Compressors and Modern Process Applications

Compressors and Modern Process Applications

Heinz P. Bloch
Process Machinery Consulting
West Des Moines, Iowa

WILEY-INTERSCIENCE
A JOHN WILEY & SONS, INC., PUBLICATION

The equipment and process descriptions used in this book represent configurations and practices in successful use at the time they were being submitted to the principal author (typically years 2001–2005). However, technology does not stand still. It will thus be to the user's advantage to seek information on new developments and to consider occasional upgrade measures that may have been either implemented or advocated for equipment and processes since then. We wish to emphasize that neither the contributors nor the publishers and/or holders of process licenses are representing this text to be a stand-alone implementation document. Also, the various photos and illustrations (see credits page xi) may not always represent the most up-to-date equipment and component configuration. Accordingly and before proceeding with the purchase of equipment and processes, the reader must consult with the other parties and obtain their written updates, comments, and concurrence.

Published by John Wiley & Sons, Inc., Hoboken, New Jersey.
Published simultaneously in Canada.

Library of Congress Cataloging-in-Publication Data:

Bloch, Heinz P., 1933–
Compressors and modern process applications / Heinz P. Bloch.
p. cm.
Includes index.
ISBN-13: 978-0-471-72792-7 (cloth)
ISBN-10: 0-471-72792-X (cloth)
1. Compressors. I. Title.
TJ990.B54 2006
621.5'1—dc22

2005035955

Printed in the United States of America.

10 9 8 7 6 5 4 3 2 1

Contents

Preface

WHY THIS BOOK WAS COMPILED

The segregation of job functions into project, maintenance, administrative, operations, purchasing, and so on often leaves gaps in the sum total of the understanding needed to optimize the safety, reliability, and profitability of modern process plants. Yet, each of these job functions is known to have an influence on the business-related goals of an industrial facility. Therefore, a measure of knowledge of capital equipment and processes must be imparted to the various groups involved in the facility's. Intelligent discourse must exist and certain basic facts must be accepted by the various parties.

Needless to say, compressors represent a multimillion dollar investment for many plants, and profitability can be neither reached nor sustained by organizations that neglect this critically important asset. This is clearly brought out in more detailed compressor texts; these are available and listed in the references. However, whereas these more detailed texts have often been recommended for and by machinery reliability professionals, a condensed overview of compressor design, operation, and maintenance is desired by other job functions and will thus be given in this book. This material will assist the very wide spectrum of readers whose process involvement brings them into contact with large process compressors. As an example and to run a smooth organization, terminology must be unified and misconceptions dispelled wherever they creep into our thinking. That is one of the aims of this book.

Along these lines, our overview of process gas compressors dealing with positive-displacement machines, and especially the later chapters concerned with centrifugal compressors, are intended to equip the reader with the essentials. As is done in certain encyclopaedias, the beginning chapters are self-contained and will be very useful to readers interested in how and why the physical equipment works. The later chapters allow more advanced readers to see the underlying analyses that make possible the design of virtually all compressors. We must keep in mind, however, that dealing with essentials may not leave enough room for discussion of every detail, and an expert will certainly appreciate this fact. So, for all details not to be within the scope of this text should come as no surprise; squeezing decades of compressor design, maintenance, and operating know-how into an introductory overview text is simply not possible. An expert will also agree that persons performing some job functions have neither the need nor the time for reading hundreds of pages of compressor details. On the other hand, a text that is aimed at imparting an understanding of critically important parameters and interactions is both feasible and worth the effort. This, too, is what we have set out to achieve here.

The entire text will prove useful to managers, engineers, and technicians who need overview-type access to compressor-related issues that range from specification, through fabrication and field erection, to compressor operation and maintenance. Recognizing the mechanical/process interaction, the text in Part II endeavors to transition to process-flow schemes in three major segments. These three segments deal with petroleum refining, petrochemical, and gas processes. Each segment identifies application and location of gas compressors and turboexpanders in over 300 processes that utilize these machines. Many compressor essentials are described in conjunction with the first two segments of Part II, dealing with refining processes and petrochemical processes, respectively. Turboexpander details are highlighted in conjunction with the third segment, explaining what the hydrocarbon processing industry collectively calls "gas processes."

Finally, we have made a conscious effort not to duplicate either the approach or the content of other books by the lead author or, for that matter, any compressor and compressor application text known to be available today. We are certain this book will be useful as a stand-alone text or as a companion text to other books. Also, many university students will appreciate the value of this book as they embark on and progress in careers in the process industries. We would be gratified to have been of assistance.

ACKNOWLEDGMENTS

The narrative of the centrifugal compressor text was coauthored by Heinz Bloch and Arvind Godse. At the time of contributing the bulk of Chapters 3 through 8, Mr. Godse was engaged in providing project-technical and equipment reliability engineering services in Muscat, Oman. Because this material forms the essence of these important chapters, Arvind derserves special mention and thanks.

Almost all of the material on positive-displacement machines originates with a professional colleague and highly esteemed mentor, Ralph James, Jr. His untimely death in 1979 is still recalled and his contributions to the practical side of reliability engineering will not be forgotten by those who knew him. Ralph was once challenged to perform the torsional analysis of a machinery train from memory. He did it, right then, on the back of an envelope.

Indeed, we rarely accomplish much without additional contributors. In this instance, we wish to acknowledge that most of the compressor illustrations were provided by A-C Compression, Appleton, Wisconsin, and Demag Delaval (now Mannesmann Demag Delaval), with major plants in Duisburg, Germany, and Trenton, New Jersey. At A-C, we were ably assisted by Larry Miller and Christopher Starszak. At Demag Delaval, now Mannesmann Demag Delaval, we are very much indebted to David Kiser. Dave went the extra distance with his competent support.

Virtually all turboexpander illustrations and details on applications in the United States and overseas were provided by Mafi-Trench (Santa Maria, California). Duane Bergmann and Randy Dirlam were most helpful in collecting the needed data.

Through these contributors, we also obtained worldwide application and location information for the compressors designed and manufactured by their respective companies. But consulting companies and individual reliability professionals were also very responsive when we asked their permission to incorporate elements of their work in our text.

Manjul Saxena compiled data on dry gas seals and Mike Saunders of Houston-based Flexelement Texas collaborated on the overview material dealing with couplings. The material on computerized machinery data evaluation we owe to Ken Atkins of Engineering Dynamics (San Antonio, Texas) and Chet Farabaugh of Kingsbury, Inc. (Philadelphia, Pennsylvania) compiled the segments on magnetic bearing technology.

Gulf Publishing Company's HP publisher Gene Swantek (now retired) graciously allowed us to use process schematics from Hydrocarbon Processing's different refinery, petrochemical, and gas processing handbooks. Uri Sela (Walnut Creek, California) gave us excellent suggestions on how to organize this book so as to best serve our target audience.

Again, we are expressing sincere appreciation to everyone for extending cooperation and permission to both use and further develop the information contained in this text.

HEINZ P. BLOCH

Illustration Credits and Photo Acknowledgments

Aerzener Maschinenfabrik: See "Maschinenfabrik Aerzen," below.

A-C Compressor Corporation, Appleton, Wisconsin: Reciprocating and Rotary Displacement Compressors, Dynamic Compressors, (3-16, 3-20, 7-1, II-1 through II-7)

Ameridrives Couplings, Erie, Pennsylvania: Power Transmission Products, (4-1, 4-2)

API/American Petroleum Institute, Alexandria, Virginia: API Standards (1-5, 2-4, 2-5, 3-5)

Babcock-Borsig, Berlin, Germany: Reciprocating and Rotary Displacement Compressors, Dynamic Compressors, Turboexpanders, Steam Turbine Drivers, etc. (1-7, 1-11, 6-1, 7-9)

Burckhardt Compression, Winterthur, Switzerland: Reciprocating and Rotary Displacement Compressors, Labyrinth Piston Compressors, (1-1, 1-3, 1-13, 1-14)

Cooper-Bessemer (now Rolls Royce), Middlesex/ UK and Houston/Texas: Centrifugal Compressors for Pipeline Services (7-5, 7-6)

Demag: See Mannesmann Demag Delaval, below.

Dresser-Rand Corporation, Olean, New York: Dynamic Compressors, Turboexpanders, Steam Turbine Drivers, etc. (Tables 3-1 and 3-2)

Dresser-Rand Corporation, Painted Post, New York: Reciprocating and Rotary Displacement Compressors, (1-2, 1-4, 1-6, 1-9, 1-12)

Flexelement Texas, Houston, Texas: Couplings (4-3 through 4-5)

John Crane Company, Morton Grove, Illinois: Sealing Products (3-30, 3-32)

Kaydon Ring and Seal, Inc., Baltimore, Maryland: Gas, Oil, and Water Seals, Face-type and Circumferental geometries: (3-24 through 3-26)

King Tool Company, Longview, Texas: Gas Filtration and Moisture Separation Equipment, Reverse-Flow Coalescers, etc. (2-10 through 2-13)

Kingsbury, Inc., Philadelphia, Pennsylvania: Hydrodynamic Radial and Thrust Bearings, Magnetic Bearings (3-23, 4-6 through 4-16)

MAN-GHH, Oberhausen, Germany: Reciprocating and Rotary Displacement Compressors, Dynamic Compressors, Turboexpanders, Steam Turbine Drivers, etc. (2-2, 2-6)

Mannesmann Demag Delaval, Duisburg, Germany: Reciprocating and Rotary Displacement Compressors, Dynamic Compressors, Turboexpanders, Steam Turbine Drivers, etc. (3-2 through 3-4, 3-8 through 3-13, 3-15, 3-17, 3-19, 3-21, 3-27, 5-8, 7-2, 7-3, 7-4, II-8 through III-15)

Mafi-Trench Corporation, Santa Maria, California: Turboexpanders (I-1 through I-12)

Maschinenfabrik Aerzen, Aerzen, Germany, also Coatesville, Pennsylvania: Blowers, Rotary Displacement Compressors, (2-1, 2-3)

Mitsubishi Heavy Industries (MHI), Tokyo, Japan: Reciprocating and Rotary Displacement Compressors, Dynamic Compressors, Turboexpanders, Steam Turbine Drivers, etc. (3-28, 7-12)

Nuovo Pignone, Florence, Italy: Dynamic Compressors, Turboexpanders, Steam Turbine Drivers, Reciprocating and Rotary Displacement Compressors, etc. (1-8, 1-10, 2-7, 2-8, 7-7, 7-10, 7-11)

Sulzer Turbo, Winterthur, Switzerland: Dynamic Compressors, Turboexpanders, Steam Turbine Drivers, etc. (3-1, 5-1, 7-8)

Zurn Industries. See Ameridrives Couplings.

PART I

Chapter 1

Positive-Displacement Compressors

Although a fair number of types of large compressors can be found in modern process plants, they can, certainly for our purposes, be separated into positive-displacement and dynamic machines. In the positive-displacement category, reciprocating compressors and twin-helical screw machines are of primary interest. Reciprocating compressors are certainly the "older" of the two. It's on that basis that we will provide an overview of reciprocating compressors before moving on to other process compressor types of interest.

1.1 RECIPROCATING COMPRESSORS*

Some refining and gas-processing facilities (e.g., hydrocracking) are best served by piston-type reciprocating compressors. Accordingly, an overview of these machines is an important complement to this text.

Reciprocating compressors (see Figs. 1-1, 1-2, and 1-5), are the oldest and most widely used process compressor type. They are manufactured in a variety of different configurations, including vertically oriented and skid-mounted ("packaged") (see Fig.1-2). Their application ranges are quite wide and include both lubricated and nonlubricated cylinder models, as shown in the coverage chart of Fig. 1-3. So-called secondary compressors often use plain plungers instead of the more typical pistons shown in Fig. 1-4.

An outstanding reference document, API Standard 618, describes many recommended features of reciprocating-process gas compressors. Fig. 1-5 shows a typical compressor cross section and the customary nomenclature. Other important components are also depicted on this cross-sectional view.

*This section was originally contributed by Ralph James, Jr., and is gratefully acknowledged.

Compression is produced by the forced reduction of gas volume by the movement of a piston or plunger in a cylinder. Suction and discharge valves (see Figs. 1-6 and 1-7), are spring-loaded and work automatically from pressure differentials generated between the compressor cylinder and piping by the moving piston [1].

Reciprocating compressors are manufactured as both air-cooled and water-cooled models. Since we are discussing the compressor as related to petrochemical and oil refining processes, we shall concern ourselves only with the cooled, liquid-jacketed type. This type compresses gas on the forward stroke and also on the reverse stroke of the piston.

Reciprocating compressors have a wide range of applications. Crankshaft speeds may range from 125 to 1,000 revolutions per minute. Piston speeds range from 500 to 950 feet per minute, the majority being 700 to 850 feet per minute. The nominal gas velocity is usually in the range of 4,500 to 8,000 feet per minute, and operational discharge pressures may vary from vacuum to 60,000 pounds per square inch (psi).

1.2 MAJOR COMPONENTS DESCRIBED

1.2.1 Crankcase

The crankcase (Fig. 1-8) is a U-shaped cast iron or fabricated steel frame. The top is left open for installation of the crankshaft. To prevent the top from opening and closing as a result of the forces of the throws, it is held together with torqued bolts and spacers or, alternatively, keyed spacers. These are placed directly above the main bearings. The main bearings, spaced between each throw, have removable top covers for ease of assembly and ready removal of the babbitted bearing-liner shells. The keyed spacers are, therefore, preferred since their removal is easiest for access to bearing covers. Main

Compressors and Modern Process Applications. By Heinz P. Bloch.

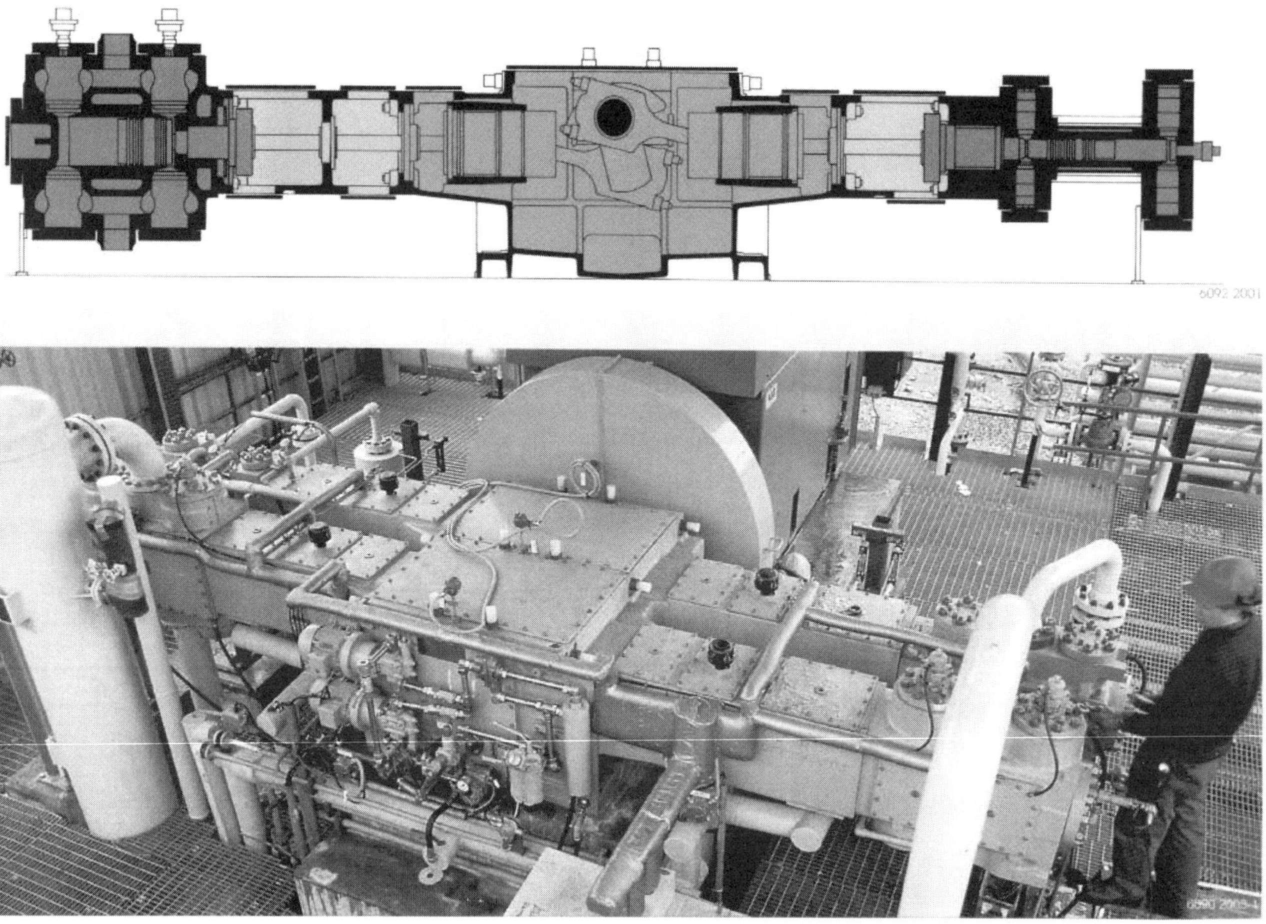

Figure 1-1. Typical horizontal balanced, opposed reciprocating compressor. (Source: Sulzer-Burckhardt, Winterthur, Switzerland.)

bearings, however, in the types of compressors used in petrochemical plants, are so overdesigned for stress that they seldom require removal for rebabbitting.

1.2.2 Crankshaft

The crankshaft (see Fig. 1-8) is the heart of the machine, and usually the most expensive component. Each throw is forged and counterweights are bolted on to balance the reciprocating mass of the crosshead and piston. If the crankcase moves on the foundation, it will cause the throw to open and close through each revolution, resulting in fatigue and breakage. For this reason, the dimension at the open end of the throw must be taken periodically while barring the crankshaft through 360 degrees. This procedure is called "taking crankshaft deflections" and is recommended as an annual check.

1.2.3 Connecting Rod

The connecting rod (Fig. 1-9) is provided with pretorqued bolts fastening cap to body at the crank end. The split is shimmed and shims can be removed as wear progresses. The wrist pin is free-floating and held in place with caps in the crosshead, which allows the connecting rod to find its own center.

1.2.4 Crosshead

The crosshead (Figs. 1-9 and 1-10) runs between two guides with about one mil per inch (0.001″ per inch) of diameter clearance. It is often weighted so that the mass inertia of all reciprocating parts is sufficient to reverse the stress on the wrist pin, even when one end of the piston is under pressure. If this is not done, the wrist pin will wipe all the oil from the side under stress and will bind or "get stuck."

1.2.5 Lubrication

Lubrication of the frame is accomplished either by a pump driven from the crank end or by a separately mounted pump. The pump takes oil from the crankcase sump and pumps it through a cooler and filter, usually 25 micron, then through

Figure 1-2. Completely packaged five-stage, high-pressure air compressor for a U.K. chemical plant. (Source: Dresser-Rand Company, Painted Post, New York.)

piping to the main bearings. The crankshaft has holes drilled from the main bearing surface through to the connecting-rod bearing face. From here, the oil passes up through a hole in the connecting rod to the wrist pin and from there, through holes to the crosshead sliding faces. Oil scraper rings in the frame end prevent oil leakage out along the piston rod. Because of this tortuous passage of oil, prelubrication is required before start-up. This is accomplished with an auxiliary lube pump.

Crankcase oil heaters are specified for outdoor compressors to keep the oil at required viscosity and to prevent condensation with resultant corrosion. Oil, however, is a poor conductor and local overheating and carbonization have occurred with these heaters. Therefore, when using crankcase heaters while a compressor is not in operation, the auxiliary lube pump should be continuously run.

1.2.6 Cylinder Materials

Up to 1,000 psi, cylinders are normally made of cast iron. Above this working pressure, materials are cast steel or forged steel, at the manufacturer's discretion. Nodular iron castings are sometimes specified in preference to cast iron.

API-618 specifies that all cylinders must have replaceable liners. These are usually of cast iron because of its lubricating and bearing qualities. Liners should be honed to a finish of 10 to 20 microinches.

1.2.7 Cylinder Sizing

Cooling jackets, lubricant, and packing and ring material limit cylinder temperatures. A compressed gas discharge temperature of 275°F is set as ideal maximum, with 375°F as the absolute limit of operating temperatures. Using a common thermodynamic equation with the above limits, the design ratio per cylinder can be set. The design compression and tension load on the piston rod must not be exceeded and, therefore, rod load must be checked on each cylinder application.

Rod load is defined as follows:

$$\text{R.L.} = P_2 \times A_{\text{HE}} - P_1 \times A_{\text{CE}}$$

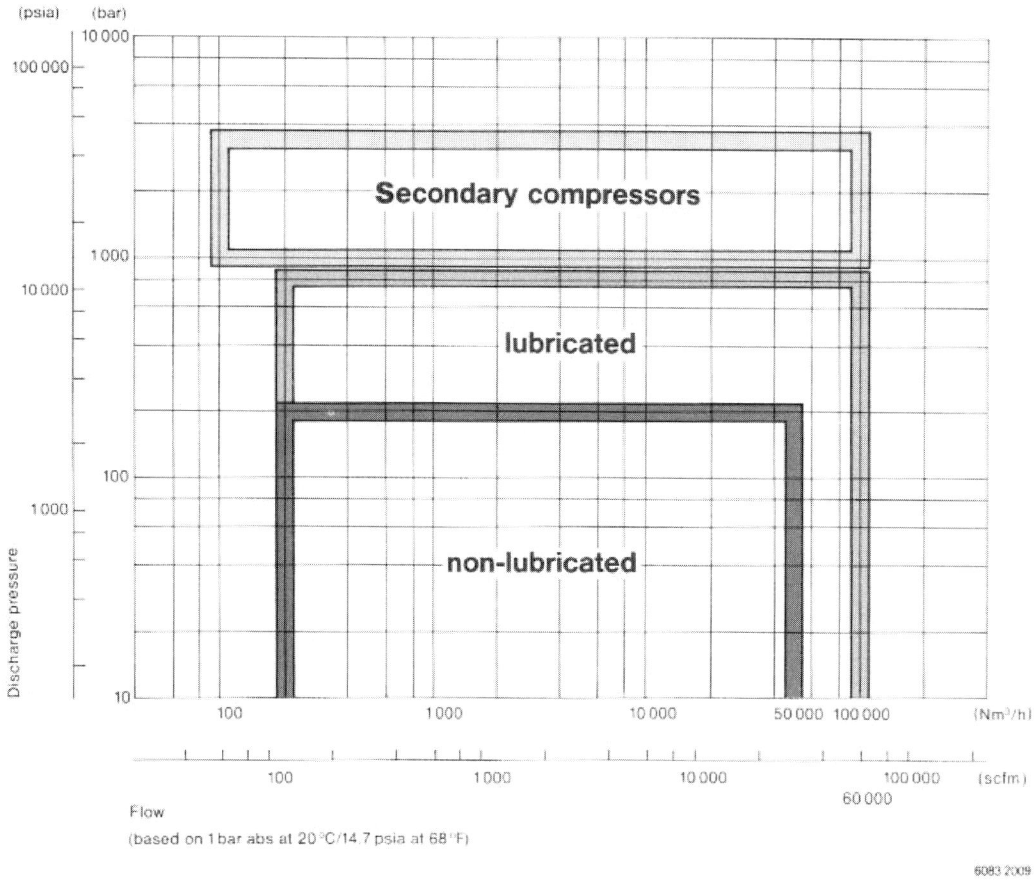

Medium

N_2, Ar, He, H_2, CO, CO_2, NH_3, CH_4, C_2H_4, C_3H_6, C_nH_m, LPG, LNG, CNG, Cl_2, HCl, H_2S, SO_2, Synthesis and other mixed gases

Applications

Chemical
Hydrocarbon processing industries
Polyethylene (ldPE, lldPE, hdPE)
Oil and gas production
Refinery
Enhanced oil recovery
Synfuels
Corrosive and toxic gases
Low-temperature and cryogenic applications
LPG on-shore installations

Figure 1-3. Typical application ranges for reciprocating compressors. (Source: Sulzer-Burckhardt, Winterthur, Switzerland.)

where

R.L. = rod load in compression in pounds
A_{HE} = cylinder area at head end in any one cylinder
A_{CE} = cylinder area at crank end, usually $A_{CE} = A_{HE}$ – area of rod
P_2 = discharge pressure, psia
P_1 = suction pressure, psia

The limiting rod loads are set by the manufacturer [2]. Some manufacturers require lower values on the rod in tension than compression. In this case, the above limits are checked by reversing A_{HE} and A_{CE}. Note, however, that the supplemental specifications of reliability-focused user companies often impose a limit on the maximum allowable stress acting on the net thread root area of piston rods. This limit is largely based on field experience and relates to rod attachment limitations rather than stress failures of the rod itself.

Compressors below 500 psig, such as air compressors, are usually sized by temperature considerations. Rod load usually becomes the limiting factor in applications above these pressures.

1.2.8 Cylinder Cooling

In process compression, higher pressures are normal. Therefore, compression ratios are low. Low compression ratios

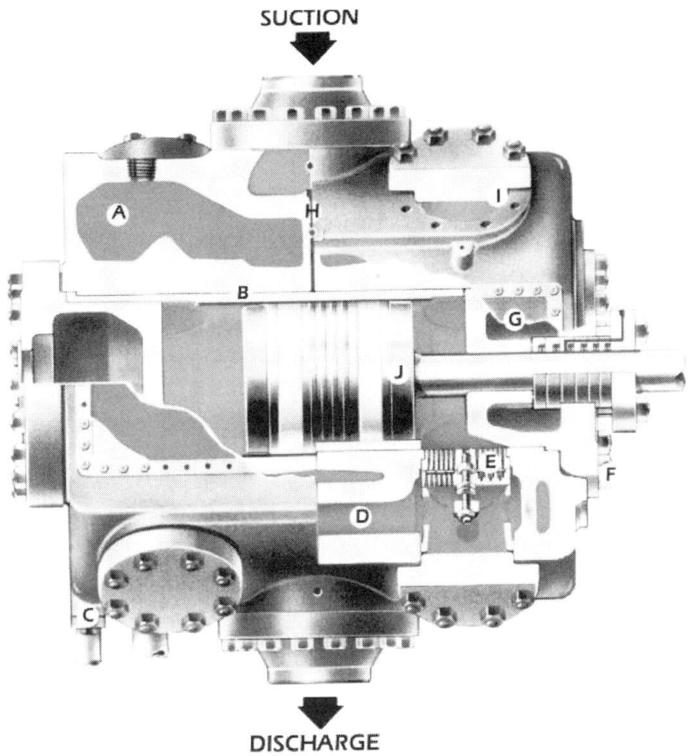

A Uniform cooling
B Dry liner
C Support pad
D Generous gas passages
E High-efficiency valves
F Rolled external threads
G Large cover plates for inspection/maintenance
H Reliable lubrication
I Superior sealing
J Multi-thru bolt rod design

Figure 1-4. Typical reciprocating compressor cylinder for long-term reliable operation. (Source: Dresser-Rand Company, Painted Post, New York.)

give low temperature rises. Thermosiphon cylinder cooling is, therefore, specified, with a discharge temperature limit of about 200°F. Thermosiphon cooling consists of filling the jackets with an appropriate liquid such as water or a suitable radiator fluid and letting the heat radiate from the outer cylinder walls. The purpose of filling the jackets is to obtain an even heat distribution throughout the cylinder.

Above 200°F, coolant circulation is applied. Raw water is to be avoided because it leaves deposits in the cooling jackets that are extremely difficult to clean out. A closed system is specified, therefore, which consists of a reservoir, circulating pumps, and heat exchanger.

The old saying "cooler is better" is really incorrect here. Overcooling of the cylinder to a temperature below the dew point of the compressed gas must be avoided to prevent cylinder corrosion, or liquid build-up that results in a slug of liquid. Therefore, the coolant must be bypassed around the exchanger under temperature control. It is also recommended that a thermostat and heater be mounted on the reservoir and the coolant circulated when the compressor is stopped on standby, to maintain the cylinders at a temperature above the gas dew point.

The coolant is usually 50% ethylene glycol and water, to prevent freezing. This has a lower specific heat than water. Therefore, the manufacturer must always be asked to size heat exchangers on this basis, even if the initial coolant is intended to be only water.

Compressor valves are the most critical part of a compressor; they generally require the most maintenance of any part. They are sensitive both to liquids and solids in the gas stream, causing plate and spring breakages. When the valve lifts, it can strike the guard and rebound to the seat several times in one stroke. This is called valve flutter and leads to breakage of valve plates. Light-molecular-weight gases such as hydrogen cause this problem mostly, and it is controlled in part by restricting the lift of the valve plate, thus controlling valve velocity.

API valve velocity is specified as:

$$V = D \times 144/A$$

where

V = average velocity in feet/minute
D = cylinder displacement in cubic feet/minute

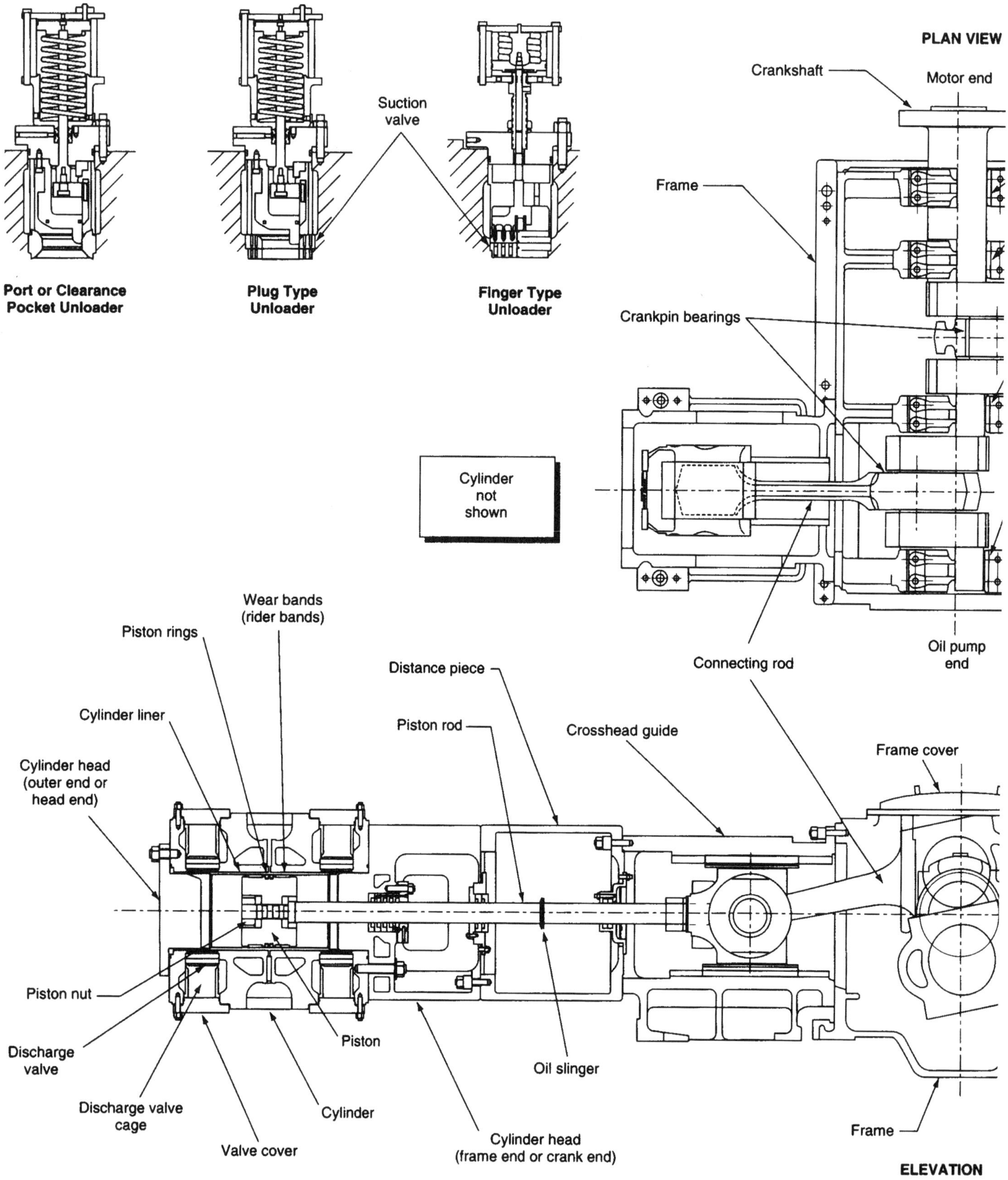

Figure 1-5. Reciprocating compressor cross section and nomenclature, per API 618.

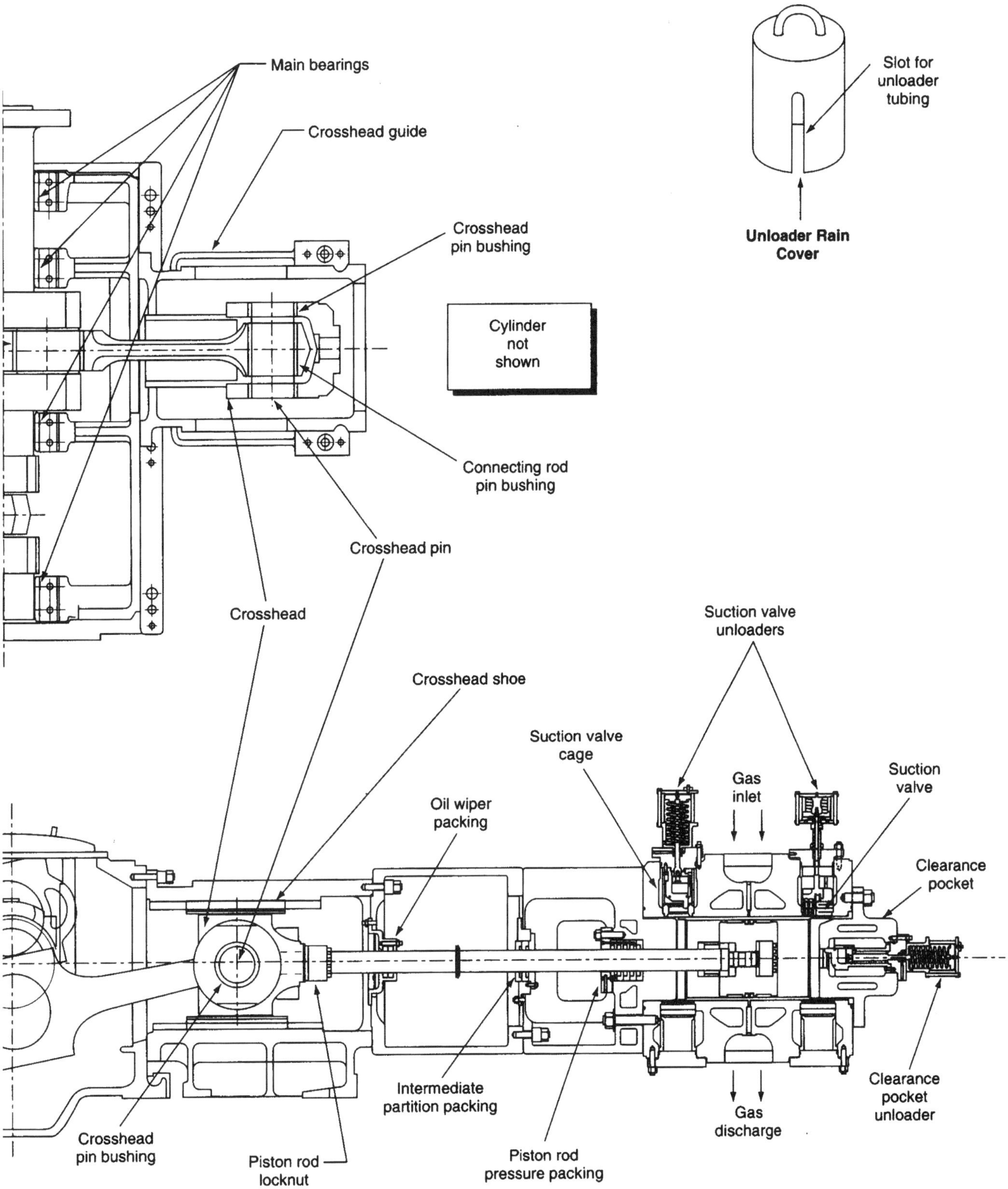
Slot for unloader tubing
Unloader Rain Cover
Main bearings
Crosshead guide
Crosshead pin bushing
Cylinder not shown
Connecting rod pin bushing
Crosshead pin
Crosshead
Suction valve unloaders
Crosshead shoe
Suction valve cage
Gas inlet
Suction valve
Oil wiper packing
Clearance pocket
Intermediate partition packing
Gas discharge
Clearance pocket unloader
Crosshead pin bushing
Piston rod locknut
Piston rod pressure packing

VIEW

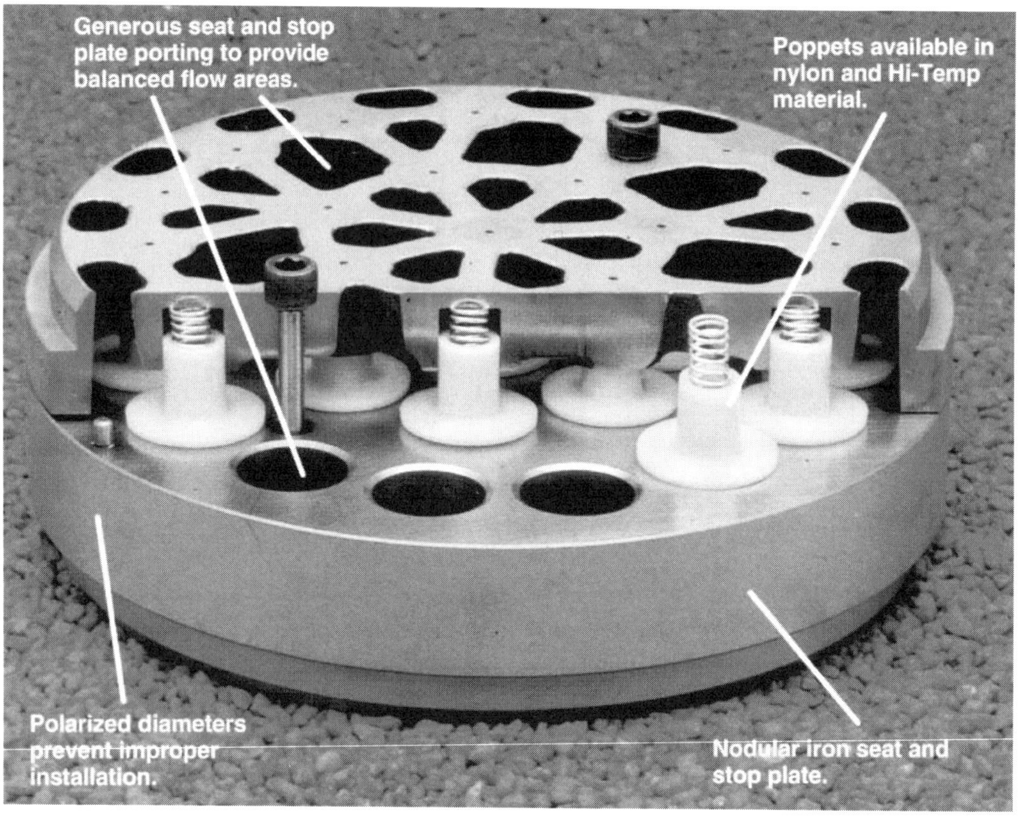

Figure 1-6. Poppet valve. (Source: Dresser-Rand Company, Painted Post, New York.)

A = total inlet valve area per cylinder, calculated by valve lift times valve opening periphery, times the number of suction valves per cylinder in square inches

Compressor manufacturers sometimes object to the above specification, since it gives a valve velocity for double-acting cylinders one half the value compared to equivalent single-acting cylinders. Therefore, manufacturers' data on double-acting cylinders often indicate a valve velocity double the API valve velocity and care must be taken to find out the basis upon which valve velocity is given. For heavier molecular weight gases (~20), API valve velocities of about 3,580 fpm are selected, and for lighter molecular weight gases (MW = 7), 7,000 fpm.

Manufacturers often use interchangeable suction and discharge valves. This can lead to putting valves in the wrong port, which can result in massive valve breakage or broken rods or cylinders. Reliability-focused users specify that valves must not be interchangeable. However, this feature can be lost or broken off, so correct valve placement should always be checked.

1.2.9 Pistons

Pistons (Fig. 1-9) are usually cast iron and are often hollow, to reduce weight. This space can fill with gas and is an explosive hazard when removing the piston from the rod. One should, therefore, specify that an easily removable plug must be supplied to vent this space before handling. Larger pistons are made of aluminum to reduce weight. These pistons have large clearance in the bore to allow for thermal expansion, on the order of 20 mils per inch of diameter. Rider rings are often supplied on cast iron pistons and are necessary on all aluminum pistons. These, or the whole piston, are rotated 90° about once per year to reduce wear. Bearing loads of pistons and rider rings are based on a unit stress calculated from half the rod-plus-piston weight in pounds, divided by the diameter of the piston or rider ring, times the width in square inches. Normal limits are 5 psi for Teflon®, 12 psi for cast iron, 14 psi for bronze, and 22 psi for Allen metal. Teflon compression rings are specified and are useful up to 500 psi ΔP. Above these loadings, copper-bearing material or Babbitt metal are used for wearing rings and bronze for compression rings. Teflon compression rings are often offered with steel expander rings underneath. These should be avoided because when the compression rings wear, the expander rings can score the cylinder. Designs of pistons and rings are available that will hold the compression ring out against the cylinder without expander rings.

1.2.10 Piston Rod

The piston rod (Fig. 1-10) mounts into the crosshead, and must be locked, either by a locknut or a pin, to prevent back-

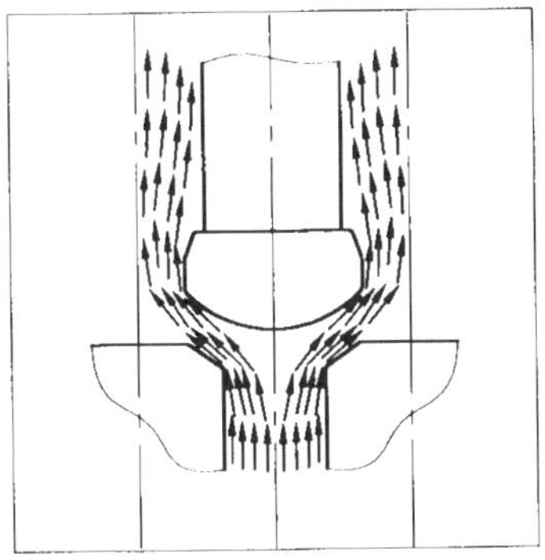

Aerodynamisches Strömungsprofil
Aerodynamic flow profile

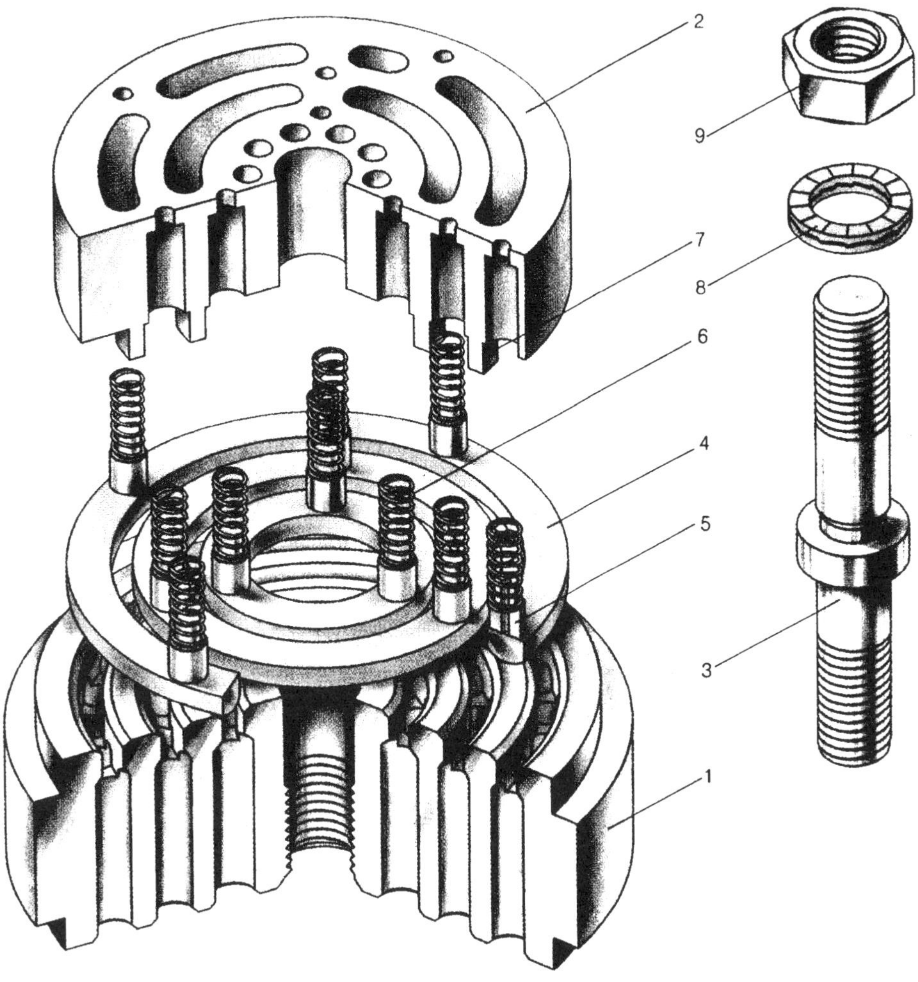

Explosionsdarstellung des Hochdruckventils

1 Ventilsitz	2 Hubfänger	3 Bolzen
4 Ventilring	5 Kappe	6 Feder
7 Zentrierung	8 Keilsicherungs-scheibe	9 Mutter

Exploded view of a high pressure valve

1 Valve seat	*2 Valve guard*	*3 Screw bolt*
4 Valve ring	*5 Cap*	*6 Spring*
7 Centering	*8 Safety washer*	*9 Nut*

Figure 1-7. Typical high-pressure plate valve. (Source: Babcock-Borsig, Berlin, Germany.)

ing off. The rod is adjusted in the crosshead to equalize the end clearance of the piston in the cylinder. This is checked by barring over the machine, crushing a piece of soft lead, and measuring the remaining thickness. This is called the bump clearance.

The rod must be hardened where it passes through the packing. Some rods are chrome plated, but problems have occurred with them, especially on high-pressure machines with a high heat buildup. This can easily cause spider web cracks in the chrome which, in turn, can flake off and destroy the packing. The best arrangement is to purchase a flame-hardened rod. As wear occurs, the rod could be plated with tungsten carbide, which should last the life of the machine. High-pressure machines often have the rod extended through the piston and out the cylinder head to balance the pressure load (rod load) on the piston. Such an extended-through rod is called a tail rod. Tail rods [1] have been known to break off and be ejected from the cylinder like a missile. Reliability-focused users specify that all tail rods must be housed in a capture box strong enough to contain the tail rod should breakage occur.

1.2.11 Packing

Compressor packing (Fig.1-11) is made up of two rings in pairs, mounted in steel or cast iron cups with the open end of the cups facing away from the pressure. The cups are bolted together and have vent, oil, and drain holes drilled in them where required. These must be correctly aligned each time the packing is opened. If Teflon is specified for packing for pressures above 500 psi, generally an additional metallic backup ring is used to prevent the Teflon from extruding out of the cups.

Compressor manufacturers supply a distance piece (Fig. 1-5) between the cylinder and the crankcase for access to the packing. This is usually good for a three- or four-cup set. In process applications, however, especially above 2,000 psi, packing sets run to 8 or 18 cups or more. Extra long distance pieces must be specified in order that sufficient space is available to dismantle the cups and change the packing rings. If any doubt exists, the extra long distance piece should be specified, as the extra cost is minor. It is generally not possible to change to a longer distance piece after the machine is built.

Figure 1-8. Crankshaft being lowered into compressor crank case. (Source: Nuovo Pignone, Florence Italy.)

1.2.12 Gaskets

Cylinder end covers, valve covers, and valves are gasketed to the cylinder. Metallic gaskets are specified. Manufacturers often offer nonmetallic gaskets, which have led to considerable trouble with leaks, especially on low-molecular-weight gases.

Often, the gasket seats must be lapped to obtain a good seal. Soft iron or metallic V-ring sets have given the best service. O-rings confined four ways on valve cover plugs have also given good service, but should never be used where they can be crushed when pulling down the cover. Plug holes must be chamfered to prevent cutting of O-rings upon entry.

1.3 COMPARISON BETWEEN RECIPROCATING AND CENTRIFUGAL COMPRESSORS

In the final analysis, economic factors, including repair and failure projections, will govern the selection of a compressor [2, 3]. Both the user and compressor manufacturer should

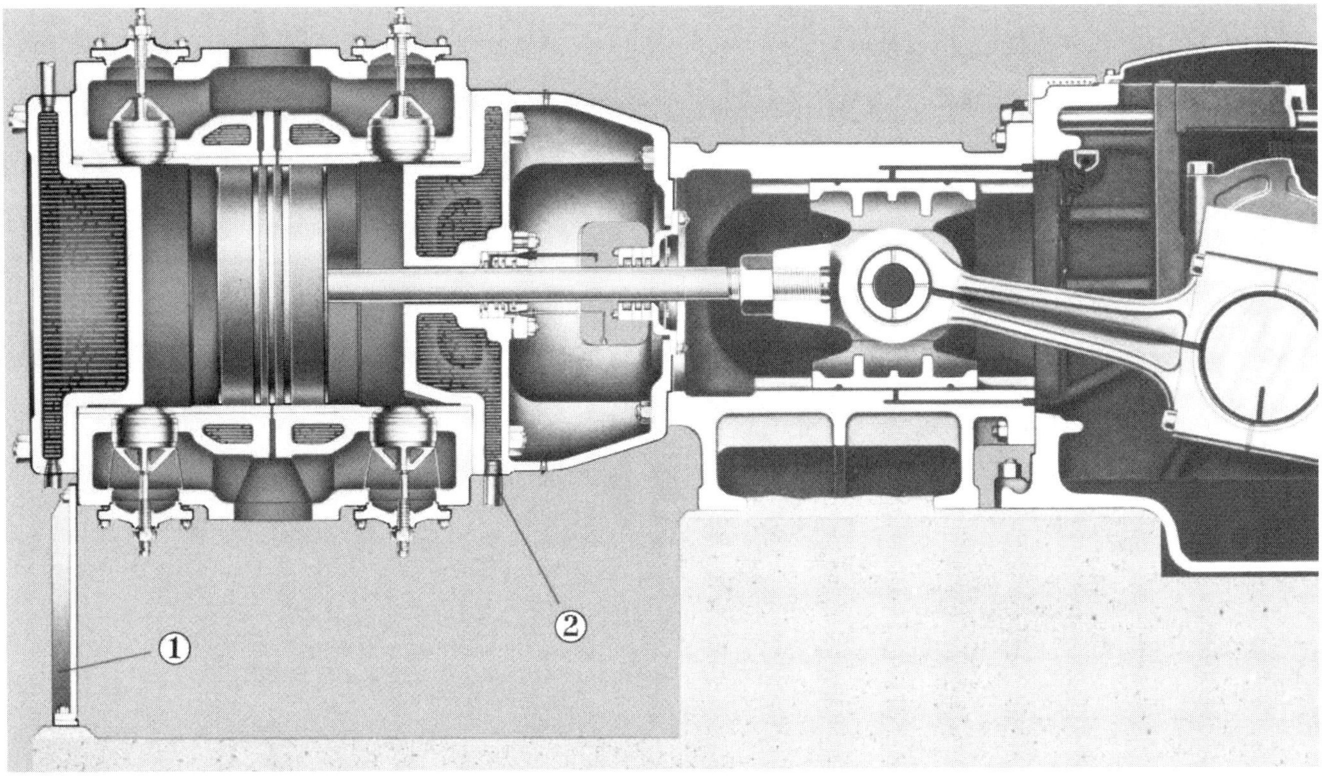

Figure 1-9. Partial cross section showing "cold" cylinder support (1), crank-end cylinder head and drain (2), piston and piston rod, sliding crosshead, and connecting rod. (Source: Dresser-Rand Company, Painted Post, New York.)

Figure 1-10. Compressor distance piece with sliding crosshead (left) and piston rod (right). (Source: Nuovo Pignone, Florence Italy.)

have full knowledge of operating conditions and factors governing selection. As energy costs increase, the higher efficiency of the reciprocating and rotary screw (double helical screw) compressor becomes increasingly important. To what extent energy considerations influence equipment selection is, interestingly enough, also a cultural issue. In a house in which lights are habitually left on in every room, or even the unoccupied rooms are cooled in the Summer and heated in the Winter, the introduction of energy conservation is neither a popular nor high-priority subject.

Obviously and in some instances, certain tangible and even intangible factors dictate the use of one or the other machine, without question. In other cases all factors must be analyzed carefully in order to make a choice between the two types. Occasionally, a reciprocating and centrifugal compressor operating in series is indicated.

1.3.1 Gas Properties and Process Conditions

Gas Analysis

A complete and accurate gas analysis is the starting point in compressor selection. Percentage by volume of component gases, entrainment of liquids and solids, and percent water vapor should be recorded. Even minute quantities of contaminants should be reported in the analysis. Trace amounts of

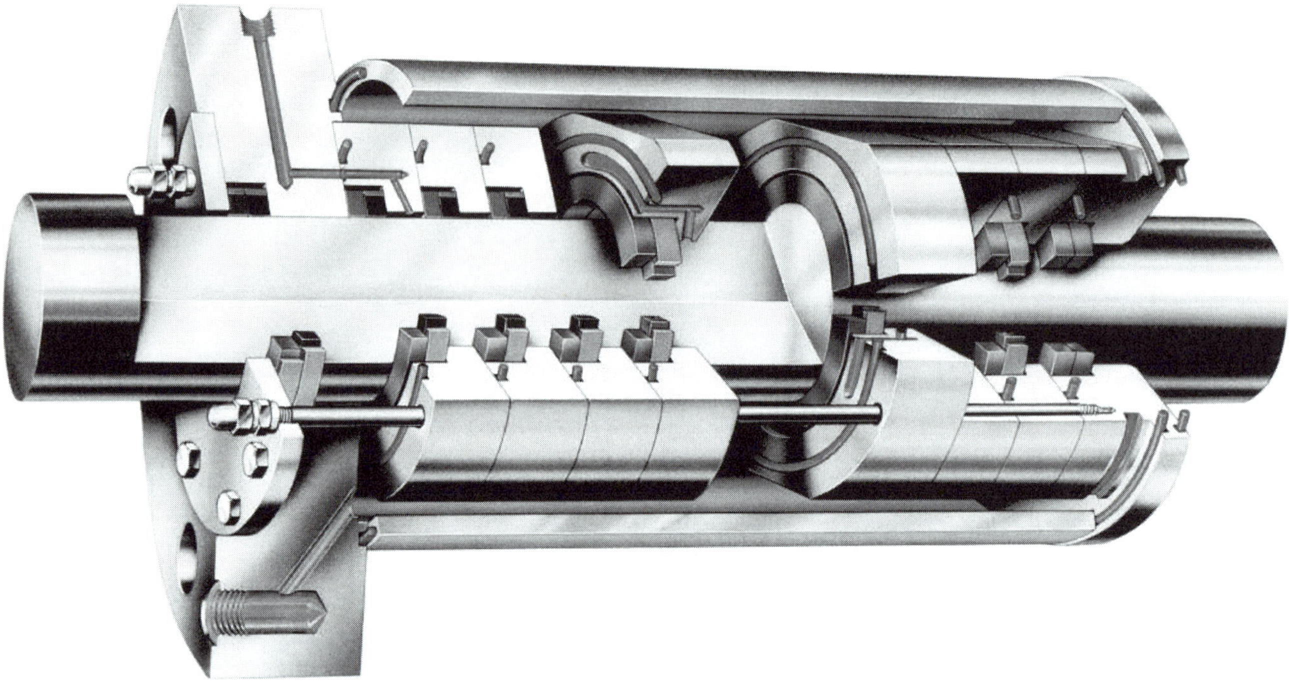

Figure 1-11. Cutaway view of packing area. (Source: Babcock-Borsig, Berlin, Germany.)

sulfur compounds and chlorides can cause corrosion or other mechanical difficulties. Even slight corrosion can produce failure of cyclically stressed parts of either type of compressor.

In a reciprocating compressor, solid particles will cause high maintenance by accelerating wear of valves, pistons, cylinders, piston rods, and packing. Solids passing through a centrifugal compressor may erode impellers and casings severely. If at all possible, any solid particles should be removed from the gas stream before the gas reaches the compressor. In some cases, a wash fluid may be used to carry fine solids through a compressor.

Water vapor or other vapors entrained in a gas use up compressor capacity. This volume must be allowed for when sizing the compressor.

Molecular Weight

Except for a change in efficiency due to valve-related losses, the reciprocating compressor is not affected by the molecular weight of the gas. Periodic changes in gas composition will have little effect on compression horsepower and pressure. On the other hand, the pressure developed by a centrifugal compressor at a particular speed is directly proportional to the gas density or molecular weight.* Also the internal flow passages of a centrifugal compressor are designed to best handle the change in density of a particular gas as it passes through the compressor.

For equal impeller peripheral speeds it will take many stages compressing hydrogen to equal the pressure created by a single stage compressing a dense gas such as butadiene or methyl chloride.

If the percentage of constituents in a composite gas varies from time to time, so will the molecular weight. A centrifugal compressor, to generate the desired pressure when handling the full flow of such a gas, would have to be designed for the lowest expected molecular weight. Excess pressure resulting from increased molecular weight would have to be reduced by throttling, change in speed, or by some other means. In other words, it is difficult to design a centrifugal unit that will economically and practically handle changing gas densities.

Polytropic Exponent

As a gas is being compressed, the polytropic exponent determines the pressure–volume relationship and temperature change. If the polytropic exponent is not known, the average ratio of specific heats may be used for calculating the theoretical adiabatic compression temperatures, volumes, and horsepower.

Temperatures are important in a reciprocating compressor. Temperatures within a conventionally lubricated cylinder should not exceed 350°F. Theoretically, a gas with a low exponent can be compressed with much higher ratios of compression per stage, yet hold temperatures within desired lim-

*Centrifugal force = (mass)(tangential velocity)/(radius). Note also that the impeller of a centrifugal compressor will develop the same head for various gases but the pressure is a function of gas properties

its. One drawback to this, however, is that volumetric efficiency decreases with high ratios, and a low exponent may make it uneconomical to compress to high ratios.

The compression exponent also influences the design of centrifugal compressors. Pressure developed by an impeller will be less for a high-exponent gas than for a low-exponent gas of equal density at inlet conditions.

Flow Rate

The centrifugal compressor is essentially a large-capacity machine. If, say, 1000,000 cfm of gas were to be compressed from near-atmospheric suction conditions, a centrifugal compressor would probably be used for the lower stages of compression. If the final discharge pressure is high, the last stages of compression could be handled by reciprocating compressors.

In general, reciprocating compressors have a cost advantage up to perhaps 10,000 cfm capacity. Another rule of thumb for process applications is that comparison of centrifugal compressors with reciprocating and rotary screw machines is recommended if the volume of a gas stream at discharge pressure is over 600 cfm. Also, custom-built centrifugal compressors are generally a parity investment compared with reciprocating and/or helical screw compressors if the absorbed power is in the vicinity of 1,500 kW.

If the flow rate varies widely, reciprocating compressor and certain screw compressors can operate with a reasonable sustained efficiency. Flow variation may be handled by suction valve unloaders and/or clearance pockets (Fig. 1-12) or, in the case of certain screw compressors, speed changes.

A suction valve unloader operates by depressing the valve feathers or plates to hold the valve open during both suction and compression strokes. So, on the compression stroke, the gas is not compressed, but flows back into the suction manifold. The unloader can be operated either manually or by remote automatic means.

Clearance pockets are volume chambers built into the cylinders or heads, or may be attached to the cylinder by a piping connection. Ordinarily, a built-in pocket is fitted with a plug valve so that the additional volume may be added or removed at will, either manually or automatically.

Generally, clearance pockets are more flexible in application than are valve unloaders. They can be sized to accomplish any degree of per-stage unloading. Two or more smaller pockets commonly are placed in one cylinder end. Some

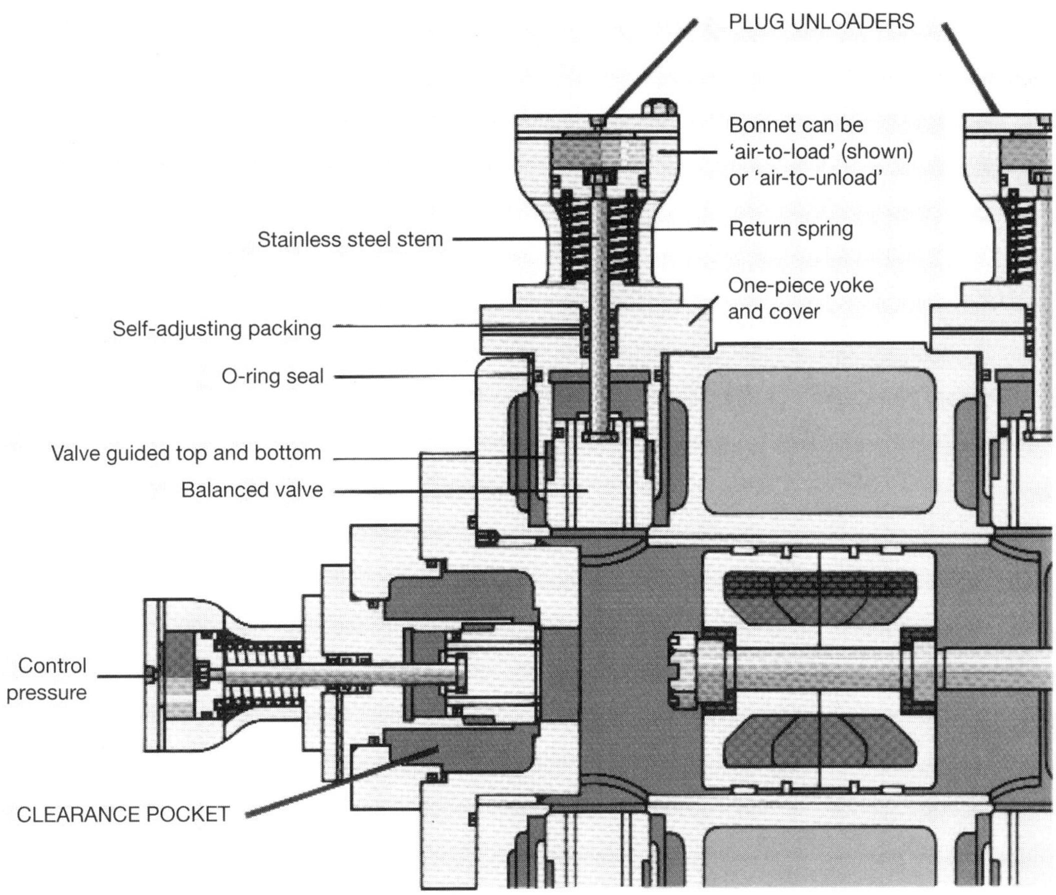

Figure 1-12. Plug unloaders (top) and clearance pocket (left). (Source: Dresser-Rand Company, Painted Post, New York.)

compressor manufacturers are able to offer variable-volume clearance pockets. Variable-volume clearance pockets plus suction valve unloaders permit continuous (infinitely variable) capacity control with minimum power consumption. (As will be seen later, with centrifugal compressors operation is not feasible below the surge point. This flow limit ordinarily is between 50% and 75% of rated capacity.)

Inlet and Discharge Pressure

In a multistage reciprocating compressor, a variable suction pressure while maintaining a constant discharge pressure may necessitate pockets and/or valve lifters in order to maintain satisfactory operation. Lowering the suction pressure will lower the overall horsepower, lower the differential pressure on all stages except the last stage, increase the differential on the last stage, and, very often, the horsepower on the last stage. Raising the suction pressure will raise the horsepower of the complete machine, raise the differential pressure on all stages up to the last stage, and probably lower its horsepower.

In centrifugal compressors, an increased suction pressure will raise the discharge pressure and increase the horsepower. If the suction pressure is lowered, the centrifugal machine will not compress to the desired discharge pressure.

Temperature

In both reciprocating and centrifugal machines, the compressor recognizes only capacity at the actual inlet conditions. Therefore, the inlet temperature must be specified and compressors are usually rated at suction conditions. Conversion to standard conditions are made for reference and comparison only.

As regards mechanical operation of equipment, centrifugal compressors are less affected by high or low temperature extremes than reciprocating compressors. Centrifugal compressors have been used to circulate gas at 800°F. With conventional lubricants, such temperatures are impractical in a reciprocating compressor. Very low temperatures also cause lubrication problems. However, reciprocating compressors have been operated at suction temperatures below minus 100°F.

A nonlubricated reciprocating compressor developed by a Swiss company in the 1940s—the Burckhardt Labyrinth Piston Compressor, Figs. 1-13 and 1-14—will often solve both of these lubrication problems. A grooved piston that does not contact the cylinder walls is used. Close clearances are employed to control leakage between the cylinder and piston. Outstanding results have been obtained with this design.

A centrifugal compressor designed for a given capacity at a specified inlet temperature will fail to deliver the required discharge pressure if the suction temperature is increased significantly. Or, if the suction temperature is lowered, the discharge pressure will be increased. Consequently, the centrifugal compressor must be designed to deliver the desired capacity and pressure under the maximum inlet temperature conditions.

Heat Balance

Process heat balance sometimes has a bearing on the selection of a driver. In turn, the driver selection may determine the compressor type. For example, if a backpressure steam turbine driver is selected because of low-pressure steam requirements, a centrifugal compressor would be a logical choice. Steam turbines are ideally suited for centrifugal compressors for several reasons. First, the rpm match between centrifugal compressors and turbines permits a direct drive. In the size range of centrifugal compressors, turbines are one of the least expensive drivers. They are mechanically efficient and lend themselves well to speed control by which compressor pressure and gas flow may be indirectly controlled. Moreover and if properly selected, they rate favorably on the list for high reliability and low maintenance cost. They may be used in continuous service of perhaps three or more years. In clean services and at best-practices user companies, uninterrupted run lengths of eight years have sometimes been achieved. Steam turbine drivers are limited, of course, to use in plants where steam is available at reasonable cost.

If economics or other reasons dictate the use of electric-motor or gas-engine drivers, reciprocating compressors are sometimes the most logical choice. One exception is that gas engines drive large single-stage centrifugals through speed-increasing gears in some gas pipeline stations. This particular combination is less expensive than gas-engine reciprocators of the same size. A second exception is in applications that favor a step-up gear between the electric motor driver and centrifugal compressor.

Under certain conditions, such as low fuel cost plus unavailability of moderate-pressure steam, a combustion gas turbine driver may be selected. Credit for steam produced by a waste-heat boiler will generally offset the higher first cost of the gas turbine. In any event, driver selection is governed by economic factors that may vary greatly from site to site.

1.4 SERIES AND PARALLEL OPERATION

Within certain limits, compressors can be operated in series or parallel. In positive-displacement machines operated in series, the discharge flow rate of the first must equal the inlet flow rate of the second machine. If not carefully matched, a series system may produce an appreciable vacuum or excessive pressure between the two compressors. With turbocompressors in series, a flow mismatch will cause surging of one or the other, and flow instability.

Positive-displacement compressors can be operated in parallel if the discharge pressures of the two machines are about the same; flow through one compressor has little or no

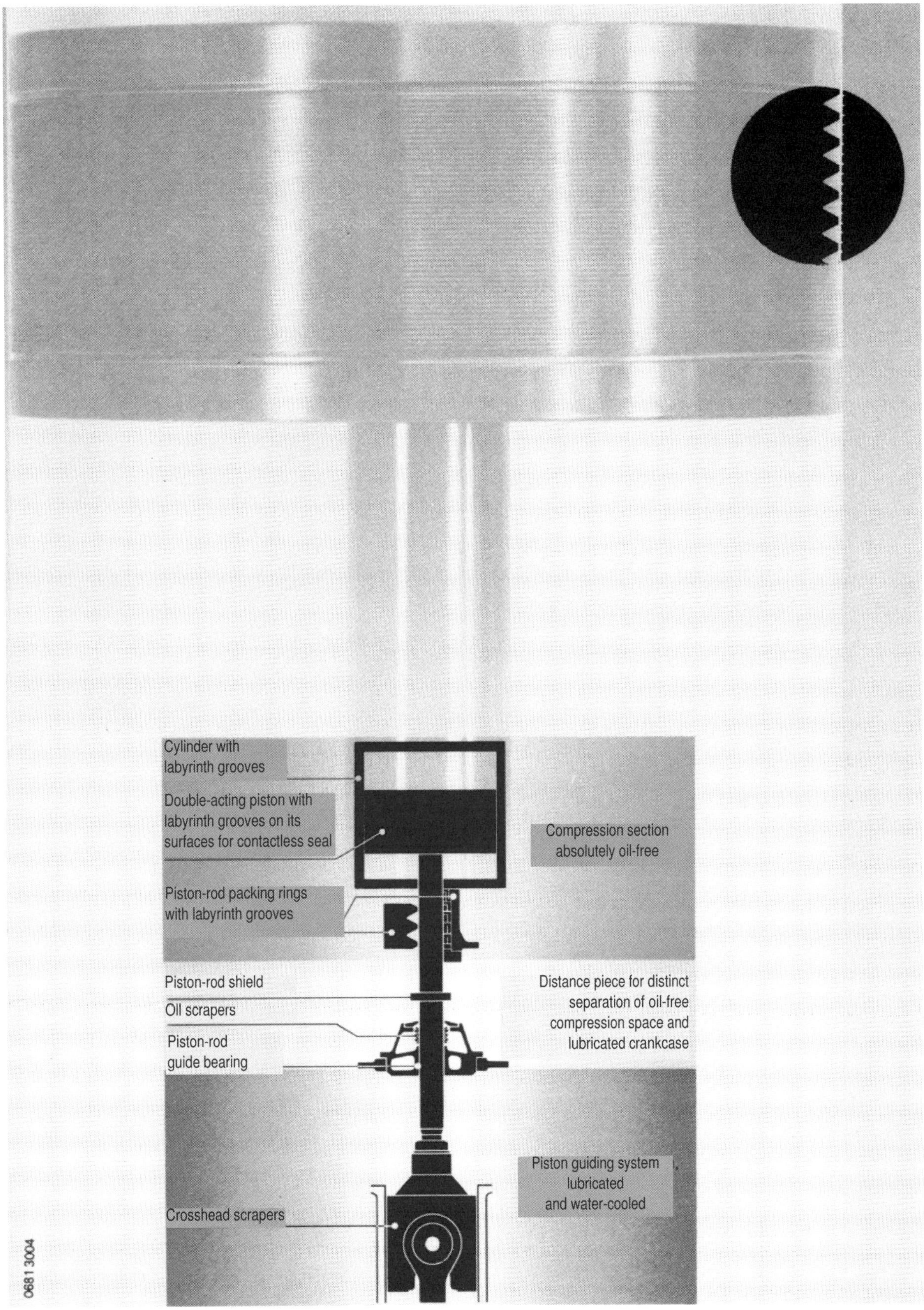

Figure 1-13. Principle of labyrinth piston compressor. (Source: Sulzer-Burckhardt, Winterthur, Switzerland.)

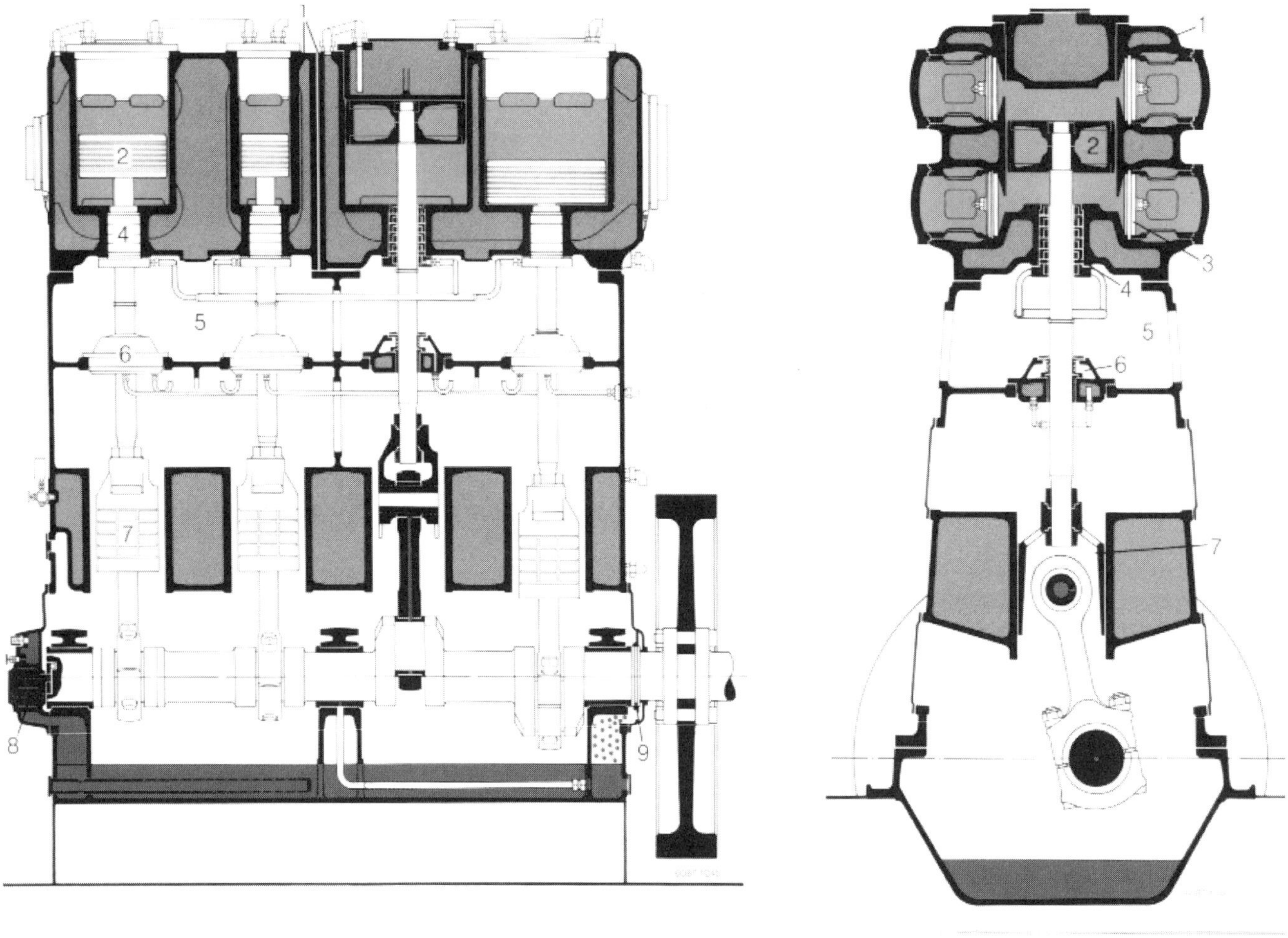

Design features

1. Cylinder with jacket for heat exchange medium such as cooling water
2. Labyrinth piston
3. Compressor valve
4. Labyrinth-piston rod gland
5. Extra long distance piece separating the oil-lubricated crankcase from the oil-free cylinder
6. Oil-lubricated guide bearing with oil scrapers
7. Crosshead guide surrounded by jacket for heat exchange medium such as cooling water (supersedes the oil cooler on small compressors)
8. Crankshaft-driven lubricating-oil pump for forced-feed lubrication of bearings and crossheads
9. Crankshaft seal

Area of gas compression and gas flow

Area of nonflowing gas

Heat exchange medium (e.g. cooling water)

Lubricating oil

Figure 1-14. Four-throw labyrinth piston compressor. (Source: Sulzer-Burckhardt, Winterthur, Switzerland.)

effect upon the other. However, turbocompressors operated in parallel do present a matching problem; one compressor will frequently pick up the entire flow while the second idles or becomes unstable. Consequently, an automatic flow-control system is used by reliability-focused users.

Turbo and positive-displacement compressors can be combined; a turbocompressor can be used in series with a positive-displacement machine, or vice versa. A turbocompressor also could be used in parallel with a positive-displacement compressor as a flow booster.

CHAPTER 2

Rotary Compressors as a Category

The term *rotary* describes a class of compressors that operate on the positive-displacement principle and employ rotary motion to transfer energy, that is, to compress gas. A number of different types and construction styles have been developed within this class of compressor and they employ different mechanical principles and configurations. The various rotary compressor types are often applied in petroleum and chemical plant services, but in smaller numbers than are reciprocating and centrifugal compressors. The somewhat typical rotary compressor population is found in relatively low discharge pressures and low power level applications. Some rotary types have decidedly lower maintenance requirements than reciprocating compressors, others, depending on service conditions and design features, do not have low maintenance requirements. Generalities aside, proven and highly successful twin-screw rotary compressors with capacities up to perhaps 60,000 acfm (100,000 m^3/hr) are in use today. Power inputs in excess of 7,000 hp per machine and years of uninterrupted service in "dirty" coke gas services are no longer the exception (Fig. 2-1).

2.1 HELICAL SCREW COMPRESSORS

Recalling that our topic deals with process applications, the scope and purposes of this overview text are best served by concentrating entirely on twin helical screw compressors. The terms rotary screw machine, rotary screw compressor, and helical screw machine are often used interchangeably.

Helical screw compressors have been marketed since the late 1930s. The high-pressure, oil-free helical screw compressor has been successfully applied in a wide variety of air and gas services since the 1950s and is an important economic competitor in the flow and pressure range for which it was developed [4]. High-pressure, oil-free rotary compressors are the most significant rotary compressor type applied in major process gas and utility air services. Preferred embodiments are written up in a widely used, comprehensive API Standard, API-619. This particular standard is primarily intended for rotary compressors that are in continuous duty and are generally nonspared. For every one of the thousands of process screw machines, many more oil-flooded machines are in refrigeration service.

2.2 OVERVIEW OF OPERATING PRINCIPLES AND BASIC CONSTRUCTION

Modern helical screw compressors used in process plants in the early 21st century are positive-displacement-type compressors. Figure 2-2 shows assembly in progress for a typical large helical screw compressor. Compression is achieved by the intermeshing of two helically formed rotors: a male rotor with convex lobes and a female rotor with concave flutes. Gas is admitted through an inlet port and fills the helically shaped flutes along the length of the female rotor. The discharge end of the rotor cavity is confined by the compressor end plate. As the rotors turn, a lobe of the male rotor progressively enters the flute cavity in the female rotor, decreasing the cavity volume and raising the pressure of the trapped gas. The compressor is designed so that a built-in compression ratio is attained at the point in the rotation cycle at which the discharge port in the discharge end plate is uncovered. The process is repeated for each interlobular space. For a given size of compressor, designs are available with a range of built-in compression ratios. The built-in compression ratio is selected by the vendor to optimize compression efficiency on a service-by-service basis. Figure 2-3 is helpful in visualizing the volume–displacement relationships that are part of these operating principles. As the screws rotate, a volume of

Compressors and Modern Process Applications. By Heinz P. Bloch.

Figure 2-1. Large helical screw compressor train. (Source: Maschinenfabrik Aerzen, Aerzen, Germany; also, Aerzen USA, Coatesville, Pennsylvania.)

Figure 2-2. Screw compressor being assembled. (Source: MAN-GHH, Oberhausen, Germany.)

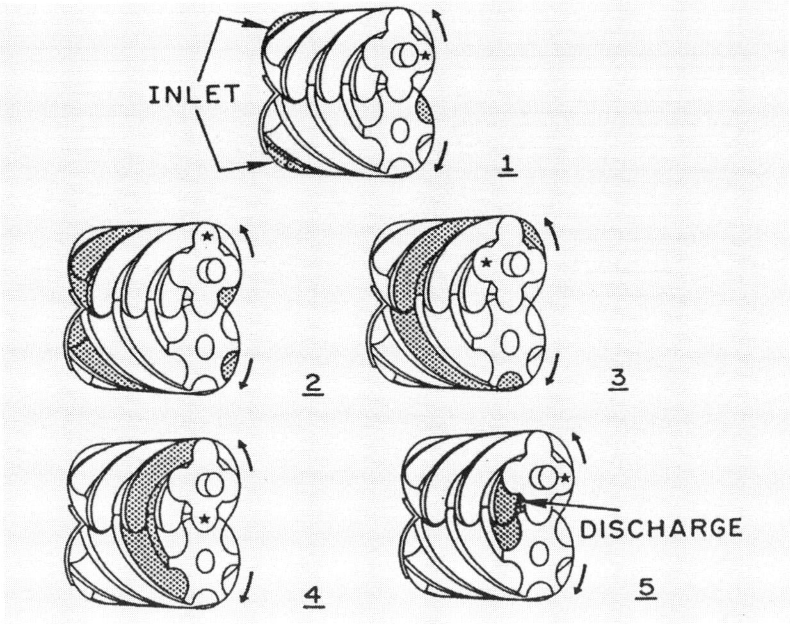

Figure 2-3. Principles of compression in double helical screw compressors. (Source: Maschinenfabrik Aerzen, Aerzen, Germany; also, Aerzen USA, Coatesville, Pennsylvania.)

gas moves from the inlet (1) successively to stations (2) through (5). Its pressure rises as it undergoes a volume reduction.

Regardless of whether the screw compressor is executed for dry compression or oil injection, the gas is compressed in chambers progressively decreasing in size that are formed by the intermeshing action of the helical rotors and by the housing wall. As mentioned earlier, oil-injected (oil-flooded) compressors (see Fig. 2-5) do not incorporate timing gears. Instead, the driven male rotor interacts directly with the female rotor without use of timing gears. Oil injected into the compressor cavity provides intensive lubrication, and a large portion of the compression heat is absorbed. At the same time, the clearances between rotors and cylinder (casing) walls are filled with oil. This prevents the reverse flow of compressed gas and increases the overall efficiency of compression.

In dry-screw compressors (Figs. 2-2 and 2-4), the male and female rotors operate dry in the process gas without the presence of oil as a lubricant. In order to prevent rotor contact and wear, and in the absence of lubricant, the male and female rotors are kept in a synchronized rotational speed relationship by timing gears, identified in Fig. 2-4 as item 12. A dry-screw compressor thus uses no liquid for sealing the rotor clearances and driving the noncoupled rotor. The rotor-to-rotor relationship is maintained by timing gears on each rotor, and the noncoupled rotor is driven by the coupled rotor through the timing gears. No rotor-to-rotor contact occurs in the dry-screw compressor. Wet-screw (oil-flooded) machines, which will be further discussed later, do not use timing gears. A cross-section view of a wet-screw compressor is shown in Fig. 2-5.

In all helical screw compressors, the compressor inlet gas volume handling capability and compression ratio are determined by rotor length, helix angle of the lobes, and the location and shape of the discharge port. Normally, there are four lobes on the male rotor and six flutes on the female rotor [4]. Consequently, the male rotor rotates 50% faster than the female rotor. In general, four shaft seals, two per rotor, are required to prevent compressed gas leakage to the bearing, timing gear, or outside atmosphere regions. Some of the many possible configurations are illustrated in Fig. 2-6.

In large machines, each rotor is supported by two pressure-lubricated, hydrodynamic-type radial bearings and a hydrodynamic thrust bearing is provided for each rotor. Rolling element (steel, antifriction) bearings are used on light-duty designs. The lube and seal oil systems for many (but not all) of the large screw compressors found in process gas service are designed and manufactured in accordance with API 614 (Figs. 2-7 and 2-8; also see [5]).

On small rotary screw compressors, the housing is vertically parted on the suction side. Cylinder (casing) and discharge side plate are frequently combined in one housing. The housings of larger machines are often parted horizontally for easy assembly. Rotors and shafts are milled out of one piece of either forged or stainless steel. Some manufacturers provide rotors with synthetic coatings. Depending on service conditions, this may lead to a rapid drop in compressor efficiency due to loss of coating on the rotor edges.

Process gas machines incorporate direction of flow from the top to the bottom, thus facilitating liquid removal from the compression space whenever liquid is injected into the rotor chamber for cooling or cleaning during operation. On-

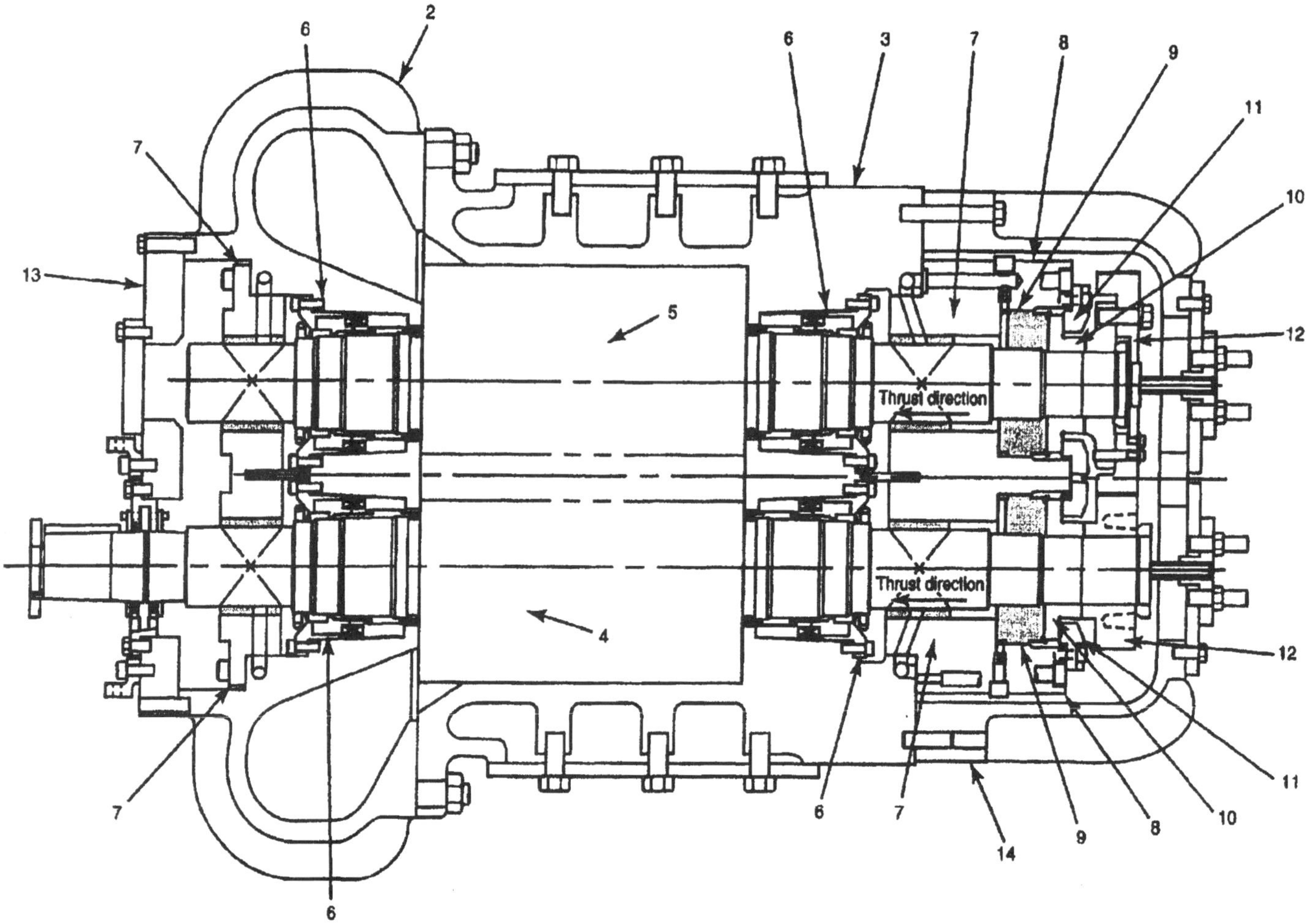

Figure 2-4. Cross-sectional view of dry screw compressor. Key: 1. compressor assembly, 2. housing inlet, 3. casing, 4. male rotor, 5. female rotor, 6. shaft seal, 7. radial bearing, 8. thrust bearing housing, 9. thrust bearing, 10. thrust collar, 11. thrust plate, 12. timing gear, 13. suction end cover, 14. discharge end cover.

stream cleaning, which is a term describing the injection of a liquid to either cool the gas or to wash the objectionable substance off the compressor internal surfaces, is highly advantageous in services in which gases are contaminated or tend to polymerize.

As long as a liquid-injected machine has timing gears, it is still called a dry-screw machine. In other words, liquid injection does not make a compressor a "wet-screw" type. A wet-screw (also called flooded screw compressor) compressor is a rotary, helical-lobe compressor into which a lubricant (compatible with the process gas) is injected into the rotor area after the closed thread position of the rotor. This lubricant helps seal rotor clearances and establishes an oil film between rotors. One rotor drives the other in the absence of timing gears.

As a further aid to keeping these separations in mind, the reader may wish to consult the Additional Reading section given at the end of this text. These references further explain selection criteria pertaining to rotary screw compressors used in general gas compression services that are beyond the scope of this overview.

2.3 CONSIDERATIONS FOR SCREW-COMPRESSOR STAGING

A stage of compression is the compressor casing region between the gas inlet and gas exit nozzles within which a progressively increasing pressure occurs. Whether or not more than one stage of compression is needed is determined by discharge temperature and differential pressure limitations. If there is no limit to discharge gas temperature due to gas considerations such as polymerization ("gumming up"), the discharge temperature of a compressor stage should be limited to 350°F (175°C). Discharge temperatures approaching 450°F (230°C) are feasible when oil-cooled rotor designs are employed. Although, in this instance, oil would be forced to circulate through the rotors for cooling, this does

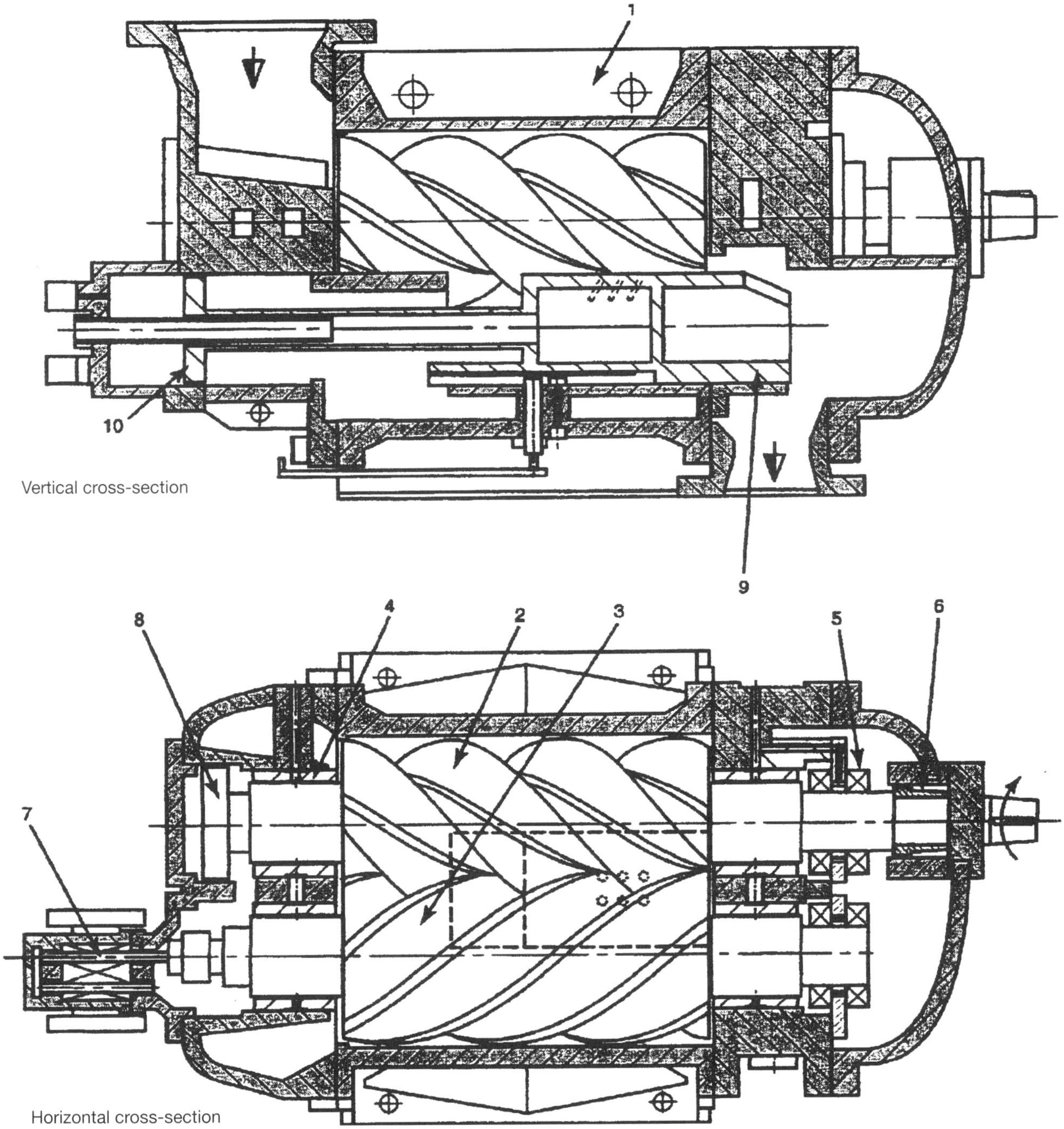

Figure 2-5. Cross-sectional views of liquid-injected screw compressor, per API 619. Major components: 1. housing, 2. male rotor, 3. female rotor, 4. radial bearings, 5. axial bearings, 6. mechanical seal, 7. oil pump, 8. hydraulic thrust compensating piston, 9. capacity control slide valve, 10. double acting hydraulic piston.

not constitute an oil-injected or wet-screw (flooded) compressor.

The pressure differential (discharge pressure minus inlet pressure) is also limited since pressure differences cause rotor bending that can result in casing contact. A differential pressure of 10 to 100 psi (0.7 to 7.0 bar) is readily obtained in a single stage, and single-stage operation is typical for the differential pressures of most oil-flooded machines in use as of 2006.

Nevertheless, surprisingly high differential pressures are often achieved by competent vendors; their designs use rotors with low length-to-diameter ratios.

Multistage (multicasing) arrangements are thus not uncommon and can accommodate pressure ranges from approximately 0.1 bar absolute, to 40 bar. Interstage cooling is used in many of these applications. Depending on compressor size, speeds from 2,000 to 20,000 rpm can be encoun-

Sealing system

Shaft sealing is provided on both the suction and discharge sides of the machine between the compression chamber and the bearings. The seals are designed to reduce gas leakage from the compression chamber to an absolute minimum, and prevent lube oil entering the compression chamber.

The gas working pressure is, usually, reduced with the aid of several floating rings (carbon or white-metal seal rings); in addition, a labyrinth bushing is provided on the outboard side to prevent lube oil entering the compressor. Where positive shaft sealing is required, mechanical contact seals can be employed.

The various seals systems are described below:

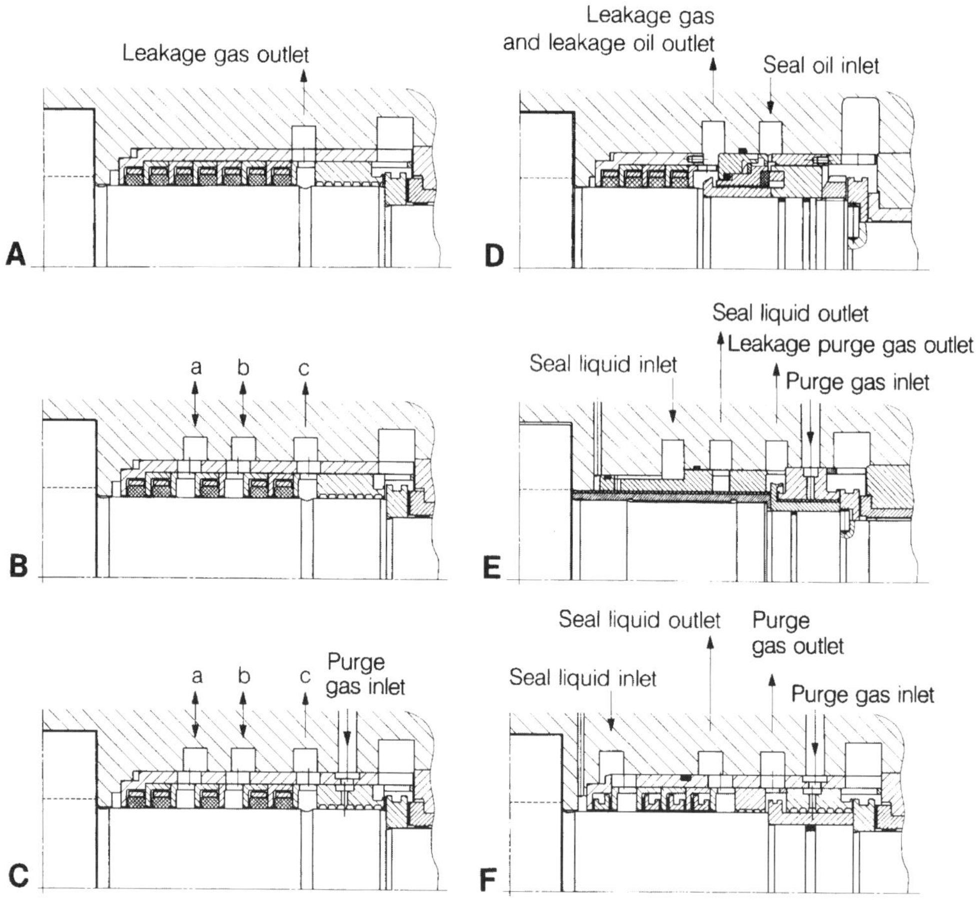

Dry seal systems:

A Chosen for clean, harmless gases, where the leakage gas can be vented to atmosphere.

B Selected for all those cases where the process gas cannot be vented to atmosphere, with the following possible connections:

i)
a: Letdown to lower pressure;
b: Extraction using an ejector;
c: Atmospheric vent connection.

ii)
a: Letdown to lowe pressure;
b: Seal-gas admission;
c: Vent to atmosphere or flare.

iii)
a: Seal-gas connection;
b: Closed;
c: Vent to atmosphere or flare.

C As under **B**, but with an additional gas purge (usually air or nitrogen) between sealing and bearing space to ensure that process/seal gas does not penetrate the bearing space and thus into the lube oil.

Liquid seal systems:

D Mechanical contact seal used in combination with dry carbon ring seal gives positive sealing under all conditions, including standstill. As the amount of leakage oil from the contact seal is only small, the oil in contact with the gas can be discarded. A combined lube/seal-oil system can be employed.

E Labyrinth seal with liquid injection; in this case no leakage gas accumulates, but seal liquid enters the compression chamber. As a rule, water is used as the sealing liquid (no high demands on quality). On account of the comparatively large clearances, the amount of water required is high but remains nearly constant as hardly any wear occurs. The leakage liquid is isolated from the lube oil by a gas purge (usually air or nitrogen).

F Floating white-metal seal rings; no leakage gas accumulates, but seal liquid enters the compression chamber. As a rule, water is used as seal liquid. On account of narrow clearances, high-quality water (demineralised) must be used. The amount of liquid required is small, but this may increase with wear. The leakage liquid is isolated from the lube oil by a gas purge (usually air or nitrogen).

Figure 2-6. Principles of sealing. (Source: MAN-GHH, Oberhausen, Germany, Publication M636e05952 br.)

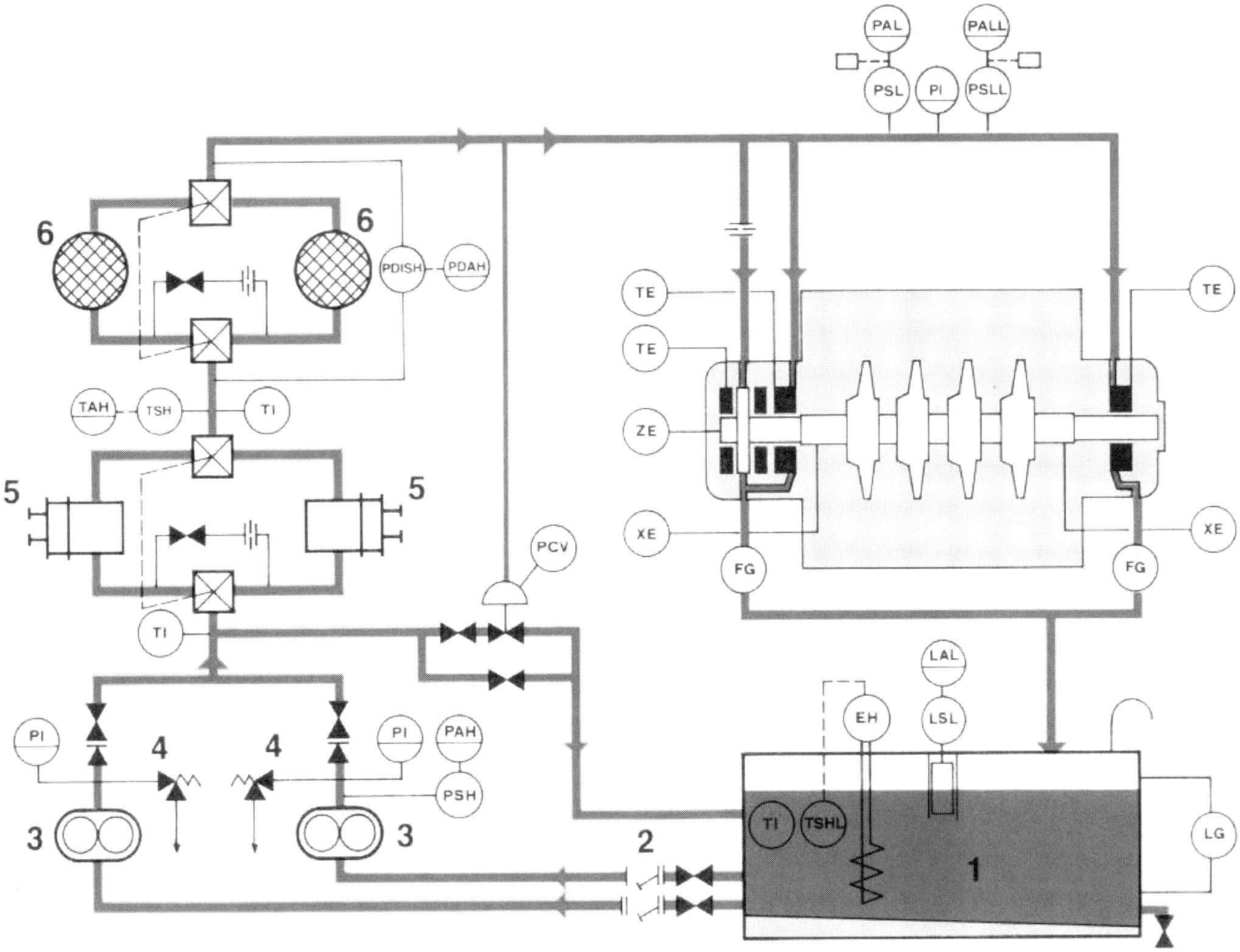

A force feed lube oil system ensures that journal and thrust bearings are properly lubricated. The system is basically designed in accordance with API 614 and is suitable for continuous compressor operation. It generally comprises:

a) oil reservoir;
b) steam turbine or electric motor driven screw or centrifugal pumps;
c) full-flow oil coolers;
d) twin filter (5/10 micron) allowing cartridges to be changed during operation;
e) automatic by-pass valve to control oil pressure at the journal bearing manifold;
f) monitoring and safety instrumentation.

LEGENDA

1. Oil reservoir
2. Suction strainer
3. Oil pump
4. Relief valve
5. Oil cooler
6. Oil filter

○ - Locally mounted instrument
○ - Panel mounted instrument
LG - Level glass
LSL - Low level switch
LAL - Low level alarm
PI - Pressure indicator
PCV - Pressure control valve
PSL - Low pressure switch
PAH - Low pressure alarm and stand-by pump start-up
PSLL - Minimum pressure switch
PALL - Minimum pressure shut-down
PSH - High pressure switch
PAH - High pressure alarm
PDISH - High diff. pressure indicator and switch
PDAH - High diff. pressure alarm
TI - Temperature indicator
TE - Thermoelement
TSH - High temperature switch
TAH - High temperature alarm
TSHL - High/low temperature switch
ZE - Axial displacement probe
XE - Radial probe
FG - Flow glass

Figure 2-7. Circulating (Force-feed) lube oil system for large screw compressors and centrifugal machines. (Source: Nuovo Pignone, Florence, Italy.)

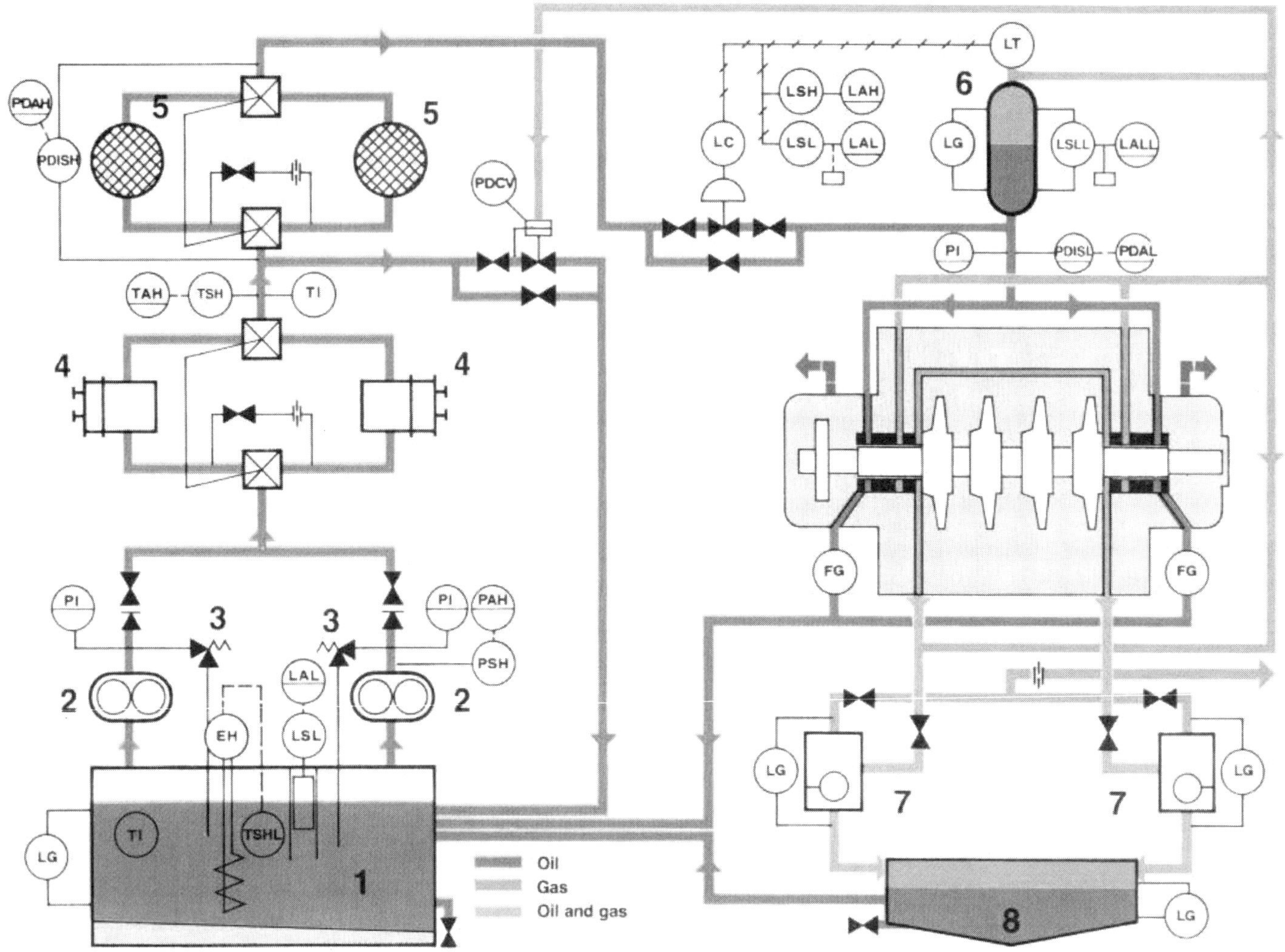

The seal oil system supplies filtered oil to the liquid film rings or to mechanical type seals at the correct pressure and temperature. The system is basically designed in accordance with API 614 and is suitable for continuous compressor operation. It generally comprises:

a) oil reservoir;
b) steam turbine or motor driven screw or positive displacement pumps;
c) full-flow oil cooler;
d) twin filters (5-10 micron);
e) overhead tank with level or pressure control;
f) automatic seal oil traps;
g) monitoring and safety instrumentation.

LEGENDA

1. Oil reservoir
2. Oil pump
3. Relief valve
4. Oil cooler
5. Oil filter
6. Overhead seal oil tank
7. Seal oil trap
8. Drain pot

○ - Locally mounted instrument
⊖ - Panel mounted instrument

LG - Level glass
LSL - Low level switch
LAL - Low level alarm
LT - Level transmitter
LC - Level controller
LSH - High level switch
LAH - High level alarm
LSLL - Minimum level switch
LALL - Minimum level shut-down
PI - Pressure indicator
PSH - High pressure switch
PAH - High pressure alarm
PDISL - Diff. pressure indicator and switch
PDAL - Low diff. pressure alarm
PDISH - High diff. pressure indicator and switch
PDAH High diff. pressure alarm
TI - Temperature indicator
TSH - High temperature switch
TAH - High temperature alarm
TSHL - High/low temperature switch
FG - Flow glass

Figure 2-8. Circulating seal oil system for large screw and centrifugal compressors. (Source: Nuovo Pignone, Florence, Italy.)

tered. The limiting factor is typically the circumferential speed of the male rotor, which generally ranges from 40 to approximately 120 m/sec, and up to a maximum of 150 m/sec for very light gases.

The maximum allowable compression ratio for one twin-screw compressor stage is that which will not cause the final compression temperature to rise above the permitted value of 250°C. This ratio and the associated temperature will to a very large extent depend on the specific heat ratio, cp/cv, of the gas to be compressed. For example, where the specific heat ratio equals 1.4, the maximum compression ratio would be approximately 4.5, and where the specific heat ratio cp/cv equals 1.2, the maximum compression ratio would be approximately 10 for one oil-free, twin-screw compressor stage.

2.4 REASONS FOR USING SCREW COMPRESSORS

As was alluded to earlier, twin-screw rotary screw compressors are a subset of the machinery group making up rotating positive-displacement compressors. Of the various machines available, rotary screw compressors are primarily used in higher-pressure air and process gas services, whereas certain other rotary piston blowers and single-screw compressors are more typically used in lower-pressure, high-volume applications. Both machines can be used as dry or wet fluid movers.

Rotating positive-displacement machines offer the same advantage as reciprocating positive displacement equipment with regard to flow versus pressure relationships, that is, nearly constant inlet flow volume under varying discharge-pressure conditions. Recall, also, that positive-displacement machines do not have a surge limitation, which is to say, there is no minimum throughput requirement for these compressors. (In this, they differ from centrifugal compressors that cannot be operated below a certain minimum throughput volume).

The rotor-tip speeds on rotary-screw and rotary-piston blowers are low; this allows for liquid injection and handling of contaminated gases. By design, the rotors are self-cleaning during operation, which is a significant advantage in dirty-gas services. But "dirt" needs to be kept away from bearings and dirt in the compression space needs to be disposed of, as will be seen later.

2.5 OIL-FREE VERSUS OIL-FLOODED TWIN-SCREW COMPRESSORS

As can be inferred from Figs. 2-4 and 2-5, rotary-screw compressors are available in oil-free or oil-flooded construction. Technically speaking, "oil-free" refers to not having oil in the compression space, but bearings still require lubrication by a clean medium, typically oil. We say "typically" because pressurized water can also be used as a lubricant, but this is a relatively new and somewhat advanced technology. Although oil-free twin-screw compressors are widely called "dry-screw" machines, a major manufacturer defines and designates as "dry screw" *any* screw compressor equipped with timing gears.

Therefore, whether or not the compression space is dry, oil-flooded, or water-injected makes no difference: *With* timing gears, the machine is defined as a dry-screw machine. *Without* timing gears, it cannot function as a dry-screw machine, because the resulting contact of mating rotors would destroy the machine. If there are no timing gears, a separating liquid must be used, and that makes it a flooded or wet-screw machine.

Fields of application for oil-free machines include all processes that cannot tolerate contamination of the compressed gas or in which the lubricating oil would be contaminated by the gas. As discussed later, dual-circuit oil-flooded machines can also be used for such applications.

Oil-flooded machines can achieve slightly higher efficiencies than "dry-screw" machines and can utilize the oil for cooling as well. The process gas must be clean for dry screw compressors to work satisfactorily in plants that desire to limit maintenance work.

Properly designed rotary-screw compressors are constructed with no metallic contact whatsoever inside the compression chambers, either between the rotors themselves or between these and the walls of the housing. As was mentioned earlier, although originally intended for air compression, rotary screw compressors are, in the 21st century, compressing a large number of process gases in the petrochemical and related industries. These include air separation plants, industrial refrigeration plants, evaporation plants, mining, and metallurgical plants.

Practically all gases can be compressed: ammonia, argon, ethylene, acetylene, butadiene, chlorine gas, hydrochloric gas, natural gas, flare gas, blast furnace gas, swamp gas, helium, lime-kiln gas, coking-plant gas, carbon monoxide gas, all hydrocarbon combinations, town gas, air/methane gas, propane, propylene, flue gas, crude gas, sulfur dioxide, oxide of nitrogen, nitrogen, styrene gas, vinyl chloride gas, and hydrogen gas can be found on the reference tabulations of experienced manufacturers.

One manufacturer alone has dozens of oil-flooded twin-screw compressors in successful service for at least three years of uninterrupted service. Other screw compressors have continuously stayed on line for six or more years. Uninterrupted service is the key here, as is vendor experience with a particular gas composition.

2.5.1 Bearings

Although air machines are often equipped with hardened-steel, rolling-element bearings that have very little tolerance for water and H_2S, the majority of compressors for process gas applications are furnished with journal bearings, sometimes called sleeve bearings or sliding bearings. Journal bearings are shown in Fig. 2-4 (item 7) and Fig. 2-5 (item 4). Here, as in most larger screw compressors, the thrust bearings are typically of the Babbitt-lined sliding or multisegmented pad type commonly found in centrifugal process gas compressors. The service life of these bearings is practically unlimited as long as proper lubricating and operating procedures are in force. These bearings are also far more tolerant of H_2S and water than are steel rolling-element bearings [3]. However, rolling-element bearings are acceptable for (relatively) light loads and where oil cleanliness is assured. Needless to say, selecting a separate closed-loop bearing-oil circuit is a powerful step in the direction of maintaining a clean bearing environment. (See also Fig. 3-22 for typical compressor bearings).

2.5.2 Shaft Seals

Rotary compressors have to be sealed against the atmosphere and the sealing area is equipped with connections for sealing-medium supply and relief. In principle, it is possible to apply a cooling medium to the cylinder wall, but noncooled cylinder housings (or casings) can be used as well.

The compressor main shaft seals are located at both ends of each rotor. As mentioned earlier, in the dry or oil-free helical-screw compressor there are four seals, two per rotor. Three types of seals are commonly used. These are the labyrinth, restrictive ring, and mechanical (contact) type oil seal [6]. Self-acting gas seals also available but have limited experience and are not considered an economic choice for the low operating-pressure levels experienced in screw compressor services.

2.5.3 Internal Seals

Internal seals (see Fig. 2-6) are not to be confused with the seal(s) in Fig. 2-5, item 6, that seal against atmosphere. In many rotary-screw compressor applications, it is necessary to provide separate circuits for the liquid that takes up a portion of the compression space and the oil that gets sent to the bearings. With separate liquid or oil circuits, a sealing barrier is needed between the process gas and the bearings. A number of different seal types are feasible and their selection is largely governed by vendor experience [2]. Although disclosure of general vendor experience is common, disclosure of competing selection criteria and definition of certain proprietary configurations and parameters is a function of more detailed engineering studies. The vendor initiates these studies upon issuance of purchase documents. In any event, the vendor selects from

- Carbon ring seals
- Barrier-water, floating-ring seals
- Double-acting mechanical seals with stationary springs
- Combined floating rings and mechanical seals

At the compressor input shaft, manufacturers often opt for either labyrinth seals or double-acting mechanical seals with rotating springs.

Carbon ring seals with connections for the injection and eduction (evacuation) of inert, clean gases are used in cases in which leakage gas, even in connection with sealing gas, may enter into the bearing areas or into the atmosphere. The gas pressure is relieved across floating carbon rings at the beginning of the seal chamber.

With barrier-water, floating-ring seals, barrier water enters the seal chamber and a small amount of water reaches the compression space. Most of the water is returned to the barrier-water system for cooling, filtration, and reuse. Barrier-water seals are able to fully prevent gas leakage and can provide valuable cooling and scrubbing duties.

A double-acting, stationary-spring mechanical seal and a combination mechanical and floating-ring seal are primarily used for compression with high differential pressures.

2.6 SCREW COMPRESSOR VOLUME CONTROL

Operating at a certain speed and incorporating defined sizing parameters or internal geometries, positive-displacement compressors provide a certain volume of gas per unit of time. This gas is usually referred to as "acfm," or actual cubic feet per minute. However, the receiving process may, occasionally, demand less than this built-in volume capability. In that case, one needs to limit, or control, the compressor throughput.

In principle, it is necessary to consider the problems of volume control for dry-running and oil-injection-type screw compressors separately. First, then, let us deal with the control of the volumetric output of dry-screw compressors.

Control by Variable Speed

In consequence of the fact that screw compressors displace the medium positively, the most advantageous method of achieving volume control is that obtained by varying the speed. This may be done in any of thc following ways:

- By variable speed electric motors
- By use of a torque converter
- By steam turbine drive

Speed may be reduced to about 50% of the maximum permissible speed. Induced flow volume and power transmitted through the coupling are in this manner reduced in approximately the same proportion. The allowable turndown depends on the adequacy of bearing lubrication at low speed and compressor discharge temperature. More than 50% reduction is possible in special cases. As mentioned earlier, there is no surge limit (a minimum flow below which the gas would alternate between forward and reverse flow) for these positive-displacement machines.

Bypass

Using this method, the surplus gas volume is allowed to flow back to the intake side by way of a compressor-discharge pressure controller. An intermediate cooler brings the surplus gas volume down to intake temperature.

Full-Load/Idling-Speed Governor

As soon as a predetermined final pressure is attained, a pressure controller operates a diaphragm valve that opens a bypass between the discharge and suction sides of the compressor. When this occurs, the compressor idles until pressure in the system drops to a predetermined minimum value. The valve will close once again on receiving an impulse from a pressure sensor. This brings the compressor back to full load.

Suction Throttle Control

This method of control is suitable for air compressors only. As in the case of the full-load/idling-speed control method, a predetermined maximum pressure in the system, for example in a compressed air receiver, causes pressure on the discharge side to be relieved down to atmospheric pressure. Simultaneously, the suction side of the system is throttled down to about 0.15 bar absolute pressure. When pressure in the entire system has dropped to the permissible minimum value, full load is once again restored.

2.6.1 Volume Control for Screw Compressors Equipped with Oil Injection (Oil-Flooded Compressors)

Suction Throttle Control

Since the final compression temperature is governed by the injected oil, a greater range of compression ratios, such as may arise when the induced volume is throttled down, can be safely coped with. This permits the main flow volume to be varied within wide limits.

Built-in Volume Governor

Large compressors are frequently equipped with an internal volume-regulating device. By operating a slide valve (Fig. 2-5, item 9) that is shaped to match the contours of the housing and built into the lower part of the housing, designed to move in a direction parallel to the rotors, the effective length of the rotors can be shortened. The range of this control mode is typically between about 10% and 100%. Compared with suction throttling, this type of control offers more efficient operation.

Slide valves can be likened to mechanical cylinders that can be moved along the periphery of the compressor casing. Suppose a casing contains a helical rotor that is 36 inches long. Suppose further that the gas inlet nozzle of the compressor normally admits the gas at the end face of this rotor. In that case, there would be 36 inches of rotor length within which the gas would be compressed from its inlet pressure to its outlet pressure. In contrast, consider a slide valve to be a hollow cylinder that allows gas inlet pressure to exist not only at the edge of the rotor, but over the first 12 inches of rotor length. Thus, gas compression can take place only in the remaining 24 inches of rotor and the expelled volume of gas is considerably less than its rated 100% amount.

To function properly, slide valves require lubrication by very clean oil. Only clean oil will ensure long-term, satisfactory operation of these close-clearance components.

2.7 SCREW COMPRESSOR AUXILIARIES

For a brief description of lube and seal oil systems and schematic representations of both refer to Section 2.2 and Figs. 2-7 and 2-8. Although these are, indeed, the most important compressor auxiliary systems, experience shows the upstream scrubbers to be perhaps the most misunderstood and most neglected.

2.7.1 Suction Scrubber and Drain Seal Drum

In the great majority of installations that employ gas compressors, suction scrubbers are placed upstream of the compressor inlet nozzle. This is where they serve a dual purpose. If a massive volume of liquid should move toward the compressor inlet, a liquid-level-sensing device will shut down the compressor. The second, and for some installations equally relevant, purpose of the suction scrubber is that of taking out of the gas stream any contaminants which, if left in the gas, could cause abrasion, wear, or "gumming-up." In the case of gas compressors provided with a finite volume of liquid (i.e., flooded, or "wet-screw" compressors), excessively high concentrations of these carryovers might jeopardize compressor component life.

As can be gleaned from the block diagram in Fig. 2-9, the suction scrubber is installed between the upstream gas supply header and the compressor inlet nozzle. This suction scrubber is not to be confused with what the compressor manufacturer or packager calls the primary and secondary separators, coolers, and so on, all of which are downstream of the discharge nozzle of a wet-screw compressor.

Any liquid collected near the base of a compressor suction scrubber (1) flows into a tank (2) (see Fig. 2-9). This "tank" is called a drain-seal drum because it "seals" like the liquid-

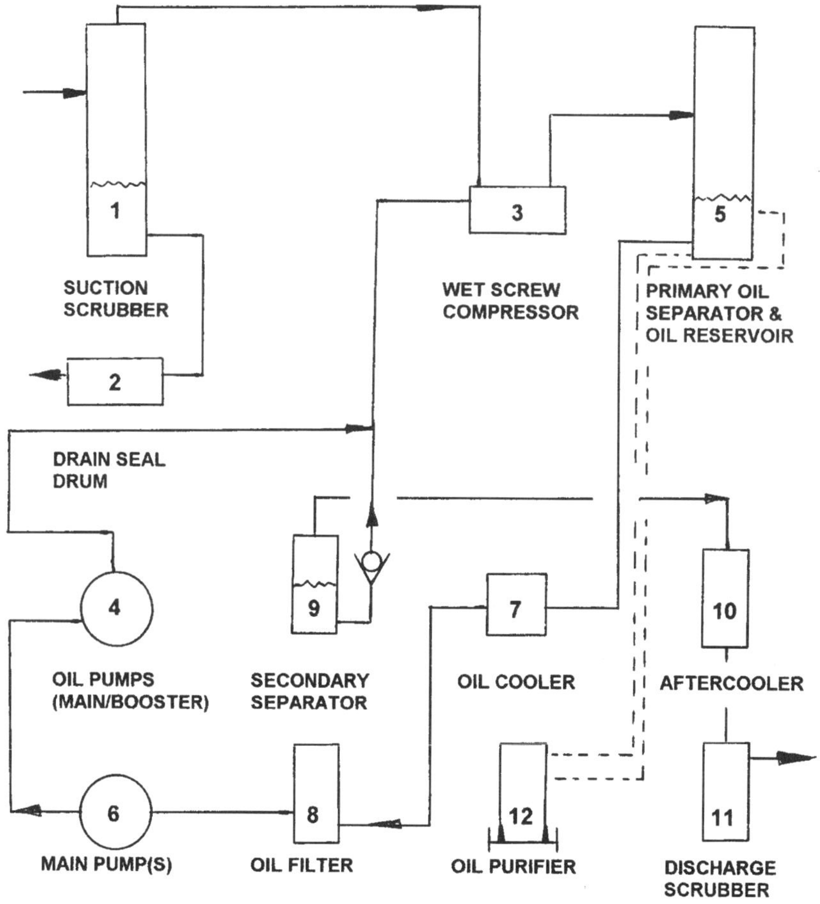

Figure 2-9. Block diagram of wet-screw compressor and auxiliaries.

filled gooseneck trap under a kitchen sink. The liquid-filled gooseneck (or seal) prevents downstream vapors residing in certain drain headers from flowing back into the interior of the scrubber (1).

The overhead gas (overhead because it rises to the top of the suction scrubber) enters the suction aperture of the wet-screw compressor (3). A pressure-sensing line from some upstream header signals the compressor slide valve (Fig. 2-5, item 9), causing it to make adjustments so as to maintain a required pressure ahead of the scrubber. A certain amount of oil is injected into the compressor; its quantity and pressure are supplied by one or more oil booster pump(s) (4).

In essence, the compressed gas is now mixed in with oil and moves on to the primary oil separator and oil reservoir (5).

2.7.2 Primary Oil Separator and Oil Reservoir

These two separators (primary and secondary) contain coalescer elements. They should not be confused with the suction scrubber which, unfortunately, is occasionally but erroneously called a separator. Primary and secondary separators are a necessary components located after wet-screw gas compressors. The intermingled liquid and gas phases being advanced from the compressor suction to the compressor discharge must be separated after their arrival at the compressor discharge.

At the wet-screw compressor discharge flange, a combined stream of gas and oil exits through a check valve to the oil reservoir where most of the oil is separated from the gas. The remaining oil is separated in a further downstream separator (the secondary separator), and residual oil amounts of typically 5 parts per million (ppm) continue to remain in the gas stream. Oil carryover can be further lowered by downstream cooling and final moisture separation. The primary and secondary oil separators have to be properly maintained and the pressure drop across the separator cartridges taken into account to determine the overall performance of the compressor package. It should also be recognized that the efficiency of oil separation depends on the degree of contamination of the separator elements. A clogged filter is obviously not efficient.

Principle of Oil Separation

The principle of oil separation is shown in API 619. In a standard configuration, gas and oil leaving the compressor discharge next enter a combined oil reservoir/oil separator. From here gas, with a considerable amount of entrained oil, flows upward and toward (typically) vertically oriented coalescer cartridges. Oil droplets agglomerate on the outside of these cartridges and mostly gas flows to the inside of the cartridges. Any residual oil is expected to drop into the bottom inside of the cartridges and small tubes lead from this oil layer back to the compressor suction. The main amount of oil is pumped from the oil reservoir/oil separator through an oil cooler and an oil filter into the slide valve and suction aperture regions of the wet-screw compressor.

The lower part of a primary separator (5) (see Fig. 2-9) constitutes an oil reservoir and the collected oil is now fed, by one or more oil pumps (6), through an oil cooler (7) and an oil filter (8) to the oil booster pump(s) (4).

2.7.3 Secondary Separator

The partially cleaned gas moves on to the secondary separator (9) (see Fig. 2-9), which is intended to take out the rest of the oil. Any oil taken out here exists at compressor gas discharge pressure and will thus readily flow toward an injection nozzle at or near the compressor suction.

2.7.4 Oil Cooler

Oil is cooled in an oil cooler (7) (Fig. 2-9) so as to regain its required viscous characteristics and restore its capacity, upon being reused and reinjected, to absorb heat from the gas being compressed. This then limits the outlet temperature at the compressor discharge nozzle. Moreover, the portion of the oil coming into contact with the bearings will perform even more critically important lubrication duty and must be cooled to have adequate film thickness and protective properties.

2.7.5 Compressor Aftercooler and Discharge Scrubber

The compressor aftercooler (10) (see Fig. 2-9) removes heat from the compressed gas before it enters the compressor discharge scrubber (11), assuming, of course, that such scrubbing is required by the downstream process. It can be assumed that the scrubber materials or scrubbing media require that incoming gas temperatures stay within a certain range or limits.

2.7.6 Oil Purifier or Oil Conditioner

An oil purifier or oil conditioner (12) (see Fig. 2-9) is either a portable or installed device that can be used to decontaminate or purify a circulating oil stream that has been slowly degraded by an influx of contaminants. Such a device will not, normally, require shutdown of the compressor. Since the finite volume of oil used in wet-screw (oil-flooded) machines will slowly but surely become contaminated by any particulate matter or water that finds its way into the compression space, such a device is sometimes required.

The need for oil purification is a function of the effectiveness of the compressor suction scrubber (1) and the allowable frequency of changing the elements in the oil filter (8). However, purifier connections are generally specified as a minimum requirement.

Again and as mentioned below, information on the effectiveness of the upstream compressor suction scrubber (1) can be of extreme importance. To ensure a high-reliability, low-maintenance installation, these data must be supplied either by a competent compressor manufacturer or by knowledgeable manufacturers of self-cleaning, reverse-flow filter–separators for use upstream of positive-displacement compressors.

2.8 ISSUES WITH H_2S IN ALL GAS COMPRESSORS

It can be said that a separate lube oil circuit would more fully protect the bearings against intrusion of contaminants, including the longer-term detrimental effects of H_2S on, especially, steel rolling-element bearings.

Based on experience amply documented by NACE (National Association of Corrosion Engineers) and various compressor manufacturers and users, a combination of hydrogen sulfide, water, stress, and steel hardened to RC 25 or higher will place the steel used in rolling-element bearings in the 100% failure range [3].

2.9 CONSIDERATIONS FOR UPSTREAM SEPARATORS

All too often, problems with positive-displacement compressors are tackled by confining the problem analysis to the compressor. In many instances, the root causes of machine distress are centered on the upstream separation equipment. This issue is so important and so frequently overlooked that it clearly merits being mentioned even in this "overview" text. Take, for example, the protection of positive-displacement compressors in the hydrocracking process.

2.9.1 Process Example

Many major refineries operate reciprocating compressors in their hydrocracker units. These machines are often arranged in parallel; two machines are typically operating, while a third compressor is either down for repairs or is on standby. In some facilities, the overall reliability has been poor, with two-week runs occasionally the rule rather than the exception.

In the late 1990s and at just one refinery, maintenance cost outlays for three such machines (about 4,500 kw/6,000 hp each) exceeded $900,000 in a 12 month period. Comparable industry data put a realistic MTBR (mean time between repair) at six months. Moreover, yearly expenditures in excess of $100,000 per compressor would be considered above average [9].

At the two week MTBR problem plant, valve repairs predominated and an examination of these valves showed serious solids buildup. Other components affected included piston rings, rider bands, unloader components, crossheads, and cylinder liners.

2.9.2 Reviewing the Problem and Outlining the Solution

If hydrocracker machines suffer from unreliability, check first for evidence of liquid carryover into the compressors. This carryover often leaves a residue that has the appearance and viscosity of black shoe polish. In the majority of cases, the residue contains high percentages of asphaltenes and iron sulfide. The defect condition of suction and discharge valves may tempt analysts to consider switching to another valve type; either poppet or concentric-ring valves made from PEEK high-performance plastic have given good service at facilities that protect the cylinder internals from liquid and particulate matter.

Well-performing plants often use PAG (polyalkylene glycol) lubricants in cylinder and packing areas [5]. These lubricants are not very soluble in the hydrocarbon stream and will do an adequate job in this difficult service. It should be noted that the insoluble asphaltene content can still condense in the compression system and cause all of the aforementioned hardware problems.

Experience at dozens of locations worldwide points to self-cleaning, reverse-flow mist coalescers with integral slug interceptors* (KTCs, for the purpose of this segment of our text) as probably the most effective and reliable method for removing iron sulfide, and also asphaltenes, from gas streams. However, these coalescer vessels must be configured and sized to accommodate the particular gas conditions encountered at a given process unit. In other words, they must be "engineered" and not just "fabricated."

Occasionally, a plant with side streams protects only the first stage inlet of its hydrocracker reciprocating compressors with this vessel. Such a facility will often enjoy relatively good maintenance experience only for the first stages of their compressors. Should a contaminated side stream enter the subsequent stages without first being routed through a KTC, problems can still occur in these stages. Asphaltenes are among the main culprits here.

*Designation and configuration employed by King Tool Company, Longview, Texas.

Asphaltenes and Where They Come From

Asphaltenes are present in petroleum in its original state. Heavy-fraction asphaltenes are also prevalent in resid fuels, simply because refineries extract ever larger amounts of light fractions out of their crudes. They do this in different cracking and visbreaking units.

Handbooks define asphaltenes as condensed polynuclear aromatic-ring systems with alkyl side chains. They are believed to be held in solution within oil by the presence of resins and aromatics. Resins act as deflocculating agents, protecting the asphaltene particles and preventing their flocculation and precipitation in the presence of aliphatic hydrocarbons.

Process engineers often assume that dispersants have to be used to keep the asphaltenes well dispersed to prevent their flocculation or aggregation. A dispersant functions by adsorbing on the surface of materials that are insoluble in the oil, and converting them into stable colloidal suspensions.

Contaminant Removal

Although asphaltenes may respond to the addition of dispersants, experience shows that where KTCs are used, asphaltenes will be removed even though no dispersants are added to the gas stream. Well-engineered KTCs will also efficiently remove iron sulfides, ammonium chlorides, salts, oils, treating chemicals, and regular hydrocarbon liquids.

As most major producing and refining plants know from experience, iron sulfide is one of the most difficult contaminants to filter out. Ordinary impingement separators and filter coalescing separators do not perform nearly as well as KTCs. The latter are needed for two reasons:

1. Iron sulfide is a light solid that is readily sluffed into small particles by the forces of the gas flow. As a result, large particles initially trapped on a filter will eventually sluff down to a small enough size to pass through even a 0.2 micron absolute filter paper and be reentrained right back into the gas stream. Iron sulfide also readily attaches itself to liquid, although it is not easily wetted by water or hydrocarbon.
2. Based on experience with similar services, the iron sulfide appears to readily pass through conventional filters and is generally found in sludge-like deposits on wetted surfaces at the compressor suction and discharge valve ports.

Design Peculiarities of Self-Cleaning, Reverse-Flow Mist Coalescers

Perhaps the most significant difference in the proprietary design of KTCs as compared to many of the more conventional filter–separators is in the gas flow direction as it exits from the filter elements. With a conventional flow filter–separator (Fig. 2-10), the gas enters the first-stage fil-

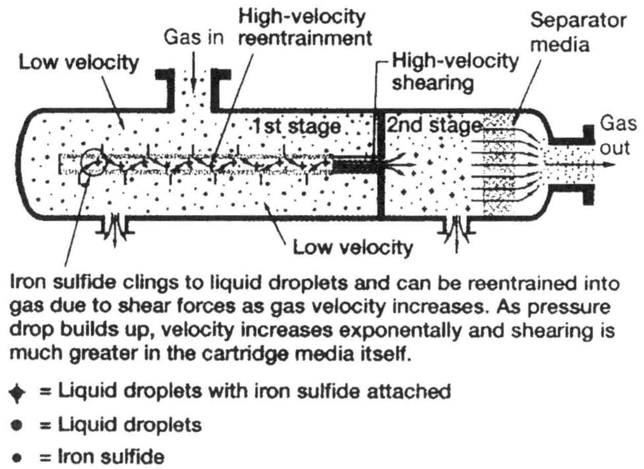

Figure 2-10. With a conventional flow filter–separator, the gas enters the first-stage filter elements. (Courtesy King Tool Co., Longview, Texas.)

ter elements and its velocity is slowed as it passes through the enlarged filter-element area. Initially, the larger iron sulfide particles are caught by the filter, but the gas forces gradually sluff it to a particle size that will pass through the element. The gas, particulates, and any liquid coalescing on the inside of the filter element tend to be reaccelerated and reentrained in the collector tube, and subsequently fed, at an appropriate velocity, into the second-stage separator section. Since the wire mesh or vane-type separator often used in this section generally passes fine mist and particulates smaller than 8–10 microns, some of the liquid and particles are left entrained in the gas.

In the KTC (Figs. 2-11 through 2-13), the gas first passes through the separator section, then through the collection tubes to the filter elements. The front end constitutes a slug-free liquid knockout or "deentrainment" section; it is sized to reduce the gas velocity out of the filter elements. At this lower velocity, any iron sulfide particles that are forced through the filter can simply either fall out or attach themselves to the coalesced liquid droplets that fall into the sump. This design is effective in removing essentially all entrained particulates and mists, as well as free liquids and large agglomerated materials.

In order to achieve the lower velocities and accommodate the range of densities, viscosities, and surface tensions likely to be represented in the gas stream, KTCs are usually larger than a conventional filter–separator and, consequently, more expensive. Obviously, the designer must apply uncompromising sizing criteria that have been validated by field experience.

Conventional Equipment Examined

In a cost-conscious environment, reliability professionals should anticipate questions about the suitability of less expensive separators to remove harmful contaminants from compressor inlet gas streams. The following observations touch on this issue.

Potential buyers have access to a large number of well-documented case histories in which KTCs have either been installed at the inception of a project, or have replaced conventional filter–separators to solve particulate and asphaltene carryover problems. In hydrocracker units, many of these problems involved iron sulfide.

It takes knowledge and technology to effectively remove asphaltenes, iron sulfides, and other contaminants from process gases. Some equipment may only be able to offer filter elements having smaller nominal or absolute ratings; with those, however, filter failures abound. Moreover, the shear forces involved with higher-pressure-drop cartridges may lead to solids carryover.

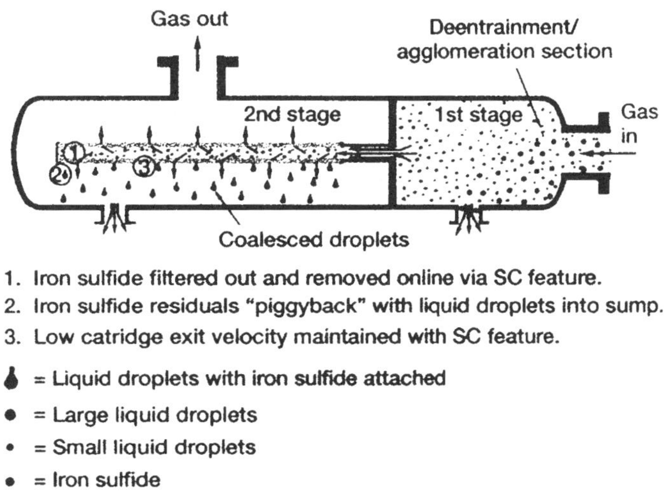

Figure 2-11. In an SCC, the gas first passes through the separator section. (Courtesy King Tool Co., Longview, Texas.)

Figure 2-12. Reverse-flow filter–coalescer installation. (Courtesy King Tool Co., Longview, Texas.)

Figure 2-13. Reverse-flow filter–coalescer package ready for shipment. (Courtesy King Tool Co., Longview, Texas.)

Water wash units often stir up the particles; they can create unique corrosion problems and byproducts, and occasionally fail to wet all the particles that in turn escape along with the gas. In these systems, unless an ultimate separation is made upstream of the compressor inlet, considerable amounts of fine mist will be present. Such carryover may well reintroduce iron sulfides and other constituents that are trapped in the droplets, float on the droplet surface, or are dissolved in the water. Dissolved solids are detrimental inasmuch as they may coat hot compressor components after the water flashes off.

This brings us back to the central issue, namely, what approach to use in solving hydrocracker compressor problems at your plant. We found that it makes good sense to make thorough comparisons, examine case histories, develop cost justifications based on long-term economics, and often budget for the installation of a KTC to begin with.

Case Histories Involve KTCs and Claimed "Equivalents"

Some time ago, a major oil refinery purchased a KTC as a replacement for a conventional-type filter–separator for a hydrogen compressor project. The filter–separator was simply unable to prevent costly compressor maintenance. From the time the KTC was commissioned, maintenance requirements decreased to the expenditures experienced by best-of-class facilities. An additional six KTC's were purchased by this refinery and word spread to affiliated refinery sites and offshore installations. Every single one of these reported the same beneficial results and literally millions of dollars were saved. When in one instance a slug of liquid was intercepted shortly after commissioning of a KTC, the refinery documented a payback period of only three weeks!

In another refinery, a single hydrocracker compressor was reported (again in the 1990s) to cost $600,000 per year to repair and maintain. When a KTC solved the problem, the refiner purchased eight more units to protect not only other hydrocracker machines, but reciprocating compressors in hydrogen booster, makeup, and recycle services as well.

The "equivalents" often look good on paper, since they inevitably cost less at the initial proposal stage. Unfortunately, there have been incidents of coalescing filter cartridge collapse within a few hours of equipment startup. A combination of liquid and hardware debris ingestion into the reciprocating compressor resulted, and the users went back to KTCs.

Retrofitting self-cleaning, reverse-flow coalescer/slug catcher vessels is generally straightforward. The owner–contractor will have to find a place to install the equipment and provide safe venting to a flare system for the brief discharges of gas during on-stream cleaning of the cartridges.

Gas Analysis and Its Value

Some plants initiate gas testing in efforts to identify and quantify the source of particulates and harmful mists. A well-structured program of continuous on-stream gas analysis is not without merit; it will often find swings in gas quality and the presence of harmful components. However, in sour crude refining services it is extremely difficult to conclusively monitor the many types of compounds and byproducts such as sulfides, chlorides, hydroxides, sulfates, metallic salts, light hydrocarbon liquids, oils, varnishes, and so on, that often exist in a gas stream. Since they are either formed or deposited by the various reactive constituents in the gas, or are mixed in with entering off-gases, freshly formed particles and liquid aerosols can be very small in size and thus escape most detection and filtration methods. Keep in mind also that the fraction of solids and liquids in mist form that can cause compressor failure may be very small in relation to the total compressor throughput or even total amount of constituents being removed. In time, a small amount of carryover can result in substantial accumulations. Visualize, for example, that a single 10 micron droplet has the ability to fracture into 1,000 one micron droplets, 8,000 one-half micron droplets, or 1,000,000 one-tenth micron droplets, or solid elements in the case of iron sulfide (FeS).

In any event, tests often give nothing other than indications of the presence or lack of presence of a certain particle in the gas stream. Tests do not always give indications of whether or not a problem was solved. With so much at stake, bottom-line cost comparisons are mandatory in reliability-conscious plants. From case histories and comparison of KTC retrofit experience, it was found that life-cycle costing requires close comparison with competing approaches. More often than not, these competing approaches will fall short of providing a complete solution. Also, the owner–user will have to look at the total cost of the problem being experienced. Obviously, not all hydrocracker problems are severe enough to justify the KTC solution.

Upstream Separator Recommendations Can Be Generalized

Whenever a least-risk, lowest-maintenance-cost compressor project is planned and budgeted, a reliability-focused organization will take a number of important precautionary steps. The same is true if an installation were to lack reliability and availability, or if it constituted a drain on available maintenance resources. In either case, one would do well to consider the following points concerning upstream separators:

- If the gas leaving the liquid disengagement vessel is at or near saturation, make absolutely certain that it will not undergo cooling in the piping. Beware of adverse atmospheric exposure conditions in the piping connected to the compressor inlet; beware also of cylinder coolant temperatures being lower than incoming gas temperatures. When in doubt, heat trace the piping. Even the perfect KTC will not be effective if these concerns are not addressed and liquid enters the machine.

- Continuous gas analysis will detect changes in gas composition, flow rate variations, liquid entrainment, and so on. These data allow competing equipment manufacturers to use uniform design input for internal configuration and sizing decisions.
- It would be advantageous to clean the entire system before implementing a KTC installation. Such cleaning should be pursued at the earliest convenience, perhaps during a unit shutdown or plant turnaround.
- Investigate user experience. Keep in mind that a comparison of before versus after maintenance expenditures will be one of the ways to prove the effectiveness of KTC. A return to on-stream reliability, downtime avoidance, and unit throughput increases can make a compelling case for capital outlays.
- Ask about the ultimate purpose of asphaltene dispersants; they may or may not be advantageous. Is it acceptable to simply "wash" the contaminants on through the compressor or should the asphaltenes be removed?
- On self-cleaning, reverse-flow coalescers for hydrocracker units, be sure to incorporate an integral slug interceptor/deentrainment section on the front end to precondition the gas entering the filters.
- Recognize that piping between KTC and compressor inlet may have to be heat traced to avoid condensation.
- Favor low-initial-pressure-drop equipment, equipment that can remain online for long periods of time by virtue of built-in means of on-stream cleaning.
- Aim for flexibility and maintainability. It should be possible to change fouled elements at operator convenience.

Experience shows that "problem plants" that follow these recommendations will see drastic increases in unit uptime and compressor reliability. Maintenance cost reductions of 70% and more are not at all unusual and rapid payback is the rule rather than the exception.

Chapter 3

Understanding Centrifugal Process Gas Compressors

3.1 WHERE CENTRIFUGAL COMPRESSORS EXCEL

Since the early 1950s, centrifugal compressors have secured a unique place in modern process plants. Centrifugal compressors are generically part of the dynamic compressor family, which also includes axial compressors. Although typically operating at high speeds, centrifugal and axial compressors are usually highly reliable machines. The capability of operating at high speeds has brought significant advantages to users worldwide.

Centrifugal compressors of the various types shown throughout this book excel where savings in space and foundation mass in addition to lower weight of the machinery are prime considerations. In most applications, centrifugal compressors require minimal maintenance only, and uninterrupted run times of eight years are now often achieved in clean gas services in major olefins plants. When supplied with a variable-speed driver, centrifugal compressors offer significant flexibility of operation without appreciable loss of efficiency.

Only in applications in which low flow rates or particular (high) pressure requirements make it uneconomical to use centrifugal compressors, the process designer will probably opt for the reciprocating and rotary screw machines discussed earlier in this text. Conversely, axial turbocompressors (Fig. 3-1) or axial blading followed by centrifugal compressor impellers mounted on the same shaft (Fig. 3-2) are applied in extreme high flow applications. Integrally geared high-speed packaged centrifugal air compressors (Fig. 3-3), first introduced in the late 1960s, have made major inroads where the larger, 500 kW and higher, multistage reciprocating models had served for decades. Figure 3-4 shows a coverage chart for the various types of compressors and Tables 3-1 and 3-2 (see pages 44–45) convey typical size comparisons, pressure capabilities, and speed for horizontally versus vertically split centrifugal compressors.

The components of all centrifugal compressors, regardless of whether in "packaged" air or "conventional" process gas service, must be expertly designed and manufactured. However, engineering expertise is not only limited to the components identified in the nomenclature of Fig. 3-5, which include thrust and radial bearings, shaft sealing arrangements, and so on. Expertise is also needed for important auxiliary support systems associated with lubrication and sealing systems, or compressor-control and monitoring elements. All of these must be absolutely dependable and reliable; they will be highlighted later in this text.

Rotor dynamics issues, including vibratory behavior and sensitivity to imbalance, must be considered; even quality control and shop testing must be flawless [8]. The typical rotor behavior at the limits of a compressor performance curve (Fig. 3-6) and the rotor performance changes in response to changing gas conditions are of interest as well. Gas conditions or process gas characteristics are manifested by physical parameters such as pressure, temperature, and weight flow rate. Their quantity relationships can be quickly observed on charts such as Fig. 3-7.

Notice how these charts are of great value in scoping studies. Entering the mass flow rate on the vertical scale and intersecting at the diagonal for a given molecular weight, one would progress to another diagonal representing suction pressure, traverse to the suction (inlet) temperature, and drop down to the inlet flow "seen" or recognized by the compressor. Similar charts require input of this suction volume and a desired compression ratio to allow quick determination of absorbed power and resulting discharge temperature.

Figure 3-1. Axial turbocompressor with top of casing removed. (Source: Sulzer Turbo, Winterthur, Switzerland.)

3.2 CENTRIFUGAL COMPRESSORS, FANS, OR BLOWERS?

Associated with every type of machine are unique terminologies and concepts. Understanding the various terms and concepts will be important in successfully guiding equipment selection, installation, operation, and, finally, effective troubleshooting in the event of malfunction.

There is no clear demarcation between fans, blowers, and compressors. Various types of fans belong in the very low compression ratio region. Fans are used to produce static pressures up to a maximum of 30 inches of water column. Blowers usually handle gases at compression ratios not exceeding 1.4 per stage and, in multistage machines, a maximum ratio of 4:1. The higher-pressure machines are generally called compressors, but Fig. 3-4 is not too unusual in using the term turbo-blower for machines with pressure ratios up to 2.2:1.

3.3 CENTRIFUGAL COMPRESSOR CONFIGURATIONS AND COMPONENTS

3.3.1 Horizontally Split Compressor Casings

Compressor casings house the compressor rotor, stationary gas passageways, bearings, and seals. Quite obviously, these are important components affecting the layout of the machine. Materials of construction range from cast iron to a variety of steels with different amounts of alloying constituents. Stainless steels containing high percentages of nickel and even virtually pure titanium are available for special applications.

Gas composition and pressure and the number of gas side streams entering and/or leaving the compressor as dictated by the process are among the main factors to be considered. These largely influence materials of construction, method of fabrication (whether welded, cast, or forged), and making the decision between horizontally or vertically split casing configurations. The preferred, or even mandated, casing split orientation is governed by either the physical properties of the gas and/or its pressure.

Considering pressure only, horizontally split construction, Fig. 3-8, is generally acceptable at pressures below 55 bar (~790 psi). Low molecular weight gases might mandate lowering this guideline value further and by substantial margins. As the actual volume flow at the inlet increases, and regardless of molecular weight, the required casing size increases to the point where the relatively large sealing areas of horizontally split machines make it difficult to contain the gas. Large horizontally split compressors are thus rarely used at pressures exceeding 20 bar (~285 psi).

Cast iron casings are strictly limited to centrifugal compressor applications in which the gas is nonflammable. Here, maximum allowable working pressures and discharge temperature should not exceed 28 bar (~400 psi) and 260°C (500°F), respectively. The compressor should also be located in a safe zone, away from hydrocarbon process units. The main concern is that even during plant emergencies not directly involving the compressor, its casing may be exposed to

Figure 3-2. Combined axial-centrifugal turbocompressor (Source: Mannesmann Demag Delaval, Duisburg, Germany.)

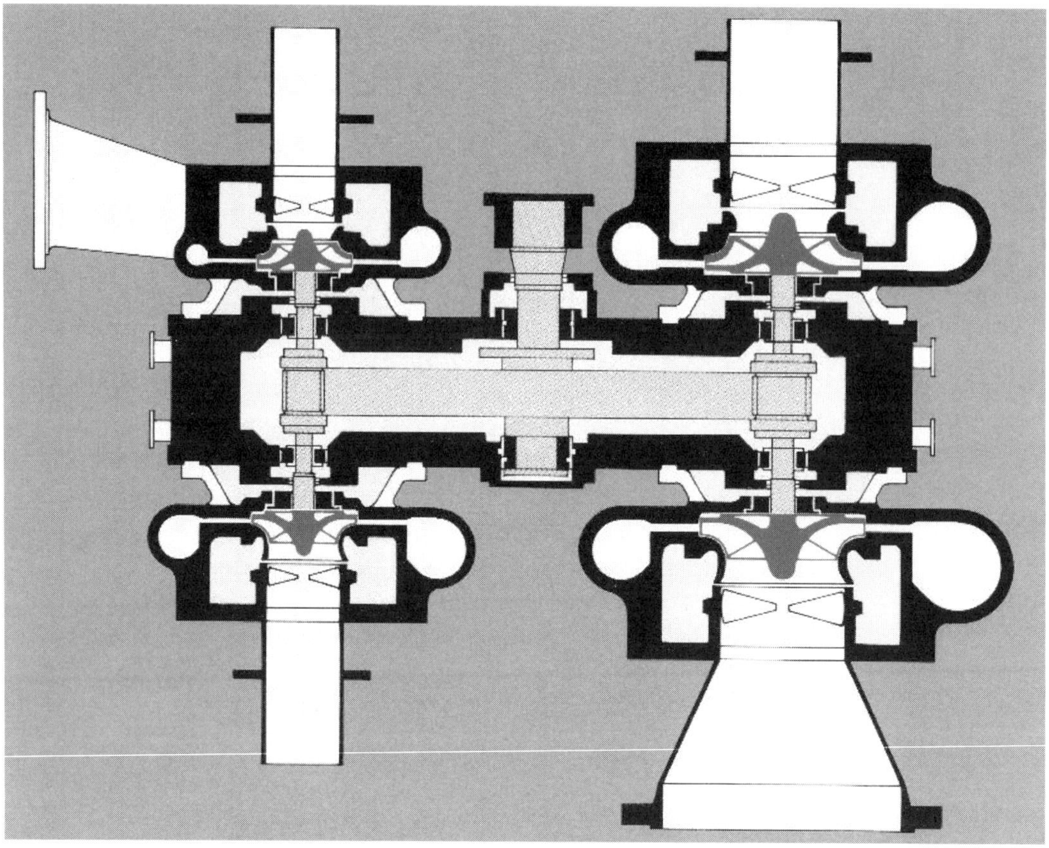

Figure 3-3. Integrally geared, high-speed "packaged" air compressor. (Source: Mannesmann Demag Delaval.)

heat from a neighboring conflagration. Spraying water on a hot cast iron casing may cause it to fracture into shrapnel. When in doubt, prudence dictates steel casings on all except reciprocating compressors. Reciprocating compressor cylinders, up to perhaps 1,000 psi (~70 bar), and reciprocating compressor frames are generally still made of gray or nodular cast iron.

As can be seen in many of the compressor illustrations in this text, horizontally split compressors have their gas inlet, outlet, or sidestream nozzles located on the lower half of the casing. Horizontally split construction is simple and allows for easy maintenance access after lifting off the top half. With nozzles located only on the lower casing half, the piping need not be disturbed for rotor access during maintenance. This access is needed to check the relative component locations and assembly clearances. In these compressors, the horizontal joint is designed as a metal-to-metal joint. The two parting surfaces should be carefully lapped before assembly. Since the parting areas are large and often of irregular surface roughness, horizontally split compressors are not suitable for low molecular weight gases. Experience shows that light gases can escape in the event of uneven tightening of the casing bolts.

Both tightening sequence and the amount of torque applied require careful adherence to proper procedures and workmanship. Any undue compromise could certainly be costly, especially on the larger horizontally split compressors. The required accuracy of assembly can be achieved relatively quickly with proper use of high-torque hydraulic tensioning devices. Tightness checks using pressurized inert gas and a soap bubble solution are needed before allowing the machine to be returned to process duty. Again, any oversights or compromising on this issue may seriously jeopardize equipment reliability. With progressing age and after the casing has been opened several times, barely perceptible deformation of casing split surfaces will inevitably have taken place. Gas leakage, both internal and external to the compressor casing, will prove either costly or hazardous, or both. In those situations, it may be necessary to use approved and chemically compatible, room-temperature vulcanizing compounds (RTVs) at the casing split in order to salvage an expensive casing.

3.3.2 Vertically Split Compressor Casings

Vertically split construction, also called barrel construction, is typically used above the pressure and molecular weight limits of horizontally split designs (see also Table 3-2). Vertically split construction (Fig. 3-9) is typically recommended

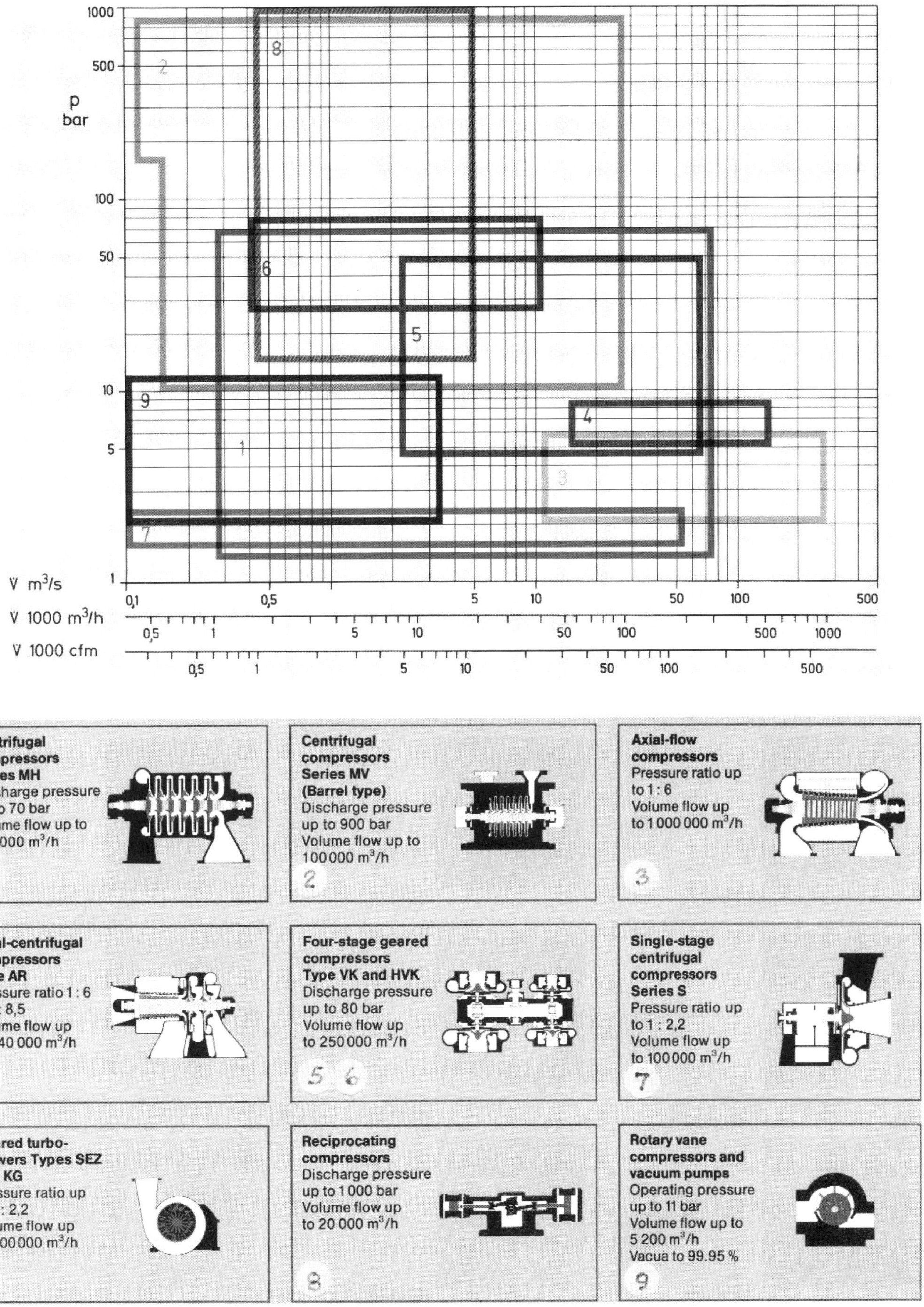

Figure 3-4. Coverage chart. (Courtesy of Mannesmann Demag Delaval.)

Table 3-1. Typical size comparisons, flows, pressures, and speeds for horizontally split compressors. (Source: Dresser-Rand, Olean, New York)

	Case model	Max ACFM/PRESS/SPEED
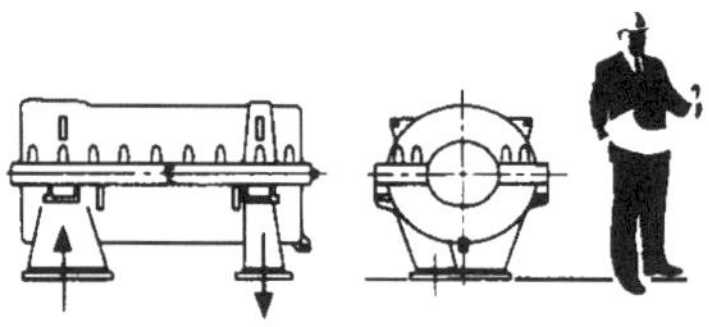	1M	4900 825 psig 14,550 RPM
	2M	7050–9500 700 psig 11,000 RPM
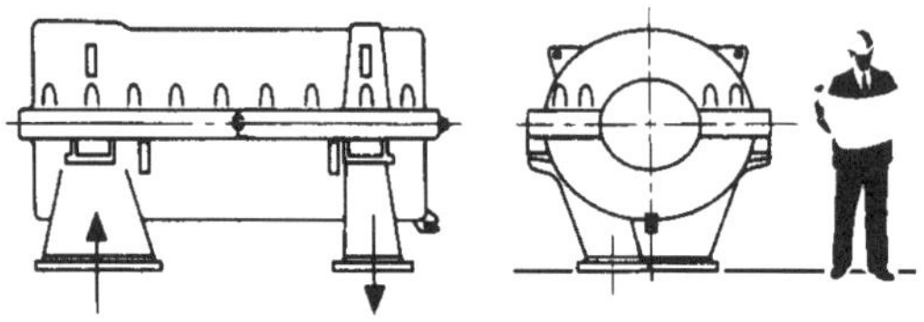	3M	14700–26500 625 psig 9390 RPM
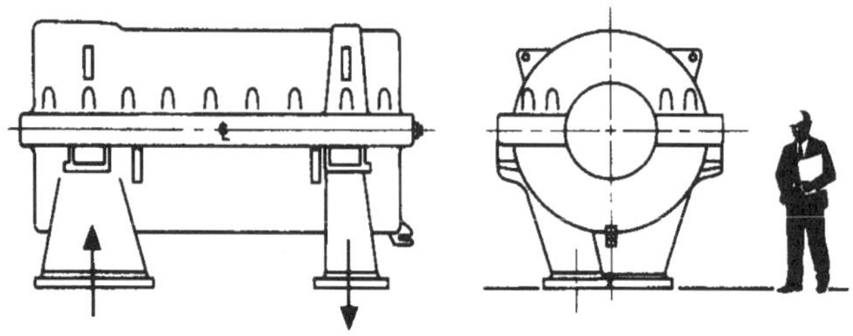	4M	41000–81000 450 psig 6500 RPM
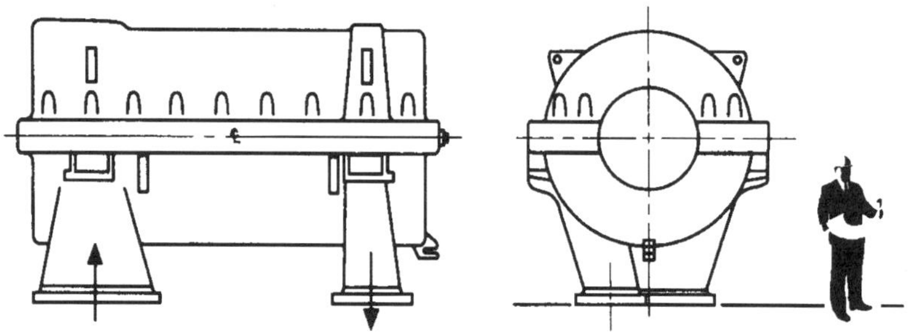	5M	60000–10800 350 psig 5000 RPM
	6M	180000 250 psig 4000 RPM

in low molecular weight applications. One example would be a hydrogen-rich service in which the partial pressure of hydrogen exceeds 200 psi (~14 bar). Aside from the need for structural strength of the equipment, operational safety is important whenever high-pressure toxic or flammable gases are being compressed. Barrel construction of the casing has extended centrifugal compressor capabilities to discharge pressures of approximately 10,000 psi (~700 bar). Applications for these machines can be found in the offshore and fertilizer industries. Vertically split or barrel construction utilizes a forged steel casing with one or both steel end covers inserted and secured with stud bolts (Fig. 3-10, left) or shear rings (Fig. 3-10, right).

Shear-ring construction is primarily and exclusively used in high-pressure services. Vertically split construction for the lower range of pressures utilizes cylindrical barrel-shaped casings ("barrels") of either forged or rolled (fabricated) steel.

On multistage barrel compressors, the impellers are initially assembled as part of a separate bundle in a horizontally split inner casing (Fig. 3-11) and with the assembled rotor sitting in the diaphragm assembly. The full bundle is then inserted in the barrel and the end covers secured in place. Since the inner bundle is to be removed for overhaul by being pulled out along the axis until it clears the main barrel, vertically split compressors are typically placed at the end of a tandem train; the compressor(s) closest to the driver are likely horizontally split. Also, vertically split compressors require special guides or rails bolted to the barrel to facilitate removal of the bundle during maintenance events (Fig. 3-12). Guide rails and other special tools (Fig. 3-13) should be supplied by the compressor manufacturer.

Table 3-2. Typical size comparisons, flows, pressures, and speeds for vertically split compressors. (Source: Dresser-Rand, Olean, New York)

	Case model	Max ACFM/PRESS/SPEED
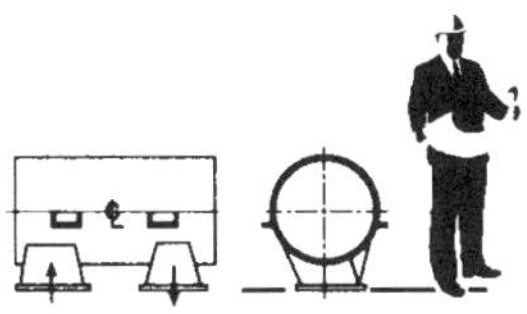	.8B	1806–4087 6500 psig 25,200 RPM
	1B	1067–6857 10,000 psig 16,000 RPM
	2B	2200–8940 5300 psig 14,000 RPM
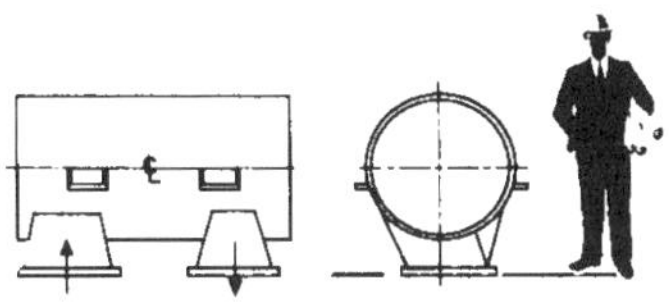	3B	3675–11,000 4000 psig 12,000 RPM
	4B	15,904–19,500 2500 psig 10,000 RPM
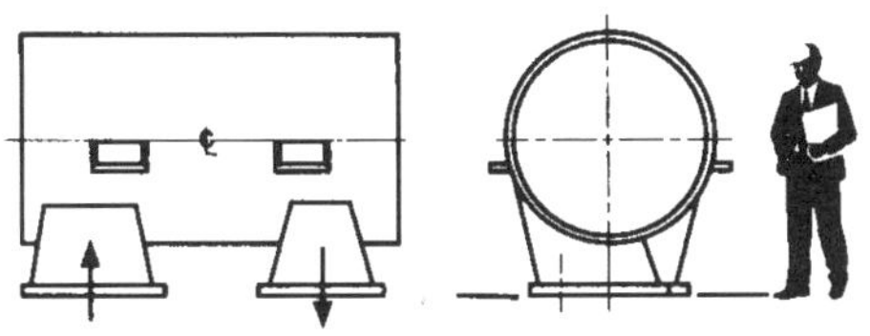	5B	23,758–26,000 1800 psig 8500 RPM
	6B	23,758–41,282 1000 psig 6500 RPM
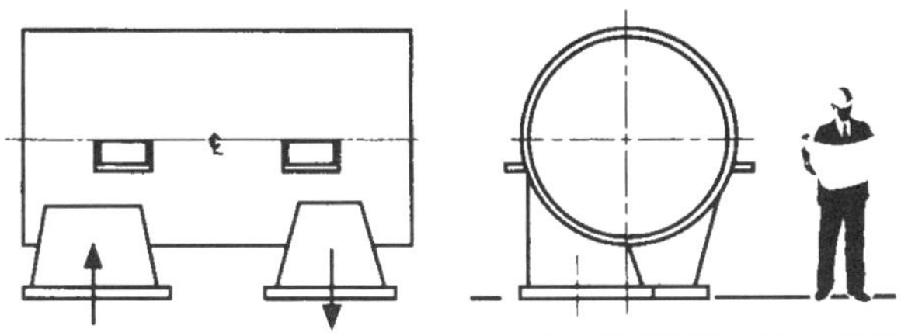	7B	60,132–82,516 1200 psig 6500 RPM
	8B	82,516–108,434 1200 psig 5000 RPM
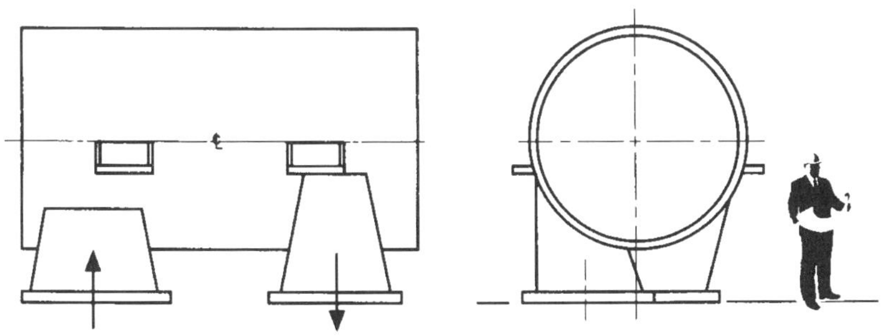	9B	180,000–360,000 300 psig 4000 RPM

At this point, it would be appropriate to compare cast construction with fabricated (rolled plate, welded) construction. In general, there is the risk of castings being rejected due to nonmetallic inclusions, blowholes, and porosity. Although special repair procedures may be available, delivery delays may result. Quality-related risks increase with the greater complexities inherent in irregular-shaped casings. Moreover, achieving uniform casing thickness is not always easy.

There are thus advantages to using either a forging or rolled steel plate [9]. Better uniformity, overall quality, and dimensional accuracy are possible in comparison with castings. Fabricated casings open up additional possibilities with respect to the desired orientation of suction, discharge, and intermediate nozzles. Compactness and, hence, more favorable overall weight are key features of fabricated construction. The issue takes on added importance with space-limited offshore installations, and also with considerations involving

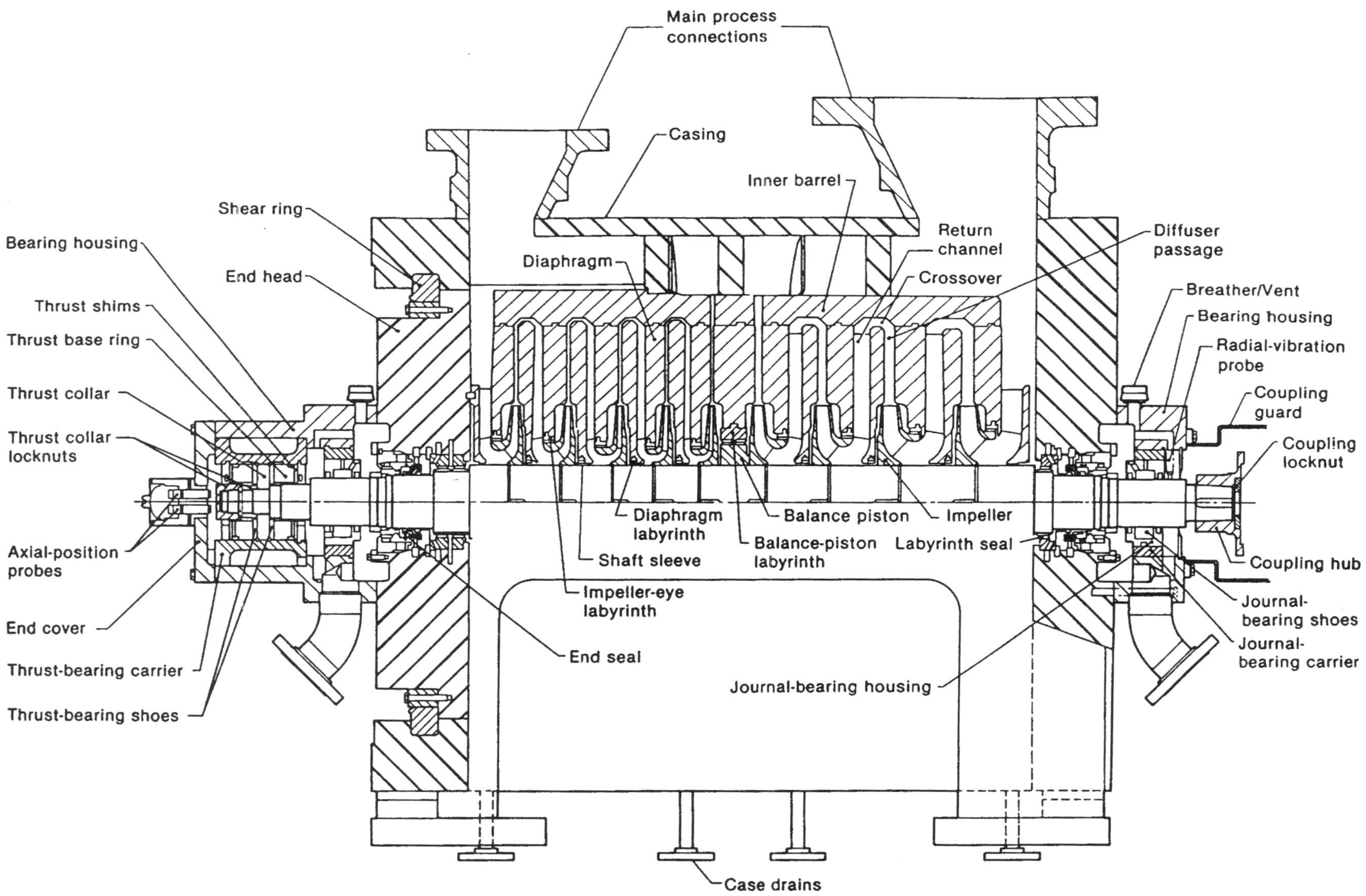

Figure 3-5. Centrifugal compressor nomenclature. *Note:* Some compressors may use bolted-head construction.

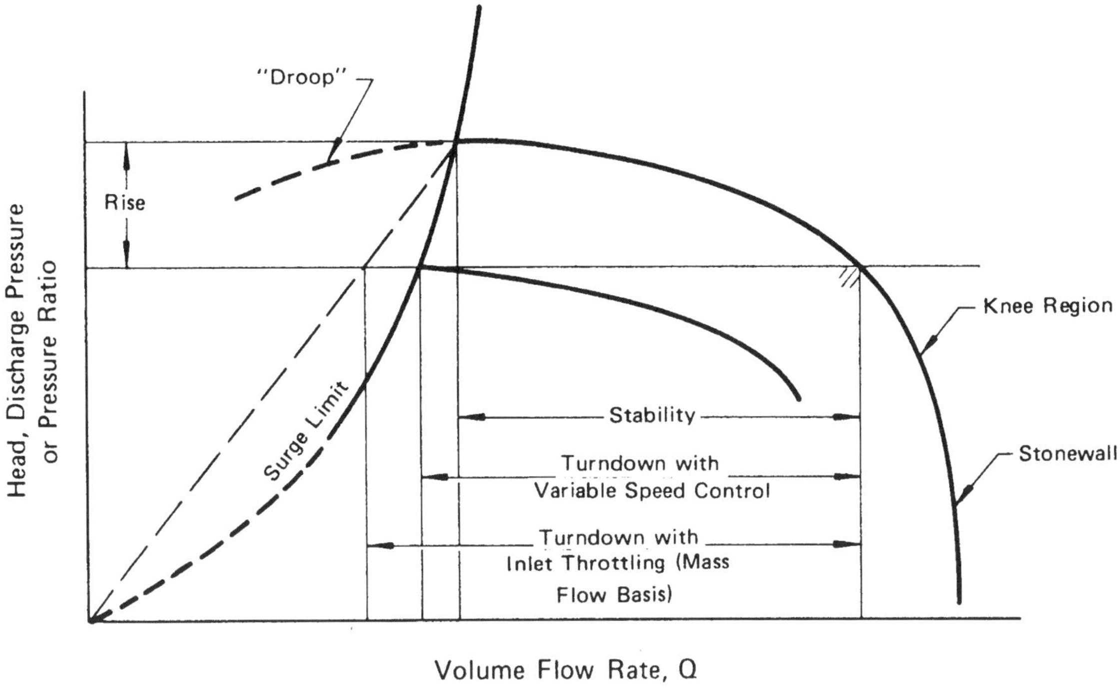

Figure 3-6. Centrifugal compressor performance curve and associated nomenclature.

time, cost, vendor's experience, availability, and other factors for competing compressor geometries.

Some of the many available casing layouts and train arrangements are indicated in Fig. 3-14. A large number of sidestream and cooling options are available and can be custom designed.

3.3.3 Compression Stages Versus Sections

An explanation of the nomenclature "stage" and "section" will be useful. For single-shaft machines with single inlet and outlet nozzles, each impeller represents a stage of compression. Hence, the number of impellers equals the number of stages. However, centrifugal machines also are occasionally required to accommodate the entry of side streams (through entry ports or inlet nozzles) and, at times, gas exit ports (discharge nozzles) for cooling and reentry of gas, or just a partial exit port (or nozzle—the terms are used interchangeably) for the gas stream. In all of these cases, the one or more impellers located between adjacent casing flanges (nozzles) constitute a "section." In other words, each section can consist of one or several stages.

3.3.4 Compressor Impellers

Two of the many types of compressor impellers are shown in Fig. 3-15. Open, semiopen, and closed types are found in industry. Semiopen and closed are the most prevalent versions of impellers used today. Impellers should be chosen based almost entirely on performance. As will be seen later, their respective head and flow parameters ("coefficients") are selected for optimum efficiency over the desired flow range; accordingly, an economical solution may differ from one application to the next. Virtually all gas compression processes deserve careful selection and could end up with either a single-stage, overhung design (Fig. 3-16) or perhaps a single-shaft version of a sidestream compressor with a semiopen first stage (Fig. 3-17), followed by as many as twelve closed impellers. There are also train configurations with multiple casings driven by a common gear, or involving a stack gas expander and motor–generator, gas and steam turbines, gear speed increases, and fluid clutches. The variations are again endless and Fig. 3-18 gives barely a glimpse. It is also possible to have every stage designed for optimum performance by choosing different speeds and impeller diameters.

Fully open impellers, as the name suggests, do not have shrouds and back plates. Usually, the profile of an open impeller is radial and its stability range is limited. Since both of its sides are open due to the absence of the shroud and back plate, an open impeller generates very little axial thrust.

Manufacturing of impellers is a labor-intensive exercise and frequently requires precision machining. Blade contours are generally milled from solid hub ("disc") material by means of multiaxis, numerically controlled machine tools. Milling is followed by vacuum brazing for narrow impellers or welding of the shroud ("cover") for the wider impellers.

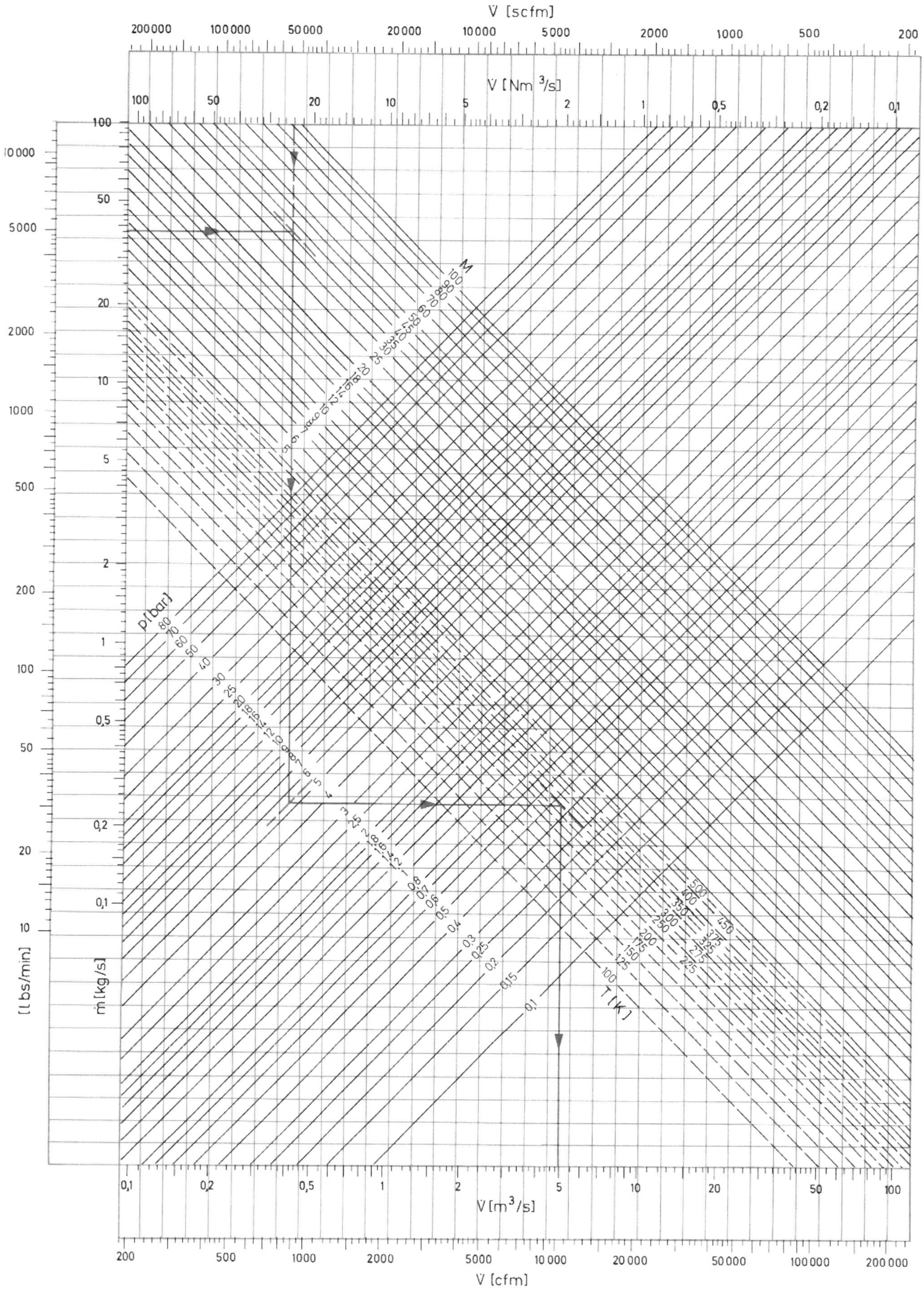

Figure 3-7. Diagram illustrating quantity relationships.

Figure 3-8. Horizontally split compressor casing. (Source: Demag Delaval, Duisburg, Germany.)

Large-size blades are bent to the desired contour and later joined to the hub by welding. Cast construction is also feasible for large-size impellers; sand castings are economical. Mass-produced, small impellers can be made using precision investment casting processes. Impellers are individually subjected to balancing and overspeed spin testing before nondestructive examination and final acceptance.

Impeller Geometry

Closed-impeller construction with backward-leaning vanes offers a wide operating range. Two versions are available: a two-dimensional ("2 D") and a three-dimensional ("3 D") one. The backward lean in a 2 D version has the same curvature throughout the blade width. In the case of a single-shaft compressor with multiple stages, the volume at the inlet to the next stage is being reduced and this lowers the stage efficiency. A compromise comes to mind: reduce the impeller diameter to gain efficiency. However, reducing the diameter would reduce head produced and more stages would be needed. Now the shaft would have to be made longer, which would affect the dynamic stability or susceptibility to undesirable vibration behavior of the compressor.

All of these factors must be given consideration for good compressor design practices. The same machine will not incorporate significantly different impeller diameters on the same rotor. That said, 2 D impellers have optimum performance up to flow coefficients of about 0.06. This coefficient-based terminology and its more precise meaning will be described a little later.

Three-dimensional (3 D) impellers with contoured ("twisted") blades seem to be better adapted to varying flow conditions. Decreasing the diameter will increase the flow coefficient and a given flow rate that demands more width. This means that the axial width of a 3 D stage exceeds that of the equivalent 2 D impeller. Since the axial dimension is wider compared to the 2 D version, the number of wheels that can be installed is restricted due to stability considerations. At optimum performance, the flow coefficient of a 3D impeller is in the range of 0.09. As will be seen later, efficiency drops markedly as a flow coefficient of 0.15 is approached.

Flow capacity and the head developed by a compressor stage can be increased by using an "inducer"-type impeller, consisting of an axial flow element similar to a feed screw auger that transitions into a radial flow element. This design is primarily used in overhung, single-stage compressors but includes also the thousands of packaged air machines (Fig. 3-3) and modern, similarly configured process gas machines (Fig. 3-19). Individually designed to combine high head and high flow, inducer-type impellers are mounted at the overhung ends of each pinion shaft. A common bull gear drives up to six sets of pinions at optimum speeds, speeds

Figure 3-9. Vertically split ("barrel") compressor casings. (Source: Demag Delaval, Duisburg, Germany.)

Figure 3-10. Vertically split compressors with end covers secured by stud bolts (left) and shear rings (right).

that are generally governed by the mechanical tip speed limitations of the impeller material selected by the designer. The operating range of these impellers can also be improved by using a measure of backward lean for the radial portion.

3.3.5 Impeller Arrangements on Compressor Shafts

Compressor impellers generate axial thrust, and mixed-flow impellers create higher axial forces than equivalent closed impellers. Arranging the impellers in back-to-back fashion will substantially reduce net thrust. However, high-pressure applications require special design attention to avoid subsynchronous vibrations.

Impeller performance and surge-point location are related; the surge point determines the useful operating range and head capability. Comparing three- dimensional impellers with backward-leaning, radial inlet impellers is of interest. The more highly contoured three-dimensional geometries often feature improved operating and flow range.

Locating the thrust bearing at the outboard side of cen-

Figure 3-11. Inner bundle of a high-pressure, vertically split centrifugal compressor. (Source: Mannesmann Demag Delaval, Duisburg, Germany.)

Figure 3-12. Withdrawing inner bundle from a vertically split compressor. (Source: Mannesmann Demag Delaval, Duisburg, Germany.)

trifugal compressors (in the foreground of Fig. 3-9) will facilitate maintenance access and is thus preferred. Economic considerations seek to limit the size of these thrust bearings. With possibly substantial net thrust acting on an impeller toward its suction, balance pistons ("balance drums") are provided to reduce this thrust load by creating suitable opposing pressure regions. However, a residual amount of thrust must be absorbed by a thrust collar that, in turn, loads up the bearing pads. The attachment method chosen for the thrust collar must be such that the collar lock nut will not "unthread" itself during operation.

3.3.6 Diffusers

Diffusers are the stationary passages in the compressor whose primary function is to "diffuse" or slow down the gas velocity. As the impeller discharges flow into the diffuser passage (Fig. 3-5), the diffusion process converts velocity energy into pressure energy. Diffusion can be achieved by means of parallel-wall diffusers or volutes, depending on the compressor design. Single-shaft, multiimpeller compressors, with or without sidestreams, employ parallel-wall diffusers formed by diaphragms. Compressor diaphragms are stationary inserts that must direct the process gas into successive impellers. It should be noted that diffusers, return bends, and return passages all constitute volumes created by the walls of

Figure 3-13. Mounting the end cover on a barrel-type compressor. (Source: Mannesmann Demag Delaval, Duisburg, Germany.)

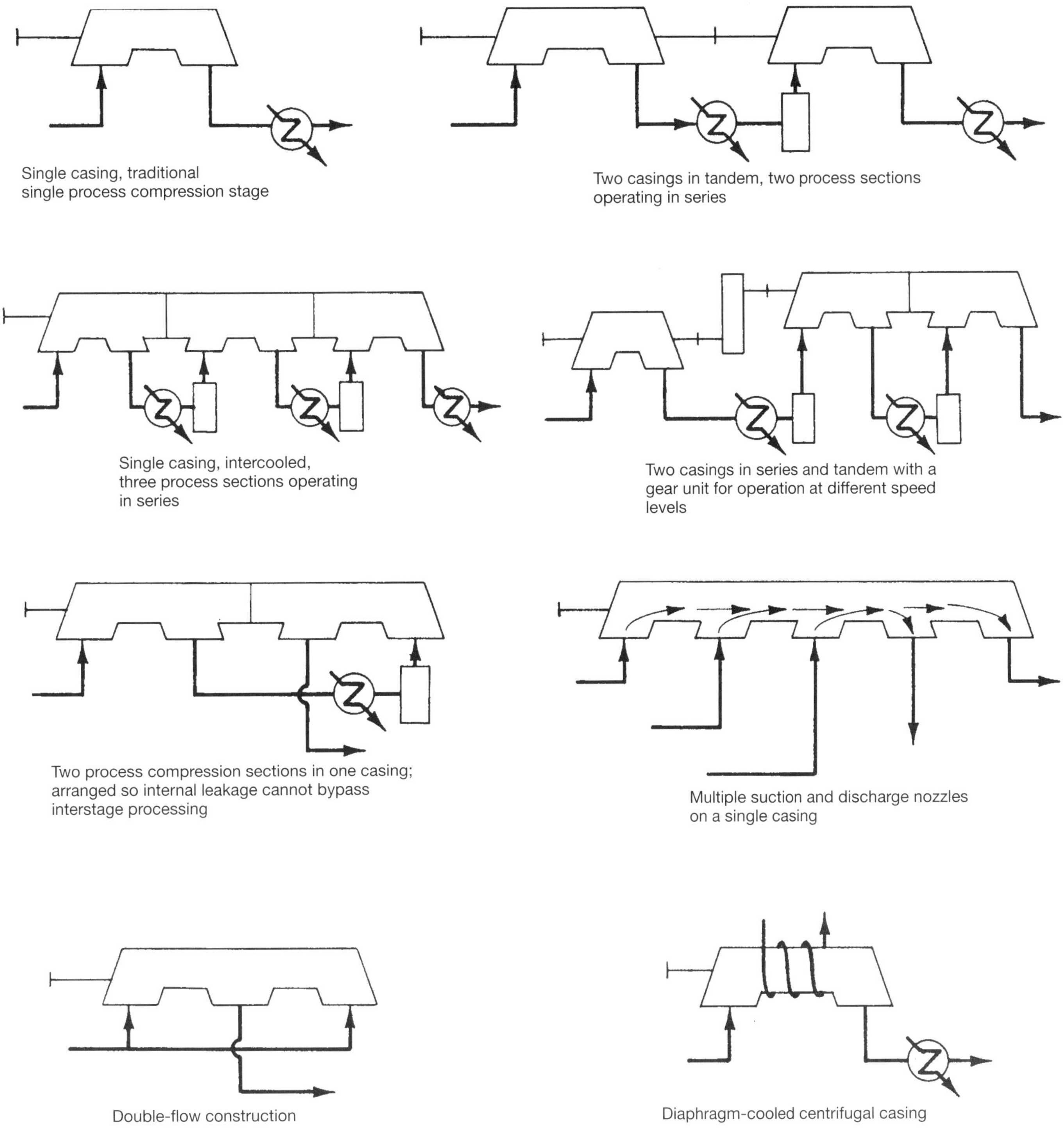

Figure 3-14. Some of the hundreds of available casing layouts used by all industries.

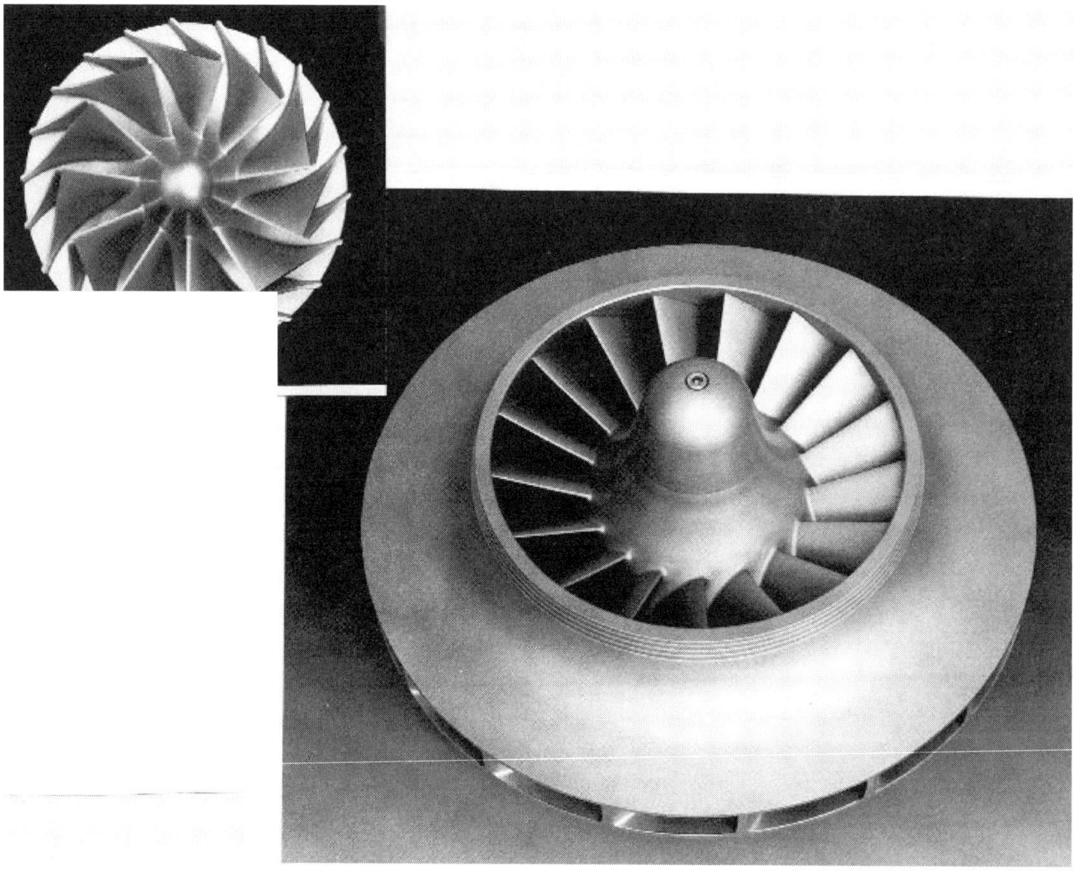

Figure 3-15. Semi-open and closed impeller designs. (Source: Mannesmann Demag Delaval, Duisburg, Germany.)

the diaphragms and casing. Although most of the pressure rise takes place in the impeller itself, the remaining pressure rise takes place in the diffuser.

There are two types of parallel diffusers: "vaneless" and "vaned." In a vaneless diffuser, the gas travels at the same angle as it leaves the impeller. In vaned diffusers and as the name implies, vanes provided in the diffuser contribute to achieving a particular performance. Within a certain flow range, the performance of the vaned diffuser is excellent; outside this range the gas flow approaches at less than optimum angles, which reduces operating efficiency. The choice between vaned diffuser and vaneless diffusers is thus governed by the operating conditions anticipated for the compressor. Vaneless diffusers may be the best choice if anticipated operating conditions include prolonged and wide swings in gas flow rate. Applications involving single-stage machines favor a volute that itself is part of the casing. The pressure rise will take place in the discharge nozzle. As do vaneless diffusers, volute casings offer a somewhat wider stable operating range than vaned diffusers. Volute casings and vaneless diffusers are thus preferred in applications in which efficient operation at part load is more important than the somewhat higher efficiencies that might be possible with vaned diffusers at peak load conditions.

Adjustable guide vanes and adjustable diffuser vanes are used to adapt to varying process demands. Adjustable guide vanes (Fig. 3-16, item 11) are, typically, a standard feature of single-stage machines. In multistage machines, variable guide vanes may cover only the first two or three stages. Inlet guide vanes alter gas velocity and direction without incurring the more pronounced pressure drops caused by suction throttling. Variations in inlet guide vane angle provide good part-load performance in applications in which considerable changes in head requirement accompany changes in throughput. Aside from contributing to power savings, varying guide vane angles also affect the surge point (Fig. 3-20), to be discussed later. Variable or adjustable guide vanes tend to increase the initial purchase cost for compressors and must be properly serviced during planned inspection, maintenance, and repair downtimes.

Although not normally supplied on most process machines, single-stage compressors are sometimes offered with adjustable diffuser vanes that control the angle at which the compressed gas travels in this stationary passageway. Compressors equipped with these features can maintain reasonable efficiency while responding to rather wide fluctuation in pressure ratio and throughput conditions. Needless to say, the linkage elements (Figure 3-21) associated with variable

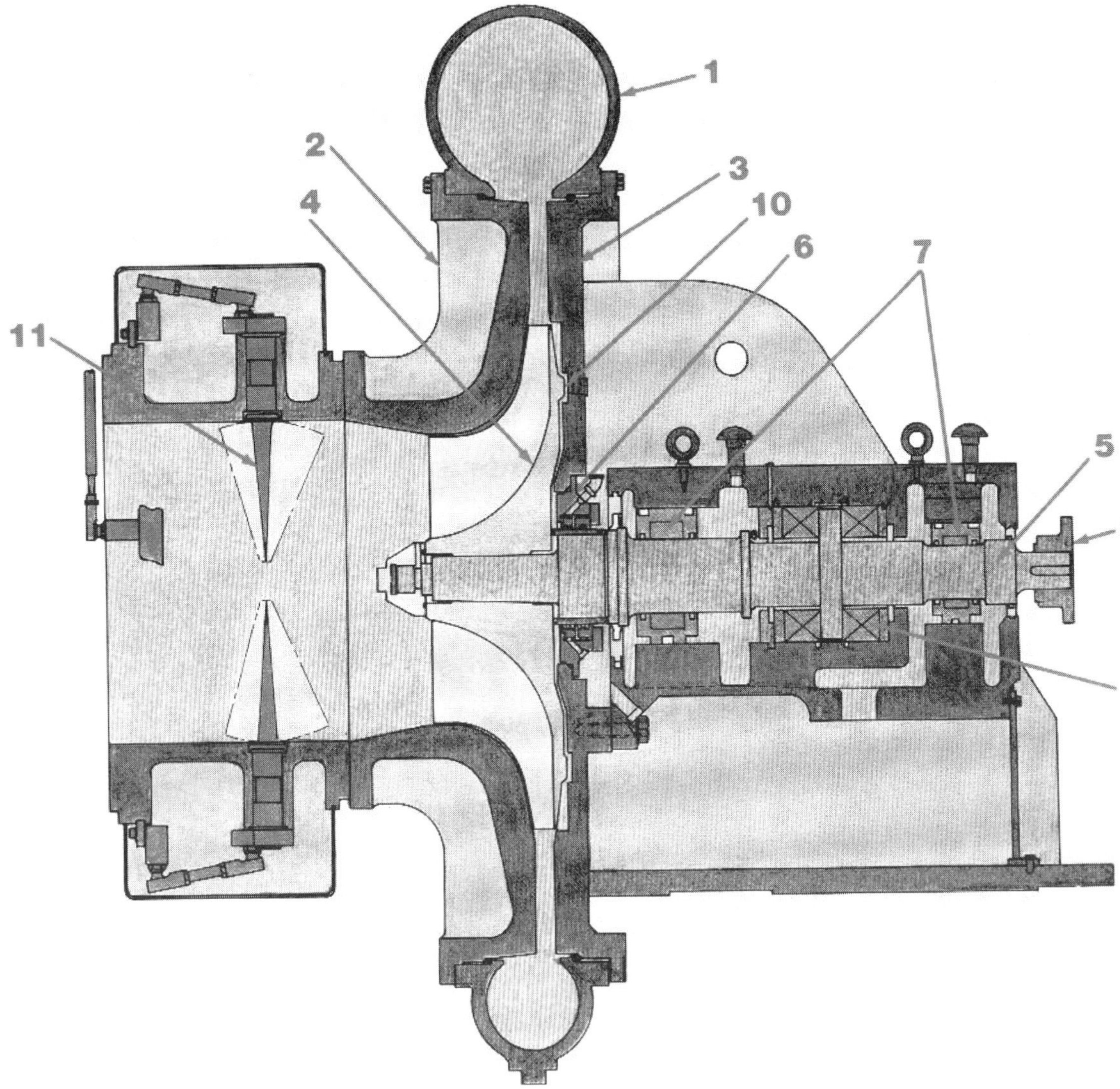

Figure 3-16. Single-stage, overhung compressor showing (1) casing, (2) inlet cover, (3) back plate, (4) impeller, (5) shaft, (6) seal, (7) journal bearings, (8) thrust bearing, (9) coupling hub, (10) balance plane, and (11) variable inlet guide vanes. (Source: A-C Compressors Appleton, Wisconsin.)

guide vanes and/or variable diffuser vanes must function smoothly if compressor damage is to be avoided. Maintenance neglect has, more than once, caused costly failures. This fact must certainly not escape a reliability-focused user.

3.3.7 Internal Labyrinths

Internal labyrinths, specifically identified in Fig. 3-5 but also visible as impeller-eye labyrinths and shaft labyrinths in Fig. 3-11, are used along the gas path to minimize leakage between stages, along the balance piston, and ahead of the seals separating compressor internals from compressor bearings. Internal leakage from regions of higher pressure to regions of lower pressure contributes to compressor inefficiency. In high-pressure applications such as CO_2, synthesis gas, and in offshore facilities with reinjection applications, it is especially important to improve overall performance by minimizing losses. Since the gas volume at the last stage of compression is substantially reduced and an elevated pressure gradient exists at the discharge end, balance piston leakage will greatly influence overall compressor performance.

The exact quantification of leakage and improved designs for holding leakage to a minimum are of major interest to compressor manufacturers and users. Labyrinth materials and profiles must be optimized and high-performance polymers (HPPs) such as PEEK and Vespel® CR-6100 (a Teflon®/oriented carbon composite) that permit tighter clearances are among the noteworthy developments. Abrad-

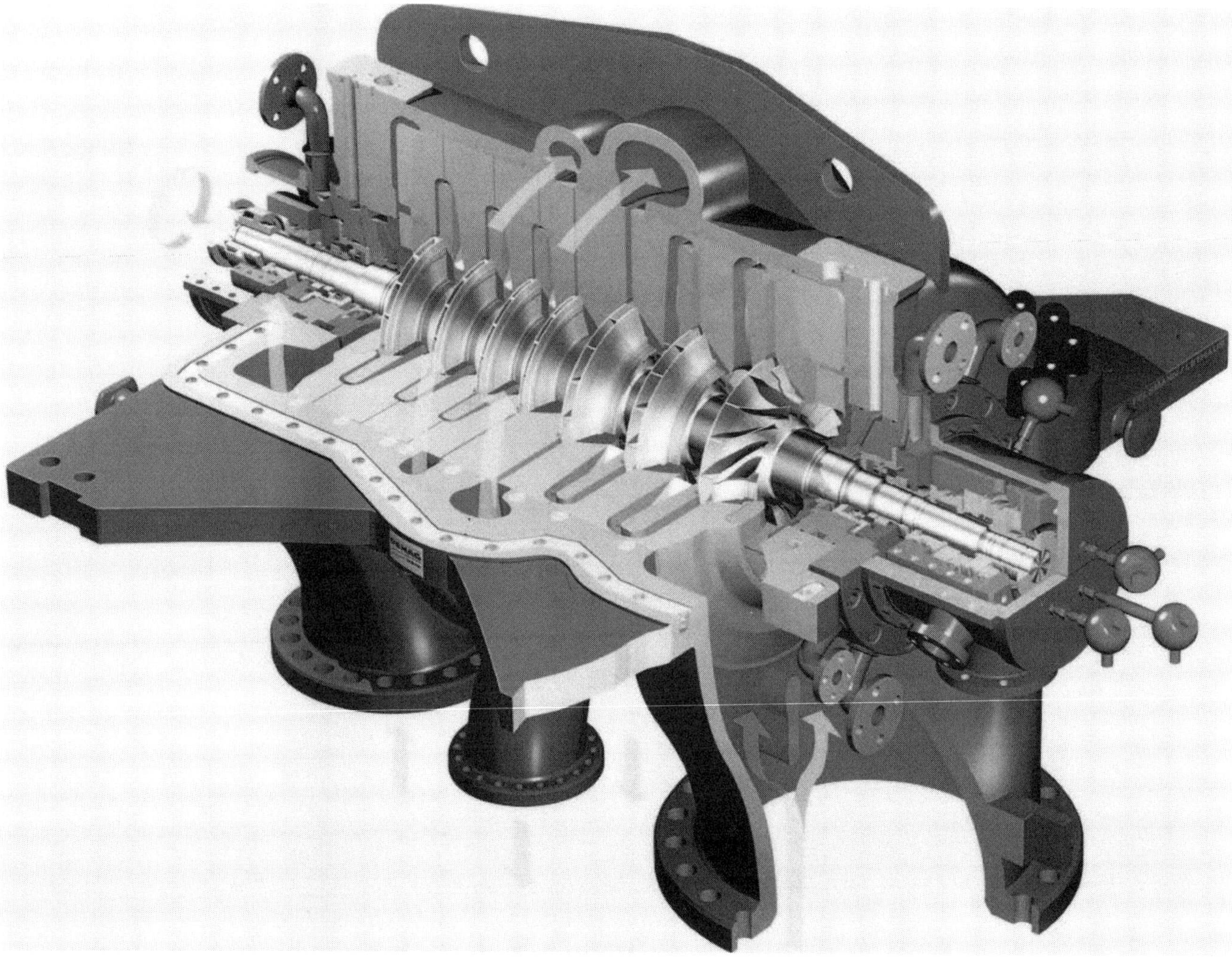

Figure 3-17. Horizontally split sidestream compressor with semiopen first stage and closed impeller stages 2 through 7. (Source: Mannesmann Demag Delaval, Duisburg, Germany.)

able ("frangible") seals and honeycomb designs are often selected to minimize stage-to-stage leakage. In any event, materials technology advances have pointed to high-performance polymers and these can be of great value in both OEM (original equipment manufacturers) or upgrade situations for existing machines. Of course, HPPs should be used within their respective physical limits and manufacturer's experience. On the plus side, proper leakage control also contributes to reduced fluid-dynamic excitation of the compressor rotor.

As will be seen later, separating labyrinths are provided in the sealing areas to avoid mixing of sealing and lubricating oil streams found in certain sealing systems. Some separating seals are configured with a "wind-back feature" that attempts to curtail gas leakage flow by producing counterrotation of the gas attempting to escape. In other compressors, inert gas (occasionally called "barrier gas" or "buffer gas") is injected so as to prevent migration of oil from the seal environment to the bearing areas.

3.3.8 Bearings

Over the years, compressor manufacturers have furnished different types of hydrodynamic bearings, ranging from plain sleeve bearings (lower part of Fig. 3-22) and pressure-dam sleeve bearings to very precisely machined tilting-shoe (tilting-pad) bearings, shown in the upper part of Fig. 3-22. These bearings differ little, if at all, from the bearings discussed in conjunction with our earlier overview of helical screw compressors. As can be seen in Fig. 3-22, process gas compressor bearings are "split" to permit installation without the need of coupling removal. However, although sleeve bearings are often quite satisfactory, they have, collectively, been somewhat susceptible to subsynchronous oil whirl and also high-frequency oil whip. All of these are problems creating rotor instability. Whirling is orbiting of the rotor center in the bearing. The orbiting frequency is generally 0.42–0.47 times the running speed. Axially grooved bearings reduce instability risk, but require more oil flow to feed all grooves.

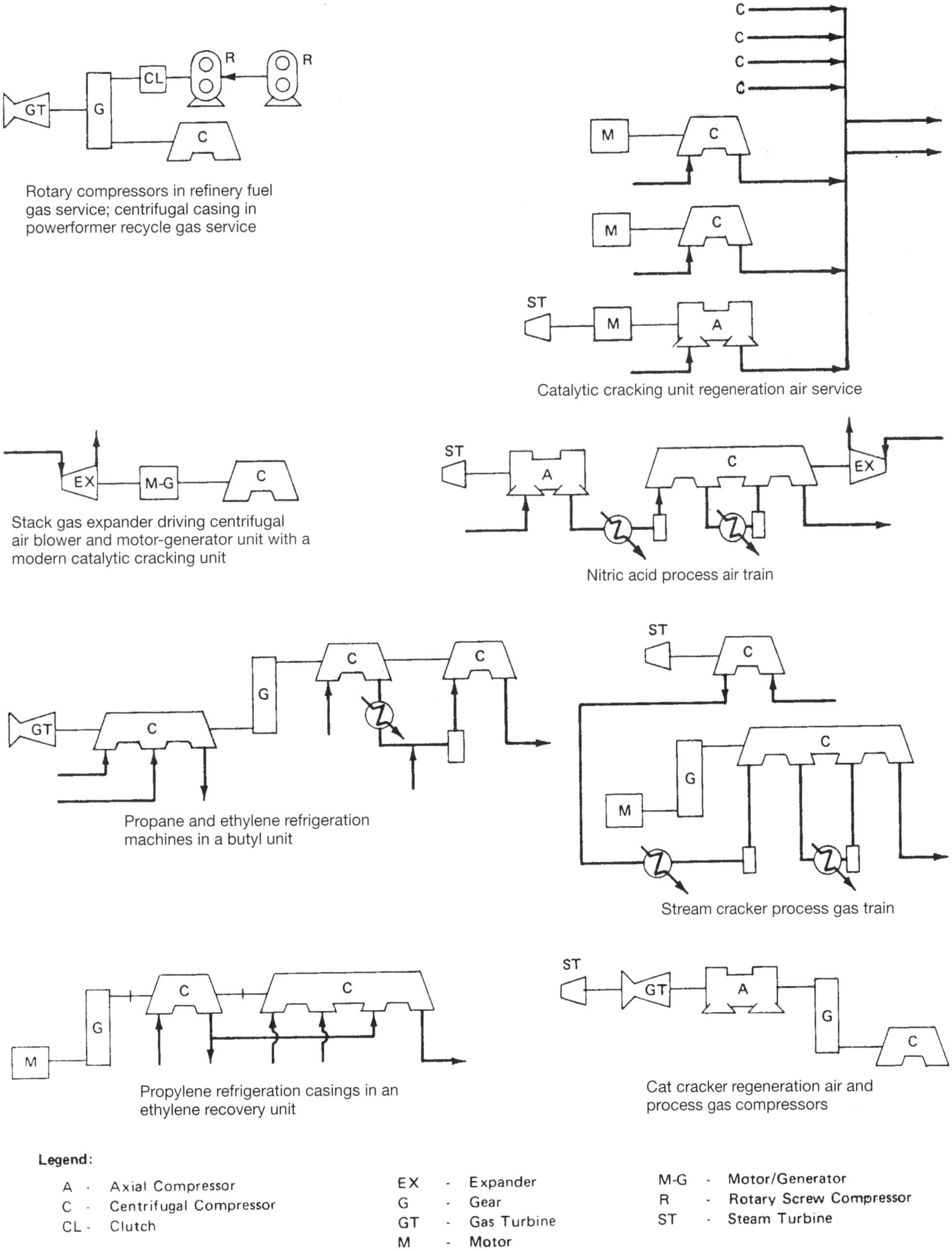

Figure 3-18. Compressor train arrangements found in the hydrocarbon processing industry.

Figure 3-19. Integrally geared multistage process gas compressor. (Source: Mannesmann Demag Delaval, Duisburg, Germany.)

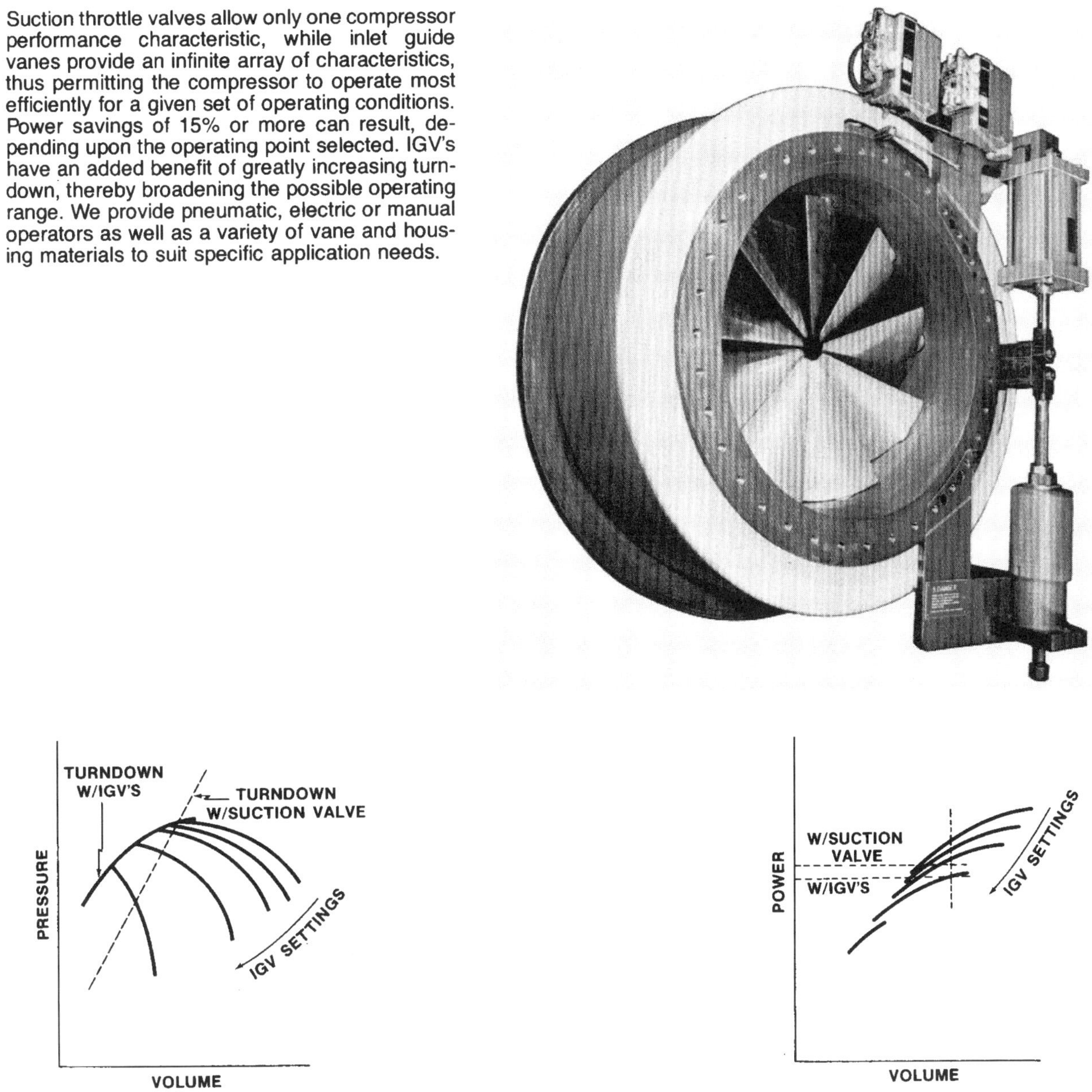

Figure 3-20. Adjustable inlet guide vanes and their effect on surge and power. (Source: A-C Compressor Corporation, Appleton, Wisconsin.)

They also have slightly reduced load-carrying capacity compared to full sleeve bearings. However, axially grooved bearings normally run cooler than plain bearings.

Lobe-type bearings with symmetrical and asymmetrical bore patterns represent an improvement over axially grooved, cylindrical bore bearings. Their noncircular bores allow the oil film to form a wedge. They were precursors of the present-generation tilting-pad bearings. Tilting-pad bearings also called tilting-shoe bearings, have largely overcome the whirl and whip problems and are favored due to their ability to not only support, but also overcome small amounts of line-bore inaccuracies at the bearing housing locations of compressor casings. Tilting-pad bearings (upper part of Fig. 3-22) are available in different geometries, with different numbers of pads, pad pivot locations, pad curvature, and different axial-width-to-bore-diameter ratios. Each particular geometry or configuration can have an effect on rotor instability, which is essentially its propensity to vibrate when exposed to certain loads. In most cases, the instability risk can be calculated and optimized geometries chosen to mitigate the risk.

Putting it another way, different bearing geometries, internal clearances, oil film characteristics, and their controlled interactions are all needed to avoid rotor instabilities. Rotors can become unstable, or prone to vibrate excessively, if supported by the wrong bearings [10]. This tendency to become

Figure 3-21. Linkage elements associated with variable-inlet guide vanes. (Source: Mannesmann Demag Delaval, Duisburg, Germany.)

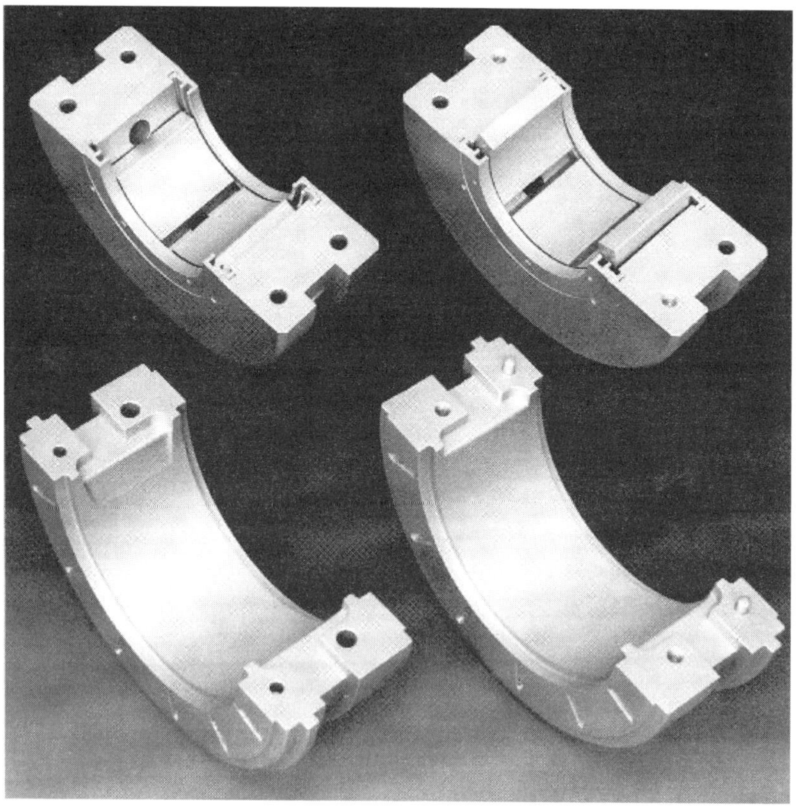

Figure 3-22. Compressor sleeve bearing bottom, and tilt pad bearings top.

unstable is a function of rotor geometry, gas conditions, and gas properties, but also bearing curvature, pivot location, and axial width.

All of these parameters can affect the "critical speed" of compressors. Centrifugal compressors are generally designed to operate above their first critical speeds. Critical speed is somewhat analogous to the resonant frequency at which a rotor assembly would vibrate if struck by a hammer.

Although it would be technically desirable to have the pivot point of the bearing pads off-center and with the leading edge portion somewhat larger than the trailing edge portion to enhance stability, the resulting bearing would be unidirectional. However, centered pivots are found to predominate in tilting-pad bearings since the bearing is now suitable for shaft rotation in either direction. Regardless of overall geometry, most hydrodynamic bearings with steel backing have a Babbitt lining of about 0.8 mm (0.032 inch) thickness, and clearances of 0.0015–0.002 inch per inch (mm per mm) of journal diameter. They utilize oil supply pressures in the relatively low range of 1 to 2 bar (~15–30 psi). Shaft rotation and bearing geometry cause higher pressures to be self-generated within the bearing. An oil film varying in thickness from 0.0001 to 0.001 inches prevents metal-to-metal contact. Since shear action on the oil produces heat, the lubricant must be cooled. Circulating-oil systems are best suited to accomplish the necessary cooling and filtering.

Thrust bearings are used to locate the rotor axially and at the same time absorb any axial rotor thrust. Thrust bearing configurations include flat land, tapered land, and tilting-pad models. Their respective load-carrying capacities range from 50 psi to 250 psi. However, the most popular configurations comprise self-equalizing, leading-edge, spray-lubricated tilting pads with permissible specific loads ranging from 200 to 400 psi (Fig. 3-23). Bearing material options include tin/lead-base Babbitt and various copper-bearing alloys (bronzes) to suit specific applications. In each case, the compressor designer desires not to exceed 50% of the bearing manufacturer's allowable load rating. Bearing designs also call for suitable conservatism whenever the possibility exists of additional thrust being transmitted from or through multiple casings.

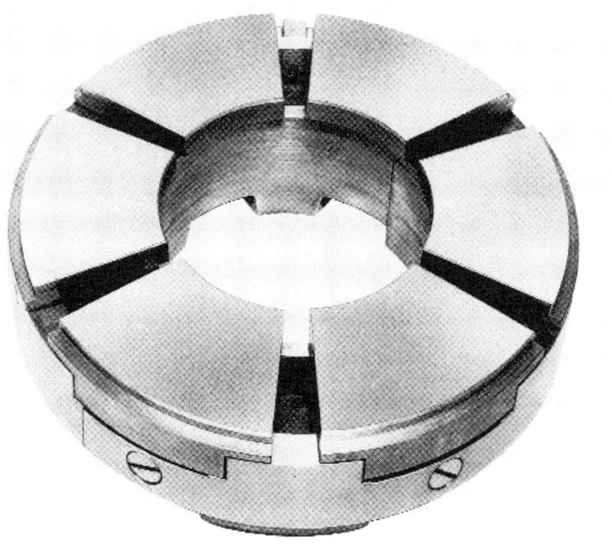

Figure 3-23. Hydrodynamic, equalizing, pivoted-shoe thrust bearing. (Source: Kingsburg Corp., Philadelphia, PA.)

Thrust bearings must have the correct axial clearance, typically 0.008 to 0.010 inches (0.2–0.25 mm), to perform properly. During maintenance-related shutdowns and before further dismantling, the exact rotor float from its mid position toward the active and nonactive sides of the thrust bearing should be checked. When the bearing is removed, the pads should be examined for wear and other surface damage. Also, the rotor free-float should again be checked before the compressor is fully assembled at site. Free-float shows the feasible axial impeller movement within the casing before the thrust bearing is being installed. Final readings are taken and recorded after installation of the thrust pads. In essence, the correct positioning of the rotor in its casing and observing thrust collar locations relative to the active and inactive thrust bearing pads require considerable care and patience. Axial position probes (proximity probes) are used for on-stream monitoring of rotor location.

In desert conditions in which air-cooled lube oil coolers are provided, extreme temperatures are of concern. Oil-supply header temperatures are high due to prevailing ambient conditions and, possibly, dust-clogged cooling fins affecting exchanger efficiency. This, obviously, risks reaching higher bearing metal temperatures and could impose limitations on allowable compressor loads.

Although bearing oil supply, oil drain headers, and bearing metal temperatures are typically monitored, bearing metal temperatures are the more important of the three. Alarm settings of 110°–120°C (230°–248°F) are generally used for bearing temperatures, whereas high header temperature (bulk oil) alarms are often set at 80°C (~176°F). Tin-based Babbitt is preferred over lead Babbitt due to its superior corrosion resistance and better bonding with steel backing. The tin-based material is, however, not as "forgiving" of dirt inclusions in the lubricating oil as is the lead-based version.

In summary, manufacturers attempt to achieve higher compressor efficiencies by improving impeller design, and numerically controlled machine tools are often used in the manufacturing cycle. But power losses can also be reduced and downtime risk reduced through innovations. Flexing pivot bearings and dry gas seals are but two of many recently proven bearing and seal designs. Dry gas seals have been successful in a large number of retrofits on older centrifugal compressors and are now accepted as part of the original manufacturer's (OEM's) supply in very many installations. Similarly, stage (internal) labyrinth clearances can be reduced and small, but worthwhile, efficiency gains

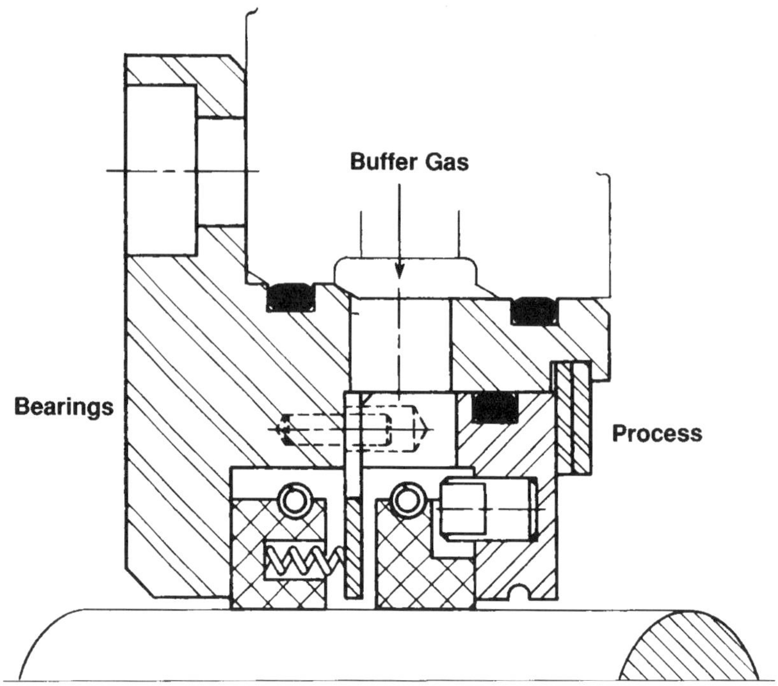

Figure 3-24. Circumferential barrier seal. (Source: Kaydon Mfg. Co., Baltimore, MD.)

achieved whenever superior and thoughtfully selected high-performance polymers (HPPs) are properly applied.

3.3.8 Shaft Seals

The primary purpose of compressor shaft seals is to avoid gas leakage into the atmosphere. Both safety and loss of revenue are at issue here. The American Petroleum Institute, in its Standard API-617, describes five generic types of seals: labyrinth, restrictive ring, mechanical contact, liquid film, and "dry" gas seals. Mechanical contact (Fig. 3-24) and liquid film seals (Figs. 3-25 and 3-26) employ liquid as the sealing medium. Process gas or a buffered inert gas medium alone is used in the other three seal types.

Labyrinth seals (Fig. 3-27, also Fig. 3-28a) represent the

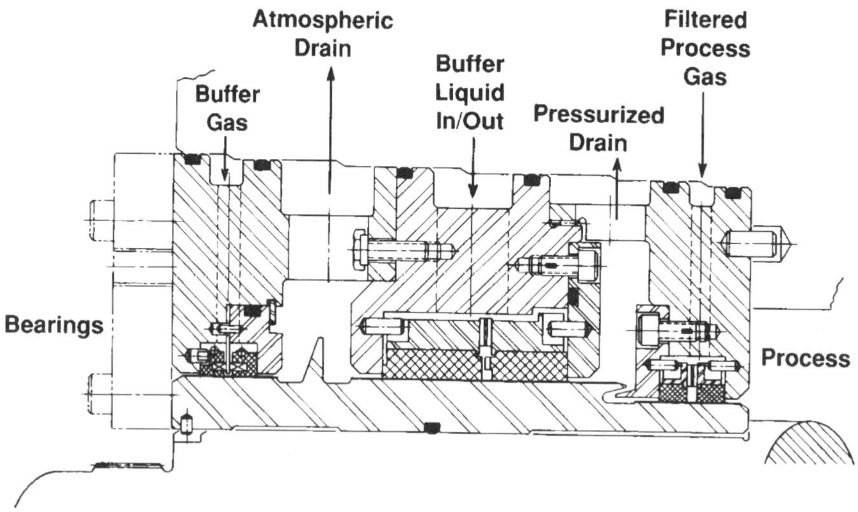

Figure 3-25. Liquid-buffered bushing seal. This arrangement is typically used in ethylene glycol process compressors with water buffer. Filtered process gas injected between floating rings minimizes seal liquid transfer to process. Gas buffered barrier seals isolate bearings from atmospheric drain. (Source: Kaydon Mfg. Co., Baltimore, MD.)

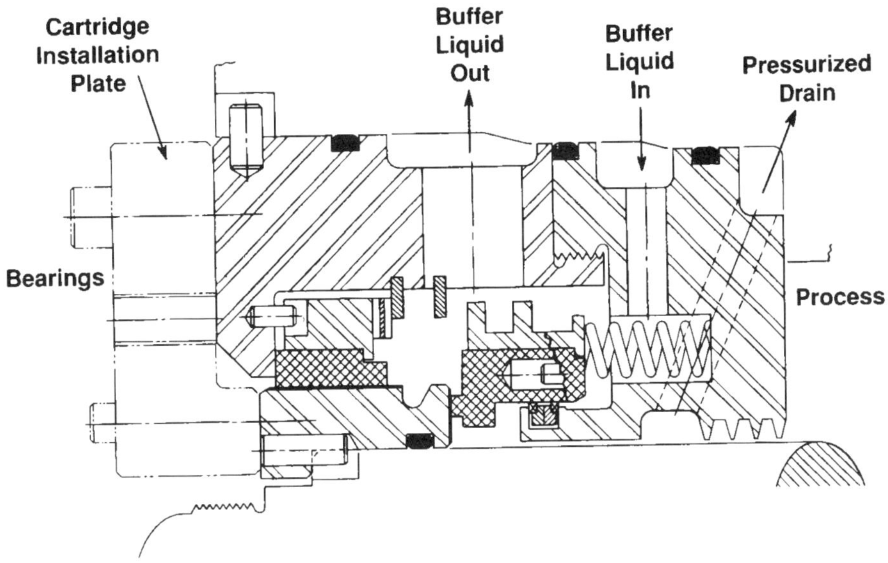

Figure 3-26. Liquid-buffered face and bushing seal. (Source: Kaydon Mfg. Co., Baltimore, Maryland.)

most straightforward and simple construction. They are also the least expensive but are suitable only for applications such as air compression, in which limited loss is acceptable. This type of seal takes up more axial space than other seal types, again placing constraints on its usage. However, labyrinth seals are generally used in sealing duties associated with balance pistons (sometimes called balance drums) and seal gas-equalizing regions (Fig. 3-29). Gas leaking past the last compressor stage enters a space that is pressure balanced or "equalized" back to compressor suction via balance-line piping. The gas so returned will be at substantially higher temperature and slightly higher pressure than the gas entering at the compressor suction nozzle at normal inlet conditions.

Aluminum labyrinth seals are generally cost-effective. For moderate operating temperatures and the majority of nontoxic applications, these seals consist of fins inserted into a stationary backing component. More expensive stainless steel or tough-pitch copper can be used where higher temperatures are involved. Since the seal does not guarantee 100% sealing, it may be necessary to consider buffer gas injection to prevent the process gas from reaching the bearing cavity.

In some cases, restrictive carbon ring seal designs are selected. Restrictive carbon ring seals have virtually zero shaft clearance and present a more difficult escape path for the process gas. The rings are segmented and are available in a number of configurations. They are usually surrounded by a garter spring that keeps the individual segments assembled and also serves to apply the desired contact pressure against the shaft surface. However, zero clearance invites wear and more frequent carbon ring replacement. Again, buffer gas is recommended for difficult applications.

Liquid film or contact seals (Figs. 3-25 and 3-26, also Fig. 3-28 b and c) should be considered whenever sealing requirements are not met by either labyrinth- (Fig. 3-28a) or restrictive-type seals. Liquid film and contact types of compressor seals have been widely accepted in all kinds of hydrocarbon services at high and low pressures, but are specifically *excluded* from such nonhydrocarbon services as high-pressure air and oxygen compression. Compressors handling charge gas, process refrigerants such as ammonia/propane, hydrogen-rich gases, and fertilizers have generally used liquid film or contact seals.

Under normal operating conditions, these seals prevent leakage in either direction. They incorporate seal rings (Fig. 3-28c) arranged with the high-pressure (HP) side toward the process and the low-pressure (LP) side toward the atmosphere. Seal oil pressure is maintained in the space between LP and HP rings at a predefined constant differential pressure, or ΔP, over the process gas pressure. A small amount of this oil, perhaps 12 liters (~3 gallons) per day (!) is allowed to flow through the very small diametral clearance between shaft and HP seal ring toward the process gas. This oil film acts as a barrier and will prevent the process gas from contacting the oil contained between inner (HP) and outer (LP) seal rings.

Although these approximately 12 liters per day become "sour" seal oil due to contact with the process gas, many times this amount will flow through a larger diametral clearance in the LP ring and will remove considerable heat as it passes to the "sweet" oil cavity, a space between the LP seal ring and atmospheric pressure maintained between the LP seal ring and the adjacent bearing. The entire setup prevents process gas from flowing to the atmospheric side of the seal and prevents the "sweet" oil from escaping to the atmosphere. The "sour" seal oil contaminated by process gas enters one or more traps, degasifiers, or oil purifiers to remove the process gas. The "sweet" or uncontaminated oil is returned to

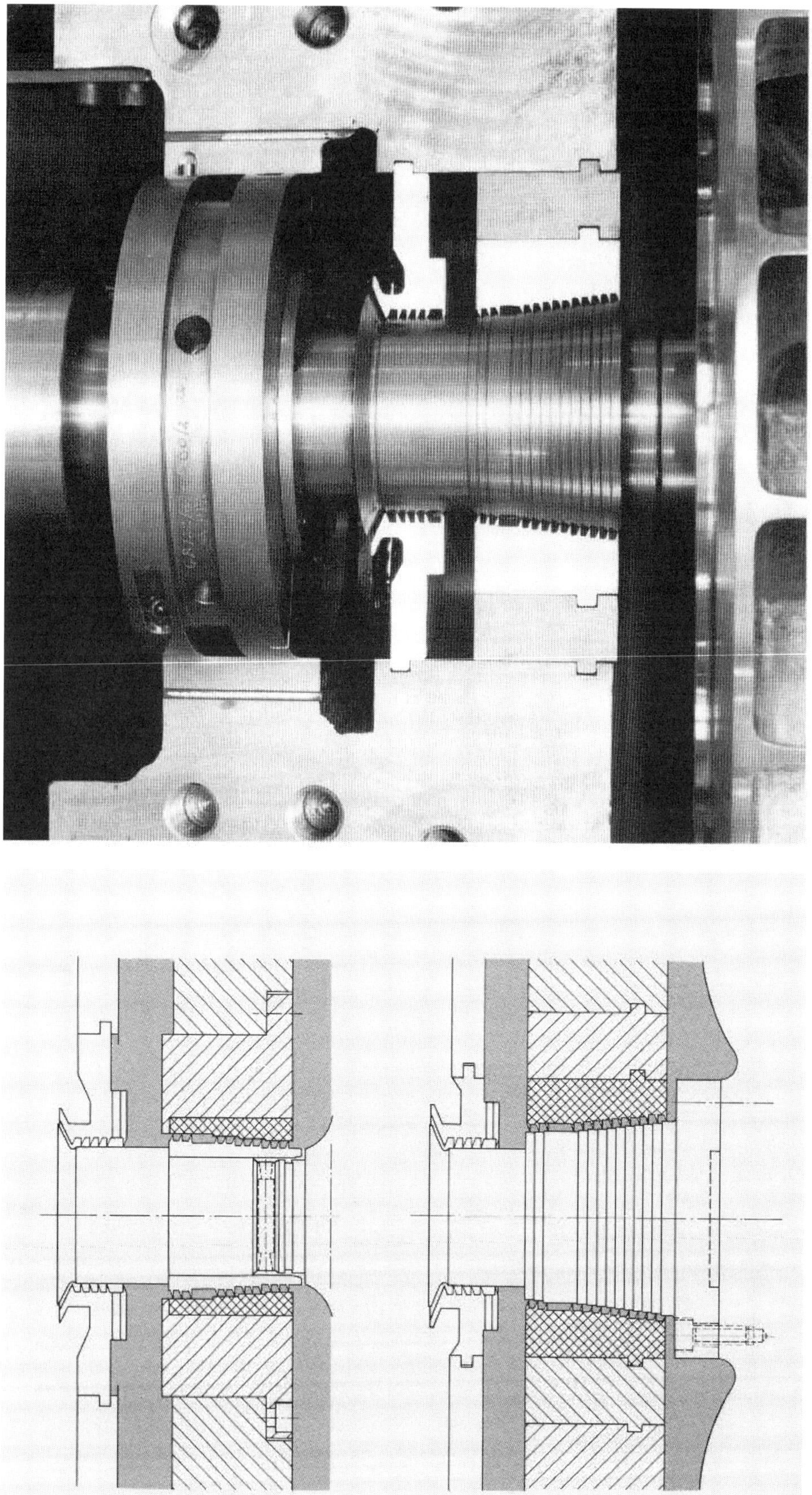

Figure 3-27. Labyrinth seals. (Mannesmann Demag, Duisburg, Germany.)

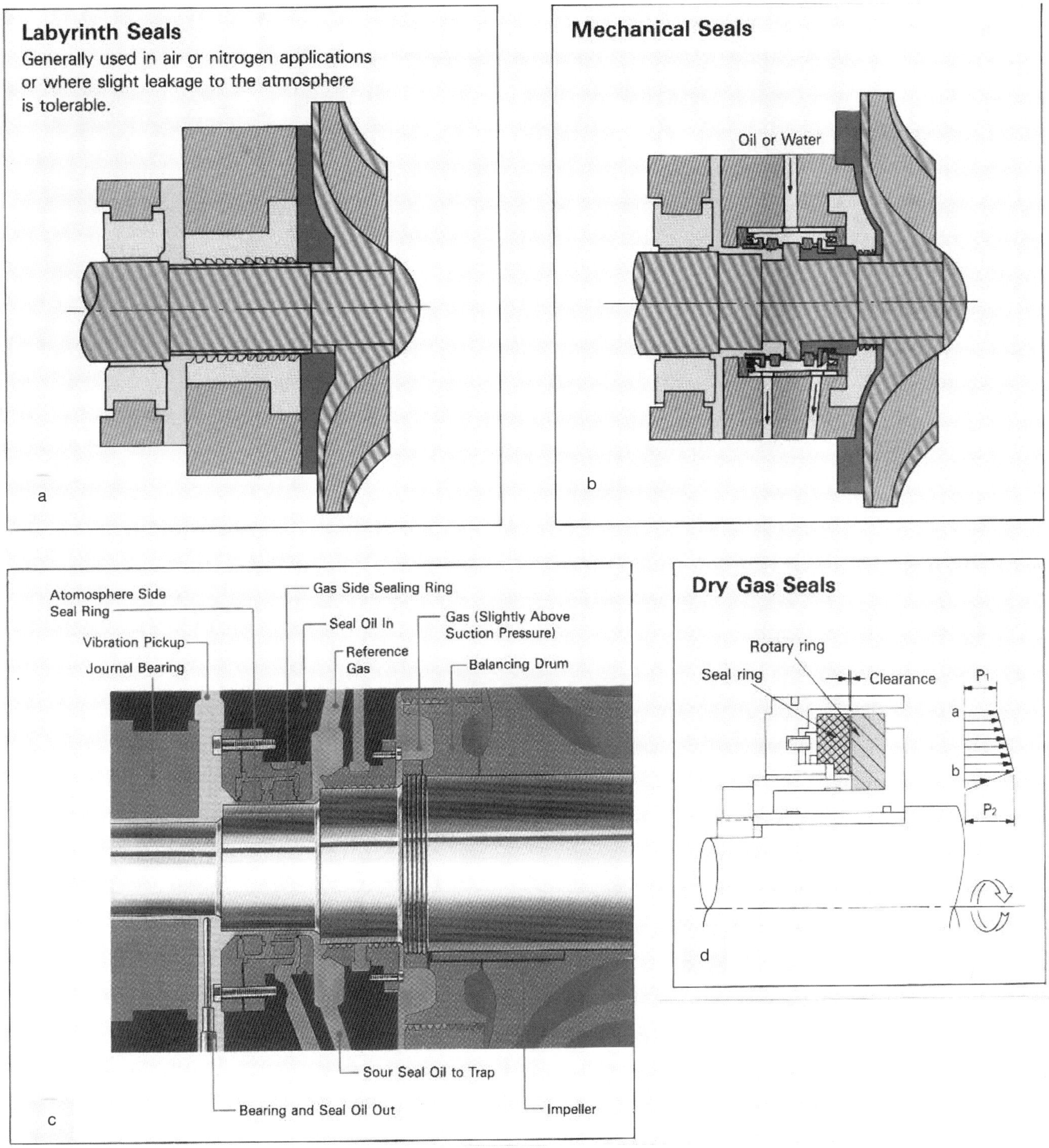

Figure 3-28. Compressor seals. (Source: MHI, Tokyo and Hiroshima Japan.)

the seal-oil reservoir or, in combined lube and seal-oil systems, to the common reservoir.

An efficient and reliable seal system should comply with the minimum requirements of API-614, described earlier in Figs. 2-7 and 2-8. Such seal systems can involve elaborate instrumentation as well as high procurement cost for pumps, coolers, degassing system, heaters, and coolers. High oil-flow rates may be needed, and future long-term operating costs must be factored in as well. Combined lube-and-seal-oil systems can often save money and floor space. When combined or "common" lubrication and seal-oil systems are feasible, there will typically be a separate seal-oil booster pump with its suction branching off from the lubricating-oil header. Alternatively, separate oil pumps are connected to

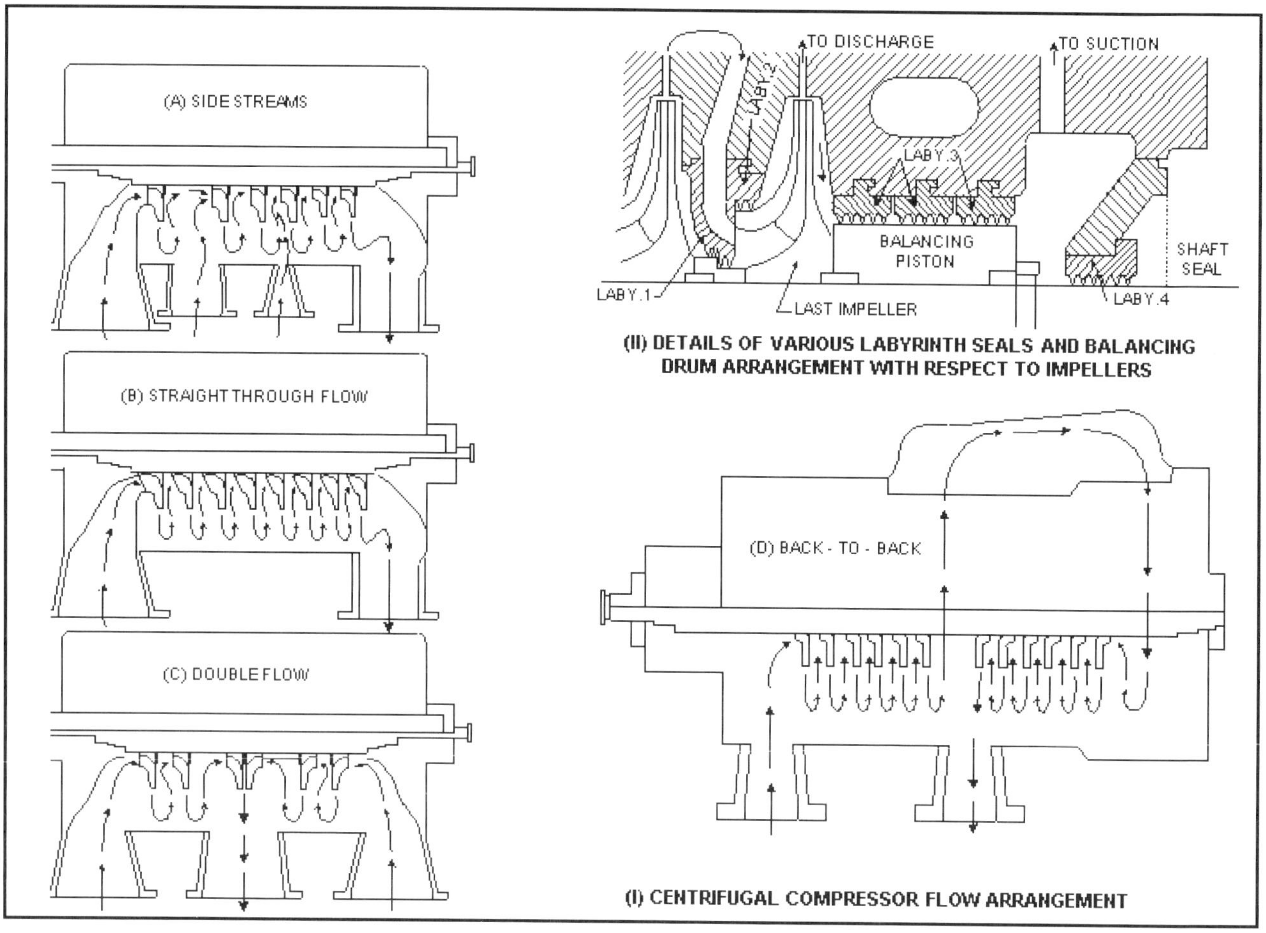

Figure 3-29. Typical compressor flow options (A through D) and details of labyrinth seals and balancing drum arrangement relative to impellers.

the common reservoir. However, where the process gas contains H_2S in excess of 50 ppm by volume, the stipulations of API-614 require seal-oil systems independent of lube-oil systems.

Cost and reliability concerns have prompted industry to steadily move toward application of dry gas seals (DGS, see Fig. 3-28d), which promise such advantages as reduction in power consumption, space savings, reduced control requirements, generally cleaner environment, and long-term, uninterrupted operation. As was mentioned earlier, both new and retrofit applications are frequently of interest and may be economically justified.

Dry Gas Seals and Support Systems: Benefits and Options*

Since their introduction in the mid-1970s, dry gas seals have revolutionized sealing of axial and, in particular, centrifugal compressors. Today, dry gas seals are being installed in the majority of new compressors and numerous retrofit opportunities exist as well. Compared to traditional liquid or "wet" sealing, dry gas seals offer several important advantages:

- On shutdown, pressurized containment of process gas is possible, reducing plant flaring.
- Properly engineered, dry gas sealing prevents oil contamination of compressed gases.
- Reduced power consumption (conservatively in the vicinity of 1%).
- Increased reliability/availability of the process train.
- Predictable rotor characteristic with consistent stability.
- Reduced space (real estate) requirements.
- Prevents reservoir-oil contamination and associated bearing failures in sour or saturated gas applications.

Additionally, plant safety is often improved by typically eliminating the following problems associated with wet seal systems:

*From articles contributed by Manjul N. Saxena, Clough Engineering Limited, Perth, Australia; and Joe Delrahim, John Crane Conpany, Morton Grove, Illinois.

- Seal-oil pump failure with the consequent possibility of gas blowout.
- Seal blowout caused by an oil-trap float failure, a clogged coalescer, or a sticking buffer-gas regulator.
- Thrust bearing failure resulting from excessive buffer-gas differential pressure.
- Possibility of a lube-oil tank explosion. Gas escaping from the seal return oil can increase the flammable gas concentration above the lower explosive limit (LEL) when the oil-reservoir breather is clogged or overwhelmed due to excessive gas leakage at seals. One compressor supplier has reported 26 such reservoir explosions.

Life Cycle Cost Comparison of Dry Gas Versus Wet Sealing Systems

If a technical review confirms the suitability of dry gas seals for a given application, the selection process must consider a number of factors for a comparative life cycle cost analysis between the dry gas and wet seal systems. These would include:

- First cost
- Plant availability and its impact on throughput
- Cost of lost process gas
- Operating utilities and maintenance costs

First cost. As was explained elsewhere in this text, wet seal systems consist of seal cartridges, seal-oil pump, blower, degassing unit, seal-oil reservoir, overhead seal-oil vessel, connecting piping, and a control and monitoring panel. A dry gas seal system requires only seal cartridges, filtration, and a control and monitoring panel.

For a land-based plant, a wet seal system would cost on the order of \$500,000 compared to perhaps \$250,000 for a dry gas seal system in a single-casing compressor with a 2 to 3″ (50 to 75 mm) diameter shaft. The high cost of real estate and maintenance in an offshore application would automatically justify a dry gas seal application.

Availability and plant throughput. In process plants with a planned availability of perhaps 340 days per year, any trip of the main process compressor (usually nonspared) translates into a loss of production. In general, wet-seal systems issues are believed to cause 75% of motor and 30% of gas/steam turbine-driven compressor downtime. With a realistic availability of 98% or even 99% for motor and turbine-driven compressors, respectively, this equates to days of production downtime due to wet-seal systems.

In contrast and historically, dry gas seals have exhibited some "infant mortality," with a subsequent 99.9% availability. Performing an accounting calculation involving a given plant's variable versus fixed production cost ratio and specific rate of return on investment will be revealing. Even a single day of extra production due to increased reliability of dry gas seals can have notable beneficial effects on the bottom line.

Compared with a negligible frictional loss of two to three kW in dry gas seals, up to a full percent of compressor shaft power is often consumed in wet gas seals. Again, both power cost and lost production may be rather significant.

Lost gas. Dry gas seals typically leak 5 to 10 scfm/seal (8 to 16 Nm^3/hr per seal) compared to an aggregate leakage rate of perhaps 100 scfm/seal (161 Nm^3/hr per seal) to flare for many wet seals. For natural gas service at an assumed price of \$6/MMBTU* (1000 BTU/scf*), 180 scfm of saved gas per compressor body results in a daily saving of \$1,650. Similar calculations should be carried out for other gases.

Operating and maintenance cost. Dry gas seals use pressurized nitrogen/air as a buffer and separation gas. Wet-seal systems employ a seal-oil pump, degassing blower, and, in a sour gas application not employing continuous on-stream oil purification, seal-oil change-out is required every 4,000 to 8,000 hours. Consider 50 kW of power consumption by a seal-oil pump and blower as the major operating cost difference per compressor casing. Then, at \$0.06/kW-hr, dry gas seals would save \$24,000/year in utility cost. Labor and material savings, based on two days of reduced maintenance support per year by a team of six persons at \$1,600/day (including materials) equates to \$19,200 annually.

To the extent applicable and supportable by plant statistics, a reviewer might also evaluate such potential savings as:

- Impact of a six year operating cycle between overhauls for a dry gas seal compared with a three year cycle for a wet seal.
- Savings from a process loop inventory not flared on the spurious compressor trips (say, one trip per year) due to pressurized hold capability of dry gas seal.

Assuming equipment life of 20 years and a predefined internal rate of return, the present value of the benefits derived from a dry gas seal installation may well multiply the total annual savings.

Principles of Dry Gas Seal and Construction Features

The sealing mechanism of a dry gas seal is generally understood to consist of two rings: a stationary (plain face) and a shaft-mounted rotating face with "etched-in" grooves. On rotation, the groove continuously pumps the sealed gas into the reducing cavity. The high pressure thus generated creates a thin gap between faces, allowing minute leakage of 5–10 scfm (8–16 Nm^3/hr). The separation of the stationary and rotating seal faces, also called "lift off," occurs between 150 to 350 rpm. Manufacturers have developed numerous rotating

*1 MMBTU = 1.055 Giga joules; 1 BTU/scf = 37.25 kJ/m^3.

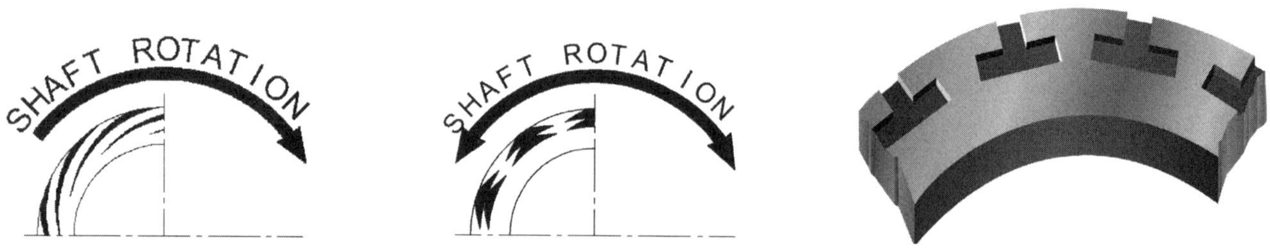

(a) (b) (c)

Figure 3-30. Dry gas seal face configurations: spiral groove (a), bidirectional groove (b), and T-groove (c).

seal face patterns: radial, spiral, wavy face, V-groove, T-slot, and combinations thereof (Fig. 3-30) to achieve the desired operating and sealing properties.

A number of factors determine the lift-off speed, such as seal tip speed, depth of groove, design of the rotating face pattern, dynamic and static sealing pressure, stationary face spring-closing force, molecular weight of compressed gas, slow- or high-speed application, slow roll after machine shutdown, and so on. Normally, seal tip speed is about 90 fps (27.5 m/s), although seals up to 650 fps (198 m/s) are operating successfully.

Based on the seal groove geometry, the working gas film thickness (or gap) between the rotating and stationary seal faces varies from 0.0001″ to 0.0003″ (3 to 8 microns). Research continues to improve seal-face geometry, to maintain film stiffness and to prevent face contact under transient operating conditions. Ideally, rotating faces are perpendicular to the axis of rotation. On upset, if the axial gap reduces, the pressure between the two seal faces rises rapidly. The hydrostatic and hydrodynamic forces stabilize, thereby restoring the seal-face gap. Similarly, film stiffness restores the gap if there is radial misalignment or thermal deformation. Seals generally tolerate ± 0.12″ (± 3 mm) of axial and ± 0.024″ (± 0.6 mm) of radial displacement.

The normal dynamic seal-leakage rate varies from 5 to 10 scfm (guaranteed at 25 scfm) depending on the shaft diameter, operating speed, and sealing pressure. Gas turbulence and viscosity increase with temperature and reduce leakage rate. However, cooler operation is preferred to ensure long life of the seal elastomers. The static leakage is less than dynamic leakage rate. On shutdown, good seals with automated compressor isolation can hold system pressure for up to 24 hours, thus allowing operators to investigate any problem before depressurization. The consequent reduction in flaring is environmentally and economically desirable.

Post-shutdown, "slow roll" of the rotor in gas-turbine-driven compressors should take into account the seal vendor's recommended minimum speed requirement, typically >100 rpm for a 4″ seal. A stationary carbon ring can accommodate slow roll at speeds of 5 to15 rpm on steam turbine shutdown.

Refer to Table 3-3 and Fig. 3-31 for different categories of dry gas seals and their principal features and applications:

Silicon carbide and tungsten carbide are commonly used materials for rotating seal faces. Their good thermal conductivity and low deformation at high temperature assist in maintaining the seal-face gap. Silicon carbide or special grade carbon is used for the stationary seal faces. Sleeves and

Table 3-3. Dry gas seal types, features, and applications*

Seal type	Feature/application
Single	• With inboard labyrinth, usually applied in air, N_2, CO_2 or other nonhazardous services.
Tandem	• Two seals in-line with the same orientation. • The second backup seal is designed for full sealing pressure.
Tandem, with intermediate labyrinth	• Two seals in-line with an intermediate labyrinth to safeguard and reduce the sealing pressure of the secondary backup seal. • Most widely used in hydrocarbon gas applications.
Triple seal in tandem	• Three seals in-line with the same orientation. • The primary and intermediate seals share the high-pressure (usually above >1,750 psig or 12,000 kPa) sealing load.
Double, opposite faced	• Two opposite, face-to-face seals with clean gas injected in between at higher than the sealing pressure. • Used in applications in which absolutely no leakage to the atmosphere is acceptable (e.g., toxic gases). • Preferred in screw-compressor applications due to limited axial sealing length.

*Dry gas seals are installed with the labyrinth on the gas side and the carbon ring/labyrinth on the outer (or atmospheric) side. Quad seals with four seal stages have also been successfully installed and operated.

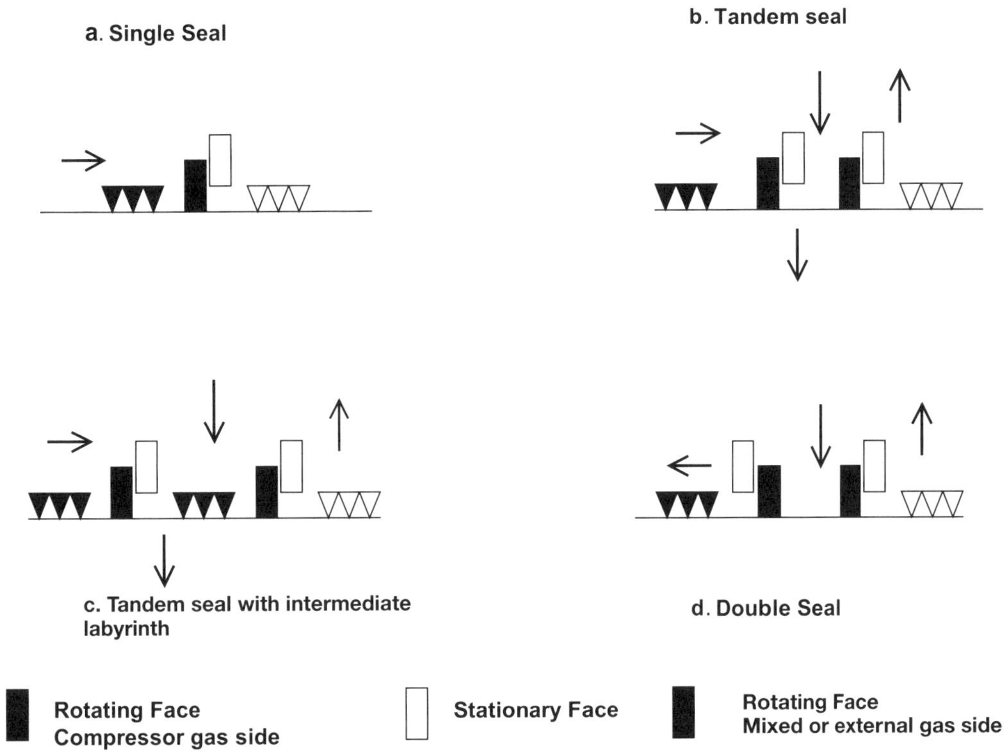

Figure 3-31. Different categories of dry gas seals.

other cartridge parts are generally made of stainless steel. Review and minimize screwed, multiple-sleeve construction for improved reliability. If H_2S is present, seal materials should be NACE compliant. Consider duplex stainless steel parts, Hastelloy springs, and silicon carbide faces if wet chlorides are present in the process gas.

Secondary sealing rings. Under pressure, permeable O-rings (fluoroelastomer) absorb gas. On shutdown, a high rate of depressurization can damage these O-rings (explosive decompression) since absorbed gas exits. To prevent explosive decompression, suppliers recommend controlled depressurization. The depressurization rate reduces as the sealing pressure increases. For sealing pressures above 900 psig (about 6,000 kPa) consider an O-ring of harder (but lesser compression/sealing ability) material, such as, perhaps, Aflas®. Alternatively, consider using polymer (e.g., Teflon®) seals with a position-retaining step. The O- or V-ring material is also affected by gas composition and should be discussed with the seal supplier. Generally, a water-saturated gas with 5% CO_2, 1 % H_2S, or the presence of methanol or ethanol will require review of the secondary seal material.

Face contact on reverse pressurization results in instantaneous seal failure. Reverse pressurization can occur in low-pressure service due to high plant-flare pressure on upset/depressurization. To prevent such incidences, the seal can be protected by suitable instrumentation or by installing a check valve in the leakage line to flare. For extended periods of high flare pressure, other solutions may be needed. During the engineering phase of a project, the maximum possible flare pressure and anticipated duration of flaring are communicated to the seal vendor.

Reverse rotation on compressor shutdown and consequent seal damage is possible under one or more of the following conditions:

- Suction valve does not close
- Recycle valve fails to open
- Malfunction of the check valve

A T-slot or other bidirectional-groove seal face will maintain the seal gap on reverse rotation. However, limited gas film area and gas film stiffness may well prevent their universal application. Reverse rotation is rare due to large machine inertia and hence not often considered a prime criterion for seal face selection.

Seal designs represent a continuously evolving technology; as manufacturers resolve field problems, new construction features are introduced. Engineers should not hesitate to evaluate new developments and select the features most suitable for their application. Some significant new features are:

- Provision of a rotating face shroud that contains seat material, in case of seal face fracture.
- Provision of an oil slinger or carbon rings between the seal and the bearing to prevent oil migration from the bearing to the seal

Table 3-4. Successful operating envelope for dry gas seals

Operating parameter	Conventional units	SI units
Maximum dynamic sealing pressure	6160 psig	42500 kPa
Maximum static sealing pressure	6160 psig	42500 kPa
Minimum operating temperature	(–) 242 °F	(–) 152°C
Maximum shaft size	13.75 inch	350 mm
Maximum compressor operating speed	70000+ rpm	
Normal allowable radial movement	± 0.024″	±0.6 mm
Maximum allowable axial movement: (size dependent)	± 0.24″	± 6 mm
Maximum surface speed at seal balance diameter	590 fps	180 m/s

- Larger return-oil passage to allow free flow of oil from the bearings to reservoir.
- Recently developed, directed-lubrication bearings with decreased oil requirements may assist further in reducing this problem.
- Centering the collar on the main seal sleeve.
- Preventing gas leakage along the shaft by providing a relief path to the flare in the seal sleeve.

As of the early 2000s, the successful operating envelope of dry gas seals was defined as given in Table 3-4. As of 2005, no single seal or manufacturer has achieved all of the above parameters in the same product. Thus, every supplier's experience should be individually evaluated. Moreover, development work is always underway and it pays to stay informed about up-to-date technology.

Dry Gas Seal Support Systems

Typically, on compressor trip/shutdown, the automated suction and discharge valves close and the recycle control valve opens. Consider blow-down of the compressed gas in this loop if the trip is due to failure of the dry gas seal, usually the primary component. However, on shutdown from other causes consider keeping the system under pressurized hold. On pressurized hold, the dry gas seals are subjected to an intermediate "settling-out" pressure, which depends on the volume of the loop. The estimated settling-out pressure should be provided as a minimum design condition for the seal and support system. In a multibody compressor train with interstage check valves, the transient settling-out pressure should be considered in the seal design. Some end users and suppliers prefer or recommend that the dry gas seal and support system to be suitable for the maximum discharge pressure.

The seal support system must carry out one or all of the following functions:

- Maintain seal, separation, and buffer gas supply at the required flow/pressure to ensure safe and reliable operation under all operating, transient, and shutdown conditions.
- Safely dispose of the normal seal leakage.
- Contain or eliminate hazardous or toxic leakage.
- Monitor seal performance.
- Activate alarm or shutdown on malfunction of either the seal or its support system.

Each dry gas seal (see Fig. 3-32 for a schematic representation of different categories of dry gas seals) needs certain utilities. In a "between-bearing" compressor, the pressure balance line reduces the bearing thrust load and the sealing pressure to compressor suction pressure. The seal gas or clean gas that is injected in a cavity ahead of the primary seal faces (see Fig. 3-32) should be noncondensing and compatible with the compressed gas. By using a differential pressure control valve (DP, Fig. 3-33) or plain control valve, the seal gas is controlled at 25 to 35 psi (175 to 250 kPa) above the sealing (reference) pressure to ensure an inrush velocity of

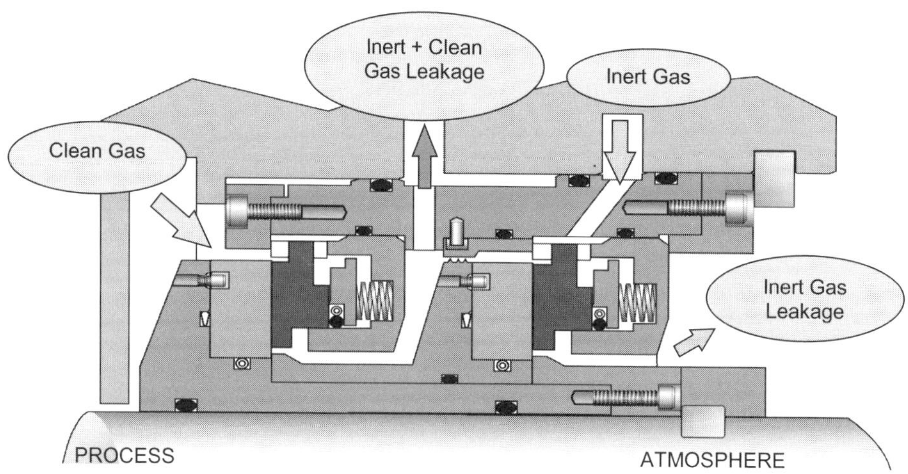

Figure 3-32. Tandem dry gas seal.

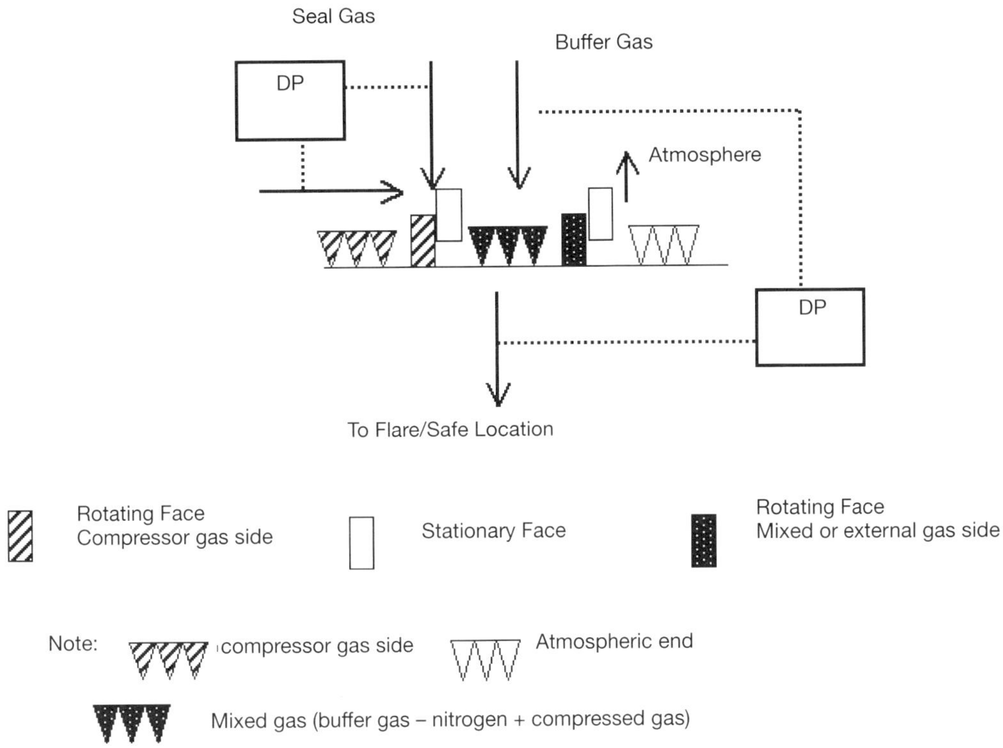

Figure 3-33. Process control on tandem seal with intermediate labyrinth.

approximately 20 ft/sec (6 m/s) when rotating and about 10 ft/sec (3 m/s) at static conditions. Some suppliers recommend flow control of the seal gas.

Although the labyrinth clearances at the top end of shaft are larger when the machine is stationary, the closing force of the seal spring reduces the leakage and, hence, a lower inward velocity is acceptable. Typically, the total dynamic seal-gas flow rates range from 50 to 150 scfm (internal and external leakage combined). Solids or magnetically charged particles larger than 2 microns should be removed from the seal gas, preferably by using coalescing filters.

Condensed liquids (water or C6 plus hydrocarbon) in the sealing gas tend to produce a sticky substance that adversely affects the elasticity of O-rings and springs, resulting in seal hang-up and machine trip from increased leakage rates. The effect is exacerbated under the stationary/pressurized hold condition when no windage heat is generated between the seal faces. In high-pressure applications, the traditional 20°F (11°C) margin of the seal gas temperature above dew point is insufficient to prevent condensation of water/heavy hydrocarbons as the leakage gas pressure decreases to near atmospheric before discharge to a safe location or the flare. Check the seal gas-phase envelope for the entire operating range before determining the superheat margin. If required, consider heat tracing of the seal piping.

Seal vendors recommend that machine starts/stops be minimized. With turbine-driven compressors, inform the seal vendor if the compressor is required to slowly turn in the hot standby condition. When seal gas is taken from the compressor discharge on hot standby, insufficient discharge pressure may result in dirty compressed gas reaching the seal faces, with resultant seal failure. Selection of suitable seal-gas source(s) to meet all operating/transient conditions is of prime importance for reliable operation.

Buffer gas is filtered to 2–5 microns prior to injection into tandem/triple seals between the primary and secondary seals. Nitrogen at 50 to 100 psig (~350 to 700 kPa) is commonly used as buffer gas and regulated above the backpressure in the primary leakage line. This line is generally connected to the plant flare system. Though not as desirable as nitrogen, offshore applications often use pressurized air when it is the only available utility. If air is used as buffer gas, use sufficient volume to ensure that the air and hydrocarbon mix does not reach the lower explosive limit (LEL). The LEL depends on the gas composition and should be evaluated on a case-by-case basis. Minimize the length of the atmospheric vent line to keep the backpressure below 5 psig (35 kPa).

Separation gas (or inert seal gas, Fig. 3-33) is injected between the seal and bearing to prevent lube oil migration toward the seals. Carbon bushings and/or an oil slinger assist in preventing this oil migration. Generously sized bearing lube oil return lines and tank breathers reduce backpressure and, consequently, lube oil migration toward the seals.

The seal support panel is often located at the compressor shaft centerline or higher. This will assist in draining liquids from the low point. The seal monitoring and control panel provides the following principal functions:

- Measures seal gas and sealing differential pressure to ensure that seal gas pressure exceeds the pressure of the adjacent process gas.
- Measures, routes, and monitors primary seal leakage to the flare or a safe location.
- Routes and monitors secondary seal leakage to the atmosphere or a safe location.
- Monitors separation-gas injection.

Dry gas seal failures are generally instantaneous, leaving little opportunity or time for operator intervention. Most failures occur early in the seal operating life. Once technical issues are resolved, failures are rare. Refer to Table 3-5 for commonly used alarms and trips in the seal control panel.

Reliable Auxiliaries Are Important for Dry Gas Seals

The seal or buffer gas filter is generally stainless steel with a coalescing, duplex filter and continuous transfer valve. The filter is typically sized for a minimum of three times the normal flow, or the required flow rate for the seal failure case, whichever is greater. For a saturated or heavy condensing seal gas, consider installing the filter downstream of the pressure throttling device to maximize the removal of condensate. Consider designing the seal gas filter in accordance with the applicable pressure vessel code.

Leakage from each seal should be measured separately. When the primary seal leakage is routed to the atmosphere, consider installing a flame arrestor in the vent line.

Testing. Dry gas seal cartridges are tested for both dynamic and static seal leakage in the seal supplier's shop. The test is carried out, using ambient air or nitrogen, at various operating points up to the maximum sealing pressure and operating speed. In tandem construction, both primary and secondary seals are tested for the maximum sealing pressure. Dynamic and static seal leakage is measured at various pressures and speeds. The test procedure is jointly agreed on. For cryogenic applications, it is possible to replicate low temperatures during testing. Manufacturers usually provide both a normal leakage and guaranteed maximum leakage rate.

Field Installation. Most seal failures occur early in seal life. For critical or large applications, consider hiring the services of the seal vendor's representative for first start-up in the field. Dry gas seals are customized and purchase of a spare cartridge set is recommended for quick maintenance turnarounds.

Retrofit. Consider the following issues when retrofitting dry gas seals on compressors with existing wet seals:

- Changes in piping/valve arrangement, including blowdown to take advantage of pressurized holding capability.
- Review design pressure of upstream equipment that may be subject to settling-out pressure
- Impact on train rotor dynamics. Wet oil seals increase rotor critical speeds, reduce amplification factors, and improve damping. However, as seals wear over time, they may cause subsynchronous vibration and adversely affect rotor stability.
- Installation of dry gas seals will decrease rotor lateral critical speed and increase the amplification factor. However, the rotor response is predictable and remains unchanged over time. Careful evaluation is required when retrofitting dry gas seals on flexible rotors. If preliminary review warrants it, a third-party rotor dynamic analysis is a justifiable expense.

Table 3-5. Commonly used alarms and trips in dry gas seal systems

Initiating condition	Single seal	Tandem seal	Double seal
Primary seal leakage, high	Alarm (1)	Alarm(1,2)	Alarm(1)
Primary seal leakage, high-high (failure)	Trip(1)	Trip(1,2)	Trip(1)
Seal gas differential pressure, low or high	Alarm(3)	Alarm(3)	Alarm(3)
Seal gas differential pressure, low-low	Trip(3)	Trip(3)	Trip(3)
Seal/buffer gas filter differential pressure, high	Alarm	Alarm	Alarm
Buffer gas flow, high—secondary seal failure	Not applicable	Alarm	Not applicable
Buffer gas differential pressure, low	Not applicable	Alarm	Not applicable
Separation gas pressure/flow, low	Alarm	Alarm	Alarm

1. Trip level flow is usually 5 to 8 times the normal leakage. Alternatively, differential pressure may be used to actuate a trip.
2. For tandem seal, normal flow is seal leakage plus buffer gas flow.
3. To prevent dirty compressed gas reaching seal faces.

The full cost of a retrofit should be considered when carrying out a life-cycle benefit analysis.

Additional Applications. In addition to centrifugal compressors, some other common applications for dry gas seals include:

- Dry screw compressors (normally, double-pressurized seal configurations due to space limitations)
- Steam turbines (successful only if steam condensation can be prevented)
- Cryogenic pressure-recovery service
- Agitators (using low-speed dry gas seals)

Advanced computing methods including FEM/FEA/stress and torsional analysis continue to increase our understanding of the hydraulics and mechanical rotor dynamics of large machines installed with dry gas seals and magnetic bearings. As experience increases, commercial considerations for remote and unmanned operations may force their application to be the norm.

Stricter environmental regulations and the cost of energy are reflected in developments that will, hopefully, lead to further dry gas seal and support system improvements. Some of these are quite obviously underway and maintaining contact with principal vendors is always beneficial.

Experience shows that a continuous supply of clean and dry buffer gas is one of the most important requirements for trouble-free operation and long seal life. Unfortunately, this key requirement is often ignored throughout the various planning, commissioning, and operating processes. This shortcoming, particularly during the commissioning period, is almost guaranteed to result in multiple seal failures that cause operational loss and delays in plant start-up.

As mentioned above, common control system designs for gas seals consist of filtration, regulation, and monitoring. However, though these control systems typically offer elaborate monitoring and regulation features, the filtration issue is often overlooked. In most cases, users and contractors initially choose standard filtration for virtually every application, regardless of gas composition and/or presence of liquid, or condensation occurring in certain gas mixtures. One competent seal manufacturer's documented project histories show this shortcoming to be the root cause of the majority of seal failures, particularly during the commissioning phase, but also during the "normal" operation period.

Recognizing that most control systems feature inadequate, and indeed elementary, filtration systems, the only obvious advantage of incorporating sophisticated monitoring devices is to indicate when a seal has failed or is about to fail. At best then, these monitoring systems serve to point to the root cause of seal failures, yet do little, if anything, to prevent them.

Getting to the root of the problem begins with the proper analysis of mechanical failure. This involves a review of gas composition, commissioning procedures, and control system design, as well as the various interfaces between the seal, compressor, and seal control system. Further, it is essential to understand how the control system piping and instrument layout interact and function; this requires close examination of the associated flow schematic. Of equal importance is control-related and typically preestablished software that must allow proper logic input to ensure the safe operation of the unit.

In recent years, gas conditioning units (GCUs) have greatly enhanced the reliability performance of dry gas seals by solving critically important gas supply issues. Unlike conventional gas panels that only incorporate coalescing filters, a modern GCU features a knock-out filter/coalescer vessel that removes solid particulates as well as free liquids and aerosols. A heater–controller also monitors and maintains gas temperature. Maintaining gas temperature above its dew point prevents condensation of aerosols in the process gas stream. Therefore, the collective features of successful GCUs must effectively manage liquids to ensure that the cleanest possible gas supply is always available.

In start-up, slow-roll, and settling-out conditions, a thoroughly engineered GCU will maintain adequate gas flow using a seal gas pressure intensifier or similar device. Typically, a flow switch signals the intensifier control, which automatically activates and deactivates the intensifier as needed. The intensifier then provides sufficient seal gas flow to prevent unfiltered process gas from working its way back to the seal faces across the inboard labyrinth. Experienced machinery engineers and failure analysts know that clogging the seal-face grooves will cause failures. With minimal customer interface connections and self-checking, self-regulated functions, modern GCU's can meet the difficult sealing challenges faced by many industrial facilities.

In summary, the key to minimizing seal failure is a thorough review of the components of the plant and their impact on the total system. In order of importance, the following factors should be considered in examining dry seal support systems for centrifugal compressors:

- *Gas composition.* Understanding the actual gas composition and true operating condition is essential, yet often overlooked. For example, it is necessary to understand when and where phase changes and condensation will result in the sealing fluid.
- *Commissioning procedures.* Is clean and dry buffer gas available? Is the seal protected from bearing oil? How is the compressor pressurized or de-pressurized? How is the machine brought up to operating speed? Are all personnel fully familiar with the compressor maintenance and operating manual? Is the full control system included and adequately described in these write-ups?
- *Control system design.* Is clean and dry buffer gas available at all times? Key elements of the system de-

sign include buffer gas conditioning, filtration, regulation (flow vs. pressure) and monitoring. It is important to not only review the control system to ensure that problems do not arise, but also gain an understanding of the system's design philosophy in order to gain an appreciation for the manner in which the logic is addressed. Further, it is crucial to review the buffer gas conditions. Is a heater required? What is the temperature setting for the heater? It is necessary to analyze the connecting piping structure, size, material, and type to avoid liquid entrapment? Also, what components and systems are most suitable to withstand harsh outside environments?

- *Interface between the compressor, seals, and control system.* Review the startup and shutdown sequence, liquid removal if applicable, alarm settings, shutdown settings, and flow measurement units. Be sure to recognize signs of problems within the startup and shutdown points. Also, in terms of the logic and the interface, what control setting do you want? Remember that you cannot design one control system to fit every scenario.
- *Plant specifications, including tubing versus piping, pipe sizing, logic system, and wiring diagrams.* Sometimes, the plant specification is totally different from a supplier's recommendation, but for good reason. For example, a plant may specify tubing instead of piping, or a different type of welding procedure. Or, although a supplier may recommend a shutdown on a specific setting, the plant may opt for coordinated shutdown to avoid process upset.

In all instances, but especially in the case of dry gas seal retrofit applications, the list could be extended. Certainly from the reliability-focused user's point of view, there are additional precautions that deserve to be looked into. These include not only the previously mentioned dynamic analysis of the train with the dry gas seal (DGS) retrofit, but also preparation of the spare rotor for DGS retrofit and making plans for machining the existing, or procurement of new, end covers. Planning must include upgrading of compressor controls and incorporating requirements associated with the DGS system. Considerable thought must be given to the buffer-gas source.

Again, either the process gas must be absolutely clean or some other absolutely clean and fully compatible medium must be found. Permanent provision of a reliable source of clean purge gas must be made also for the tertiary seal to avoid ingress of lube oil into the DGS. Trim balancing is used for rotors retrofitted with dry gas seals. Subsequent onsite testing of these retrofitted rotors must demonstrate vibration severities that are equal to, or lower than, those observed when the machine was originally commissioned with a wet-seal system. The spare rotor must be included in these precautionary checks.

Chapter 4

Power Transmission and Advanced Bearing Technology

4.1 COUPLINGS FOR ROTARY POSITIVE DISPLACEMENT AND CENTRIFUGAL COMPRESSORS

Couplings represent an important link in compressor trains. They deserve nothing short of the best engineering. Their primary function, of course, is to bridge the space between drivers and driven equipment and to transmit torque from the driver to the driven machine. However, since it is practically impossible to maintain absolutely perfect alignment between the two machines, couplings must be designed to accommodate shaft misalignment and axial float as well (Fig. 4-1). Pre-1960s installations employed gear type couplings (Fig. 4-2) as an industry standard. However, since the 1970s contoured diaphragm couplings (Figs. 2-3 and 2-4) have successfully replaced gear couplings and are now often favored even by cost-conscious original equipment manufactures (OEMs). Contoured diaphragm couplings are nonlubricated and thus eliminate oil-related concerns or issues. Properly selected diaphragm couplings tolerate a reasonable measure of misalignment. They do, however, require careful installation.

There is less latitude available in the axial positioning of these couplings than in gear-type couplings. Failures, although virtually unheard of in reliability-focused plants that use rigorous selection techniques and installation methods, could occur with little or no warning.

As in the case of other machinery specifications, the coupling standard, API-671 will serve the user quite well. API-671 defines, in great detail, the minimum requirements for couplings used on critical, nonspared machines. Such a machine could be a 60,000 rpm expander drive or a more conventional 1,800 rpm motor drive; the machinery user determines which machines are critical.

4.1.1 Coupling Functional Parameters, Types, and Configurations

Gear-type couplings, with lubrication provided by either grease or a continuously applied stream of oil, incorporate conventional meshed gear teeth. Available in numerous different geometries, tooth profiles, and sizes, these couplings absorb misalignment by relative movement (sliding) of the driving gear mesh versus the driven gear (Fig. 4-2). By nature of their design, gear couplings are "wearing" component and may require periodic replacement after limited wear has taken place. Moreover, gear couplings are also prone to potential "lock-up" of the gear teeth. Such lock-up could produce reaction loads that often exceed the capacity of the thrust or radial bearings of a machine. The fact that gear tooth wear is unavoidable and that lock-up is difficult to rule out entirely makes the long-term performance of a gear coupling unpredictable. Although certain gear couplings have been in operation for five, ten, or even 15 years, the risk of excessive vibration due to lock-up causes is real and unscheduled outages are a possibility.

Single contoured-diaphragm couplings (Fig. 4-3), are of great importance to reliability-focused process plants. Although now primarily manufactured by B. F. Goodrich, they are still commonly known by the original trade name "Bendix" (which was purchased by Lucas Aerospace, then TRW, and now B. F. Goodrich). Torque is transmitted in shear through a contoured, profile-ground single disc or diaphragm. A contoured diaphragm coupling exerts a lesser bending moment on associated assemblies than an equivalently rated gear coupling. In many retrofit situations, replacing an existing gear coupling by a diaphragm coupling allows the transmission of higher power within an available space envelope. The need to change an existing shaft is thus often eliminated.

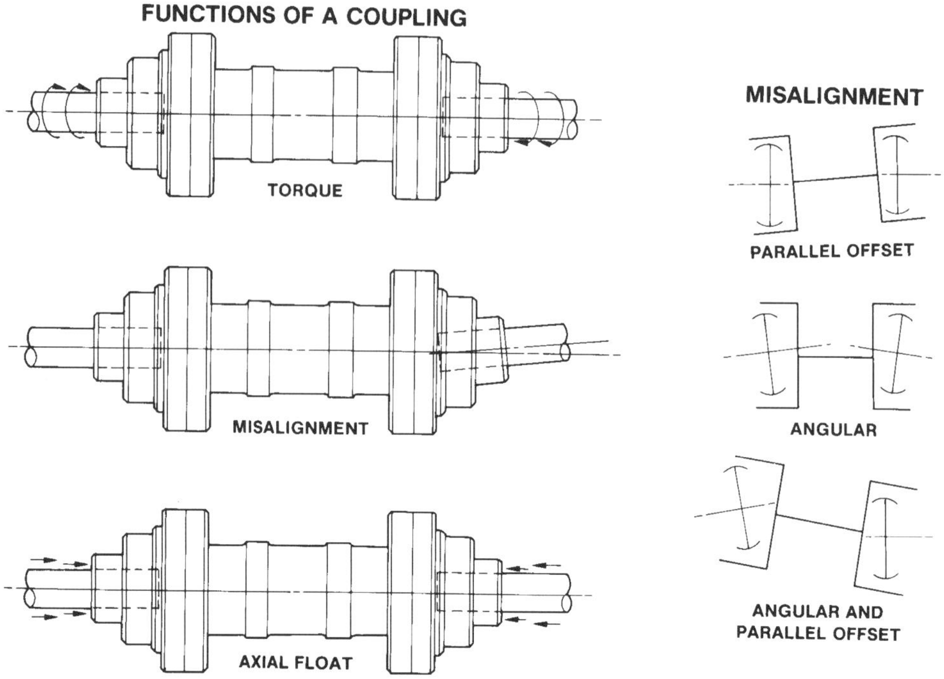

Figure 4-1. Functions of compressor couplings. (Source: Ameridrives Couplings, Erie, Pennsylvania.)

Multiple diaphragm couplings (Fig. 4-4) were originally designed and manufactured by the Zurn Corporation and are now available under the Ameridrive name. They use the same principle of torque transmission as Bendix couplings, but multiple thin, often "wavy" diaphragms produce a coupling with much lower axial and angular stiffness than the single-diaphragm coupling

Disc couplings (Fig. 4-5) were originally manufactured under the Thomas brand name and have been in use since the late 1950s. Until the 1970s, disc couplings were generally considered best suited for use in less critical, motor-driven equipment. Refinements in link geometry (optimized by using finite element analysis) as well as discs coated with PTFE have made the disc coupling an increasingly popular choice for many turbomachinery users and original equipment manufacturers (OEMs). The disc-coupling drive mechanism is significantly different from that found in a diaphragm coupling. With the disc design, torque is transmitted through tension in a series of laminated tangential links, connected alternately to a driver and driven flange on a common bolt circle. Since steel is typically twice as strong in tension than in shear (as in a metal diaphragm), the disc-pack cou-

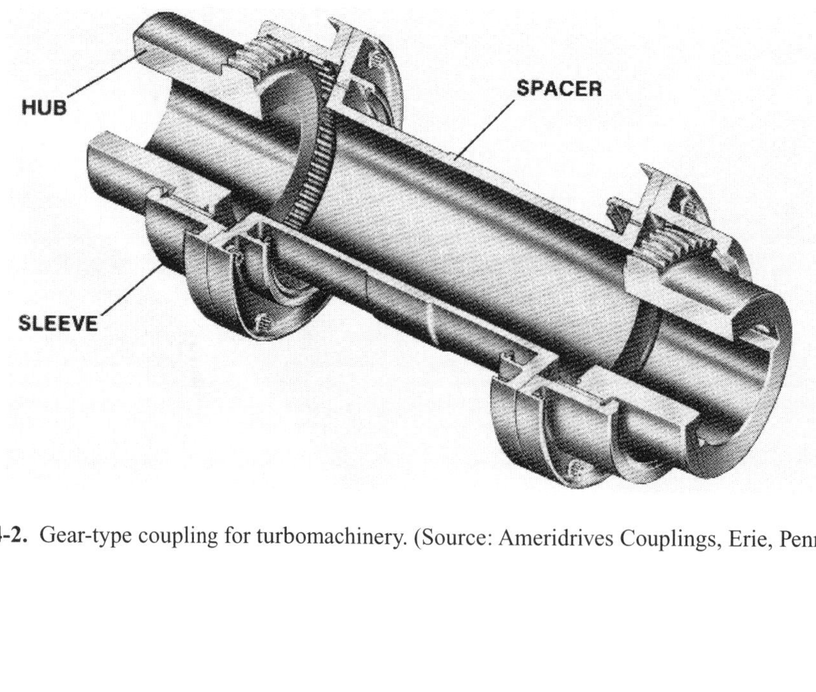

Figure 4-2. Gear-type coupling for turbomachinery. (Source: Ameridrives Couplings, Erie, Pennsylvania.)

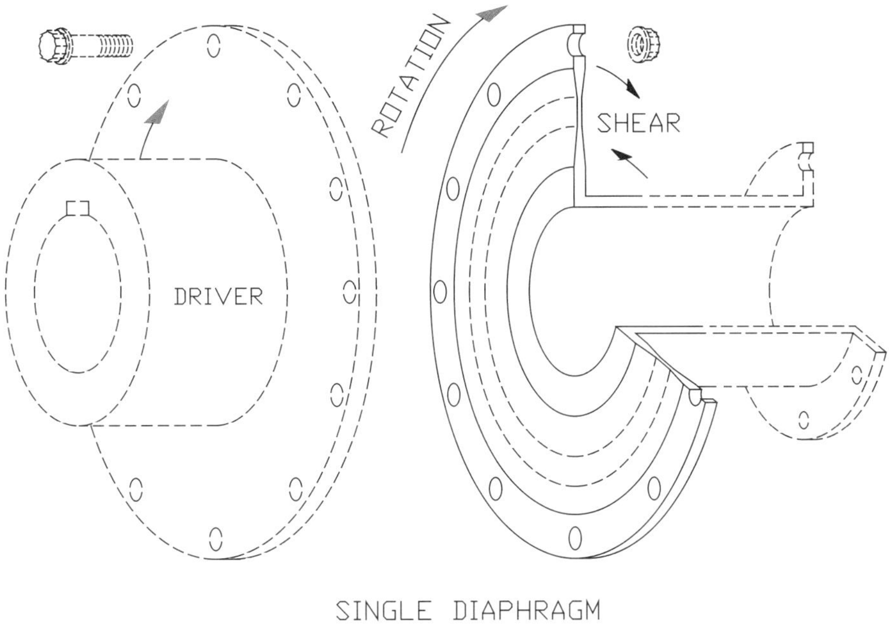

Figure 4-3. Contoured single-flexible-elements diaphragm coupling. (Source: Flexelement Texas, Houston, TX.)

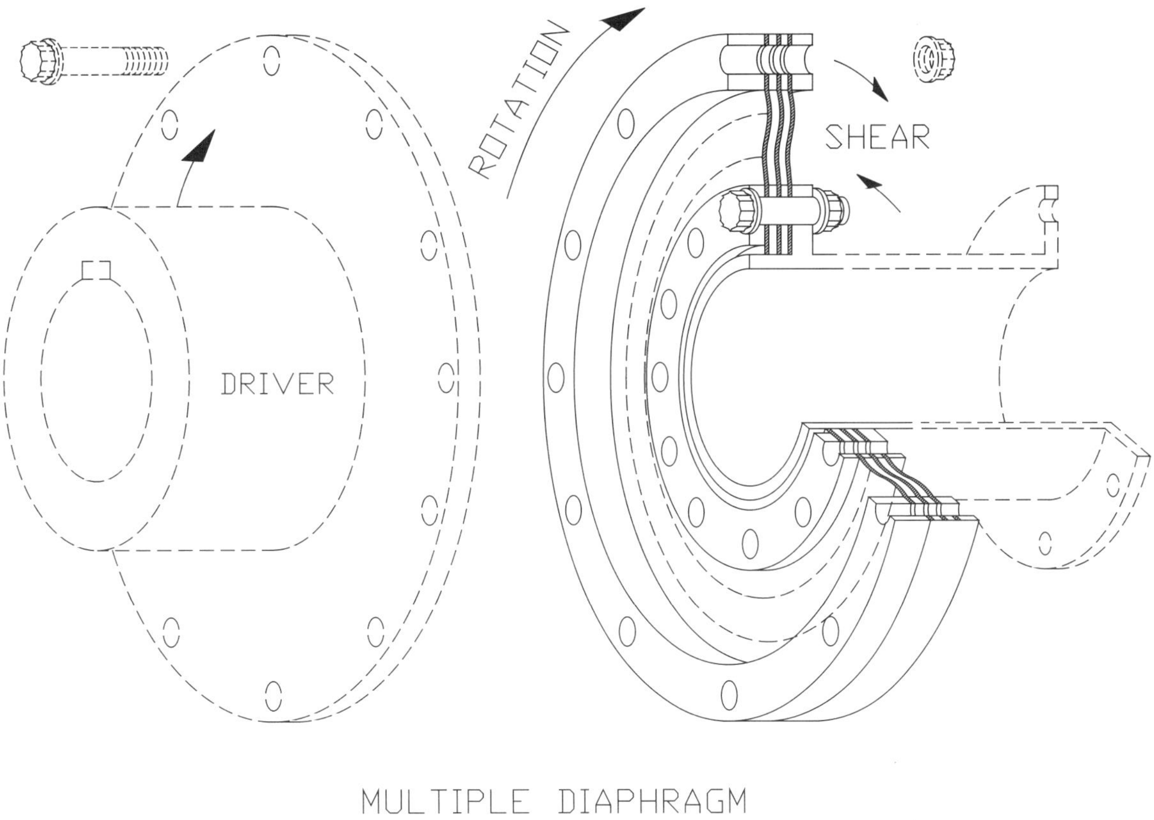

Figure 4-4. Contoured multiple-flexible-element diaphragm coupling. (Source: Flexelement Texas, Houston, TX.)

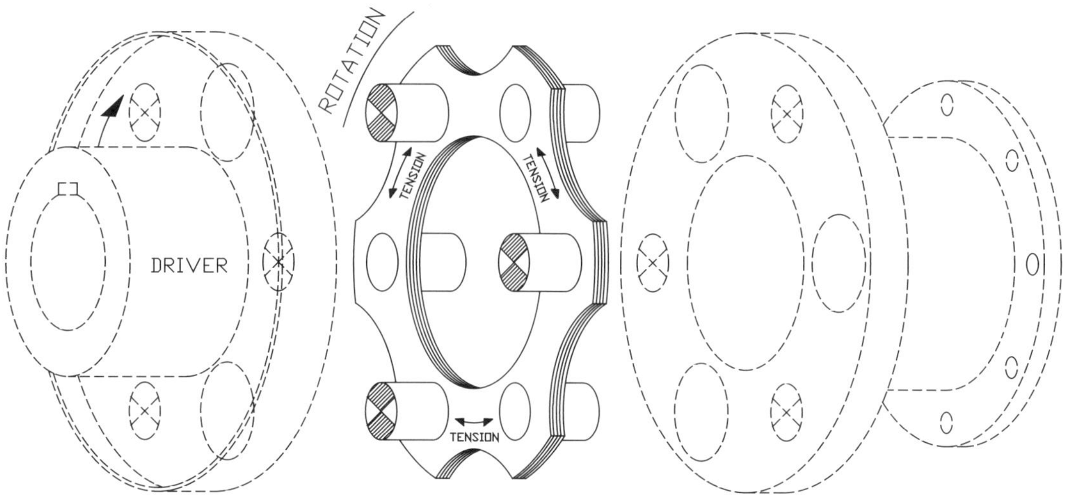

Figure 4-5. Disc pack coupling. (Source: Flexelement Texas, Houston, TX.)

pling tends to have a much smaller outer diameter than a comparable diaphragm coupling. A minimal outside diameter, combined with a large inner diameter, make the disc-pack coupling particularly well suited to low-moment applications.

Nonlubricated flexible couplings using metal diaphragms or discs to absorb stresses due to misalignment, when properly selected, have the ability to function for many years in continuous operation without the need for inspection, maintenance, repair, or replacement. The coupling should no more be considered a wearing item than the main rotor shaft. Both parts can be designed to absorb high levels of combined steady and cyclic stresses and have a useful life approaching infinity. Careful selection and application of couplings for use on high-energy machinery will result in years of safe, reliable operation with no need for scheduled replacement or maintenance beyond some simple inspection procedures.

4.1.2 Marine-Style Couplings and Reduced-Moment Geometries

A style or configuration common to all coupling types, the marine coupling takes its name from an industry in which heating a rigid hub for removal proved cumbersome and even dangerous. In marine couplings, the sliding gear-meshing components or flexing elements are integral with the spool piece. The spool pieces are easily removed and replaced without having to disturb the rigid hubs. Marine-style couplings rank among the simplest configurations from a hands-on viewpoint. However, the greater mass and different center-of-gravity location have occasionally been known to combine in a manner that makes marine couplings unsuitable for machines running above their respective first critical speeds.

Resonances are potentially lethal on high-energy machines. Lateral, torsional, and axial resonant frequencies (ideally) are calculated and avoided at the machinery design stage. Sophisticated rotor dynamic analysis programs make possible the accurate calculation of shaft or coupling resonances to advance and improve machinery designs. But rotor dynamic analysis is widely used in areas other than troubleshooting problem machines. Among other uses, it is employed in ascertaining that the right coupling is chosen for trouble-free, long-term operation of process compressors.

Again, any of the gear, disc, or diaphragm couplings can be supplied in marine-style and/or low-moment configurations that have the sliding gear teeth machined, or the flexing element assembled to the hub. The flexing plane and center of gravity are thus moved to the inboard side of the shaft end. Reduced moment geometries are used commonly in sensitive rotor systems in which excessive overhung moment can significantly change the shaft critical speed. They are also very useful in machines with very close shaft ends or little separation between shafts. Close shaft spacing can limit coupling flexibility and can create high vibration and bearing damage due to excessive reaction loads.

4.1.3 Coupling Assembly and Selection Summary

A typical coupling assembly may consist of 150 or more separable components. Many are factory preassembled so that installation personnel are not delivered a puzzle of pieces to assemble on site. A reputable coupling manufacturer has the expertise and equipment to best provide assurance of correct bolt torques, and so on.

Couplings for critical machinery should be selected, engineered, and manufactured to suit the specific operating characteristics of the connected machines. A catalog selection based only on torque, speed, shaft size, and misalignment capacity may result in a coupling not well suited for a specific machinery train. Indeed, ineffective coupling selection may be the cause of compressor unavailability, including catastrophic failure.

Gear couplings are rarely considered appropriate for reliability-focused plants. This is reflected in the sales figures of a major manufacturer producing both flexible-element couplings as well as gear couplings. In 2004, gear couplings represented only 8% of the manufacturer's high-performance couplings sales.

The initial choice of a coupling must be based upon its torque capacity (using an appropriate application factor) and its misalignment capability. A system-based review should be used to carefully examine the mass/elastic characteristics (weight, inertia, center of gravity, axial, angular and torsional stiffness) to ensure that adequate, safe margins are maintained between operating speed and lateral, torsional, or axial resonant frequencies.

Rotor dynamics analysis (RDA) techniques have been developed and programs are widely available that, when properly applied, permit a detailed review of the coupled rotor system long before the design is finalized and manufacturing begins. These same techniques, in the hands of expert machine design engineers, can be used to ensure safe and successful coupling retrofits. At times, and usually preceded by years of poor performance and high vibrations, the original coupling choice manifests itself in a bad and even dangerous mismatch between rotor geometry and operating conditions.

4.2 MAGNETIC BEARINGS FOR COMPRESSOR ROTORS

The innovative developments alluded to earlier in this text most certainly involve magnetic bearings and conventional but externally pressurized bearings. The aim of either of these innovations is to allow higher speeds and vibration-free compressor operation. In turn, this would make feasible tighter internal clearances and more efficient compression. Magnetic bearing systems represent noncontact technology, producing only negligible friction loss, zero wear, and enhanced reliability. Magnetic bearing technology uses magnetic fields to both levitate and axially restrain the rotor. Mechanical contact is made at standstill only.

4.2.1 Early Applications of Magnetic Bearings

Since the mid 1980s, magnetic bearings have found their way into many different turbomachinery applications. In compressors, the technology has followed the progress made by dry gas seals that eliminated the need for high-pressure oil seals. Acceptance of magnetic bearings grew when the merits of developing a completely oil-free "dry" machine were recognized. Early applications included magnetic bearings replacing oil bearings on pipeline compressors, with the bearings mounted outboard of the dry gas seal in an air-purged cavity. However, these units were driven by either gas turbines or electric motors with a speed-increasing gearbox that required its own lube oil system. Although the idea of a "dry" machine helped to promote the technology, it was still difficult to justify magnetic bearings commercially because of their high initial cost and the fact that a lube oil system was still required to support the driver and/or gearbox.

Much of this changed in the late 1990s and early 2000s when machine designers started to go a step further by taking advantage of the magnetic bearings' ability to operate directly in the process gas. This is particularly true for compressors in which this type of "consolidation" provides some unique technical and commercial advantages. In addition, the development of direct-drive, high-speed motors has accelerated acceptance because now the machine designer can consider using magnetic bearings for both motor and compressor. This offers the development of a completely oil-free equipment train and is the niche in which magnetic bearings are being applied today for compressors. Table 4-1 presents some general guidelines for applying magnetic bearings in turbomachinery.

Magnetic bearing technology is unique because it replaces a completely mechanical system (bearing hardware plus lube system) with an electromechanical system (bearing hardware plus control system). Using electronics instead of mechanics offers inherent advantages like higher reliability and lower maintenance, but there are also some disadvantages that must be recognized in order to correctly apply the technology for compressor applications.

4.2.2 Operating Principles of Active Magnetic Bearings Explained

Magnetic bearings, like oil bearings, provide low-friction support of a rotating shaft. The bearing loads are a result of both static and dynamic forces produced by the compressor shaft. A typical horizontal shaft arrangement, shown in Fig. 4-6, contains five degrees of freedom (DOF) that must be controlled by the bearings. Two DOFs (x and y axes) are controlled by each radial bearing on the ends of the shaft; the axial or thrust bearing controls the longitudinal DOF in the z axis.

The radial magnetic bearings react to static loads that are due to the shaft weight as well as any aerodynamic "side loads," which can come from a pressure differential in the volute. Dynamic loads are primarily due to shaft unbalance, but can also be caused by shaft/coupling misalignment, aerodynamic loads, liquid carryover, and so on. The magnetic thrust bearing must react to static loads caused by the pressure differential across the machine and dynamic loads from various sources such as surge or other aerodynamic causes.

Table 4-1. General guidelines for applying magnetic bearings

Magnetic bearing applications	Oil bearing applications
Operation within a process fluid environment such as turboexpanders or hermetic (i.e., sealless) compressors	Machinery in situations in which oil lubrication is readily available, such as gear-driven units or oil-sealed compressors
Machinery in situations in which space and weight are of concern, such as offshore platforms	Multicasing trains consisting of low-speed drive and step-up gearbox
Higher maximum peripheral speeds	Lower maximum peripheral speeds
Lower specific (psi) bearing loading	Higher specific (psi) bearing loading
Remote, unmanned, or inaccessible locations such as subsea, desert, pipelines, etc. Remote monitoring possible	Conventional land-based machinery. Remote monitoring available through third-party systems.
Technology includes integral monitoring of axial and radial shaft location, bearing load, speed, and temperature	External monitoring optional as required
Trains with integrated high-speed motor drive	Trains that can share a common lube system between machine, driver, and gearbox
Vibration- (i.e., noise-) sensitive machines or special process machines that need to withstand highly unbalanced loads	Heavy-impact loads such as rock crushers, paper pulp refiners, dredge pumps, etc.
Wide operating speed range crossing critical speeds	Limited speed range or fixed speed
Reduced oil-leak-type fire risk	Contained fire risk applications
Extreme temperature environment (cryogenic to >250°C)	Acceptable oil temperature environment and applications in which oil provides additional shaft cooling, such as certain microturbines
Available electricity for motor drive	Gas- or steam-turbine-driven equipment with gearbox

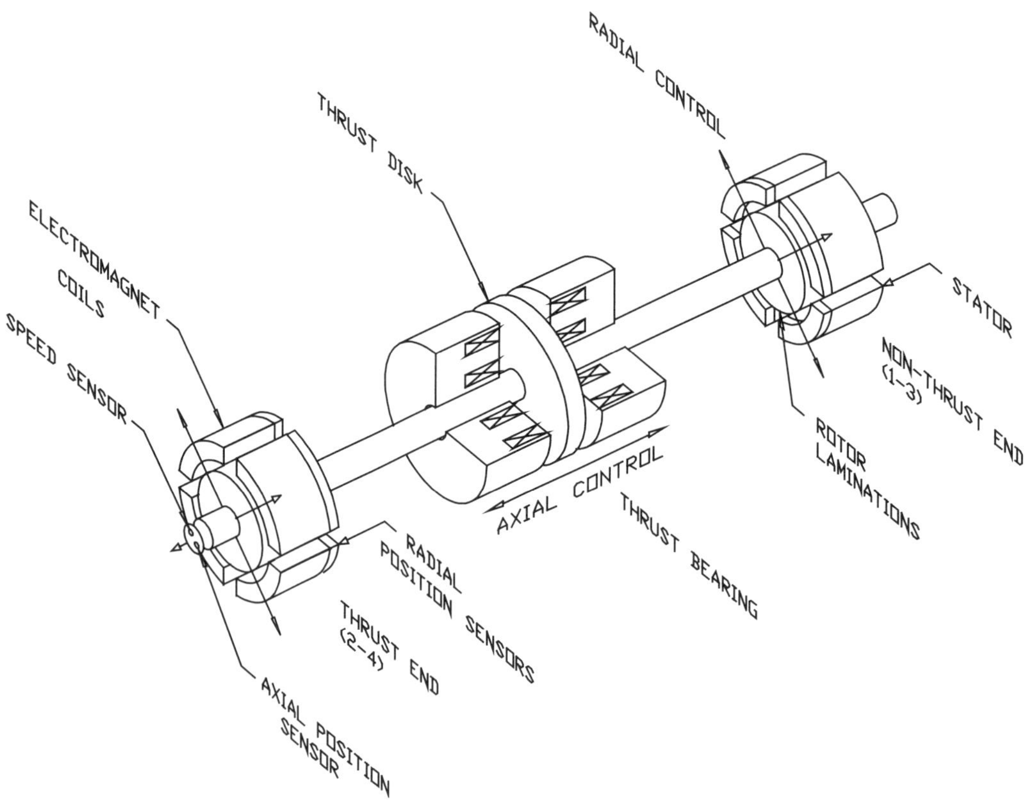

Figure 4-6. Magnetic bearing arrangement for a typical compressor shaft with five degrees of freedom.

A magnetic bearing system as shown in Fig. 4-7 is made up of electromagnets that work by force of attraction to levitate the shaft in a relatively large air gap. Position sensors monitor the position of the shaft and provide feedback to the control system. The shaft position is compared to a reference (i.e., center position). If the shaft is not centered, an error signal is generated and then conditioned by a proportional integral derivative (PID) control algorithm. The control output is a correction signal to the power amplifiers, which in turn provide correction currents to the electromagnets. The electromagnetic actuators provide the restoring force to recenter the shaft.

The control loop shown in Fig. 4-7 controls one degree of freedom, so that a typical compressor requires five control loops for five degrees of freedom. The PID control, power amplifiers, and associated electronics are housed in a control cabinet that is usually located in a nonhazardous environment away from the compressor with power and signal wires (i.e., cables) connecting the bearings to the control system.

The magnetic bearing hardware inside the machine consists of the radial and axial magnetic bearing stators (i.e., electromagnetic actuators), position sensors, rotor laminations for the radial bearings, and the thrust disc. In addition, auxiliary bearings are included to provide a safe coast-down of the shaft in the event of a failure or overload that results in a delevitation. A detailed description of the mechanical hardware and control electronics is provided later.

4.2.3 Load Capacity

Magnetic bearings have lower specific load capacity than conventional bearings and the compressor designer must provide sufficient projected area (length times diameter) to achieve the necessary load capacity. This problem is somewhat mitigated by the fact that magnetic bearings allow higher peripheral speeds and, therefore, the journal diameter can be made larger to increase the projected area or active area of the bearing. In conventional oil-lubricated bearings, the load capacity is limited by the strength of materials and the ability to develop an oil film to support the load. For magnetic bearings, the specific load capacity depends on the magnetic properties of the materials used in construction, and this capacity is limited by the saturation of the magnetic material. Once the material saturates, there is no more bearing force available and the shaft will be displaced or "delevitated." Bearing sizing must, therefore, provide excess load capacity to handle transient overload conditions.

A radial magnetic bearing is made up of four electromagnets that work as opposing pairs to control two degrees of freedom. To maximize load capacity, the magnets are organized along 45 degree axes (V and W) so that the upper two magnets can support the weight of the shaft as shown in Fig. 4-8.

The load capacity then becomes a function of the projected area (D × L) and the magnetic properties of the laminations used in the construction as shown in Table 4-2. The projected area of the bearing is the outside diameter (D) of the rotor laminations times the stack length (L) of the laminations.

4.2.4 Magnetic Bearing Hardware

Radial Magnetic Bearings

The radial bearing stator is made up of a stack of laminations that are designed and stamped to form magnetic poles to conduct the magnetic flux. The laminations have coils wrapped around the poles, forming electromagnets. This type of construction is similar to electric motor hardware. The bearing rotor is also made up of laminations that are shrunk onto the shaft. The rotor laminations are necessary to minimize the hysteresis and eddy current losses caused by alternating north and south poles on the bearing stator as shown in Fig. 4-8.

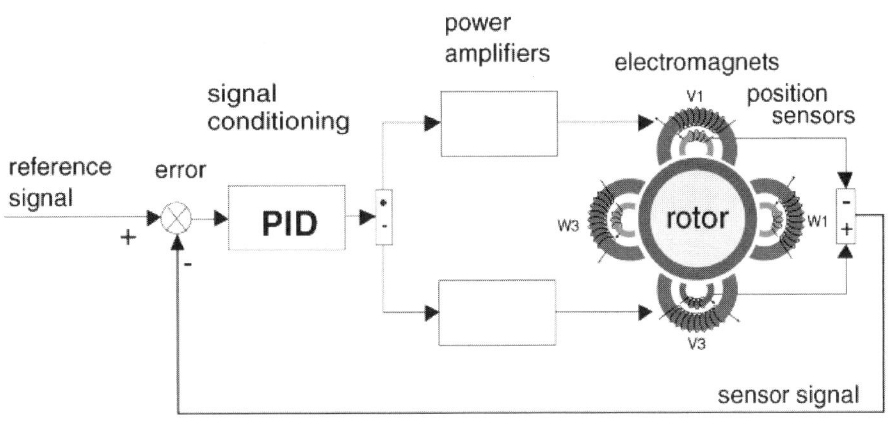

Figure 4-7. Magnetic bearing system.

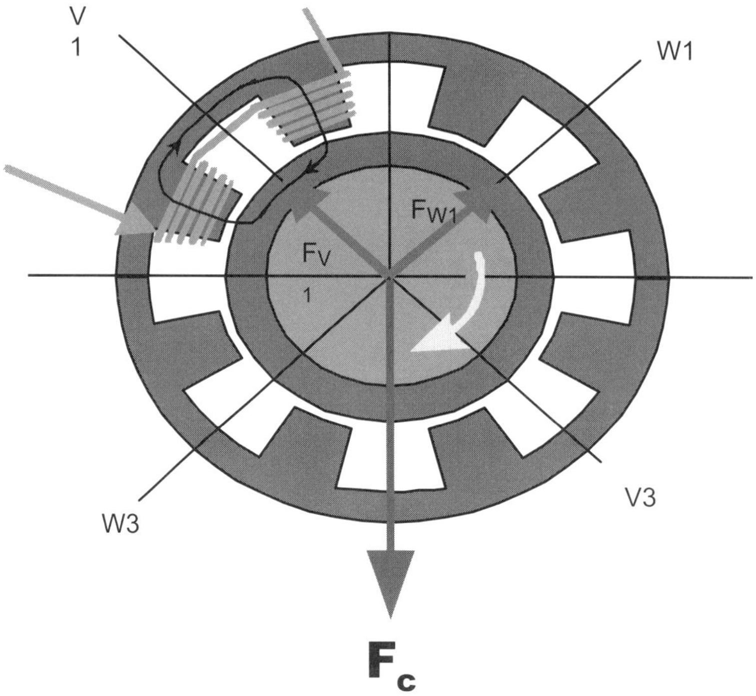

Figure 4-8. Typical arrangement for a radial magnetic bearing.

Position sensors are located close to the bearing actuators to provide continuous rotor position information back to the controller. The compressor operator must remember that the shaft is rotating with respect to this position sensor surface and not the bearing surface as it is with conventional journal bearings. This must be considered during rotor balancing.

The radial bearing stator and the position sensor assemblies are typically mounted together to ensure concentricity. Figure 4-9 shows a complete radial magnetic bearing cartridge.

The speed limit for the radial magnetic bearing is determined by the mechanical strength of the laminations mounted (shrunk) on the shaft. This provides for a relatively high peripheral speed compared to other bearing technologies.

Magnetic Thrust Bearings

The magnetic thrust bearing is comprised of a rotating thrust disc and opposing electromagnetic stator assemblies as shown in Fig. 4-10. These stator assemblies are toroid-shaped electromagnets that work by force of attraction to levitate or suspend the disc within the axial clearance.

The operating principle is identical to the radial bearing, in which shaft (axial) position is monitored and the currents adjusted in the magnets to hold or force the disc to the center of its clearance. A magnetic thrust bearing stator and thrust disc are shown in Fig. 4-11. It is important to note that the thrust disc is solid and does not have to be laminated. This is due to the thrust stator geometry, in which the disc is not passing by alternating magnetic poles as it is in the radial bearing.

The speed limit for a magnetic thrust bearing is determined by the strength of the thrust disc mounted on the shaft. The thrust disc must be made of a magnetic steel and some of the higher strength materials can provide up to 300 m/sec peripheral speed for a disc that is shrunk on the shaft. This is important because the lower specific load capacity of the magnetic bearing will require more projected area and the thrust disc diameter will be bigger than a conventional thrust bearing.

Table 4-2. Magnetic bearing load capacity

	Silicon–iron	Iron–cobalt
Force along axis, F_{V1}, F_{W1} (daN)	3.5 D (cm) × L (cm)	5.0 D × L
Combined vertical force, F_c (daN)	5.0 D × L	7.0 D × L

Auxiliary Bearings

Another important feature of the magnetic bearing system is the auxiliary bearings that provide safe coast-down of the machine in the event of a failure of the control system or an overload of the bearings. Figure 4-12 shows a cross section of a radial magnetic bearing, including the auxiliary bearings. The auxiliary bearings must provide support for both radial and axial loads during coast-down.

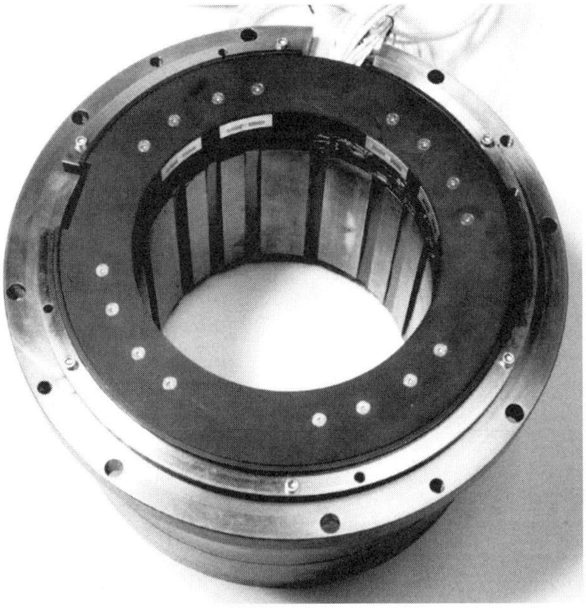

Figure 4-9. Radial magnetic bearing cartridge.

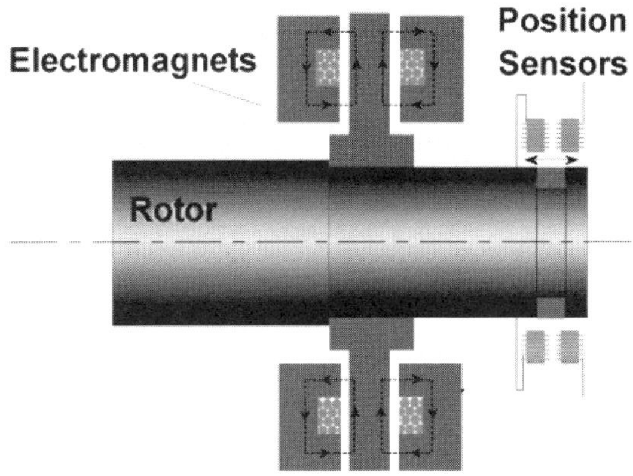

Figure 4-10. Magnetic thrust bearing.

Note that the auxiliary bearing clearance is set at about half the magnetic bearing clearance. Under normal operation, the shaft is levitated and the auxiliary bearing does not rotate, but in the event of a failure the auxiliary bearing provides a safe coast-down of the machine, preventing contact between the magnetic bearing rotor and stator. The auxiliary bearings can provide several full-speed, full-load coast-downs of the compressor before being replaced. This is sufficient for the life of most machines, considering a normal MTBF of 40,000 hours for the control system.

Magnetic Bearing Control System

Original magnetic bearing control systems used analog technology to implement the PID control. Each of the five control loops were on separate analog PC boards, and any adjustments were made by changing the components (i.e., resistors and capacitors) on the board. Eventually, microprocessor speeds improved enough and, since the mid 90s, digital control has replaced the analog systems. This provided a big savings in cost and complexity because all of the PID control that required several analog PC boards could now be done on one microprocessor. The major advantage of moving from analog to digital control is the dramatic reduction of components, which results in a higher mean time between failure (MTBF). Today's digital control systems have an actual MTBF of more than 40,000 hours or approximately five years continuous operation.

The PID control establishes the bearing characteristics by modifying the error signal with phase and gain. Phase is synonymous with damping and gain is synonymous with stiffness. Magnetic bearings are very different from conventional oil bearings in this respect. For conventional oil bearings, the bearing characteristics depend on the bearing geometry (i.e., preload, clearance, etc.) and the oil properties. For magnetic bearings, the bearing characteristics depend on the frequency of the disturbance. For example, a

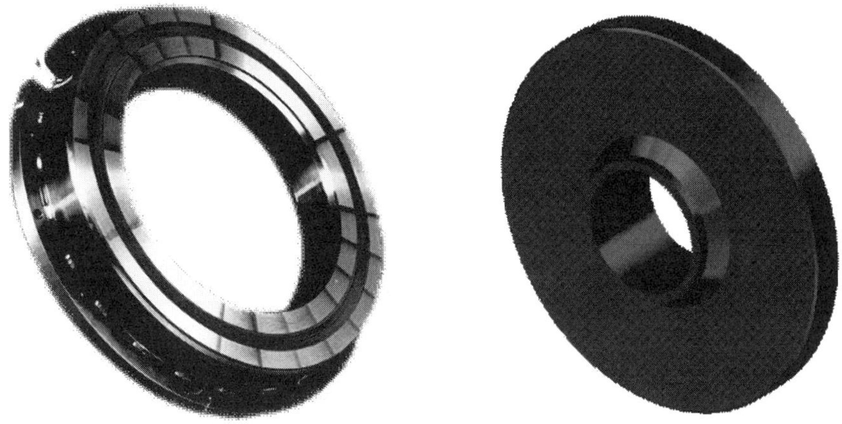

Figure 4-11. Magnetic thrust stator (left) and thrust disc (right).

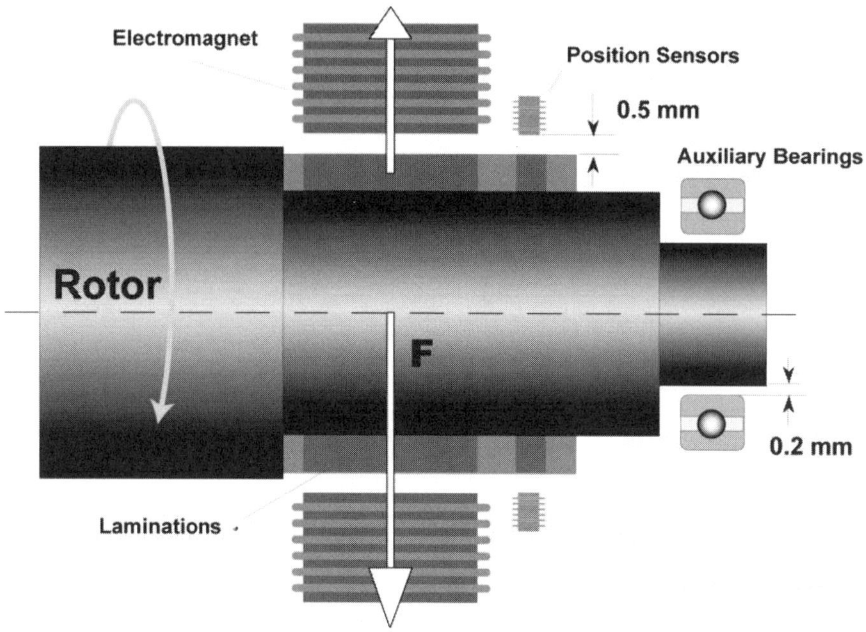

Figure 4-12. Radial magnetic bearing and auxiliary bearing.

synchronous vibration due to unbalance is controlled by the bearing based on the gain and phase (i.e., stiffness and damping) of the controller at the synchronous frequency. The bearing characteristics are set to provide a stable rotordynamic response over the compressor operating speed range meeting acceptable vibration criteria specified by API-617. Typically, a rotordynamic analysis is completed to predetermine the bearing characteristics. If the rotor model is accurate, no further adjustment or "tuning" is required. With the digital control system, any further adjustment can be made using a laptop computer interface.

Figure 4-13 shows a digital control system in its enclosure. These control systems are designed to provide remote operation of the compressor. Typically, the controller is placed in a nonhazardous location. The cable length for the bearing power and signal wires can be up to 300 meters between the controller and the machine. The user can operate the control system either locally (at the cabinet) or remotely through a MODBUS serial link. An added advantage of the digital control system is the machine operating information that is available for monitoring, trending, diagnostics, and troubleshooting. This operational data includes vibration, bearing loads, shaft unbalance, and bearing temperatures.

The magnetic bearing control needs to have an uninterruptable power supply so that levitation is not lost due to an AC power failure. The controller shown in Fig. 4-13 includes a battery backup system to maintain levitation. Normal compressor start-up and shutdown is done with the shaft levitated before rotation is requested and rotation is stopped before the shaft is delevitated.

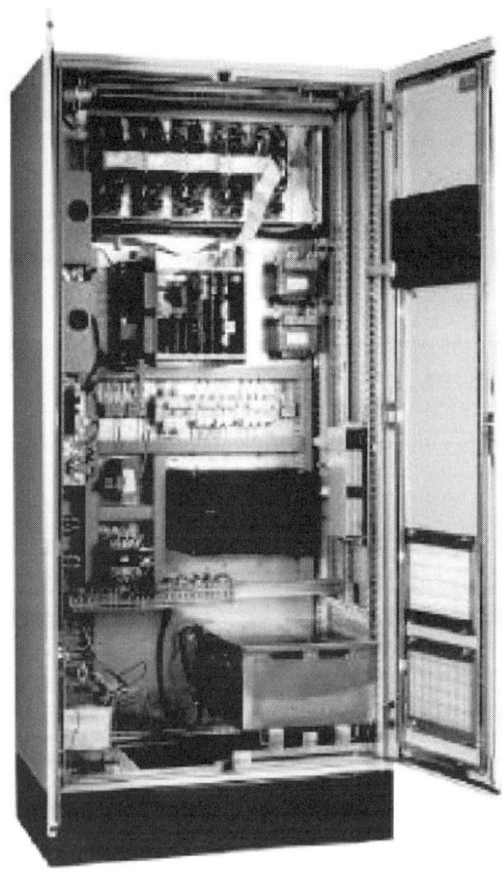

Figure 4-13. Magnetic bearing control system cabinet.

4.2.5 Technology Status and Case Histories

Magnetic bearing technology continues to grow and is finding its way into many compressor applications. The main reason for this growth comes from the OEMs who have started with a blank sheet of paper and designed a "magnetic bearing compressor" instead of "a compressor with magnetic bearings." This allows the designer to take advantage of the bearing's high speed capability and the ability to run in the process gas.

Today, there are more than 300 large turbomachines in operation with magnetic bearings and more than half of these machines have the bearings operating in the process fluid environment.

Case History No. 1—Cycle Gas Compressor Used for Polyethylene Production

These compressors are typically single-stage units with a large overhung impeller. They provide high flow with a low-pressure rise feeding gas into a fluidized bed reactor. The advantage of using magnetic bearings is to eliminate the potential for oil contamination inside the fluidized bed reactor. Table 4-3 provides a summary of the machine data.

In addition to oil-free operation, magnetic bearings also offer the advantage of providing large clearances. Large clearances can be helpful in the event of large unbalanced loads. The polyethylene manufacturing process tends to cause a buildup of polymer on the impeller and this can impose unbalanced loads on the bearings. Although the polymer buildup is normally uniform and the unbalanced loads cause minor synchronous vibration, when a piece of polymer breaks off there is a significant unbalance. Due to the large clearance, the magnetic bearings are able to react to this sudden load and continue operation. Figure 4-14 shows the compressor along with the complete shaft assembly. The shaft assembly includes the large overhung impeller and thrust disc. The radial magnetic bearing laminations can be seen on both sides of the thrust disc.

Table 4-3. Cycle gas compressor data

Machine	Single-stage compressor
Rated Power	2,000 kW
Rated Speed	3,000 rpm
Rotor weight	870 kg
Axial load	54,000 N
Driver	Asynchronous motor
Environment	
• Process	Polyethylene
• Bearing cavity	Air
• Explosion proof	No
• Control cabinet	In safe area, 300 m maximum away from the machine

Case History No. 2—Natural Gas Storage Compressor

This compressor feeds gas to an underground salt dome storage cavern. Table 4-4 provides the machine data. An advantage of the magnetic bearing system is the ability to run inside the process gas environment. This eliminates the need for seals that add cost and, by requiring increased rotor length, can adversely affect rotordynamics.

Figure 4-15 shows the cross section of the motor/com-

Table 4-4. Natural gas storage compressor data

Machine	Multistage compressor
Rated Power	5,500 kW
Rated Speed	12000 rpm
Rotor weight	
Axial load	
Driver	High-speed motor
Environment	
• Process	Natural gas
• Bearing cavity	Natural gas
• Explosion proof	Yes
• Control cabinet	In safe area, 300 m maximum away from the machine

Figure 4-14. Cycle gas compressor with magnetic bearings.

pressor train and provides data regarding the magnetic bearing sizes and load capacities. This application takes advantage of the ability of the magnetic bearings and motor to run in the process fluid. The end result is an oil-free, sealless (i.e., hermetically sealed) compressor (Fig.4-16).

A comparison of this arrangement versus a conventional machine train in this service is provided in Table 4-5. The results show a significant reduction in the number of components such as bearings and seals, both of which are typically maintenance items. In addition, the much smaller footprint and weight saving for the hermetic train provide other savings that can be realized in certain applications such as offshore platforms.

In cases in which the bearings are exposed to the process, a special varnish-impregnation treatment is used to seal the bearing windings and protect them from ingress of liquids in the gas. Likewise, the motor is given a similar treatment for the stator windings. This treatment is suitable for "sweet gas" environments according to the NACE definition of sweet gas. For sour gas applications, a canned magnetic bearing must be applied to protect the bearing from the erosive/corrosive constituents contained in the gas.

4.2.6 Application Summary

In essence, applications in which oil contamination cannot be tolerated are prime candidates for magnetic bearings. Electromagnetic suspension is achieved by a support system that applies power to the ferromagnetic material of stationary and rotating bearing components. As we have seen, the support system includes position sensors and circuitry to bring the rotor to the desired position through a power amplifier. The design sophistication of the control mechanism greatly influences the reliability of any magnetic bearing system. Published data show that conventional oil lubrication systems on typical centrifugal compressors using conventional hydrodynamic bearings might absorb, say, 100 kW. Conversion to magnetic bearings reduced this auxiliary power to around 10 kW.

Although there clearly are power savings associated with magnetic bearing systems, the principal advantages are centered on the fact that complex or costly lubricating consoles are no longer needed with magnetic bearings. They are among the contending technologies on new compressor applications.

4.3 EXTERNALLY PRESSURIZED BEARINGS

By way of recap, please note that the well-represented and widely known conventionally lubricated compressor bearings discussed earlier in this text are not really fully pressurized. Pressure in the circulating lube-oil supply system is maintained only to the extent needed to deliver lubricant to the point of application. Oil flow rates are adequate for heat removal and to form an oil film that will support the load.

It should be noted here that since the 2000s a specialty firm, Bently Pressurized Bearings, has been able to supply fully pressurized and fully lubricated bearings. These bearings, and not just the more widely known magnetic bearings, can overcome certain limitations of conventional fluid-film bearings. Externally pressurized bearing systems promote oil flow along the shaft, whereas conventional bearings have the oil flow around the shaft. Fully pressurized bearings use a quasiaxial oil wedge for rotor support; in contrast, conventional bearings use a radial oil wedge. Using controlled and

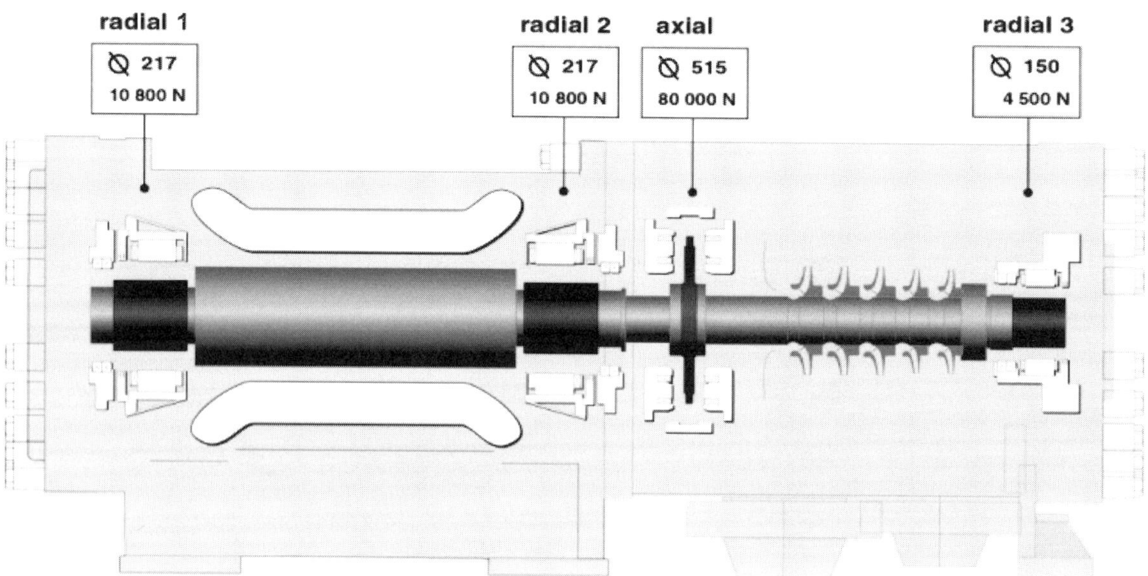

Figure 4-15. Hermetically sealed motor compressor train.

Figure 4-16. Hermetically sealed motor compressor.

controllable lubrication makes it possible to influence rotor stability as needed.

An externally pressurized bearing system operates by delivering the pressurized oil through suitably positioned recessed inlet ports. Every port directs the fluid flow to a specific bearing pocket, which in turn determines the clearance between the journal and bearing. As the rotor attempts to move in the bearing, the pressure in the adjacent pocket increases in the direction of motion and decreases in the opposite direction. This creates a restoring force as a result of the pressure differential in the opposite bearing pockets.

Conventional bearings require high eccentricity to achieve an "oil wedge" that will stably support the rotor load. If the shaft center would coincide with the bearing centerline axis, eccentricity would be zero and unstable operation would result. However, zero eccentricity in pressurized bearings would maximize space and time for the restoring mechanism to act, should the shaft move toward the bearing wall. At 70 to140 bar (1,000 to 2,000 psi), the oil supply pressure for externally pressurized bearing systems will be close to two orders of magnitude greater than that required for hydrodynamic bearings. Consequently, few commercial applications are in operation today.

Nevertheless, by properly designing and pressurizing the bearing, compressor rotors can be made to run stable at the operating speeds typically encountered in process compressors. In the event of pressure loss, the bearing will revert to conventional hydrodynamic bearing operation, allowing safe compressor shutdown. (References listed under "Additional Reading" on pp. 329 and 330 are all of interest to readers seeking more detail).

4.4 GENERAL CONDITION MONITORING OF PROCESS GAS COMPRESSORS

Their size, high speed, high capital cost, and the (typical) fact that there is no installed spare machine explain the need to

Table 4-5. Comparison of conventional compressor train versus hermetically sealed train

	Motor		Gearbox			Compressor	
	Bearings	Seals	Bearings	Seals	Coupling	Bearings	Seals
Conventional train	2	2	4	2	1	2	2
Hermetic train	2					1	

monitor centrifugal compressor condition at all times. Centrifugal compressors and their drivers—gas and steam turbines, large electric motors, and auxiliary equipment such as gear speed increasers—are generally provided with noncontacting eddy current displacement or velocity probes. Their respective signals are fed to suitable monitoring devices, with alarm and trip settings in compliance with applicable API standards.

Although the output from these monitors is, of course, routinely observed by operations personnel, modern facilities are placing importance on incorporating these online condition-monitoring systems in their generally process-oriented, distributive control systems (DCS). A modern DCS system will thus encompass operation-related important plant parameters and also the various machinery parameters linked to process compressors and their drivers. Both steady-state data and transient data related to process upsets are recorded automatically. Startup and coast-down information is now readily available. Through extensive networking, the data are typically made available to users at different locations. Displays of trend plots and real-time operating parameters contain both machine history and helpful troubleshooting information. Other displays showing piping and instrumentation schematics (P&IDs) associated with the machine along with alarm and/or trip points simplify diagnostic tasks.

Process computers are extensively used to evaluate the thermodynamic performance of centrifugal compressor trains. Vibration, temperature, and other performance data can be archived and retrieved later to make maintenance decisions and to facilitate future comparisons. Dry gas seals and contoured diaphragm couplings have generally enhanced the reliability of centrifugal compressor trains.

Occasionally, although rarely, a late model, on-line condition-monitoring package may not be available for relatively old machinery trains. Where this is the case, portable data gathering must serve instead. Some portable data loggers are surprisingly versatile and accept not only vibration data but data related to compressor start-up, coast-down, and process performance information. This electronic data collection and analysis equipment can also compare captured data to predefined alarm set points.

With the late 1990s came the emphasis on asset management approaches whereby a potential problem is identified, diagnosed, and followed by an action plan for implementing remedial measures. Unlike the prevailing prior routine of simply initiating shutdown plans, asset management is aimed at optimizing production without having to find recourse in an unscheduled shutdown. Based on risk analysis and factoring-in existing or anticipated profit margins, optimized production could mean reduced throughput, unchanged throughput, or even briefly increasing the unit throughput. Other assets have to be analyzed as part of the approach and it is necessary to collect and interpret large amounts of data. True in-depth analyses are needed to establish correlations between changes in normal values of one or more parameters, and understanding the reasons for these deviations. To be effective and reasonably accurate, a good asset management system should be able to generate definitive guidelines and must avoid generalities. Moreover, a sound asset management system should incorporate and indicate or sound early warnings well before alarm, trip, or unsafe operating conditions are reached.

4.5 COMBINING TROUBLESHOOTING AND CONTINUOUS MONITORING OF DYNAMIC DATA FOR CRITICAL MACHINERY*

In their quest for improved availability and reliability, smart operating companies are opting for continuous monitoring of key machinery assets. As can be expected, the ability to optimize the manpower needed to monitor critical process machinery is highly sought after.

Appropriate online systems can accomplish these needs. Unmanned stations are now common on many pipeline systems. Supervisory control and data acquisition (SCADA) systems or distributed control system (DCS) networks allow operators to monitor and control machinery remotely. It is common to use dial-up or cellular telephones, local area or wide area networks (LANs, WANs), or virtual private networking (VPN) via the Internet.

These systems are typically used to monitor pipeline conditions (pressures, temperatures, flow rates, etc.) and remotely adjust various parameters (load step, speed, number of units operating, etc.) to meet system demands. Common uses involve relatively slow data acquisition rates (less than 10 samples per second) of a large number of data channels (hundreds).

However, recent technology advances in both hardware and software systems have made it possible to achieve high-speed data acquisition for relatively high channel count applications as well.

4.5.1 Overview of Recent Developments

Cost-effective "packaged" systems are now available for improved machinery analysis and diagnostics. A typical installation can monitor up to 64 channels per CPU at 10 KHz or higher sampling rates. Ruggedized Class 1, Division 2 (C1D2) or Class 1, Division 1 (C1D1) data-acquisition CPUs can be mounted in the field and client applications can access the high-speed data over any available wired or wireless networking environment.

4.5.2 Innovative Data Acquisition Tools

There will always be a need for field-testing to diagnose and solve complex system dynamics problems involving all types

*This section was contributed by Kenneth E. Atkins, Sr. Staff Engineer, Engineering Dynamics Inc., San Antonio, Texas.

of pulsation- and vibration-related issues. This is where modern and comprehensive high-speed, high-channel-count data acquisition tools fill a need. These tools are now routinely used by resourceful consulting engineers to troubleshoot critical machinery-related problems where downtime is costly and where quick solutions are necessary. Desirable data acquisition systems are easily reconfigured or customized. They typically serve temporarily for only a few days of testing.

However, some problems—including transient, nonrepetitive phenomena such as slug flow or multiphase flow—often require longer-term data-acquisition efforts. In those instances, the acquisition systems can be permanently installed. This installation flexibility is needed because certain problem conditions tend to occur only seasonally, or occur due to some cause-and-effect relationships that are not obvious.

4.5.3 Early Case Study Tells the Story

In the early 1990s, one consulting company assisted a major operating company in overcoming a very complex slug-flow-related problem by installing a continuous monitoring system capable of high-speed data acquisition for 48 channels of vibration, pulsation, and strain data on a large piping network [11].

A total of three of these systems were deployed. The operating company saved millions of dollars in lost production and system modifications by being able to use the dynamic data to adjust operating conditions and avoid damaging vibration levels.

Since then, similar systems have been used by companies requiring semipermanent continuous monitoring systems for structural vibration as well as rotating and reciprocating machinery vibration.

4.5.4 Current Application Explained

A good high-speed dynamic machinery data-acquisition system should be designed to meet a wide variety of analysis requirements for a diverse range of machinery, including reciprocating compressors, screw compressors, centrifugal compressors, positive-displacement pumps, boiler feed pumps, gearboxes, and steam and gas turbines. Many different types of transducers can be employed, including accelerometers, strain gauges, pressure transducers, magnetic pickups, thermocouples, proximity probes, and so on [12].

The example installation illustrated in Fig. 4-17 uses a single C1D1 computer (requiring a nitrogen purge system) mounted locally on the refrigeration compressor deck at an ethylene plant. Two compressors are monitored: one propylene machine and one ethylene machine.

Piezoresistive pressure transducers were installed in the discharge passages near each compressor stage in two compressors. The steam turbine drivers for both compressors had recently been upgraded. There was concern about the dynamic response of the system at flow rates not previously attainable due to horsepower limitations.

The data-acquisition system was installed to capture vi-

Figure 4-17. Class 1, Div 1 data-acquisition CPU.

bration and pulsation data that could not be detected with typical DCS instrumentation.

The C1D1 computer ("headless CPU") is linked via a fiber-optic Ethernet connection to a second CPU in the plant control room several hundred feet from the compressors. The second CPU communicates with the "headless CPU" in a peer-to-peer network. It also has a wireless communications card for access via the Internet (see Fig. 4-18). The card is configured with a static IP address so that any computer with proper authentication can access the data from the Internet. In this manner, the plant's IT systems are not compromised, since no direct connection to their LAN exists.

The system is configured to continuously monitor the 43 dynamic data channels at an 8 KHz sampling rate. Detailed testing during commissioning of the upgraded turbines revealed that an acoustic resonance existed in the discharge passage of the fifth stage wheel on one of the compressors (Fig. 4-19).

This resonance was excited by blade-pass frequency (BPF), which is 17 times the compressor speed for the fifth stage wheel. The data-acquisition system was then set to track the BPF pulsation component as well as bearing housing vibration at BPF. Trends of these dynamic variables were then captured as the system was operated over its complete operating range.

Figure 4-20 shows a period of time when the 17 times compressor speed pulsation was high (up to 35 psi, p-p). Operators were able to adjust the fourth stage side-stream flow and avoid the excitation of the acoustic resonance without compromising system throughput. Figure 4-21 shows that operation in this manner greatly reduced the BPF pulsation levels.

An example of an event captured in another installation is shown in Fig. 4-22. This system was monitoring a synchronous motor-driven hydrogen compressor in an NHT unit at a refinery. Numerous unexplained trips had occurred, causing major operational upsets. It was suspected that the units were being overloaded or possibly subjected to liquid slugging because the trips seemed to correlate with higher flow rates based on the DCS historical data. Pressure transducers were installed in the compressor suction, interstage, make-up, and discharge lines. Motor supply voltage and current were also monitored along with speed and torque. The trip event captured and illustrated in Fig. 4-22 proved that the problem was electrical, and allowed the user to focus on correcting the electrical issues rather than process-related issues.

The system software architecture should be versatile, which will allow almost limitless possibilities for configuring the data-acquisition solution.

In one troubleshooting and data-acquisition system, the central function is the data acquisition server (DAQ server), which makes digitized data available using data sockets. Here, multiple client applications (running on any networked CPU) can access the data sockets simultaneously. The high-speed, time-domain data can easily be accessed by client applications to provide other useful diagnostics. For example, reciprocat-

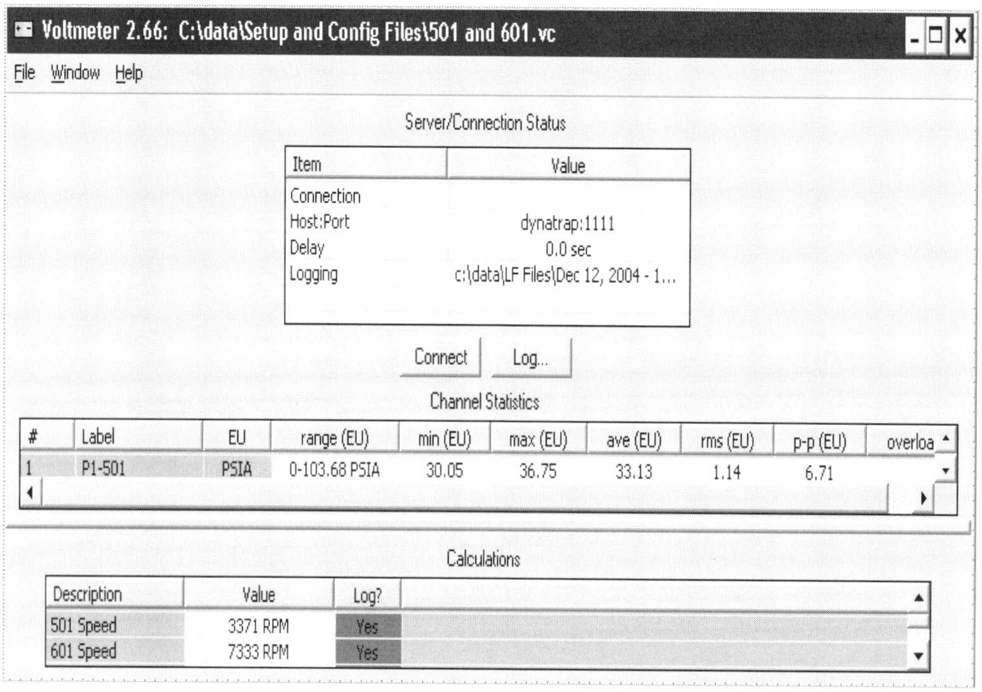

Figure 4-18. Real-time display via remote desktop Internet connection.

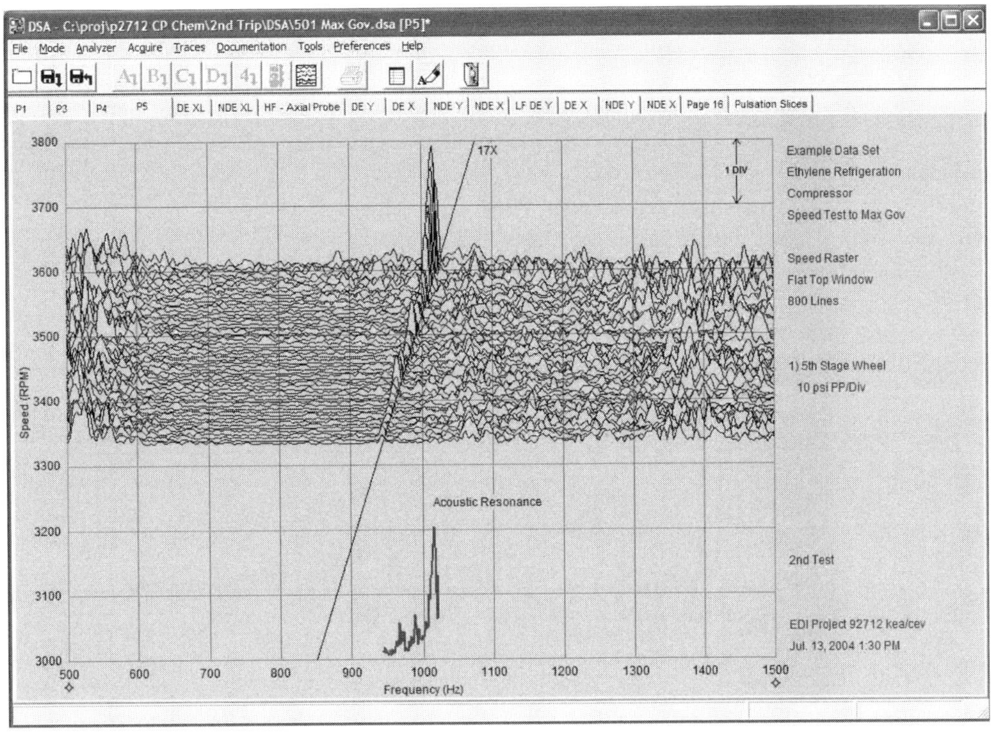

Figure 4-19. Acoustic resonance in discharge passage.

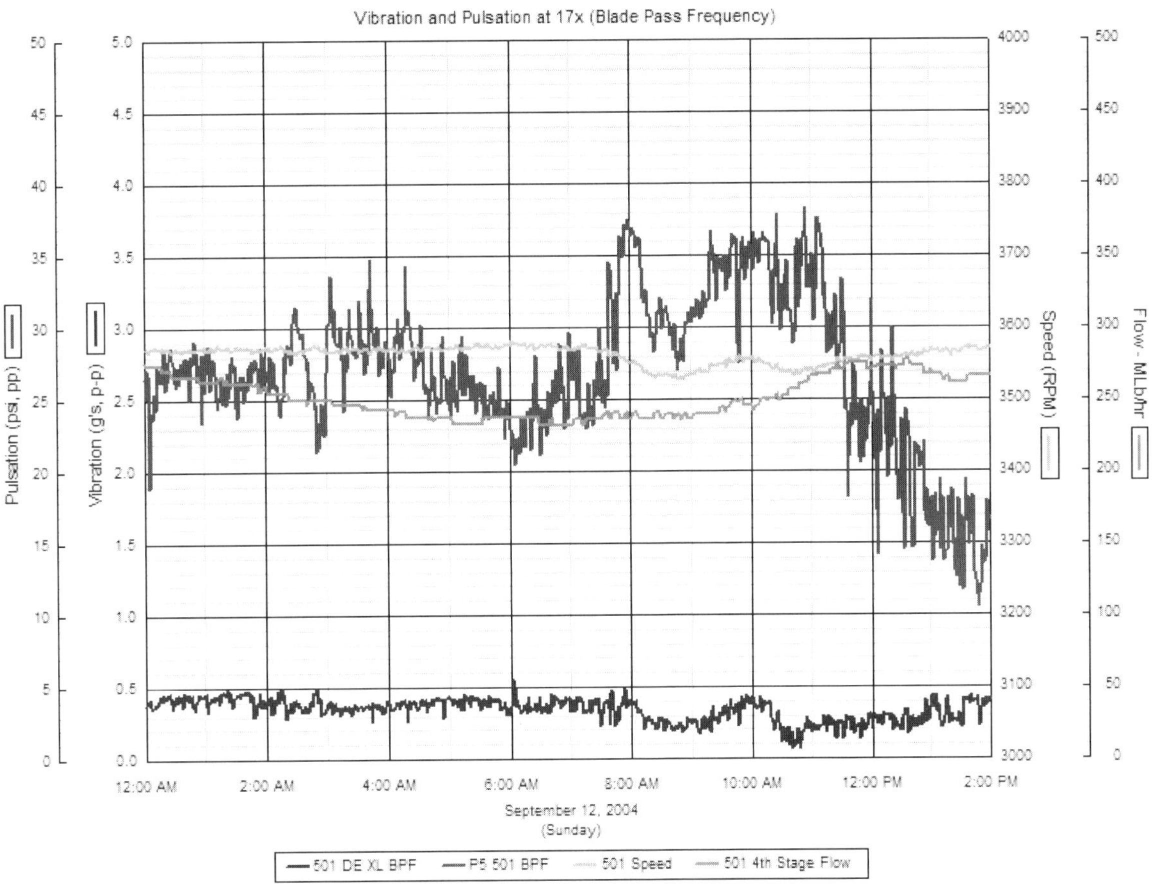

Figure 4-20. High BPF pulsation.

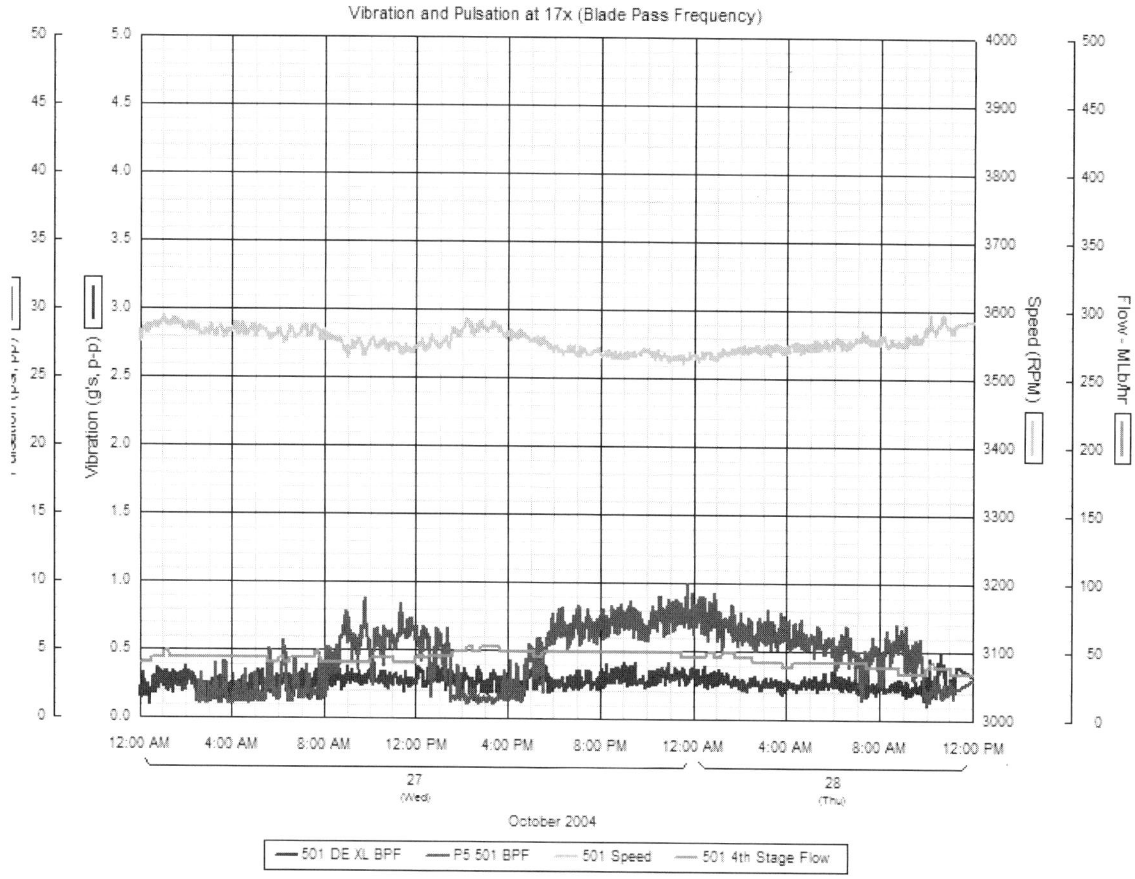

Figure 4-21. Lowered BPF pulsation.

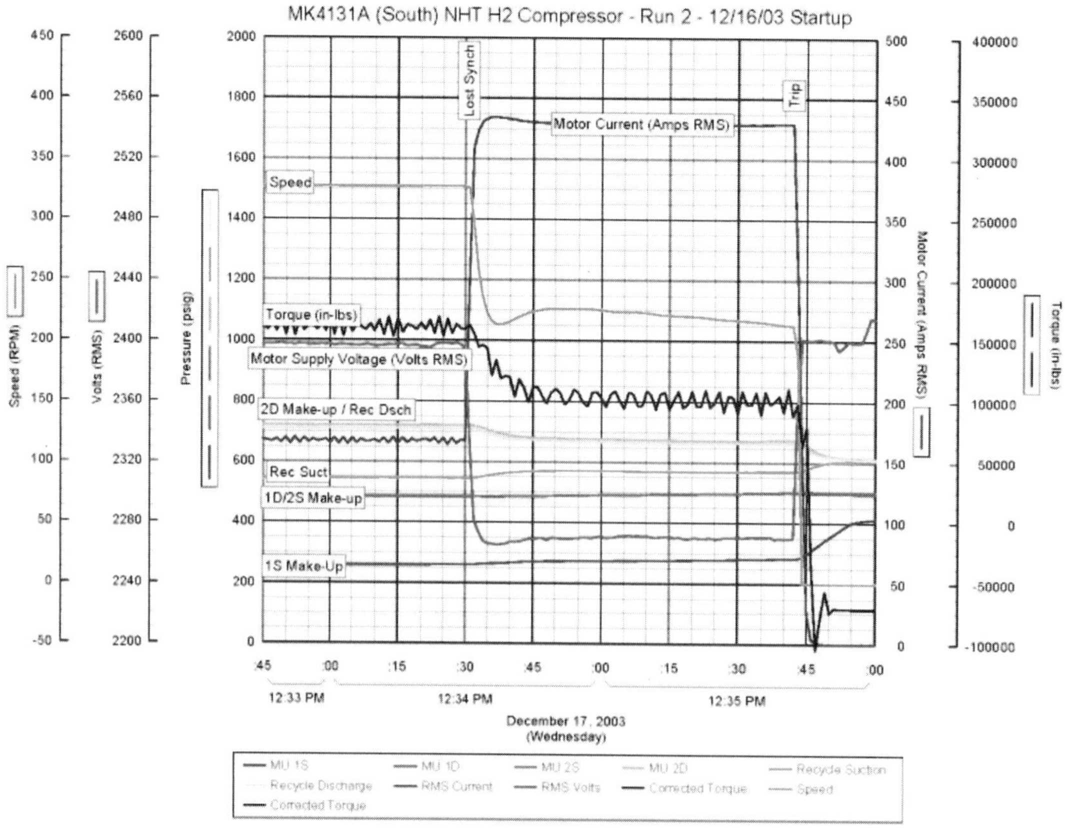

Figure 4-22. Data captured during trip event.

ing compressor cylinder pressure signals can be converted to PV cards (Fig. 4-23), or time-domain signals from proximity probes can be displayed as shaft orbits (Fig. 4-24).

The DAQ server is typically a "black box" or "headless CPU" mounted at the machinery location with no keyboard or display, and requires relatively little CPU power or data-storage capabilities. Multiple DAQ servers can be used to increase channel count as necessary. This is a significant advantage for operating companies that have broadband access to remote locations over fiber optic networks.

4.5.5 Anticipating Future Developments [13]

Developments are currently underway on client applications for other capabilities. Enhancements that will include a torsional vibration monitor and wireless access to the dynamic data using PDAs or tablet PCs will be of value to machinery engineers and process operators. These capabilities are certain to increase machinery reliability and safety in more cost-effective ways than previously found attainable.

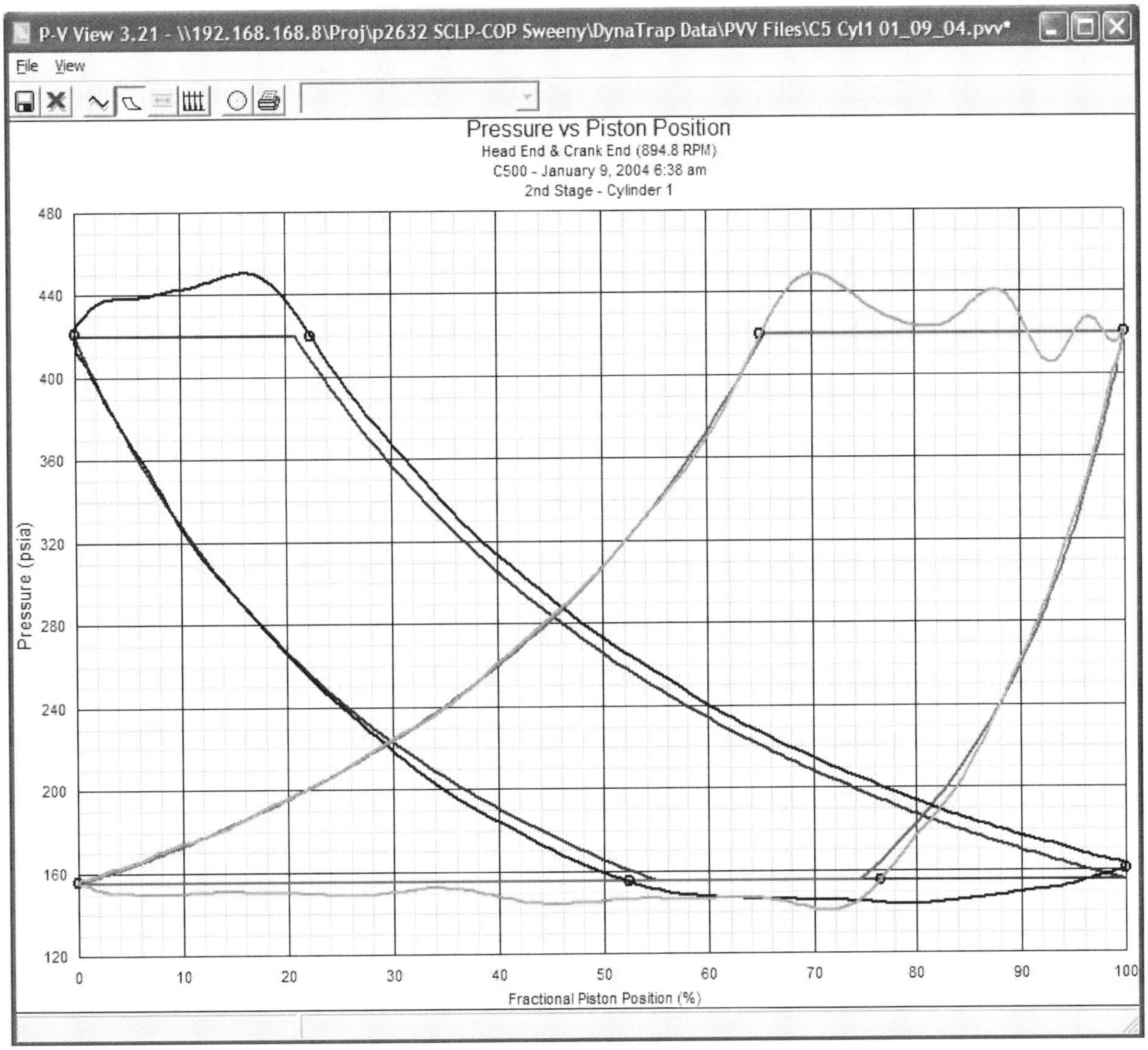

Figure 4-23. P-V card client.

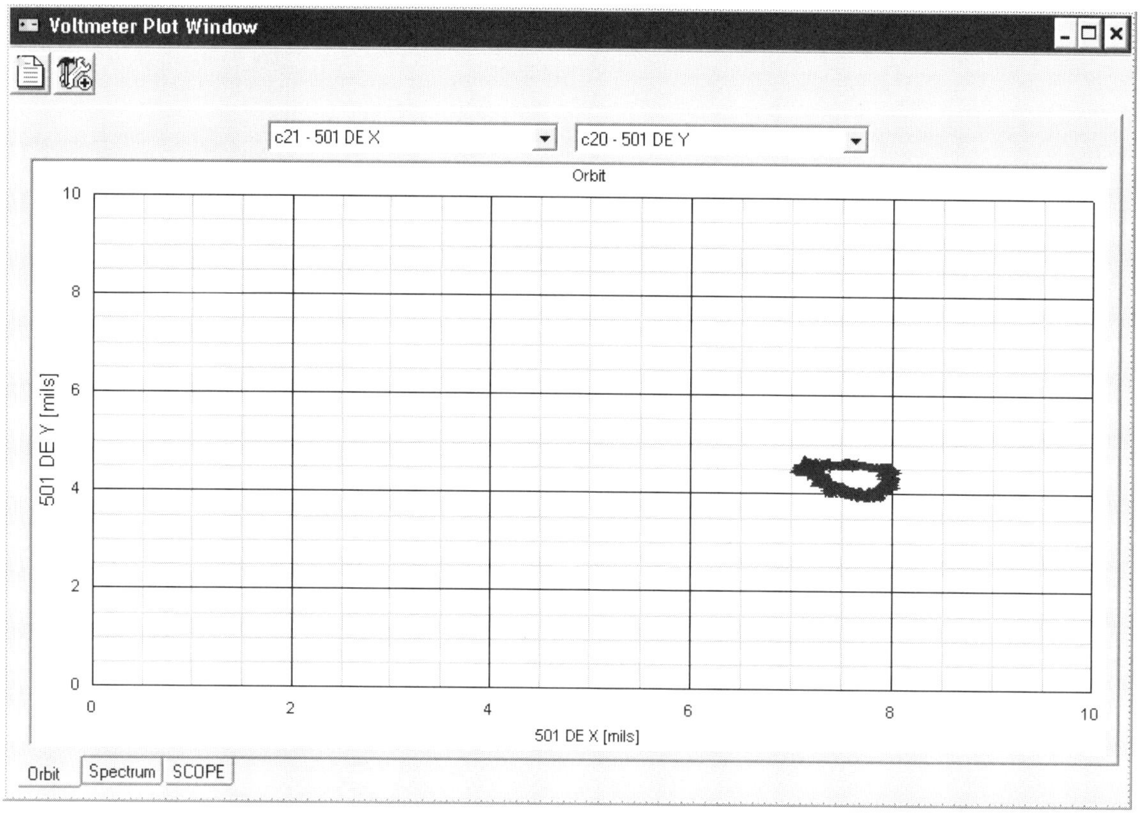

Figure 4-24. Shaft orbit client viewed via remote desktop.

Chapter 5

Centrifugal Compressor Performance

5.1 COMPRESSION PROCESSES AND EFFICIENCIES: POLYTROPIC VERSUS ISOTHERMAL

Chapter 2 allowed the reader to become acquainted with compressor essentials. Quite correctly, we left the impression that most centrifugal compressors are noncooled machines. In other words, as the gas is progressively compressed from stage to stage, it will heat up, but no cooling is provided between stages. For these noncooled, multistage centrifugals and at their particular ratios of compression, the compression process is assumed to be polytropic. The term polytropic simply conveys the factual observation that changes in gas characteristics during compression are being considered. The gas encounters friction during its passage and a certain amount of kinetic energy is delivered to the gas as frictional heat instead of being used for increasing the pressure.

This is where we encounter the term *polytropic efficiency*. This efficiency will, of course, be influenced by many factors; they include the basic design, material selection, and quality of fabrication, even workmanship. Compressor performance test results are interpreted by polytropic analysis. As will be explained later by dimensionless numbers, polytropic efficiency is independent of the gas properties of the application under consideration.

In contrast, isothermal compression is a process wherein there is no temperature increase. Although isothermal compression is clearly the most efficient process, this compression without temperature increase is rarely achieved. A limited number of manufacturers offer quasiisothermal designs with diaphragm cooling (i.e., cooling of the walls of stationary gas passageways), further supplemented by intercooling between stages. Each of these stages must operate at relatively small pressure ratios. Although it is an efficient operation, intercooling does involve additional pressure drop and, for compressor-internal intercooling, a characteristically different overall machine layout (Fig. 5-1). Good designs will balance the number of intercoolers, be they external or internal, and number of stages of compression. Such optimization trades off efficiency gains (reduced operating costs) against incremental initial equipment purchase cost.

It is certainly advantageous to understand the difference between hydraulic and polytropic efficiencies. Both need to be considered in order to choose whichever compressors will best satisfy certain specific requirements. Hydraulic efficiency considers the losses in impeller passages and eddy current losses, whereas polytropic efficiency considers, in addition to the above, the disc friction and interstage seal losses at the labyrinths between the stages.

Finally, mechanical efficiency takes into account frictional losses at bearings and seals.

5.2 SPECIFIC SPEED (N_S) AND THE FLOW COEFFICIENT (ϕ)

For decades, the manufacturers of fluid machinery have found it advantageous to make use of the concept of specific speed. Specific speed is defined by the equation

$$N_s = N\sqrt{Q}/H^{0.75}$$

where N = shaft speed in rpm and Q = flow in convenient gallons per minute for pumps or cubic feet per minute for gas machinery. N_s thus expresses the relationship of important design parameters.

5.2.1 Flow Coefficient (ϕ)

Compressor manufacturers typically use graphs of pressure coefficient (ψ) and polytropic efficiency (η_p) versus flow co-

Figure 5-1. Horizontally split "isothermal" compressor. (Source: Sulzer Turbomachinery, Winterthur, Switzerland.)

efficient (ϕ) for reasonably close estimates of compressor performance. The flow coefficient (ϕ) is a dimensionless number associated with the flow-handling capability of compressors. This flow coefficient is represented as

$$\phi = Q/[(\pi/4)D_2^2U_2]$$

where the terms D_2 and U_2 are the impeller diameter and tip speed, respectively. This equation explains that flow capability varies in direct proportion to the tip speed, whereas it varies as the square of the diameter of the impeller. When one is dealing with a single impeller, the as-designed or anticipated performance can be reliably achieved.

The flow coefficient gives us a general idea of impeller geometry and the efficiencies typically anticipated for a calculated flow coefficient. Quite obviously, a larger flow coefficient indicates a move toward mixed flow or axial compressors (Fig. 5-2). However, when referred to multistage, high-pressure compression, excessive clearance or improper design of the labyrinths can have a significant influence on the actual performance of compression equipment.

5.2.2 Pressure Coefficient or Head Coefficient (ψ)

The head coefficient (ψ) is another dimensionless performance parameter that expresses the actual head as the fraction of the maximum theoretical head at zero flow for a given impeller tip speed and geometry with radial inlet. The theoretical head, expressed in ft-lb per lb or kg-meter per kg, is obtained by using the relationship

$$H_t = (V_{2u} \cdot U_2 - V_{1u} \cdot U_1)/g$$

The conventional meaning of the velocity vectors is explained in Figs. 5-3 (a) through (e), which graphically, and for readers wanting to see the theory behind compressor design, represent the velocity triangles at the impeller inlet and outlet. Figure 5-4 shows that maximum theoretical head is obtained when the tangential component at the inlet is zero, which is the case in radial inlet impellers. The maximum theoretical head is thus equal to U_2^2/g. Therefore,

$$\text{Head coefficient } \psi = H_t/(U_2^2)/g$$

Relative to actual head, the coefficient is a cumulative representation of blade angle, slip, and efficiency. Therefore, Fig. 5-5 (a) uniquely represents the dimensionless relationships between head, flow, and efficiency for a particular impeller. The incremental reduction from a calculated theoretical head is due to eddy current and separation losses at the entrance and exit of the impeller. These are explained as obstructions in the (wrongly assumed to be smooth) path of gas travel and consequent changes of velocity and direction.

Before we close the discussion on the head coefficient, it is worthwhile to remember also the effect of the vane angle at the outlet. In the case of radial discharge ($\beta_2 = 90°$), theoretically, the head does not vary with load and the head versus flow (H–Q) characteristic should simply be a line parallel to Q on the X-axis of the H versus Q plot. If the vane angle at the discharge is inclined so as to represent a backward lean, as seen in most of the applications (i.e., $\beta_2 < 90°$), the H–Q curve shows drooping characteristics.

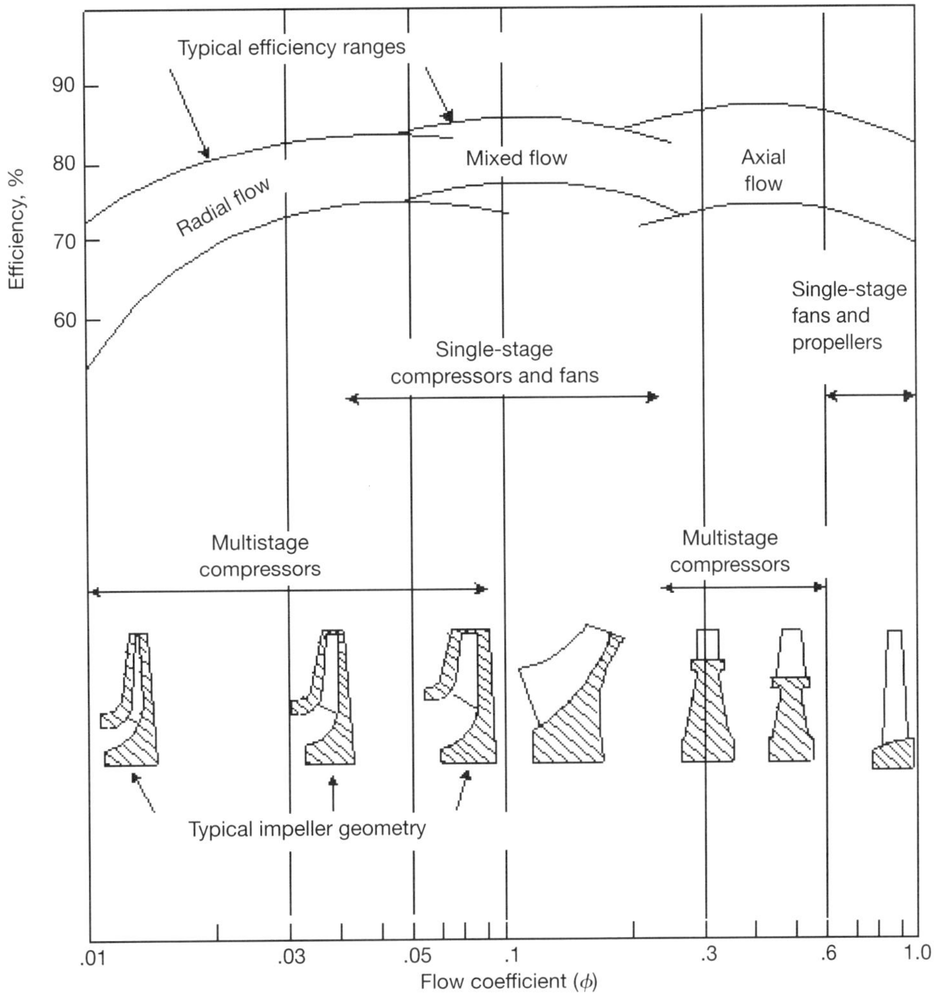

Figure 5-2. Efficiency versus flow coefficient for different impeller geometries.

Should the discharge vane angle be executed with forward lean ($\beta_2 > 90°$), the curve would have a rising characteristic. Although occasionally used on large fans and blowers, forward lean is not practical on compressors. It nevertheless illustrates that impeller blade or vane profiles influence compressor performance. Moreover, just as the discharge vane angle has an effect on the performance (*H–Q*) characteristic, inlet guide vanes also play a similar role. Inlet guide vanes (see Fig. 3-16) change the relative gas velocity at the inlet and thus the value of the gas impulse. This can be visualized from the velocity diagrams in Figs. 5-3 and 5-4.

5.2.3 Mach Number

The process gas originates at a source, say the reactor or column. As this compressible fluid flows through the suction line to the compressor inlet nozzle, it will experience pressure drop along the flow path. If the pipeline is of constant diameter, the mass flow remains constant, but the gas volume will increase. Flow volume equals flow area times flow velocity; hence, there will now be an obvious increase in the velocity. The pressure drop between the given sections of the pipe is consumed by the gain in the kinetic energy due to the change in velocity and the friction loss in the section. At a certain stage of expansion, the former becomes a predominant consumer, leaving no room for the latter. This becomes a limiting factor at some velocity close to sonic velocity, that is, the speed at which sound travels in the particular medium.

Acoustic velocity is expressed by the formula

$$C_s = \sqrt{kgRTZ}$$

where

k = isentropic exponent
g = gravitational constant
R = gas constant (= 1,544/MW)
Z = compressibility factor

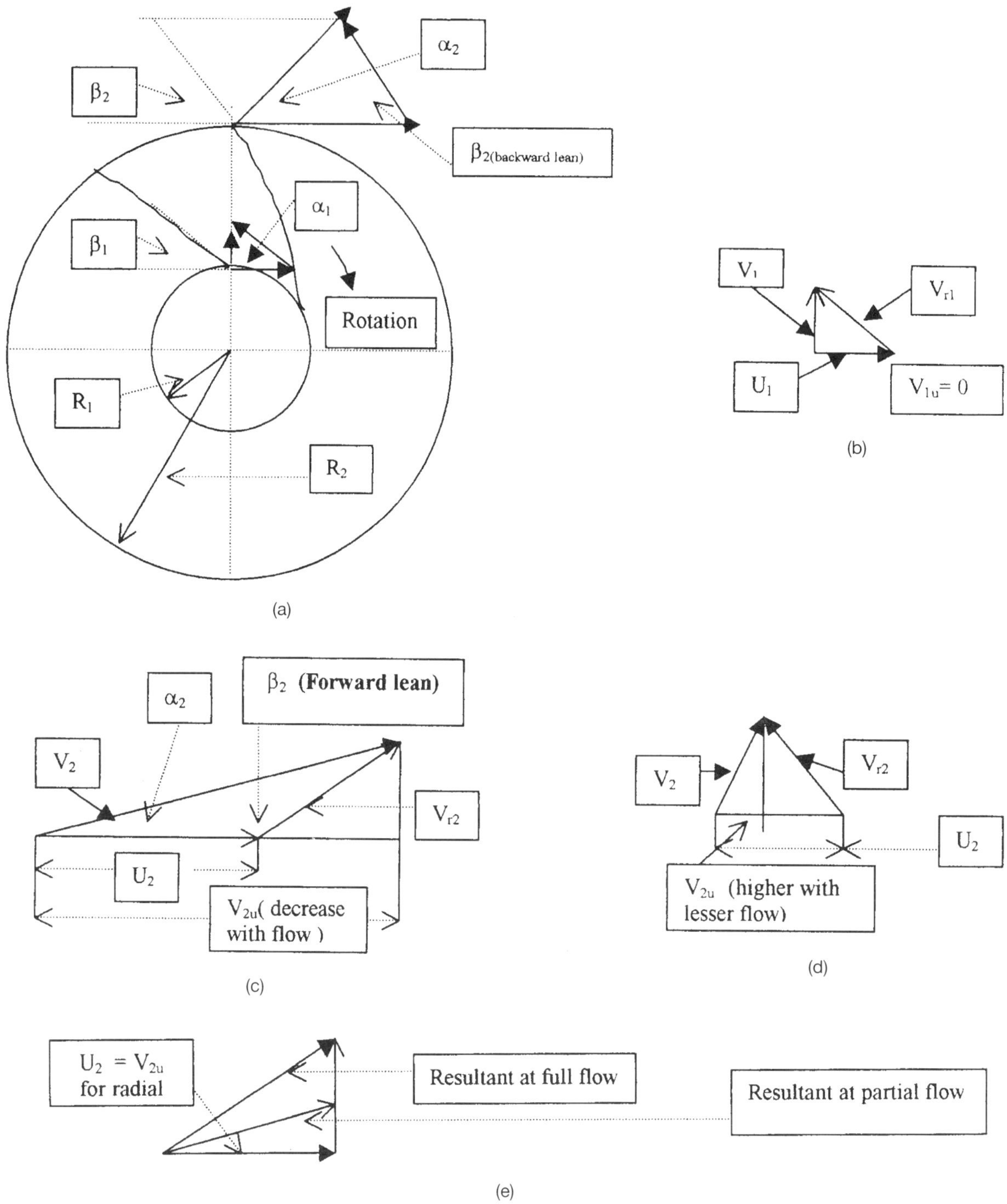

Figure 5-3. (a) Velocities at impeller inlet and outlet. (b) Velocity triangle at impeller suction with redial vane inclination. (c) Velocity triangle at impeller discharge with forward-leaning angle. (d) Velocity triangle at impeller discharge with backward-leaning angle. (e) Velocity triangle at impeller discharge with radial vane orientation.

This leads to the concept of Mach number (Ma), a term that is predominant and popular in aerodynamics. It is expressed as the ratio of the velocity of a gas (V_g) to its acoustic velocity (C_s) at a given temperature. It is linked up with the flow velocity. It does not impose limitations on the tip speed at the exit, as long as impeller geometry and higher gas temperature at the exit do not result in an unacceptable gas velocity at the discharge.

As gas flow approaches sonic velocity, shock losses in the flow channel occur and result in a "choking" condition. Under such a condition, efficiency and head reductions occur; they are accompanied by an increase in noise level. This ex-

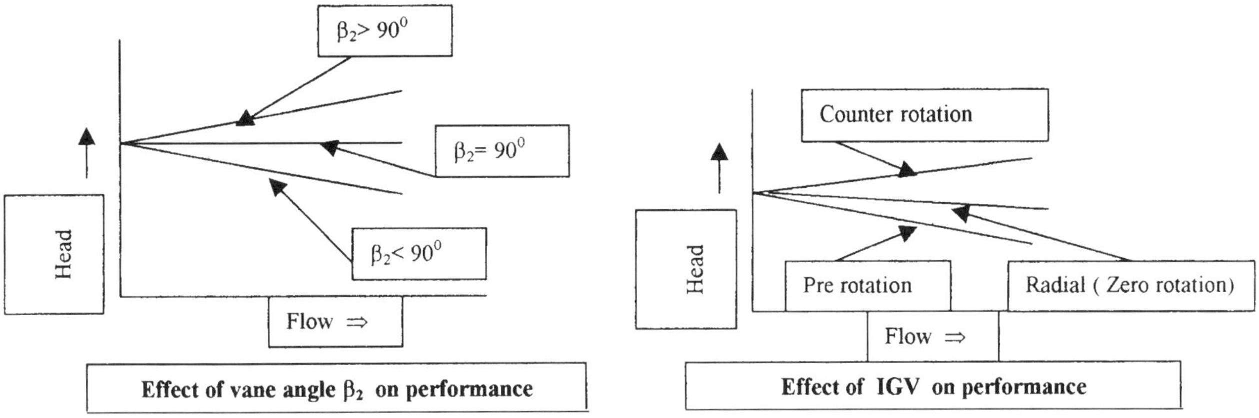

Figure 5-4. Effect of vane angle and inlet guide vane (IGV) setting on compressor performance.

plains why low velocity is important when dealing with a process gas having low inlet temperature with low value of k and high molecular weight, since all of these contribute to low sonic velocity. Refrigeration applications with heavy molecular weight gases such as propane, propylene, butane, or fluorocarbons ("Freon" gases) as refrigerants fall into this category. Piping design and compressor selection should aim to keep the Mach number as low as practical. Experience-based limits are used for relative gas velocity (V_r1).

Since the gas temperature is low at the inlet, it is important to check the Mach number effect at the inlet to the first stage of a multistage compressor. However, in compressors with side streams, checking the Mach number at the respective side stream stage is recommended. Prerotation guide vanes influence the relative gas velocity (V_r1) and serve to limit the inlet relative Mach number. Also, compressors may employ partial inducers to reduce the inlet Mach number.

5.2.4 Surge and Surge Control

Many explanations have been offered to describe the surge phenomenon. One of them considers surge as "that flow at which the head versus flow curve is perfectly flat and below which the head actually decreases." Or, "a compressor surges when the differential pressure (ΔP)—imposed across the machine by the process on which it is operating—exceeds the differential pressure the compressor can develop." However, experience shows these explanations to be of theoretical value at best [14].

It will be interesting to look at the discharge. With a reduction of flow, the angle at which the flow leaves the impeller becomes smaller; the flow now travels on a progressively wider spiral path as it moves through the diffuser. The resulting area–velocity relationship produces a downstream pressure higher than that in the diffuser, and this promotes the flow reversal.

It is important to understand the respective functions of impeller and diffuser shown earlier in Fig. 3-5. The impeller geometry controls head output, whereas the diffuser has predominant influence on the surge point [15]. Thus, flow instability leading to surge may first develop in the diffuser rather than in the impeller. Surge has a damaging effect on the machinery components and can lead to major outages if left uncontrolled. Excessive noise levels are commonly (but not always) experienced when the compressor is in surge. These (generally) dramatic increases in noise levels are due to rapid gas flow reversals and the associated mechanical vibration. Surge severity is, to a certain extent, a function of the pressure differential.

Needless to say, compressor reliability suffers whenever high vibration and high axial movement occur. Damaged internal labyrinths, thrust bearing failures, even severe rubbing of components and unacceptable power demand swings will often result. In extreme cases, even the impellers are affected.

Since most compressor applications involve two or more impellers in a single casing, as compression progresses the subsequent impellers receive reduced actual gas volumes. Moreover, they may have different characteristics [Fig. 5-5 (a)]. The surge process may thus take place in one of the higher stages rather than in the first stage. It can be visualized from the velocity triangles of Figure 5-3 that, closer to a surge event, angle α and thus relative velocity V_r become too small.

Since the mathematical expression for head includes the pressure ratio as one of the factors and with pressures and volumes linked by basic laws, it is now possible to calculate the volume at discharge conditions. A bit of mathematical manipulation brings to light an important distinction: In a given dynamic compressor, a higher molecular weight gas is compressed more than a lighter molecular weight gas. Therefore, the critical value of angle α corresponding to the onset of surge will be reached earlier—meaning at somewhat higher flow—with the heavier gas than with the lighter gas. Consequently, the useful range of a given dynamic compressor is somewhat more restricted in a higher molecular weight application. Apart from the preceding discussion, the piping downstream as well as the location of the control valves and

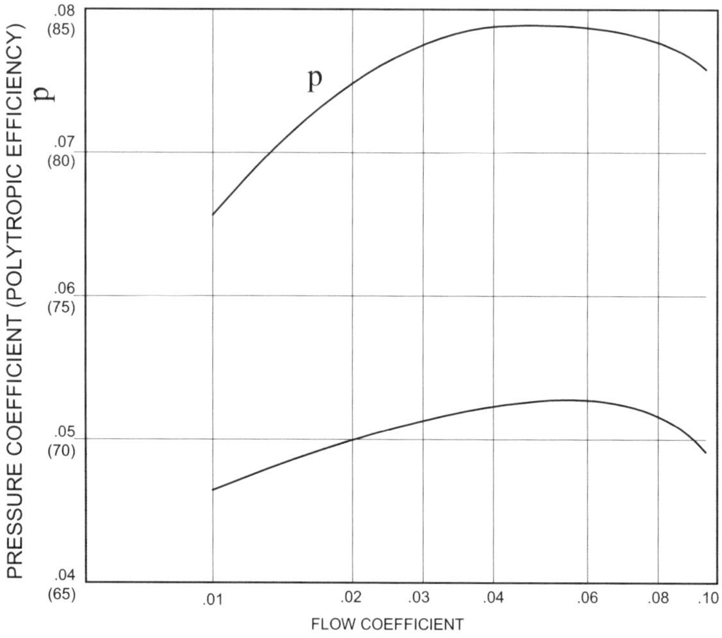

(a)

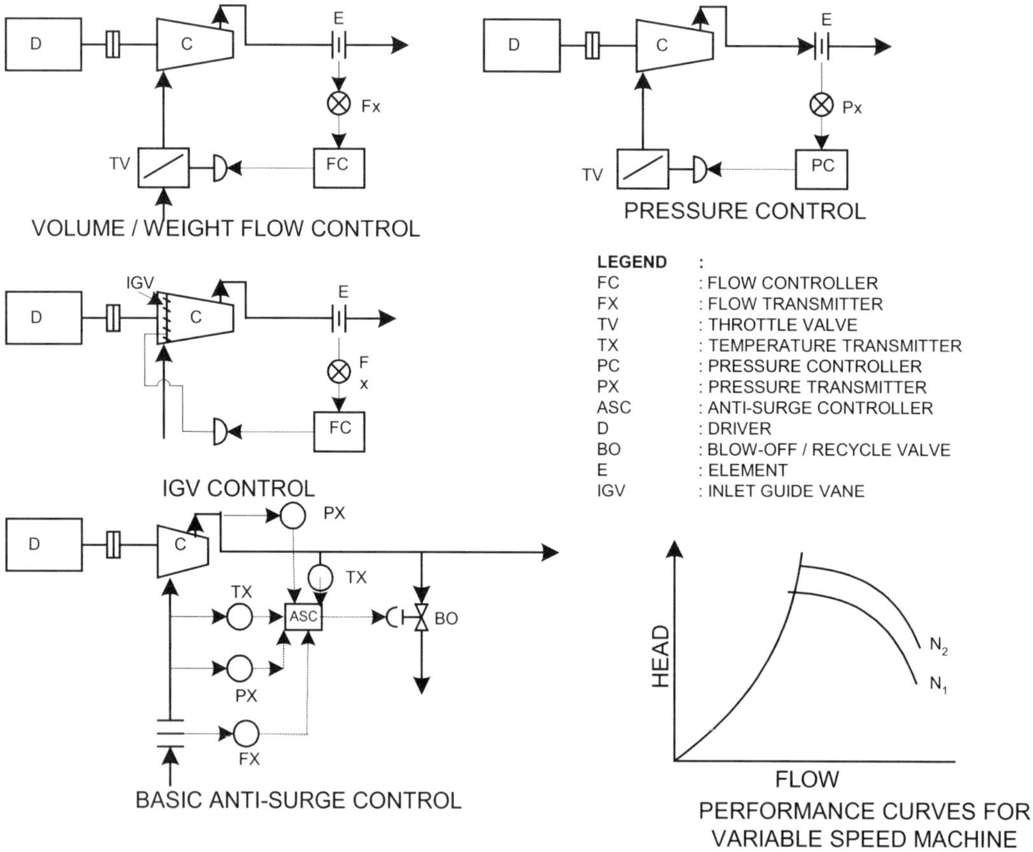

(b)

Figure 5-5. (a) A typical impeller characteristic. (b) Some typical control systems. (Continues on next page.)

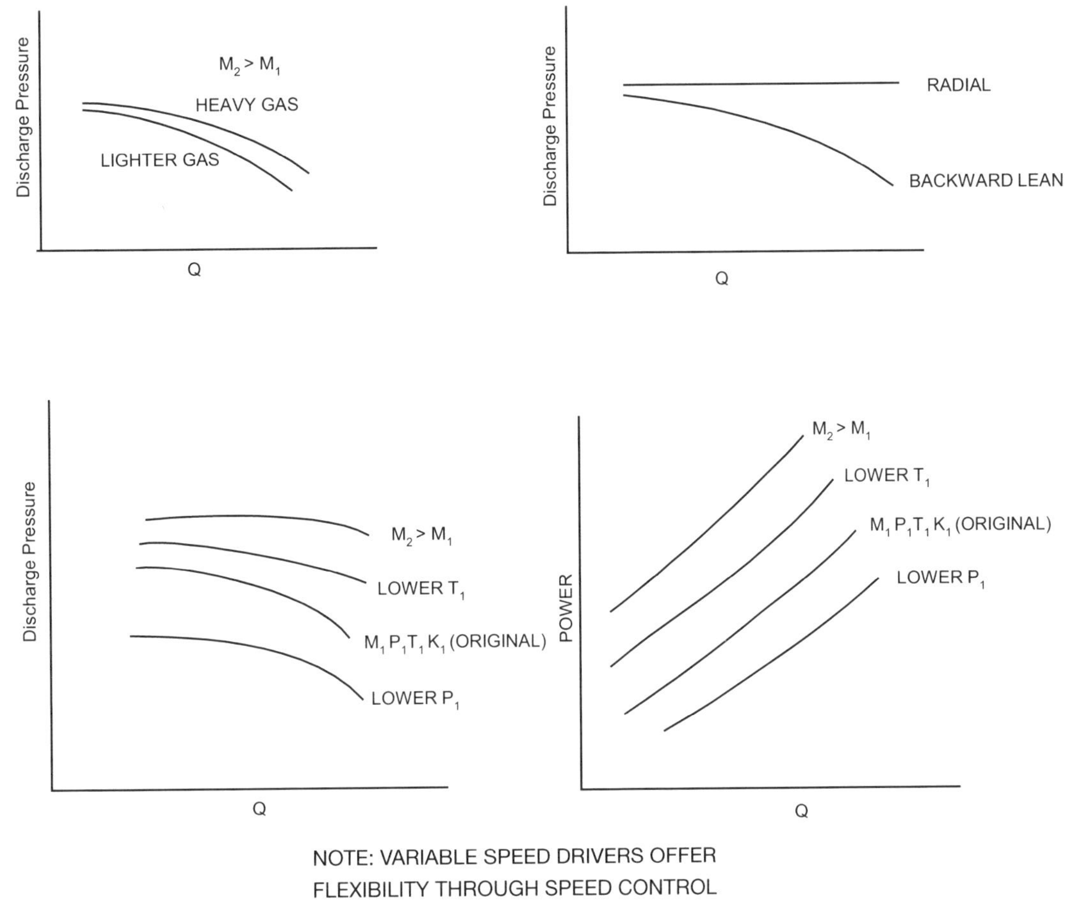

(c)

Figure 5-5. (*cont.*) (c) Changing performance patterns obtained by varying inlet conditions, and/or impeller profile.

volume contained within the connected system influence the actual surge location. Manufacturers' test stand guarantees are therefore not precise values, but rather close approximations that are considered reasonable. As can be seen from Fig. 5-5 (c), variations in gas composition tend to influence surge location [16].

Finally, there are also changes in compressor surge performance associated with any of a number of operating parameters. Figure 5-6 attempts to highlight just some of these. Collectively, the various considerations illustrate why important process gas compressors should be protected by suitable surge abatement or control systems, touched on in Fig. 5-5(b). These systems must be defined during the engineering phase and will also require field calibration and verification. The time it takes to do this work must be built into the commissioning schedule of the plant or facility.

In air compressors, the preferred method of eliminating surge includes installation of a blow-off valve in the discharge line; this then serves to maintain the needed minimum flow through the compressor. For flammable, toxic, or valuable gases, provision of a bypass from discharge back to suction is quite common. A heat exchanger is generally required in the bypass circuit so as to control the inlet temperature to the machine.

The choice of variable for operating the surge control mechanism depends on the nature of the characteristic compressor curve, that is, its pressure ratio (p_2/p_1) versus flow (acfm). It is customary to choose whichever of the two variables shows more change for a small change in the other parameter. The effects of changes in the operating parameters are shown in Figs. 5-6 and 5-7.

From a practical standpoint, it is necessary to maintain recycle valves in good mechanical condition, since the success of an antisurge system is largely influenced by how rapidly the signal is translated into smooth and accurate valve movement. It is equally important to set the surge control line not too far to the right of the actual surge line. The user's intent to operate the machine with minimum recycle should not be overlooked. The purposes of modern day sophisticated antisurge, quick response control systems will be fully served only if all of the important parameters are considered. Reliable operation with minimum recycle is to be achieved without jeopardizing the safety of the machine. To achieve this goal, investing in a well-engineered antisurge control system will be easily justified. A

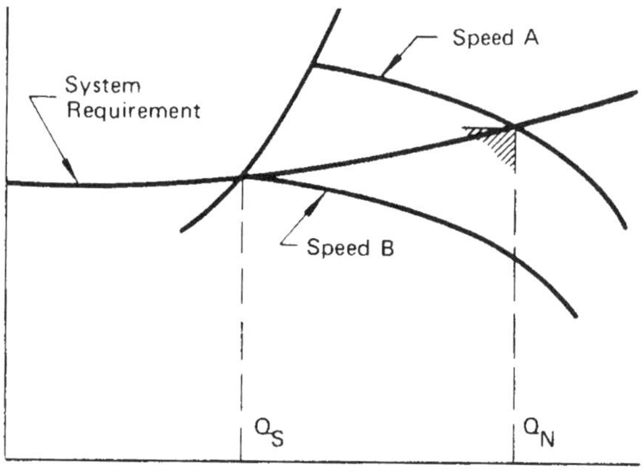

Stable flow range attainable by using variable speed control

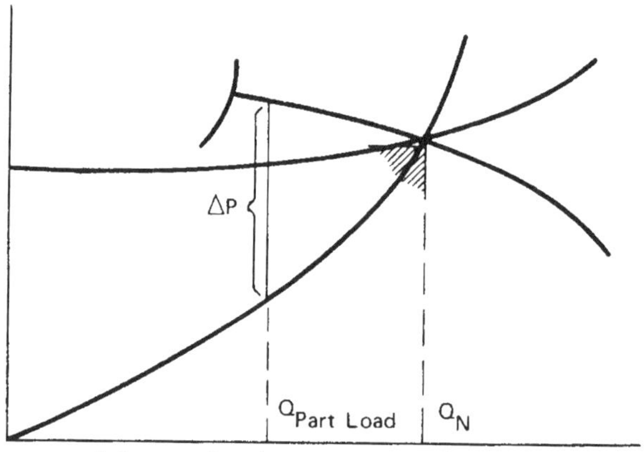

Influence of system pressure requirement curve shape on pressure drop across an inlet throttle

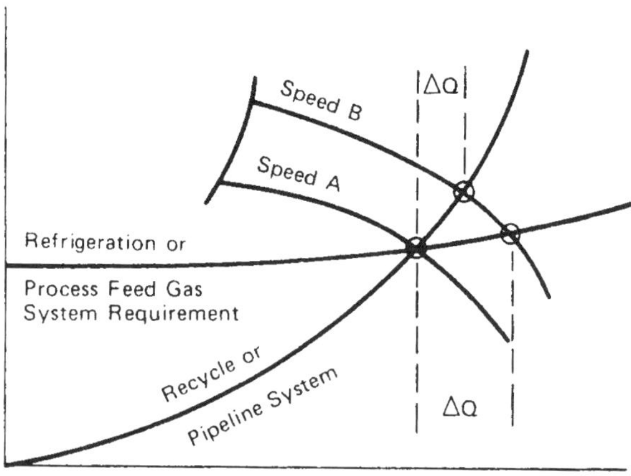

Influence of system pressure requirement curve shape on flow increase attainable through input speed increase

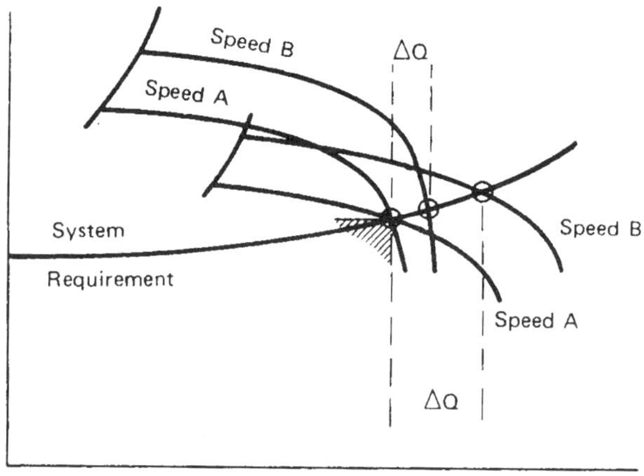

Influence of compressor curve shape on flow increase by speed changes

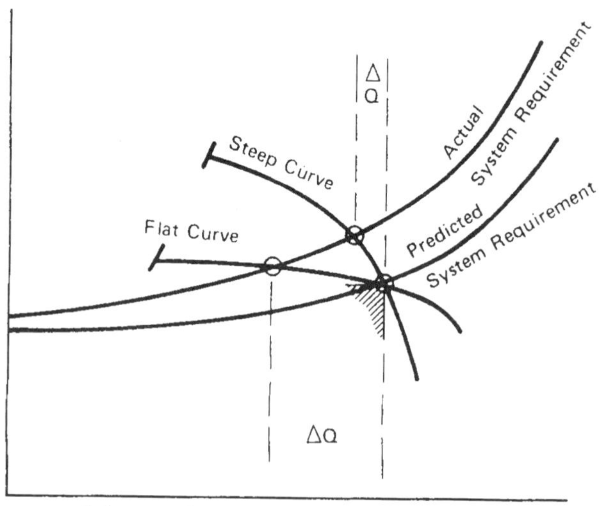

Influence of compressor curve shape on flow losses where system pressure requirements are not reached

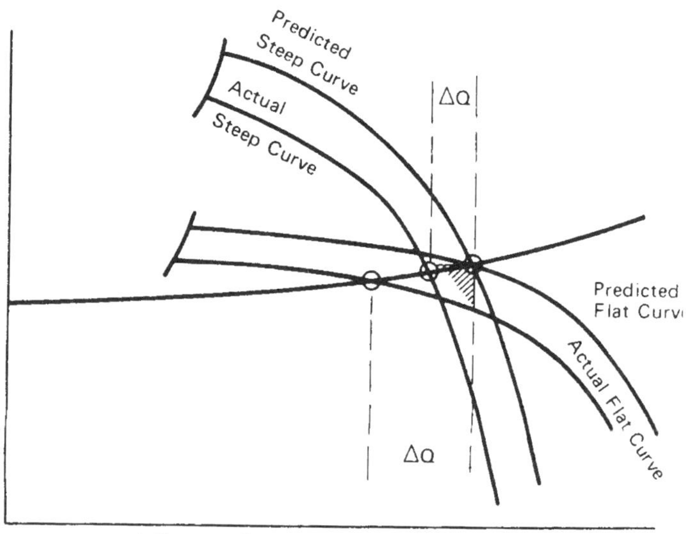

Influence of compressor curve shape on flow losses on compressor not reaching adequate head

Figure 5-6. Effect of changes in compressor operating parameters. *Note:* (1) Ordinate is Discharge Pressure, P_2, and Abscissa is Inlet Flow Rate, Q, for all graphs.

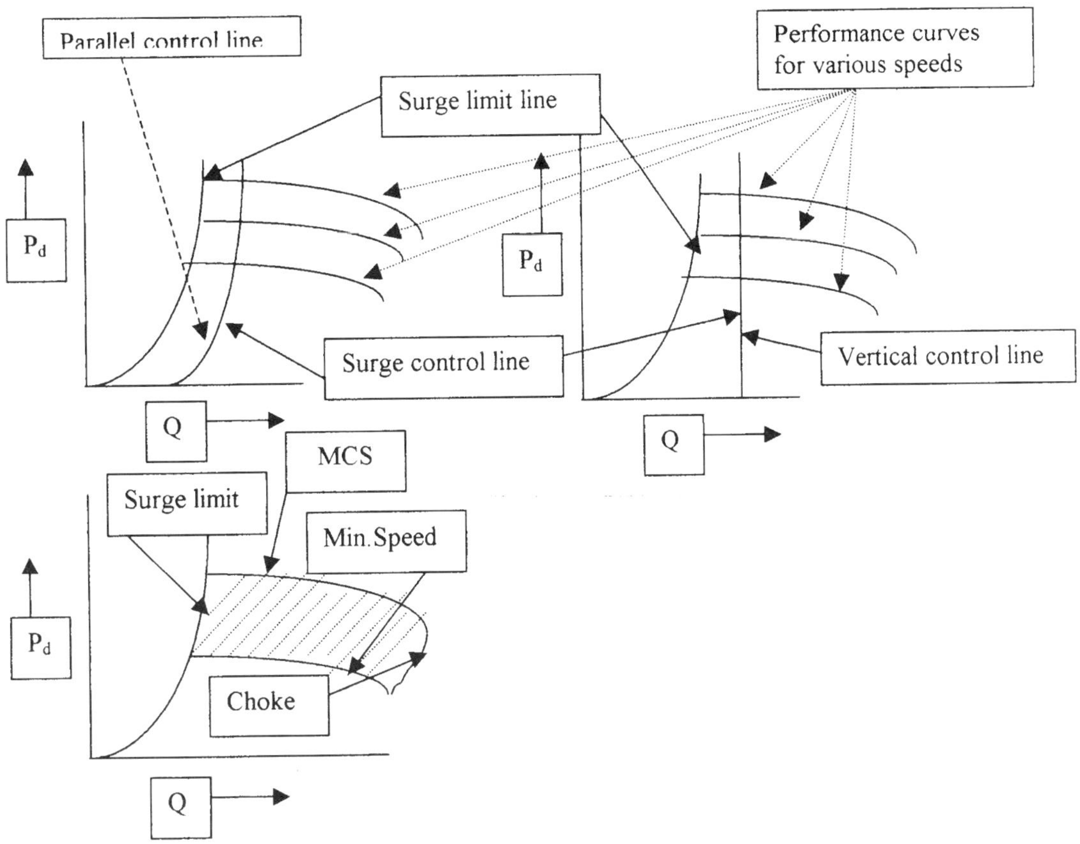

Figure 5-7. Surge control issues.

fast-acting, nonsticking recycle valve is part of the system. Antisurge valves should be fast-acting and capable of passing full flow. In other words, full recycle should be possible, with none of the flow continuing to go to the process.

As can be inferred from the above, satisfactory performance of centrifugal compressors is governed by a number of important parameters. Among others, it is defined by the limits of minimum and maximum operating speed, and the maximum pressure that a system can sustain. Similarly, the flow range from largest permissible flow (overload, or choke limit) to lowest permissible flow (surge limit) is of obvious interest. When the machine surges, flow reversals can take place within a span of 20–50 milliseconds. This shock loading taxes the machine and decreases its overall reliability. Moreover, hot gas flowing from the compressor discharge back to the compressor suction will occasionally cause substantial increases in compressor casing temperature and, again, the surge limit may shift.

The entire surge phenomenon is sufficiently elusive and its potential onset too difficult to predict to consider advocating manual intervention by operating personnel. Unforeseen process trips could occur unless properly engineered automatic antisurge systems are chosen and installed. Moreover, surge is associated with low-throughput operation. Although seemingly operating in the acceptable flow region, severe gas molecular weight fluctuations could cause a machine to go into surge. Other contributing factors might include clogging of filter or demister pads, or the collapse of filter elements. Indeed, even the actions or omissions of inexperienced operators may cause the valuable machine to go into surge during unit start-up.

A close look at the performance map of the machine (Fig. 3-6, also Fig. 5-7) will show that as the machine starts approaching surge, its *H–Q* characteristics become less steep. There will thus be big changes in flow for small changes in the pressure differential. With pressure-sensing instrumentation in control, their accuracy or "float" may be of great importance. In like manner, we conclude that probably the most important measurement in any antisurge system is the flow across a flow element. This element is normally located near the suction nozzle, but it is also feasible to place it near the discharge of once-through machines. In that case, monitoring both suction and discharge pressures will provide both the pressure differential and/or the pressure ratio needed by the antisurge controller. Monitoring suction and discharge temperatures allows computing the efficiency and will further assist the antisurge system in continuously updating all necessary information.

Recall that the purpose of the antisurge system is to move the operating point away form the surge line and toward the right side of the performance map. Manipulation of the bypass valve, labeled "BO" in Fig. 5-5 (b), ensures that flow through the machine satisfies the minimum requirement. A surge control line is located to the right of the surge line. The difference between the surge line and the surge control lines is the amount of recirculation or (blow-off) needed to prevent surge. There are various ways in which the surge control line is constructed. The preferred option is to keep it parallel to the surge line, but in the interest of maintaining high efficiency, to keep the separation between lines as small as possible. Years ago, users opted to maintain the same amount of recirculation irrespective of the operational load imposed on

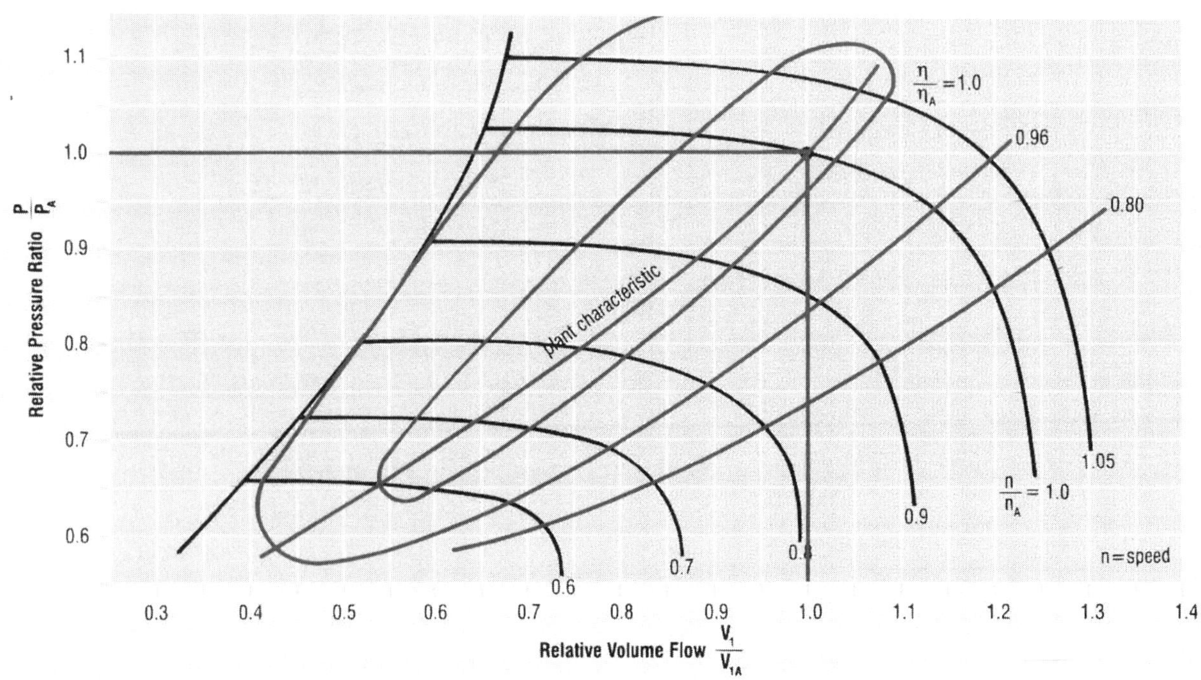

(a)

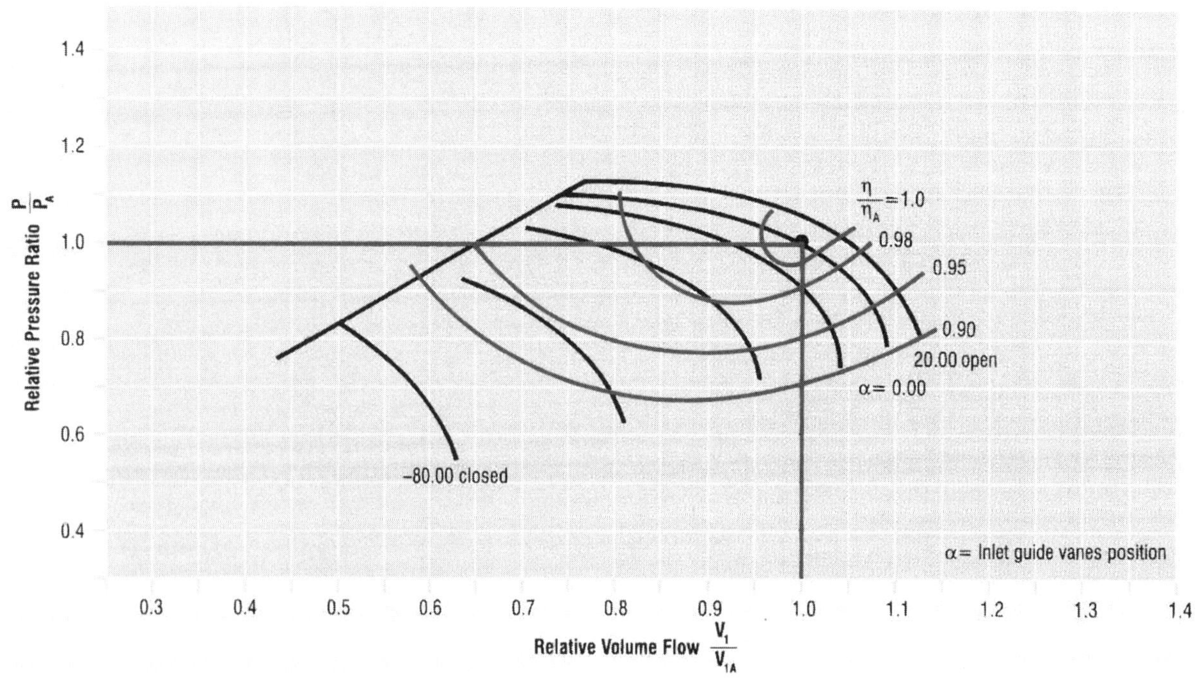

(b)

Figure 5-8. Compressor performance and surge limits with different control schemes. (a) Variable-speed performance characteristic. (b) Performance characteristic for adjustable inlet guide vanes. (Continues on next page.)

the machine. Needless to say, such inefficient operation can no longer be justified in today's cost-conscious environment. Indeed, efficiency and operating flexibility considerations have prompted some users to actively consider not only adjustable inlet guide vanes, but also adjustable diffuser vanes. Their configuration and performance characteristics can favorably affect surge limits, as illustrated in Fig. 5-8.

Since a compressor characteristic drawn as *H–Q* (head vs. inlet flow) is unique for a compressor at a given speed, its surge also is unique in character. There will be different curves for different speeds, with distinct values describing the flow at surge condition. The antisurge system designer decides what algorithms will best suit the application by choosing appropriate inputs. It is normal practice to convert compressor performance maps to equivalent Head (*H*) and "flow-squared" (Q^2) curves. This conversion eliminates the common factor (*ZRT*) at suction conditions. Many antisurge controller manufacturers use this equivalent curve. Thus the input to an antisurge controller is the pressure drop across the flow element, which represents the flow, and the head

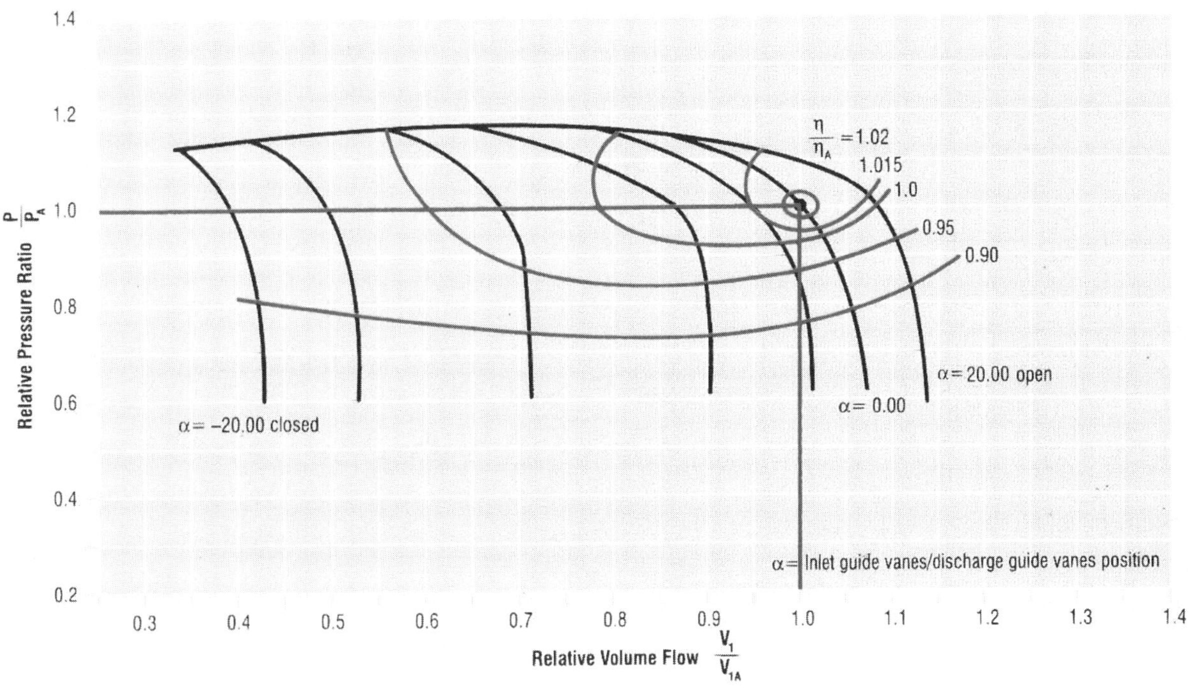

(c)

(d)

(e)

Figure 5-8. (*cont.*)Performance characteristic for adjustable inlet guide vanes. (c) Performance characteristic for adjustable inlet guide vanes and adjustable diffuser vanes. (d) Adjustable diffuser vanes. (e) Adjustable inlet guide vanes.

factor is the corresponding pressure rise across the machine. The inputs needed for calculating the equivalent head come from the suction and discharge pressures and their associated temperatures. This then paves the way for locating both the surge line and the surge control line, as well as their margin of separation. In some cases, further simplification is made when equivalent head is replaced by the pressure ratio.

We should remember that the pressure drop across the flow element is governed by the following rules for the same "inlet volume":

- If temperature at inlet decreases, ΔP across a measuring orifice increases
- If MW at inlet increases, ΔP across a measuring orifice increases
- If pressure at inlet increases, ΔP across a measuring orifice increases

Thus, if we look at an H–Q curve in any of the above cases—inlet temperature reduction, inlet molecular weight increase, or inlet pressure increase—the pressure differential (ΔP) would go up for the same inlet capacity. The mass flow goes up in all of the above situations, creating more pressure drop across the flow element. The discharge pressure in all three cases would now be higher than under original operating conditions. The critical value for surge conditions to occur at the angle at which the gas is being discharged would also be reached sooner than at original conditions. At the same time, the gas becomes denser, which contributes to the increasing pressure drop. This explains why the useful operating range of the compressor decreases with increases in molecular weight, lower inlet temperature, and higher suction pressures.

As was mentioned earlier, the manufacturer's factory testing is concerned with guarantees at only a single operating point. The establishment of surge at various speeds is a close approximation at best and will not be covered by performance guarantees.

When one is dealing with applications with compressor casings arranged in tandem for handling the same gas from one casing to another, it is customary to have an antisurge controller for every stage. Inputs are obtained from flow transmitters located either on the suction or the discharge ends of every casing; other inputs are compressor suction and discharge temperatures and pressures. These inputs help establish the operating point on the compressor with respect to the surge line. The proven high reliability of a modern antisurge system can be further enhanced by using control system redundancy. Redundant systems can bridge over such unlikely events as primary controller faults, controller output failure, or planned outages for testing and preventive maintenance. Redundant controllers remain in the tracking mode and in a state of readiness to ensure full availability of the antisurge system. A backup controller may also be part of the DCS (distributive control system) or similar plant control system, which will be called upon to take care of contingency situations if the primary controller fails. However, providing a fully redundant controller is generally preferred to the use of a DCS backup controller.

In compressors furnished with variable-speed drivers, surge can be avoided by reducing the operating speed. However, even in this instance surge abatement is usually accomplished by recycling gas around the machine. The point at which the machine actually goes into surge is called the surge line. As indicated in Fig. 5-7, the line where the machine will not be allowed to drift to its left on the performance curve is called the surge control line. The amount of recycle will be equal to the difference in flow from the point on the surge control line flow and net flow to process located to its left. It is thus ascertained that total flow to the machine will be at least equal to the flow on the surge control line. The margin between the surge line and surge control line is dependent on the kind of compressor and its capital cost. The project and reliability engineering functions cooperate to determine potential vulnerability without surge control devices. Potential consequences include damage restoration costs and overall impact on plant production. Plant downtime and time required for repairs to machines and associated piping are assessed. Conversely, the cost justification for implementing a suitable antisurge system is usually examined. In new and reliability-focused installations, the implementation of antisurge systems is certainly favored.

Any extra margin between surge line and surge control line ends up in higher power consumption when the machines are operating at low throughput. This highlights the importance of comprehensive field testing. It will be interesting to note that some operating companies assign the job of handling the antisurge system to the rotating machinery group. As the thinking goes, members of this job function typically have a better understanding of machine performance, although their cooperation and interaction with the facility's instrumentation group is always encouraged.

It is also necessary to understand the relationship between the process control and the antisurge control. A process control is designed or selected with production considerations first and foremost. Antisurge controls are governed by machine life and reliability concerns. How these two control issues are designed to interrelate will determine whether they create problems or lead to efficient, life-cycle cost-optimized machine utilization. This is best illustrated by looking at pressure control on the discharge side. If, say, less flow is demanded by the downstream units, the pressure may go up. The process control will now try to close the valve, trying to shift the system resistance. The antisurge device will interpret this as a tendency toward surge and will be asking the valve to open, thus trying to reduce system resistance and thereby promoting more flow through the machine. This is a conflicting requirement. It is thus necessary to define the nature and extent of interaction between the process control and antisurge control.

Some compressor operators have been known to place one of the controls on manual mode and allow the other one to be in the automatic mode. This approach usually leads to inefficient operation since field operators will, understandably, give priority to plant production. They are prone to place less emphasis on achieving optimized power consumption and the resulting energy savings.

Because the onset of surge and resulting machine damage can take place within seconds, quick response of the antisurge controller is of great importance. A reliability-focused plant will, therefore, insist on automatic antisurge control action. Claims to the effect that it takes time to load the machine if the antisurge is in "automatic" mode are occasionally used to justify keeping the controls in "manual," but this is where smooth operation will inevitably differ from one operator to another. However, in any such situations it is critically important that an operator be unusually well versed with the present status and condition of a plant or process unit. The bottom line is that high-quality, automatic surge controls make economic sense.

Variable-Speed Operation and Surge Issues

In situations in which variable-speed drivers are used for centrifugal compressors, plants quite often run in the manual mode of speed selection while the antisurge controller runs in the automatic mode to avoid incidents of erratic operation and surge-related damage to the compressor. As flow through the machine is being increased, it reaches a limit beyond which there is a sudden and drastic drop-off in pressure. This is called the *stonewall* or *choke* limit. As flow increases, the Mach number increases because the velocity of the gas increases. Fortunately, the end effect will not harm the compressor. Unlike the experience with surge, there is no risk of catastrophic compressor damage with choked flow.

5.2.5 Development of Head

The effect of changes in gas properties and compressor inlet condition of the compressor were illustrated in Fig. 5-5 (c). About two-thirds of the pressure rise occurs in the impeller; the remaining increase takes place in the downstream diffuser where velocity energy is converted into pressure energy.

Since a centrifugal compressor produces pressure increase by imparting kinetic energy to the gas carried thorough the impeller, impeller-tip speed becomes an important parameter. This, obviously, requires taking into account both impeller diameter and speed. The head developed by a machine is a function of the square of the tip speed, whereas flow rate handled by the machine is a function of impeller diameter and tip speed. If, for instance, the compressor is scaled to twice its previous size with tip speed remaining constant, it will produce the same head. However, it will handle four times the original flow. This is also explained by the definition of the flow coefficient described earlier. However, if the speed were doubled, the compressor would be able to handle twice the original flow and would produce four times the original head. Again, tip speed plays an important role.

5.3 ESTIMATING CENTRIFUGAL COMPRESSOR PERFORMANCE

The precise definition of compressor performance and internal configuration is ultimately the manufacturer's responsibility. However, process design and engineering companies are generally required to estimate compressor performance for a certain process or unit. Such estimates lead to preliminary decisions on power requirement, compressor speed, and driver selection. In addition, this effort will determine the number of machines, feasibility of a tandem train or "string" arrangement (see Fig. 3-18) whereby a single driver is used for more than one compressor casing, as well as overall layout and cost estimate. These initial estimates are often facilitated by data and graphical relationships furnished by the various compressor manufacturers.

5.4 CONTROLS

Much of the preceding text was aimed at imparting an understanding of what influences compressor selection. Similarly, knowing how compressors are controlled and how a compressor fits into the process are important for many job functions in a modern process plant. This understanding should go beyond the customary (for machinery engineers) initial selection and testing of the machine at the manufacturer's shop—knowledge that will, of course, assist in future troubleshooting. There have certainly been instances of perceived performance deficiencies when, in fact, compressor data were simply misinterpreted by operating personnel.

This point is best illustrated by an actual example involving two facilities separated by a long-distance gas pipeline. When the gas did not arrive with sufficient pressure at the receiving end, the destination customer blamed the originating supplier for not having a proper machine. It took some time for both parties to realize that they had not allowed sufficient time to fill the pipeline. By prematurely opening a valve at his end, the receiving customer had made it impossible for pressure to build up. Once the parties agreed to keep the valve closed for 30 hours, the pressure built up to one bar and thereafter continued to increase at a rate of approximately one bar per hour. Once the pressure reached 40 bar (~580 psi), the destination facility could be commissioned in the normal way without difficulty.

While interpreting or deciding on compressor control, it is necessary to understand how varying an inlet parameter affects machine performance. Along these lines, it is useful to recall the fundamental equation of head. Recall also the basic rule that a compressor operated at a constant "inlet volume"—the actual volume, "acfm" or "icfm," rather than the volume recalculated to normalized or standard conditions, "scfm," with speed held constant (and having adequate driver capability)—will always develop the same head. Thus, a given compressor for a given volume and speed will develop the same head regardless of the numeric value of other inlet conditions. The head developed by a given impeller is given by the following equation and is expressed in feet or meters, not in pressure terms such as psi, pascals, or bar:

$$H = (Z \cdot R \cdot T_1/x) \cdot [\{(p_2/p_1) \wedge x\} - 1]$$

where

H = ft-lb/lb
Z = compressibility factor
R = 1,544/molecular weight
T_1 = inlet temperature, degrees F + 460
$x = (n - 1)/n = (k - 1)/(k \cdot np)$, where $k = C_p/C_v$
np = polytropic efficiency
p_2, p_1 = outlet and inlet pressures, in psia

The following conclusions are thus drawn whenever the suction conditions are being changed [16]:

1. If the inlet temperature decreases, the discharge pressure will rise. The weight flow and power demand would also rise.
2. If the molecular weight increases, p_2 and the power demand will increase.
3. If the suction pressure decreases, the discharge pressure and power demand will decrease.
4. If the speed is changed from N_1 to N_2, the affinity laws will apply. The capacity will now change in direct proportion to the speed ratios, head will vary as the square of the speed ratios, and power demand as the cube of speed ratios.
5. If the k value were decreased, the discharge pressure would increase and power demand would remain constant.

To some extent, the various relationships are similar to those applying to centrifugal pumps. In pumps, the user concentrates on head versus flow characteristics and observes different pressure gauge readings with fluids of different densities. However, there are fundamental differences because compressors handle compressible fluids with distinct thermodynamic properties that result in a complex formula for head. On the other hand, pumps handle noncompressible fluids that simplify the equation for head.

It is also important to visualize the implications of changing compressor speed if a machine operates in a closed-loop arrangement. As speed is increased, one would expect a drop in suction pressure because the machine is now moving more flow, thus trying to lower the gas density on the suction side. Similarly, slowing down speed would tend to increase the suction pressure. The other process controls will sense these changes and make compensating adjustments to maintain production at the existing capacity, or readjust for the new throughput demand. However, if a compressor is handling atmospheric air, there will not be any meaningful changes in suction conditions. A small drop in pressure at the first-stage impeller inlet could probably be attributed to encountering somewhat higher frictional resistance as compressor speeds are increased.

These basic considerations must influence compressor control strategies. Even the best compressor will not be able to perform with nonoptimized controls. Process controls and machine controls must be compatible and must interact flawlessly. Buying these controls based on price alone is never a smart decision and will, sooner or later, lead to energy waste and/or declining reliability. Assigning the job to a thoroughly experienced design and manufacturing firm is clearly the course followed by reliability-focused companies.

5.4.1 Integrating Machine and Process Control Strategies

There must be full compatibility between machine and process-control strategies. In fact, the two should be coordinated by the same designer–vendor and overall responsibilities defined accordingly. As mentioned earlier, antisurge control systems are intended to save energy and prevent serious damage to the machine. They should not be selected on the basis of price alone, since malfunctioning antisurge systems can result in catastrophic damage to the compressor. Should that be the outcome, production losses are very likely to exceed any savings that might have resulted from procuring a low-cost system.

In general, process controls are configured or set to maintain one of the following: (a) constant discharge pressure, (b) constant inlet pressure, (c) constant weight flow, (d) load sharing. Relevant signals originate or emanate from sensing elements that monitor pressure and flow. This signal is then sent to the controller through a transmitter. In turn, the controller compares the incoming signal with the set point. When there is a difference between the set point and incoming signal, final action to narrow or close the gap will be achieved through actuation of either one, or a combination of suction throttle valve(s), discharge valve(s), blow-off valve(s), inlet guide vanes, recirculation (bypass) valve(s), or speed governing mechanism(s).

As was alluded to earlier in this text, controls can be separated into two types: process controls and protective controls. Process controls are aimed at satisfying production considerations. They may or may not be mandatory to safeguard human and physical assets. In contrast, protective controls are mandatory to ensure that there is no major damage to the machine due to mechanical malfunctions brought on by

- Design defects in an important component
- Fabrication deficiencies in an important component
- Material defects
- Assembly or installation flaws
- Maintenance errors
- Unintended operation
- Operator error

The reader may recognize these seven to be the only cause categories that either exist, develop, or interact to cause machines to fail [3]. Quite evidently, flaws can develop or progress during compressor operation and, unless there is intervention, there can be severe consequences. Surging, as defined earlier, is either caused by a process upset or will result in a process upset. In a modern plant, instrumentation and mechanisms that avoid damage to the compressor are part of the protective controls. The antisurge control device should identify the onset and take corrective measures as quickly as possible to open the antisurge valve or a bypass and thus safeguard the machine by keeping it out of surge. Therefore, selecting the correct valve type, its characteristics, and required time to open fully are all important. Whenever the machine is operating far to the right of the surge line, the valve should remain tightly shut.

Compressor and driver shutdowns ("trips") can be initiated by high vibration, low lube oil/seal oil pressure, high bearing temperature, high discharge or exhaust temperatures, drive motors drawing excessive current, high liquid levels in suction disengagement drums, or high differential pressure across the suction strainer/filter. All of these are among the various machine-protective controls. Similarly, there are permissive features (protective controls) that require compliance before allowing a machine to be started. These might include minimum lube oil or seal pressure in the header, inlet guide vanes (IGVs) in near-closed position at startup, antisurge valve at a predetermined opening before process gas can be fed to downstream units, and provision of adequate N_2 buffer gas supply for dry gas seals.

Although protective controls safeguard equipment life, their need is occasionally questioned because, after all, "the machine is still runnable." An experienced reliability professional will see certain parallels with the fact that an automobile tire may still hold air even though its rubber tread is completely worn away. That being the case, a rational person would not continue to operate the car at the advertised speeds and loads. Deviations from normal certainly tend to create hazards and responsible individuals will take action as soon as such hazards are detected. In centrifugal compressors, voting logic can be incorporated into the trip circuitry to avoid shutdowns due to spurious signals. Typically, the voting logic initiates a trip event after two or three relevant parameters indicate signal excursions that are deemed unacceptable. In any event, revising set points for trip conditions requires thorough scrutiny by knowledgeable individuals and should never be done without understanding and accepting the possible consequences.

5.5 THROUGHPUT CONTROL OPTIONS

For decades, gas turbines and steam turbines were among the best options that offered energy-efficient throughput variation [17]. Both driver types allow speed control over a fairly wide operating range. For centrifugal compressors, speed variations typically range from 75% to 105% of "rated" (100%) speed. Rated speed denotes a reference speed to which head and flow warranties are typically linked.

However, and for dynamic compressors, variable frequency drive motors have joined turbines as a viable option; this option merits consideration in view of the improved electronics now available. Variable speed electric motor drivers are produced in sizes up to, and even exceeding, 100 MW. They offer similar flexibility and part-load efficiency as gas and steam turbines. The trend is toward more variable speed drive motor applications because of the simplicity and relative cleanliness of these installations. Also, variable speed electric motors tend to require less maintenance than other variable speed drives. Alternatively and upon detailed analysis, process designers and equipment owners may find merit in selecting variable speed fluid drives (somewhat erroneously labeled variable speed couplings) to achieve speed variation in motor-driven compressors.

Variable inlet guide vanes (VIGVs or just plain IGVs) and occasionally even adjustable diffuser vanes (see Figure 4-7) are of interest as well. IGVs are certainly used in most high-capacity axial compressors, but also in many single-stage, overhung construction, medium pressure and flow machines. As mentioned earlier, unless properly maintained, IGVs and their linkages introduce additional failure risk. The same is true for adjustable diffuser vanes.

Suction valve throttling represents another capacity turndown option. This is mainly achieved by reducing the inlet pressure, which in turn reduces weight flow and, therefore, reduces power demand. As with the other options, precautions must be taken to keep the compressor away from surge. Suction throttling is not as energy efficient as IGVs, which can be adjusted over a wider range. Some manufacturers deserve much credit for offering adjustable diffuser vanes because these are further widening the range of operational flexibility.

Flow adjustments made on the compressor discharge side after the process gas leaves the machine are inevitably less efficient than any of the preceding options since energy was already consumed in the compression process. Therefore, operations with discharge throttling, bypassing (recirculation), and/or blow-off constitute some of the less efficient flow control modes. Their use in an energy-short and environmentally conscious environment may be difficult to justify. Suffice it to say that their applicability must be determined on a case-by-case basis.

By and large, refrigeration applications work on the principle of discharge pressure control, since this pressure determines the cooling effect for the product or medium to be cooled. The operating pressure margins of associated columns or major reaction vessels both upstream and downstream will determine compressor operating set points. Many times, compressors are placed in catalyst regeneration service. The feasibility and suitability of a given machine for this relatively strenuous service must be studied and established before purchase decisions are made.

Machinery engineers tend to confine their role to the battery limits of a compressor installation and are reluctant to acquaint themselves with relevant process information. This, of course, limits their interaction with other functional plant groups. Machinery engineers and operating personnel should become conversant with control system terminology and functionality. We restate again the intrinsic value of all plant groups knowing what certain compressor components must do. All plant groups must further understand compressor start-up and emergency shutdown procedures. The technical background of operating personnel must be sufficient to safely and effectively start process gas compressors without depending on instrumentation and control specialists, or machinery engineers, for every start-up. Having this technical background will give them confidence to routinely operate their machines. Operating technicians should call on the services of specialized groups only in the few instances in which process deviations cause abnormal machine behavior. Certainly, there are situations that genuinely require the involvement of machinery specialists.

It is noteworthy that major failures often take place while the machine is being brought on line, that is, at start-up. For a variable speed drive, say, a gas-turbine-driven compressor train, a typical start-up cycle will require eight to twelve minutes. Steam turbine start-ups require more extensive preheating of both piping and steam turbine casing. Mandatory dewatering and heat-up of piping takes time. Also, whereas the smaller rotor mass will warm up and expand relatively quickly, the much larger casing mass will take longer to heat up and expand thermally. The two total expansion amounts must coincide and the warm-up cycle must be long enough for equal expansion to occur. For large machines, the entire procedure may take three or more hours of controlled warm-up. Taking unauthorized "shortcuts" has caused major failure incidents and even loss of life on more than one occasion.

The considerably shorter gas turbine start-up sequence is generally complete when the HP shaft (connected to the axial compressor) reaches 95% and the LP shaft of the turbine (connected to the centrifugal process gas compressor) reaches 75% of the intended process operating speed. Until this point, the antisurge valves remain open and the compressor is in the recycle mode. As the speed starts to increase, the antisurge system will initiate, closing the recycle valve(s) in a predetermined manner. Gas will now flow to the process and pressure will build up in the system. Every plant is designed to allow isolation of certain process units in the event of a compressor shutdown. The plant process control scheme is designed to depressurize the system in the event of either mechanical or process-reasons trips. The process compressor may be totally isolated from the system by closing suction and discharge block valves and the trapped gas may be released to the flare. Depending on the application, the antisurge valve(s) may be left in the open condition.

An operating technician assigned to a modern process plant knows that centrifugal compressors are heavily instrumented, and for good reason. With steam or gas turbine drivers, additional supervisory and safety instrumentation becomes necessary. Whereas decades ago a plant could perhaps exist with a bare minimum of instrumentation, the late 20th and early 21st centuries are quite obviously different. With justification for automation came trends toward condition-based maintenance and more thoughtful asset management [18]. To stay competitive, plants must increasingly rely on instrumentation and controls.

Chapter 6

Testing

6.1 CENTRIFUGAL COMPRESSOR TESTING

Before the centrifugal compressor leaves the manufacturer's shop it must be tested for mechanical and sometimes even process-related performance. Since the specified process performance parameters of different compressors are rarely identical, each centrifugal compressor is likely to incorporate some custom-designed stage components. The acceptability of design and fabrication must be verified and deviations will have to be corrected before shipment [19]. As a minimum, the user–owner will specify a mechanical running test to measure shaft vibration, bulk oil and bearing temperatures, lube-oil flow rate, sour and sweet seal oil flows, systems performance in the case of magnetic bearings, and absence of gas leakage overall. Some of these measurements are needed to calculate mechanical losses and probable overall power demand. The API-617 Standard for centrifugal compressors does not allow any negative tolerance on capacity (throughput flow) at a power input within +4%. This stipulation refers to a guarantee point within a speed range kept below maximum continuous speed (MCS). Recall that MCS is 105% of rated speed.

In general, shop performance testing is highly advisable; Fig. 6-1 shows a modern test stand. Field performance testing may be feasible, but requires allocation of time, definition of instrument type, and written understandings between the various parties as to location and calibration of instruments. Performance testing is aimed at determining flow, polytropic head, pressure rise, efficiency, shaft power, surge point, and choke limit. Since shop tests are rarely done with test loops containing the actual process gas, analysis and extrapolations are needed to simulate actual site conditions. Although the well-known affinity laws are used, correction factors are applied to the calculations.

A comprehensive procedure known as ASME PTC 10 (American Society of Mechanical Engineers' Power Test Code 10) is generally used for these all-inclusive and important tests. Compressor manufacturers and their process industry clients realize the value and payback provided by final mechanical and performance tests at the factory.

Once installed, centrifugal compressor trains are expected to run continuously and without relying on "standby" or redundant equipment for a minimum of three years. However, there are major ethylene plants that now run some of their centrifugal compressors in clean gas service for as many as eight years without interruption.

6.2 TYPES OF PERFORMANCE TESTS

There are two types of performance tests: an equivalent test run on a different gas and possibly at a different speed, and a full-load test at close-to-site conditions. The need for full-load testing arose with high-pressure applications that prevail in the fertilizer industry, in pipeline boosters, crude oil production gas-reinjection compressors, underwater compressor modules, multicomponent and tandem casing trains, and in situations in which drivers such as steam turbines or gas turbines originated from sources other than the compressor manufacturer.

Plant managers must realize the importance and long-term value of compressor testing when designating the team that represents the equipment owner. At least some members of the team should be conversant and ultimately accountable for the correct interpretation of test results. Some of these individuals should be assigned to the project site during commissioning and subsequent running. It is simply not acceptable that personnel involved in the selection and testing of major machinery are no longer associated with the equipment after it arrives

Figure 6-1. Modern compressor test stand. Note the barrel compressor driven by an electric motor, through a speed-increasing gear. (Source: Babcock-Borsig, Berlin, Germany.)

at the site. It must be reemphasized that shop testing is especially important in that it can lead to timely correction of deficiencies observed during testing. Responsible personnel and an experienced work force are available at the factory, whereas the installation site might have to depend on ill-equipped local service shops and relatively unskilled workers. With increasing throughputs, the physical dimensions of compressors increase and demands are made for ever-greater reliability.

Full-load tests actually go beyond verifying the correct functioning of the compressor. The following must be kept in mind when collecting data:

- For variable speed applications the user–owner needs performance maps that include surge points over the full range of operating speeds, typically from 75%–105% of the rated speed range.

- Vibration excursions caused by aerodynamic excitation are of interest.
- Actual compressor performance must be compared with the predicted (as originally proposed) thermodynamic performance.
- Axial thrust caused by operation at off-design conditions should be determined during compressor testing.

In some cases, special test stands may have to be devised in view of the ultimate location of the machine. Some centrifugal compressors might be destined for remote or inaccessible locations. For subsea installations, it may be necessary to develop special skids that allow performance testing with the entire machine immersed in a pressurized water tank. Some offshore compressor modules deserve to be tested with all equipment assembled at site on a floating facility if, say, the manufacturer's plant is close to the sea.

6.3 THE ASME POWER TEST CODE (ASME PTC 10)

The ASME PTC 10 [20] is an extensive document dealing with the testing of centrifugal axial compressors and exhausters. Two types of tests are specified in this document. The first of the two types is labeled Type I and covers situations in which "job gas" and test gas are identical within certain prespecified limitations. Thus, in summary, one runs a near-perfect, full-load and full-speed test. Air compressors represent the best example for which Type I testing is appropriate.

Although it would be ideal to test compressors with the job gas, there are many applications in which this is simply not a viable option. In such instances, there are other possibilities; these include setting up a closed piping loop filled with a gas mixture that closely approximates the job gas. This Type II testing will permit simulation of high-pressure applications at close to anticipated site machine operating speeds and can thus be expected to give some information about the likely behavior of the machine. Included here might be data indicating how the machine will behave at and near the surge point. However, Type I tests may best indicate certain aspects of aerodynamic excitation. Aerodynamic excitations are of serious concern in high-pressure machines from even the best-known compressor manufacturers. The effect of seals on the dynamic behavior is also seen during Type I tests.

Refer to Table 6-1 for a general overview of these tests. Relevant parameters and allowable deviations are given in this table for Type I and Type II tests.

The majority of compressors in petrochemical and refinery applications require Type II testing. The parties may

Table 6-1. Parameters of interest in compressor testing

Permissible deviation from specified operating conditions for Type I tests		Permissible deviation from specified operating parameters for Type I and Type II tests			Permissible fluctuations of test readings For Type I and Type II tests	
			Limit of test values as percentage of design values			
Variable	Permissible deviation	Parameter	Min	Max	Measurement	Fluctuation
1. Inlet pressure	5%	Specific volume ratio	95	105	Inlet pressure	2%
2. Inlet temperature	8%	Flow coefficient	96	104	Inlet temperature	0.5%
3. Speed	2%	Machine Mach no. Centrifugal Axial	Refer to PTC 10 Fig. 3.3 Fig. 3.4		Discharge pressure	2%
4. Molecular weight	2%	Machine Reynolds no.	Refer to PTC 10, Fig. 3.5		Nozzle differential pressure	2%
5. Cooling temperature difference	5%	Axial compressor machine Reynolds no. at specified conditions is below 100,000	90	105	Nozzle temperature	0.5%
6. Coolant flow rate	3%	Axial compressor machine Reynolds no. at specified conditions is above 100,000	10	200	Speed	0.5%
7. Capacity	4%				Torque	1%
8.					Electric motor input	1%
9.					Mol. weight	0.25%
10.					Cooling temperature difference	0.5%

agree to either use the job gas or some other commercially available gas that is safe to handle in the factory or test stand environment. Here, the test speed can differ from the anticipated site speed. Set rules pertain for arriving at the test speed; they are based on the proportionality of velocity triangles at the inlet and outlet of the machine. During the test, the operating parameters of the test will have to be accurately maintained to validate and accurately predict the ultimate site performance. Interpretation of test results will determine what, if any, modifications are required to achieve contract performance without exceeding maximum continuous speed (MCS). Prudent purchasers prenegotiate penalties for missed performance such as excessive utility consumption.

Table 6-1 shows that permissible deviations for compressor testing per Type I and II categories differ. It is important for owner and purchaser to agree on using the latest revisions of PTC-10 and to reverify the present applicability and accuracy of the clauses and values listed in this table.

In order to have laminar flow and the desired accuracy of measurement, PTC 10 has outlined straight pipe runs upstream and downstream of potential obstructions such as valves, elbows, reducers, and similar hardware that might cause abrupt changes in gas flow in the test loop. Any flow disturbances could impair the accuracy of instrument readings. Also, the flow orifices should be certified for accuracy before performing the actual testing.

The compressor manufacturer should be asked to use the job lube-and-seal-oil system, giving assurance that these critical auxiliaries function as specified. While all compressor manufacturers are equipped to conduct mechanical running and performance testing, they may not be able to offer string tests using the job driver. Some manufacturers may therefore use a variable frequency drive motor to accommodate the different speeds needed for compressor mechanical and equivalent performance testing. If the compressor manufacturer lacks the facilities or utilities required for string testing, the compressor may have to be shipped to the driver manufacturer's facility. Alternatively, early shipment to the site and well-defined on-site testing may have to be prearranged and contractually agreed to by the various parties. Needless to say, testing and associated responsibility issues must be recognized at the bid evaluation stage.

6.3.1 Future Value of Performance Testing

Once in service and over time, performance deterioration might be encountered either because of wear or changing inlet conditions. Among the many possible causes are malfunctioning demister pads, collapse of unsuitable (assumed to be "temporary" but inadvertently left in place) suction screens, or deficient level switches that permit liquid carry-over into the compressor. If site performance tests can be arranged during later operation, they can assist in defining the extent of performance degradation. In fouling services, performance data could be used to trigger liquid injection (on-stream washing) whenever the economics favor such action. Also, if the plant is to be "debottlenecked" or expanded, compressor performance data will be of great value. Unless thought is given to this issue at the inception of the *original* specification and procurement phases, standard instrumentation may be in place. This could impose major limitations on the accuracy of results. Accordingly, piping and instrumentation deserve considerable attention during the detailed engineering phase of any major project. In any event, reliability-focused organizations have for many decades seen the justification and need for periodic on-stream testing of centrifugal compressors.

The future value of having accurate test data can also be appreciated when projecting ahead to probable "compressor revamps" that are likely to occur. Obviously, to accommodate these potential plant throughput increases, compressor performance demands may change. Upgrading may be feasible and has routinely been engineered in many facilities, including fluid catalytic cracking (FCC) units, steam crackers, and blast furnace installations that had initially employed old, conservatively designed blowers. Whenever possible, the compressor casing in an existing installation will be retained and only the compressor internals are slated for replacement. A rule of thumb calls for changing only the internals if the cost of retrofit parts does not exceed 50% of the purchase price of new equipment. The advantages of not replacing the entire compressor system are rather obvious; there will be only minimum changes made to external components such as piping, foundation, and baseplate. Also, the time required for revamp implementation will be less than the time needed to obtain delivery of a new machine and the overall work load will be reduced.

Aerodynamic efficiencies may have improved in the newer designs due to reductions in parasitic losses and the availability of dry gas seals. Compressor upgrading may thus be achieved without replacing the driver. If driver output power can be increased by modifying the firing temperatures in gas turbines, or using different steam temperatures and pressures in steam turbines, it may also be possible to uprate compressor throughput by speed changes. In many instances, speed changes can be accommodated through ratio changes in replacement gearing that might fit in the existing gearbox. Together with selective replacement of some but not all compressor stages, these options will generally reduce needed investment.

But there is a catch: To determine which options to pursue, one needs reliable test data. A small preinvestment in testing as part of the initial purchase price almost always pays off handsomely.

Chapter 7

Application Considerations

7.1 APPLICATIONS AND INDUSTRY SECTORS

This chapter is intended to highlight the indispensable and important role played by centrifugal compressors in all sectors of industry. The illustrations and capacity statistics given here are general and indicative. They may vary from plant to plant, depending on the unit capacity and process licensor.

The nominal capacities of multistage centrifugal compressors are shown in Fig. 7-1. A few interesting high-pressure applications are depicted in Fig. 7-2 for six important processes.

7.2 AIR SEPARATION PLANTS AND COMPRESSORS FOR INSTRUMENT AND PLANT AIR SERVICES

Integrally geared ("packaged") compressors driven by a bull gear that transfers power to sets of pinions are one of several successful turbomachinery innovations introduced in the late 1960s. These machines have become quite popular in air separation plants producing industrial nitrogen and oxygen. Packaged compressors feature improved efficiencies and the feasibility of custom-designed stages. They inevitably incorporate intercoolers between stages, can operate at speeds between 30,000 and 50,000 rpm—in turn allowing considerable weight reductions—and make possible speed-optimized, individual stages because each driven pinion can be designed for a different speed ratio. The high-pressure, integrally geared centrifugal compressor in Fig. 7-3 is operating in nitrogen service. It is considered small to midsize, by 2006 standards.

Since, in machines such as the one shown in Fig. 7-4, a common bull gear drives up to six pinions, twelve compression stages can be accommodated. The overall layout is compact and saves space. Reliable restrictive carbon ring seal designs are available, as are variable inlet guide vanes and sophisticated control systems that take into account ambient variations. Packaged air compressors can be designed for oil-free air, long-term uninterrupted operation between major maintenance, and utilization of two-pole motors. All of these, plus perhaps cost and other factors, made them an attractive compressor design. Today, major process plants typically use packaged high-speed compressors for plant and instrument air service.*

Power savings are also possible by fine-tuning intercooler applications and/or utilizing high-speed compressors as multiservice composite machines. In multiservice composite centrifugal compressors, each compressor casing is a compressor stage capable of handling a different process gas. A common driver is connected to the large input gear ("bull gear"). Occasionally, there is even an expander wheel as part of the associated energy recovery system.

Similarly and in multiservice reciprocating compressors, each individual cylinder or stage may handle different process gases with a common driver, usually a multipole electric motor. Air separation plants, plant/instrument air service, and nitric acid plants have used both types of these combination machines.

7.3 NATURAL GAS PROCESSING INDUSTRY

Refrigeration applications predominate in the natural gas processing industry. There are two types of plants. Natural

*Positive-displacement machines are used only for relatively low to moderate capacity needs. However, neither they nor the moderate- to high-capacity helical screw compressors discussed and illustrated earlier in Chapter 3 should be ruled out without further consideration.

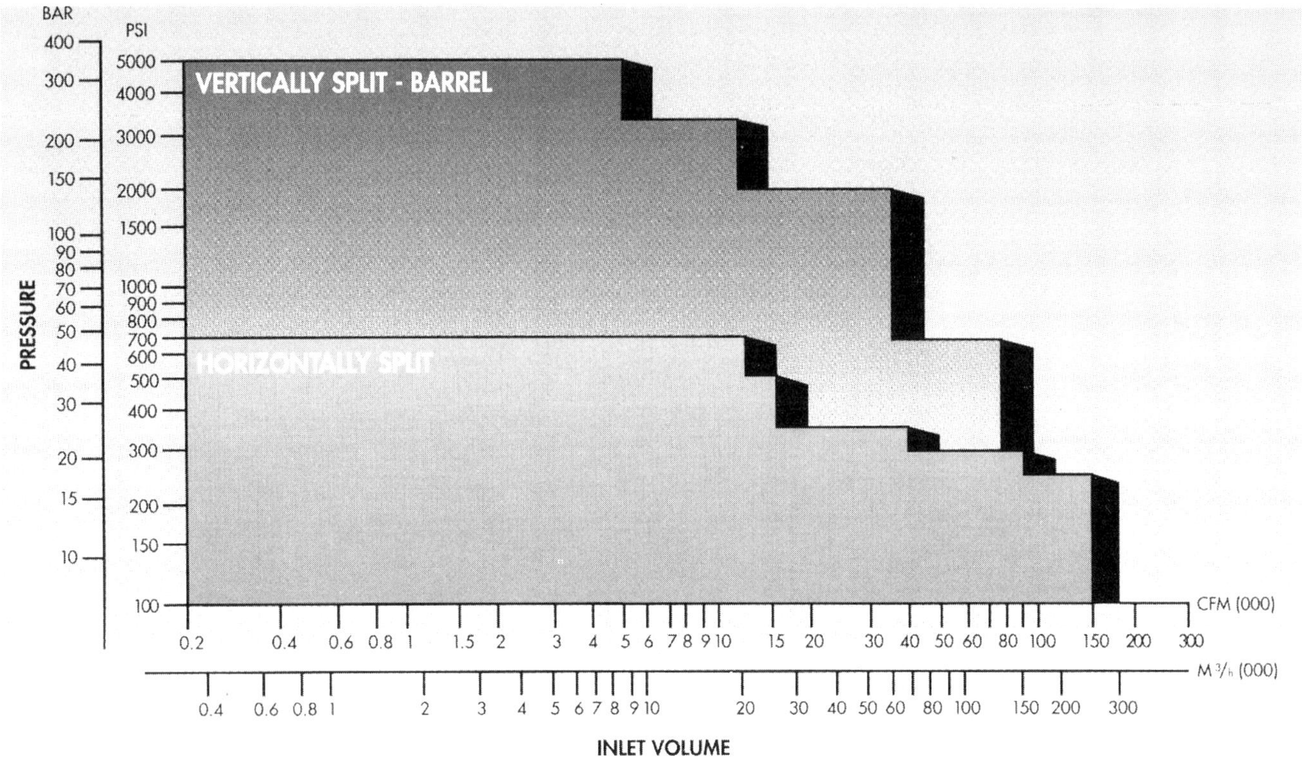

Figure 7-1. Nominal capacities of multistage centrifugal compressors. (Source: A-C Compressor Corporation, Appleton, Wisconsen.)

gas liquids (NGL) plants produce liquids such as propane and butane. Propane and butane are stored in low-pressure tanks at temperatures of –45°C (–49°F) and –7°C (~ –19°F), respectively. The various services involve process refrigeration using propane gas. The gas is liquefied in condensers and is then used for chilling the butane and propane from depropanizer and debutanizer columns. Separate loops are used because the temperature requirements for cooling the respective streams are different. Aside from the main inlet that enters the compressor at below atmospheric pressure condition, there will be a side stream at intermediate pressure. The main propane compressor typically operates at a discharge pressure of about 15 bar (218 psi).

Plants processing liquefied natural gas (LNG) have different refrigeration requirements. Temperatures of –161°C (–258°F) are needed for the liquefaction of raw natural gas drawn from gas fields at a depth of 2,000–2,500 meters. The vastly reduced volume of liquefied gas makes it possible to economically transport the fuel to global destinations.

A multistage mixed refrigerant cycle (MRC) is used to lower temperatures to these extreme values. The refrigerant gas is made up of a blend of nitrogen and hydrocarbons, consisting of methane to pentane. With an approximate composition of 4% N_2, 40% CH_4, 38% C_2H_6, and the remainder C_3H_8, the mixture is capable of evaporating over a wide range of temperatures. Representative discharge pressures of low-, medium-, and high-pressure compressor trains handling MRC are 8.5 (123), 20 (290), and 48 (696) bar (psi). Propane refrigeration is used for the purpose of precooling. The mixed refrigerant refrigeration system is staged in a cascade arrangement. Efficient plate-fin heat exchangers are used in the system. The low- and medium-pressure compressor casings are horizontally split; a vertically split casing is used for the high-pressure compressor stage.

With gas readily available, it makes economic sense to use gas turbine drivers for the LNG trains. These compression services are well suited for dry gas seals and these seals should be supplied by the compressor manufacturer. Today's mega plants often produce between two and six million tons of LNG per train. Multiple trains favor standardization of the modules, which, in turn, assists in phased and well-managed shutdowns for routine equipment maintenance.

Another interesting on-shore application concerns natural gas storage in subterranean caverns at 150 bar (~2,180 psi) and utilizing variable frequency drives. Many of these now employ dry gas seals and magnetic bearings, making them totally oil-free machines.

Finally, there are pipeline and pipeline booster compressors (Figs. 7-5 and 7-6) for services transporting natural gas across entire continents. Their casings are often constructed with suction and discharge nozzles located opposite each other to allow "in-line" connection between pipeline flanges (Fig. 7-7). Pipeline boosters are usually driven by natural gas fueled, two-shaft, variable speed gas turbines. Dry gas seals are routine for these machines and magnetic bearings are not

1 Operating range of high-pressure centrifugal compressors

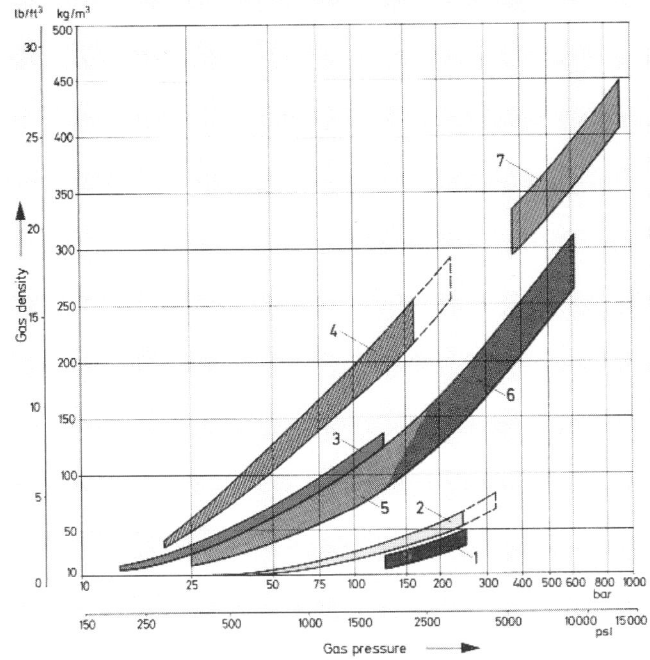

2 Typical range of the intake volume and mass flows for HP centrifugal impellers

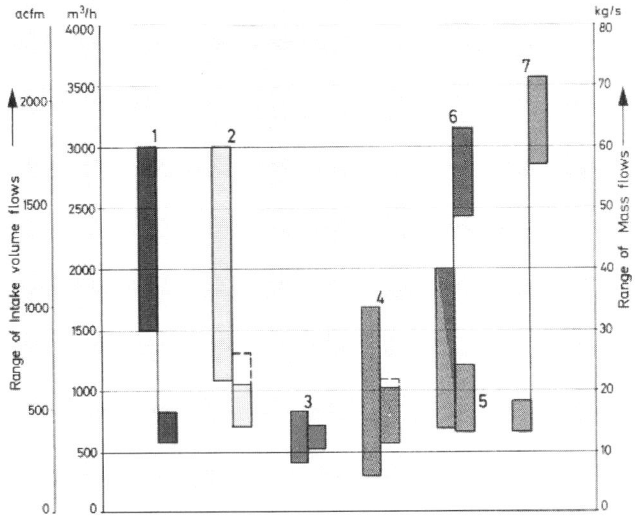

3 Pressure rise and specific compression work per stage for HP centrifugal compressors

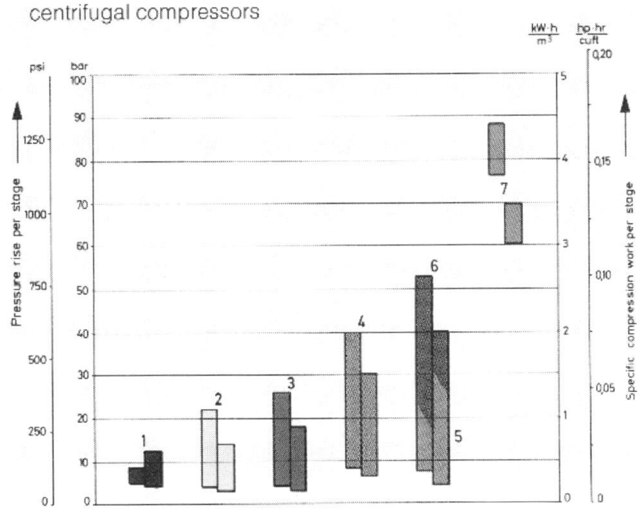

Process	Medium	Mol. weight	Max. pressure rise per casing [bar]
1 Hydrocracking	Hydrogen-rich gas	4...7	50
2 Ammonia synthesis	Syngas	8.5...9	110
3 Partial oxidation	Oxygen	32	85
4 Urea synthesis	Carbon dioxide	42...44	140
5 Gaslift	C_nH_m Mixture	20...28	90
6 Reinjection	C_nH_m Mixture	21...23	250 (400)*
7 Demag HP compressor	Nitrogen	28	500

*with interstage cooling

Figure 7-2. High-pressure applications for centrifugal compressors. (Source: Mannesmann Demag Delaval, Duisburg, Germany.)

unusual. Forged steel compressor casing and assembled packages with lube and seal-system oil systems are often furnished as packages, or modules. Delivery pressures in the range of 100 bar (1,450 psi) are quite common for pipeline boosters.

7.4 OFFSHORE INDUSTRY

Gas compressors for offshore services are found in a variety of applications. On oil production platforms, compressor modules are used for gas gathering, gas transportation through gas boosting, gas lift, and reinjection services. Their size and complexity depend on the design of the offshore platform and certain limitations of onshore processing, as well as overall economics and life of the producing field. Gas transportation involves compressor discharge pressures in the range of 2,000 psi, or roughly 140 bar. Reinjecting natural gas under the seabed is one of several measures used to boost oil production. The reinjected gas exerts pressure on the crude oil and forces it to flow into oil producing wells.

Gas reinjection services create challenging situations for compressor designers. Many experienced manufacturers had

Figure 7-3. High-pressure geared centrifugal compressor Type HVK for nitrogen. $\dot{V}$ = 8700 m³/h, p_2 = 35 bar, P = 5100 kW. (Source: Mannesmann Demag Delaval, Duisburg, Germany.)

to engineer extensive modifications to ensure sound long-term mechanical performance of their compressors. Modern geological, seismic, and satellite techniques have led to the discovery of reserves at depths of 2,000 meters and below. The resulting pressure requirements can exceed 400 and in some instances even 700 bar—over 10,000 psi. Barrel casings are used, and fabrication techniques for the thick-walled casings become critically important, as do seal design and rotor dynamics. The design must eliminate the possibility of subsynchronous vibration and its absence must be proven not only on the test stand, but also after field commissioning. An offshore module weighing approximately 1,300 tons is shown in Fig. 7-8. These barrel compressors are driven by aeroderivative gas turbines with an ISO rating of 14,780 kW. They were completed in 1981 and are evidence of mature technology.

7.5 REFINING INDUSTRY

Modern refinery processes can be quite complex. They can also be corrosive and, unless given proper attention, rather maintenance-intensive. Sour crude is a product with a sulfur content ranging from 0.5 to 1%, or 0.05 cubic ft. of H_2S per 100 gallons of crude. Aside from typically producing gasoline, jet fuel, and kerosene, modern refineries often supply a considerable slate of higher-quality and higher-profit products. Certain production decisions are driven by the need to meet waste disposal and environmental obligations. The production of hydrogen and the use of fluid catalytic cracking (FCC) and hydrocracking techniques result from judiciously weighing all relevant considerations.

A typical modern refinery processing more than 400,000 BPD (barrels per day) of crude oil may have 25 centrifugal compressors with polytropic heads in the range of 20,000–75,000 ft (~6,100–22,860 m), average impeller tip speeds of perhaps 45,000–63,000 fpm (~230–320 m/s), pressures in the range of approximately two bar (in the case of blowers) to 190 bar (29–2,755 psi) in the case of barrel compressors. A large driver may deliver 50,000 kW, although most drivers are perhaps half that size. The majority of refinery compressors are driven by special-purpose steam turbines. Their duties include recycle service in various units, process air, tail gas, hydrogen, flare gas, and wet gas com-

Figure 7-4. Multistage, direct-driven process gas compressor derived from packaged air compressor concept. (Source: Mannesmann Demag Delaval, Duisburg, Germany.)

pression with an impressive FCC train comprising a hot gas expander, an axial compressor, and an induction motor that can also serve as an electric power generator.

A typical refinery compressor platform is shown in Fig. 7-9. Outdoor installations are common in most countries with moderate temperature environments. Not having compressors located in closed buildings tends to lower explosion risk and is good for overall safety.

7.6 PETROCHEMICALS AND FERTILIZERS

Petrochemical plants with steam cracker units generally produce ethylene, propylene, benzene, and butadiene. These units are sometimes followed by downstream units that produce ethylene glycol, polypropylene, polybutadiene rubber, low-density polyethylene (LDPE), or high-density polyethylene (HDPE), among other products. Centrifugal compressors are also found in the fiber industry.

Typical ethylene plant capacities range from 80,000 to 870,000 metric tons per year. The field of application is expanding and so are plant capacities and compressor sizes. There are tandem trains with steam turbines driving three compressor casings handling process gas, with final discharge pressures in the range of 40 bar (580 psi). Their power consumptions range from 6,000 kW in small and older units, to 30,000 kW and more in some of the more recent projects. The majority of the charge gas (cracked gas) compressor population is in the 15,000 kW range (see Fig. 7-10). As of 2006, combined axial/centrifugal cracked gas trains in the

Figure 7-5. Cooper-Bessemer RB 6B pipeline barrel compressor. (Source: Cooper-Bessemer, now Rolls Royce, Middlesex, UK, and Houston, Texas.)

100,000 kW league were being considered for some new projects.

Propylene compressors are generally found in the power range from 2,000 to 30,000 kW and discharge pressures up to 17 bar (247 psi), whereas ethylene compressors with discharge pressures around 20 bar (290 psi) are found in power ranges from 3,000 to 10,000 kW. Ethylene compressors in these plants are primarily driven by electric motors that are combined with separate gear-speed increasers to optimize compressor input speed.

There are also many packaged centrifugal units, notably for refrigeration services in the fiber sector. When project engineers deal with refrigeration packages for these plants, their approach to specification, procurement, and testing may differ from the one discussed earlier in this text. Their emphasis may not necessarily be confined to the compressor, but the focus may shift to refrigeration load capability. Here, condenser and chiller are part of the package and their performance affects both temperature and overall capacity.

Large centrifugal compressors are of immense value to the nitrogenous fertilizer industry. The feedstock for fertilizer units can be fuel oil, naphtha, and natural gas. In recent years a number of developing nations have seen phenomenal growth in natural-gas-based fertilizers, including urea and ammonia. Typical urea plant capacities are 1,225, 1,550, 2,000, and 3,250 metric tons per day; ammonia plants produce 900, 1,500, and 2,050 metric tons per day. With ever-increasing demand for grain and cereals, it is reasonable to assume further expansion of this industry.

Major centrifugal compressors connected with fertilizer plants are steam-turbine-driven natural gas compressors, air and nitrogen compressors, synthesis gas compressors, and

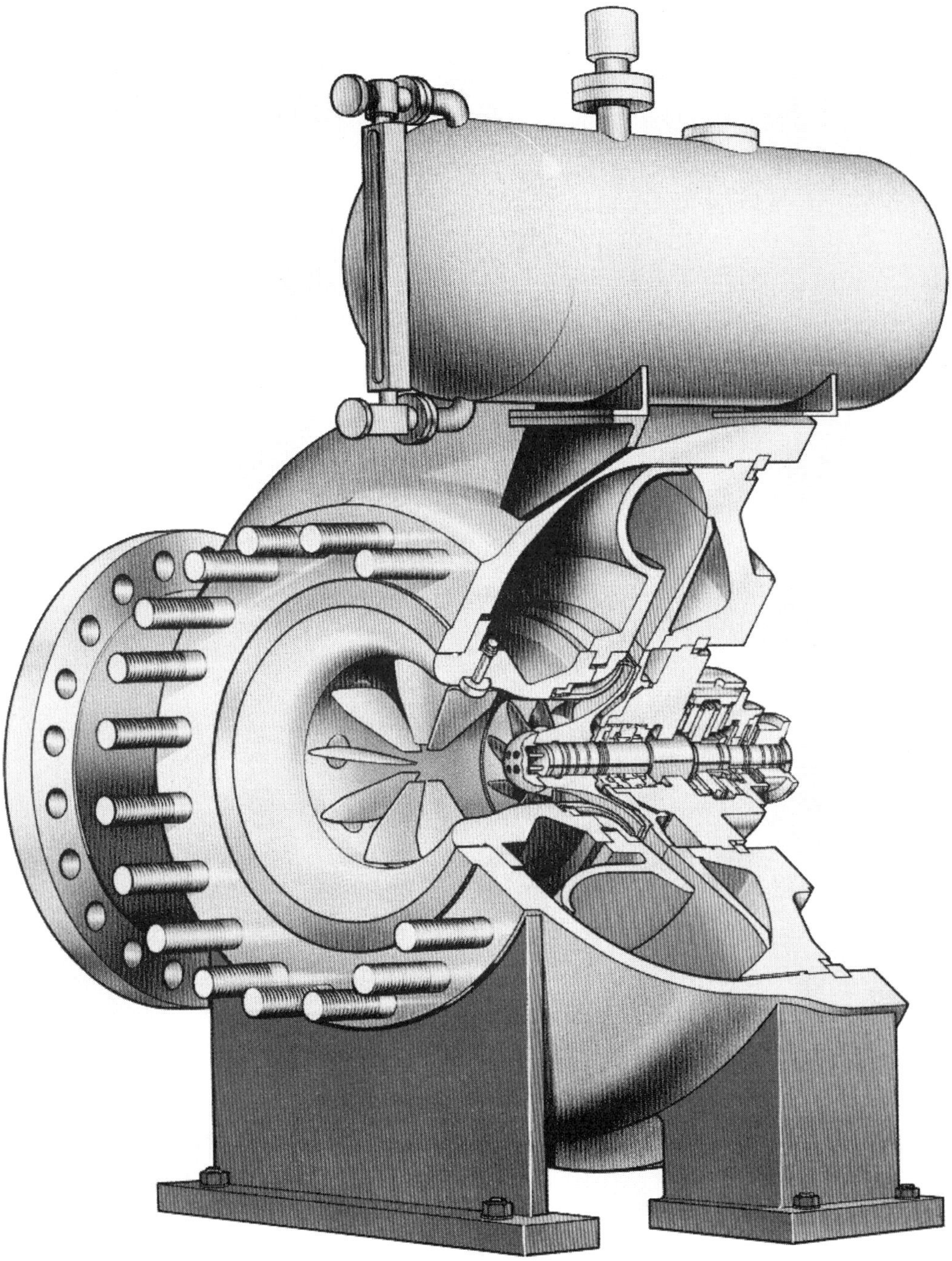

Figure 7-6. Cooper-Bessemer RFA-36 pipeline centrifugal compressor. (Source: Cooper-Bessemer, now Rolls Royce, Middlesex, UK and Houston, Texas.)

carbon dioxide compressors. Where there is an abundance of natural gas, industry has for years used gas turbines. However, and more recently, the generally favorable maintenance experience for steam turbines compared with gas turbines seems to favor steam turbines in these fertilizer plant applications. Since the various compressors are major consumers of power, high efficiency is valued at a premium. For example, typical synthesis gas compressors cover such operating conditions as suction pressures in the range of 20 to 30 bar (290–435 psi), discharge pressures in the high-pressure range around 270–342 bar (3,915–4,959 psi), in the medium-pressure range of 225–275 bar (3,262–3,987 psi), and in the lower-pressure range of 130–200 bar (1,885-2,900 psi). Typical flow capacities range from 65,000–250,000 Nm^3/hr, with a large segment of the population in the range of 110,000–150,000 Nm^3/hr. Driver ratings of each train are in the vicinity of 10,000–28,000 kW and vertically split ("barrel") compressors are used in these high-pressure applica-

Figure 7-7. Pipeline compressor during manufacture at Nuovo Pignone, Florence, Italy. Note in-line compressor nozzles.

Figure 7-8. An offshore module weighing 1,300 tons, completed in 1981. (Source: Sulzer Turbo, Winterthur, Switzerland.)

Figure 7-9. Typical outdoor compressor installation in a refinery in Australia. (Source: Babcock-Borsig, Berlin, Germany.)

Figure 7-10. Cracked-gas compressor train in a 300,000 ton/year ethylene plant. (Source: Nuovo Pignone, Florence, Italy.)

tions. Figure 7-11 shows an ammonia-synthesis compressor train.

CO_2 compressors in urea plants have typical suction pressures of one bar, discharge pressures (low-pressure range) of 17 bar (246 psi), and 143–215 bar (2,073–3,118 psi) in the high-pressure range. Capacities range from perhaps 6,500 to 30,000 Nm³/hr, with many machines in the range of 25,000 Nm³/hr. Typical driver ratings for each train are in the league of 1,000–8,000 kW. In CO_2 service, tandem compressors are customary. A horizontally split, low-suction pressure first casing is generally connected to the higher stage barrel configuration.

7.7 STEEL INDUSTRY

More than ever before, the steel industry is striving to achieve efficient operation while complying with environmental regulations. Blast furnace and coke oven gas, the formerly discarded byproducts of the steel making process, are

Figure 7-11. Ammonia synthesis compressor train. (Source: Nuovo Pignone, Florence, Italy.)

Figure 7-12. Axial compressor for a Brazilian steel mill. This compressor delivers 500,000 Nm^3 of air. A disengagement clutch can be seen in the foreground. (Source: Mitsubishi Heavy Industries, Tokyo and Hiroshima, Japan.)

now used as fuel for gas turbines driving generators in combined cycle power plants. Fly ash and other impurities are removed in scrubbers and filters before the gases are compressed to the higher pressures demanded by the gas turbine fuel supply system. A typical fuel gas compressor may use horizontally split casings with a pressure ratio in the range of 21:1 and efficiencies around 80–82% for centrifugal and 88–90% for axial compressors. An axial air machine destined for a steel mill is shown in Fig. 7-12. The machines are generally furnished with dry gas seals and the trains used for steel industry gases can demand power inputs as high as 100 MW. Components for these compressors are generally chosen from the manufacturer's standard range of products.

7.8 SPECIAL APPLICATIONS

Some compressor applications deserve special consideration. Oxygen compressors are one such example, and discharge pressures in excess of 70 bar (1,015 psi) are not unusual for certain recent plant designs. Here, the safety of an installation is of paramount importance and fire risk must be minimized or eliminated. The gas cannot be allowed to become contaminated and frictional heat generation from vibratory excursions must be ruled out. Preventing frictional heat due to high vibration requires ascertaining rotor stability and providing appropriate electrical safety, that is, grounding. Reliably grounding compressor casings will prevent the accumulation of static electricity and its potentially disastrous effects. Manufacturers often opt for Monel™ labyrinth designs with silver-coated copper inserts to rapidly dissipate heat and to provide a nonsparking environment. Oxygen compressors are kept under positive pressure in an inert nitrogen atmosphere when not in actual use.

Chlorine gas with its high molecular weight of 70 also qualifies as a special application. Selecting chlorine compressors requires close design reviews and special care. This poisonous gas must be clean and dry before entering the compressor. Maintaining dryness avoids potentially serious problems due to fouling of the process. Maximum discharge temperatures must be limited to 250°–300°F (121°–149°C) so as to limit corrosive effects.

Chapter 8

Using API-617 as a Purchase Specification for Centrifugal Compressors

Although they are not legal documents, the various standards compiled and issued by the American Petroleum Institute (API) reflect the experience and recommendations of both users' and manufacturers' groups. These groups wish to impart safety and reliability to equipment utilized in the hydrocarbon processing industry (HPI). Of the two most important standards used in procuring dynamic compression machinery, API-617 pertains to centrifugal compressors and API-614 applies to lubricating-oil systems. Virtually all API standards require users to make decisions on clauses marked with bullets (•), and understanding the standards is important for all parties. Therefore, highlighting a few clauses with explanatory remarks will be helpful in using the 7th Ed. (2002) API-617. (It should be noted that the numeric designation of a certain clause may change as the API Standard undergoes progressive updating or revisions.)

Clause(s)	Topic	API Explanation	Remarks
1.5.18 through 1.5.26	Definition of normal speed, 100% and maximum continuous speed, plus others of interest	The speed needed to attain the highest head at any operating condition. If the speed at the normal point is capable of meeting all the requirements at the other operating conditions, normal speed will be taken as 100% speed.	These definitions help to flush out any anomalies in deciding the shop performance test parameters. "MCS" or maximum continuous speed is 105% of the highest speed required to meet any of the specified operating conditions. This sets the values for the mechanical performance test and the over-speed test. It is important from the point of view of inspection and NDT examination.
1.5.25	Normal operating point	The point at which usual operation is expected and optimum efficiency is desired	
	Rated point	Obtained by the intersection of 100% speed line and the highest capacity during operation	Important to identify before ordering or before approval of inspection documents. Definition of guarantee point is required.
1.5.30	Stability	Stability is the difference in capacity between rated capacity and surge point at rated speed.	Important criterion for the design of the antisurge system and for defining operating flexibility
1.5.51	Turndown	Turndown is the percentage of change in capacity between the rated capacity and surge point capacity at rated head with unit operating at rated suction temperature and gas composition.	The amount of recycle volume and thus loss of power at low-flow operating conditions will be an important factor when it comes to a continuous mode.

Clause(s)	Topic	API Explanation	Remarks
1.5.39	Settling-out pressure	Pressure of compressor system when the machine is shut down	It is important to state the same when handling gas that has substantial change in the vapor pressure for a small change in the temperature.
1.6.1	Referenced publications	Identifies major standards covering driver, pressure vessels, instrumentation, electrical equipment, auxiliary equipment, auxiliary systems, piping, and material specifications	Look for completeness and keep deviations from applicable standards to a bare minimum. Apply piping standards without compromising good layout; access; pipe supports; upstream, interstage, and downstream piping; drains; platform; and grating. Be especially vigilant in instances in which the compressor is delivered as a "standard" package. On many occasions, comments start pouring in after contract award. Note that similar considerations should apply to instrument tubing and piping. Note cabling, junction boxes, and fire protection systems for which the user company may have developed an overall strategy based on decades of field experience. These strategies may be more comprehensive than those of the compressor manufacturer.
2.1.1.3	Acceptance criteria	No negative tolerance on the capacity and head at normal point of operation with +4% tolerance on power consumption	The process industry places emphasis on throughput with lower priority on power consumption. However, the user should identify the guarantee point at the time of purchase if performance tests are contemplated. This will avoid disputes while acceptance tests are in progress.
1.5.17 and 2.3.1.1.1; also 2.3.1.2	Maximum allowable working pressure	At least equal to relief valve setting, or at least 1.25 times the maximum discharge pressure (gauge) experienced during operation. Maximum pressure occurs when the machine operates at maximum continuous speed close to surge with anticipated combination of highest molecular weight, highest suction pressure, and lowest inlet temperature.	The manufacturer should be requested to state this, with no relationship to the actual service condition. This will be useful if future conditions should require more head. Heavier casing also guarantees less susceptibility to changes due to piping loads and its effect on machine alignment. Observe adequacy for settling-out pressure.
2.1.1.1	Head capacity characteristic curve	Rising characteristics up to surge point; may be defined in percentages	Desirable from the point of view of antisurge and general flow control since there is change in head for a change in the capacity over the entire operating range.
2.3.1.5	Axially split casing joint	Metal-to-metal joint	This is an important point to be remembered by maintenance organizations. A predefined sequence of tightening bolts, preferably with a hydraulic torque device, will avoid costly assembly errors in the future.
	Diaphragm cooling	Independent cooling passages	Precludes entry of cooling medium into compressor casing

Clause(s)	Topic	API Explanation	Remarks
2.5.3.1	Raw material for the shaft	For 8″ and above, the starting point shall be a forging. Although even below this limit it is preferred to use a forging, experienced manufacturers may be permitted to start with quality-controlled bar stock.	Heat-treated forgings are the preferred raw material from the point of view of strength and stress-concentration factors.
2.5.10.6	Balancing	Use of welding as a means to balance an impeller is not permitted.	Welding can introduce flaws, in addition to difficulties in controlling the mass. Maintenance engineers must strictly observe this rule.
2.5.4.1	Balance chamber pressure	Provision is required	Found missing in many installations. Trending this pressure gives a good indication of casing internal leakage.
2.7.1.3	Inlet oil temperature	If it exceeds 49°C, special considerations are required.	In desert atmospheres and with fin fan coolers, oil header temperatures range in the vicinity of 55°C and bearing metal temperatures above 115°C (summer conditions). Consider a closed or thermosiphon water cooling system.
2.6.2.10	Separation margin	The damped unbalanced response analysis shall indicate that the machine will meet the following SM: a. If the AF at a particular critical speed is less than 2.5, the response is considered critically damped and so no SM is required. b. If the AF at a particular speed is 2.5 or greater and that critical speed is below the minimum speed, the SM (as a percentage of the minimum speed) shall not be less than the value from Equation below or the value of 16 whichever is less. $SM = 17\{1 - [1/(AF - 1.5)]\}$. c. If the AF at a particular critical speed is equal to 2.5 or greater and that critical speed is above the maximum continuous speed, the SM (as a percentage of the maximum continuous speed) shall not be less than the below or the value of 26 whichever is less. $SM = 10 + 17\{1 - [1/(AF - 1.5)]\}$	These paragraphs represent guidelines to obtain an acceptable rotor design. Also, it calls for the verification of the calculated results on the test stand.
2.6.3	Shop verification of unbalanced response analysis	Outlines the importance of verification of the theoretical results with the ones actually obtained on the test stand	The deviation in the results beyond permissible limits helps to form guidelines. It gives the designer more insight in correcting the assumptions and thus can improve overall prediction capability.
2.6.8.1 through 2.6.8.8	Balancing	Does not necessarily specify desirable high-speed balancing	During the mechanical running test of the machine, assembled with the balanced rotor, operating at its maximum continuous speed or at any other speed within the specified operating speed range, the peak-to-peak amplitude of the unfiltered vibration in any plane, measured on the shaft adjacent and relative to each radial

Clause(s)	Topic	API Explanation	Remarks
			bearing, shall not exceed the following value or 25 micrometer (1 mil), whichever is less: In SI units: $A = 25.4\ (12000/N)^{0.5}$ where A = amplitude of unfiltered vibration, in micron (mil) true peak-to-peak, N = maximum continuous speed, in rpm. At any speed greater than the maximum continuous speed, and up to and including the trip speed of the driver, the vibration level shall not increase more than 12.7 micron (0.5 mil) above the maximum value recorded at the maximum continuous speed.
2.2	Materials	Steel casing to be used as per guidelines clauses. Disallow materials susceptible to brittle fracture at ambient temperature. In case of H_2S ("sour") environment, ferrous material selection must be limited to those having maximum yield strength of 90,000 psi and RC22 maximum hardness. Austenitic stainless steel should not be used in services likely to experience stress-corrosion cracking. For hydrogen service with partial pressures in excess of 100 psi or over 90 molal percent H_2, impeller material yield strength shall not exceed 120,000 psi and hardness RC 34.	Review appropriate guidelines for material selection.
3.1.4, 3.15, 3.1.7.	Driver sizing	110% of the maximum power required at any operating condition, inclusive of coupling and gear losses for motors and steam turbines. Determination to be made by mutual agreement for gas turbine drivers.	For steam turbines, consider future upgrade requirements. Are two-sided shaft extensions possible? Is space available for a helper turbine?
3.2	Gear couplings		Lubricated gear couplings are not normally provided on new installations. Contoured diaphragm or flexible disc couplings are preferred but must have prior experience. On older machines with gear couplings, consider, replacement with contoured diaphragm couplings.
3.5.1.4	Provision of bypass for flushing	Applicable to bearings and seals	This aspect should be carefully reviewed. Flushing with minimum effort may be difficult unless suitable provisions are "designed-in."
4.3.3	Impeller overspeed	At 115% of MCS for a minimum duration of one minute	Check the impeller for deformation and dimensional variations. Are runaway situations possible in variable speed applications?
4.3.6	Mechanical running test	Important points to note: 1. Minimum degree of filtration 10 microns or better 2. Test with contract items to include	A mechanical running test is mandatory both for the main rotor and spare rotor. It assists in determining mechanical losses. Try to use lube-oil grade equivalent to type used at

		coupling, vibration transducers, seals, bearings 3. Observation and recording of several parameters at 10% speed increments up to MCS to include oil supply rate to every line, pressure, temperature, sour seal oil rate 4. Run the machine at trip speed for 15 minutes, followed by a four-hour run at MCS 5. Collection of vibration spectra at each operating point, also during start-up and coast-down 6. Identification of lateral critical speeds, separation margins, and any testing to be done in the context of unbalance response analysis	installation site and maintain highest permissible lube-oil inlet temperature during shop testing.
4.3.7	Leakage test	After completion of mechanical running test, casing along with seal will be pressurized to maximum sealing (or seal design pressure). Also, casing shall be pressurized to rated discharge pressure (with or without end seals).	Will ensure integrity of the casing joint. Inert gas (N_2) is recommended. Helium gas is recommended for low molecular weight gases.
4.3.8	Optional tests	Various tests ranging from performance test (4.3.8.1) to dismantling and reassembly (4.3.8.8). Inspection to be defined under this optional category	All tests should be carried out at the manufacturer's facility in order to minimize site problems. This is an opportunity for owners' representatives to become familiar with equipment.
		Full-pressure/full-load/ full-speed test (4.3.8.6)	Beneficial in high-pressure applications in which seals can act as additional bearings and thus can affect the critical speed.
4.4.3.10	Preparation for shipment	Spare rotor	Preserved with rust-preventive coating and stored vertically to avoid bowing. Preferably located in humidity-controlled location.

8.1 COMPRESSOR REVAMPS

To accommodate throughput increases, compressor performance demands may change. Upgrading may be feasible [21] and has routinely been engineered in many facilities, including fluid catalytic cracking (FCC) units, steam crackers, and blast furnace installations that had initially employed old, conservatively designed blowers. Whenever possible, the compressor casing in an existing installation will be retained and only the compressor internals slated for replacement. A rule of thumb calls for changing only the internals if the cost of retrofit parts does not exceed 50% of the purchase price of new equipment. The advantages of not replacing the entire compressor system are rather obvious; there will be only minimum changes made to external components such as piping, foundation, and baseplate. Also, the time required for revamp implementation will be less than the time needed to obtain delivery of a new machine, and the overall work load will be reduced.

Aerodynamic efficiencies may have improved in the newer designs due to reductions in parasitic losses and the availability of dry gas seals. Compressor upgrading may thus be achieved without replacing the driver. If driver output power can be increased by modifying the firing temperatures in gas turbines, or using different steam temperatures and pressures in steam turbines, it may also be possible to uprate compressor throughput by speed changes [22, 23, 24]. In many instances, speed changes can be accommodated through ratio changes in replacement gearing that might fit in the existing gearbox. Together with selective replacement of some but not all compressor stages, these options will generally reduce needed investment. Coupling, seal, and driver suitability may have to be addressed as well [3, 8, 23, 24, 25, 26, 27].

PART II

Part II of this book is divided into three segments, as follows.

SEGMENT 1. GAS PROCESSING AND TURBOEXPANDER APPLICATIONS

Turboexpanders are modern rotating machines that convert the pressure energy of a gas or vapor stream into mechanical work. Since the gas expands in the conversion process, expansion turbines resemble steam turbines [4, 28, 29] in basic operating principle. To make a distinction, however, the terms "turboexpander" and "expansion turbine" are not applied to steam turbines and combustion gas turbines.

If chilling the gas or vapor stream is the main objective, the mechanical work thereby produced is sometimes considered a by-product. If pressure reduction is the main objective, then heat recovery from the expanded gas is considered a desirable byproduct. Both cryogenic and hot gas turboexpanders are discussed in [29]. In either case, the primary reason for using turboexpanders is to conserve energy.

With thousands of turboexpanders serving gas processing plants all over the world, we chose to first highlight these modern machines in a number of illustrations and then indicate their use in modern gas processes by the letter "E" placed in a circle at the relevant locations in the various process schematics. It should be pointed out, however, that our coverage in this regard is probably far from exhaustive. There are other processes in which turboexpanders are being applied today. Also, progressively escalating energy costs make turboexpander use economically more attractive as time goes on.

Figures I-1 through I-12 represent the primary operating features and a few of the very many configurations serving industry.

A typical cross section for cryogenic turboexpanders with oil-lubricated bearings is shown in Fig. I-1. The overwhelming majority of these machines are then skid-mounted together with the required lube oil support systems (Fig. I-2) or with support hardware for expanders utilizing magnetic bearings (Fig. I-3).

However, although process gas expanded in an expander/compressor provides refrigeration for cryogenic separation of propylene and other hydrocarbons from off-gas, an expander can also be used to recover shaft power in the form of electrical energy by fuel-gas pressure reduction. One such typical system is shown in Fig. I-4.

Cryogenic expanders are often used to provide the last step of refrigeration for hydrocarbon separation processes. Here again, it is common to use an expander equipped with magnetic bearings as shown in Fig. I-5; the entire expander system was shown in Fig. I-3.

A close-up of the rotor assembly for a magnetic bearing expander–compressor can be seen in Fig. I-6. The compressor wheel is in the foreground. Other important features include fully variable expander inlet guide vanes for flow and/or pressure control (Fig. I-7), an integral gas expander–generator with its expander wheel shown in the overlay (Fig. I-8), and two expander-compressors mounted on a single skid for an ethylene plant process (Fig. I-9).

The typical labyrinth seals used in expander–compressors with bearing housings pressurized by the process gas are shown in Fig. I-10. Also, multistage turboexpanders (Figs. I-11 and I-12) are making inroads in modern industry. Figure I-11 shows a two-stage, integral gear expander–generator for a cryogenic process. In Fig. I-12, the housings are removed to show the expander wheels.

Expanders Identified on Process Flow Schematics

Expanders and compressors are identified on many of the process flow schematics in Part II, Segment I of our text by either the letters "E" (for expander), or "C" (for compressor) in circles. The reader will note from the associated description if the turboexpander drives a compressor or an electric generator.

Compressors and Modern Process Applications. By Heinz P. Bloch.

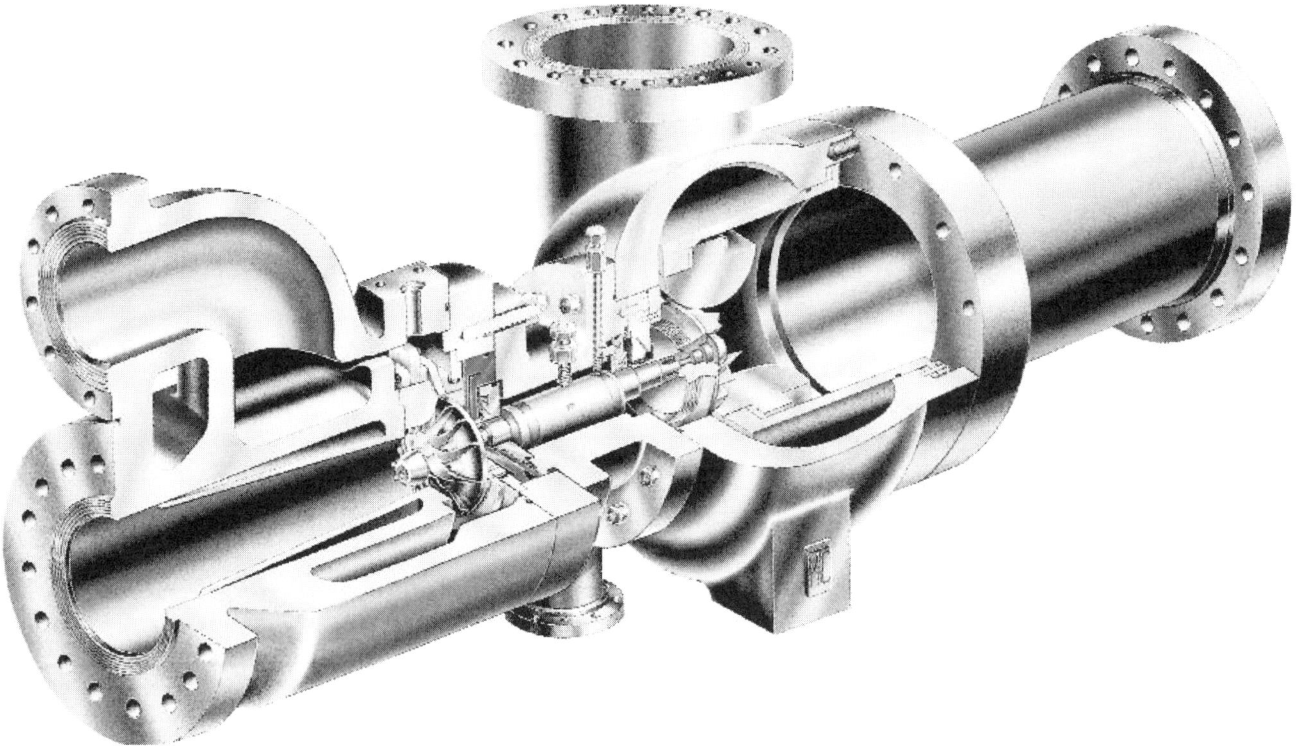

Figure I-1. Typical oil-lubricated cryogenic turboexpander. (Source: Mafi-Trench Company, Santa Maria, California.)

Figure I-2. Turboexpander package for a modem gas processing plant. (Source: Mafi-Trench Company, Santa Maria, California.)

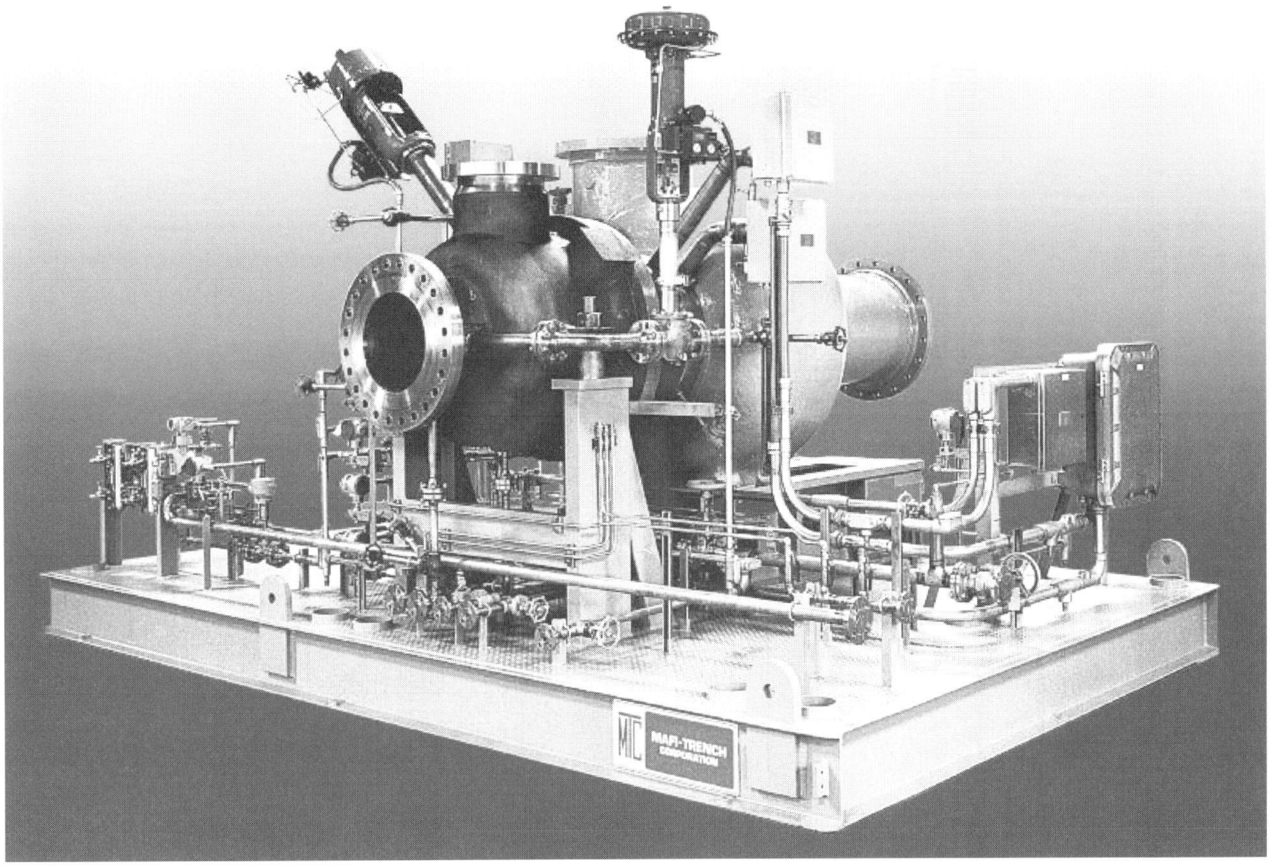

Figure I-3. Support hardware for expanders using magnetic bearings. (Source: Mafi-Trench Company, Santa Maria, California.)

Turboexpanders are often used in cryogenic air separation processes to produce high-purity nitrogen. Here, the expander usually drives a process compressor as shown in Fig. I-1.

A flashing-liquid expander (FLEX™) can be used to recover power from solvent pressure reduction. During expansion, significant amounts of vapor can be generated, which requires careful design consideration. Flashing-liquid expanders can also be used to increase refrigeration and therefore plant output. It should be noted that it is often possible to replace a throttling valve with a FLEX™ to produce additional refrigeration and liquid and thereby increase plant output.

If, as an example, an NGL or LNG process were being considered, the reader might find schematics labeled "NGL and LNG," or look it up in this segment under "Cryomax MRE" or "Cryo-Flex." Scanning the Process Index might also lead to "NGL Recover," or "LNG-Pro" to name but two.

Recall that technology does not stand still. Obviously, then, new process applications are being designed and developed even as we read this material. The reader must stay in contact with competent equipment vendors and design contractors to understand the ever-increasing cost justification for turboexpanders in modern process plants. The vendors and design contractors should be ready and willing to provide reference lists, technology updates, and application engineering support.

SEGMENTS 2 AND 3. COMPRESSORS TYPICALLY FOUND IN PETROLEUM REFINING AND OTHER PETROCHEMICAL PROCESSES

Our two main contributors, A-C Compressor Corporation (Appleton, Wisconsin) and Demag Delaval (now Mannesmann Demag Delaval, with plants in Trenton, New Jersey and Duisburg, Germany), correctly pointed out that there is substantial crossover between compressor models used in oil refining and those used in other petrochemical processing industries. This fact prompted us, in Tables II-1 through II-3, to first list the various processes and to then identify model coverage and/or arrangement. In the illustrations that follow these tables, we see the physical appearance and, in some cases, typical performance data relating to the various machines.

Figure I-4. Turboexpander connected to an electric power generator. (Source: Mafi-Trench Company, Santa Maria, California.)

Examining Table II-1

In Table II-1, A-C Compressor uses the type description "D" for the single-stage compressor illustrated in Fig. II-1. Their designation "DH" refers to a single-stage, high-pressure compressor (see Fig. II-2). "V/VS" machines are horizontally split and multistage. Two typical skid-mounted machine sets (or trains) together with electric motor drivers and speed-up gears are shown in Figs. II-3 and II-4.

The same company, A-C Compressor Corporation, manufactured the two-barrel compressors shown in Figs. II-5 and II-6. In the case of Fig. II-5, the machine is mounted on a base by itself, whereas, in Fig. II-6, the machine train shares a common base with the two driving elements.

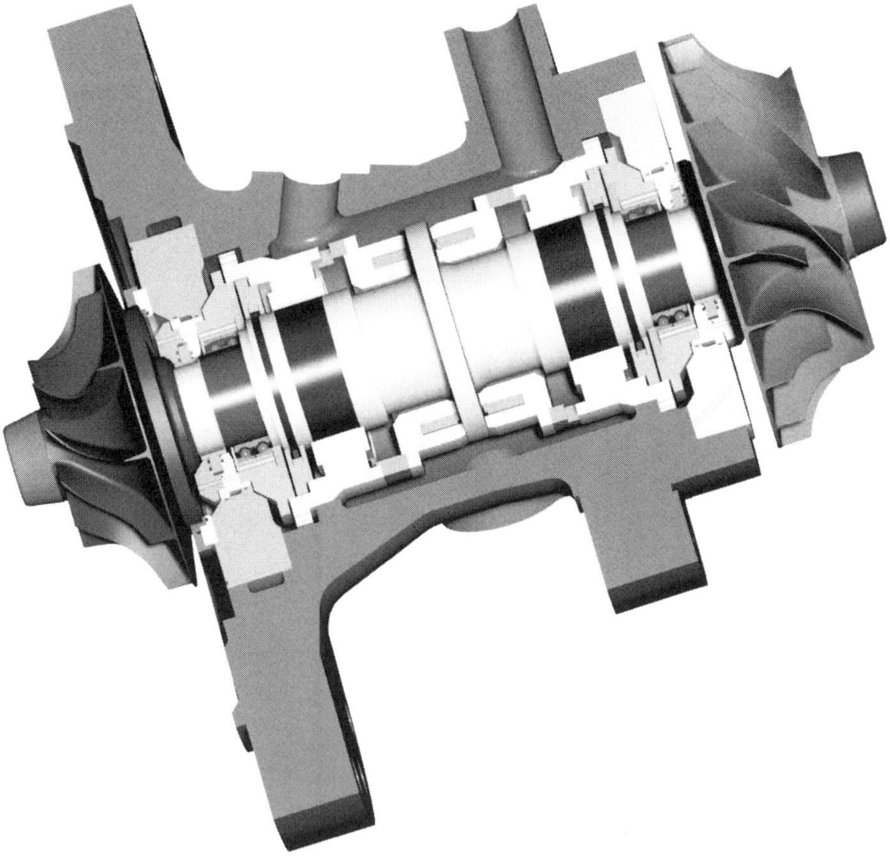

Figure I-5. Section view of turboexpander equipped with magnetic bearings. (Source: Mafi-Trench Company, Santa Maria, California.)

Figure I-6. Rotor assembly for a magnetic bearing expander/compressor with the compressor wheel in the foreground. (Source: Mafi-Trench Company, Santa Maria, California)

Figure I-7. Fully variable expander inlet guide vanes for flow/pressure control. (Source: Mafi-Trench Company, Santa Maria, California.)

Figure I-8. Integral gear expander/generator with expander wheel in overlay. (Source: Mafi-Trench Company, Santa Maria, California.)

Figure I-9. Two expander/compressors mounted on a single skid for an ethylene plant process. (Source: Mafi-Trench Company, Santa Maria, California.)

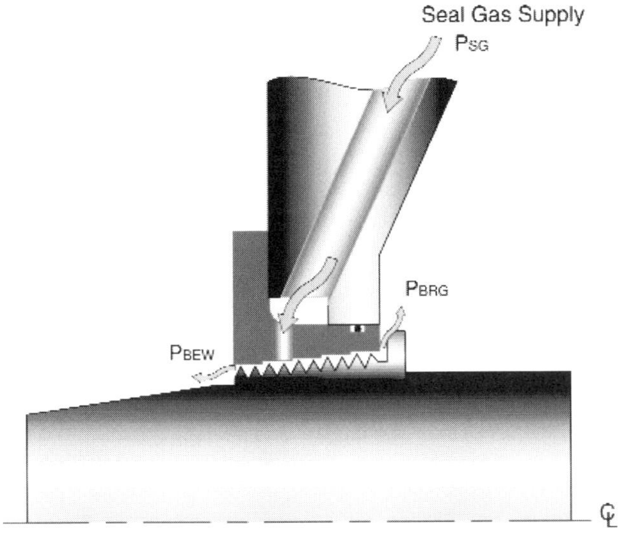

Figure I-10. Typical labyrinth seal for an expander/compressor with bearing housing pressurized with process gas. (Source: Mafi-Trench Company, Santa Maria, California.)

Finally, Table II-1 conveys that oil-free screw compressors (Fig. II-7) and axial machines are also used in certain processes covered by our text Segments II and III. Examples are oil refining processes such as alkylation, and petrochemical services in steam crackers, in which charge gas and refrigeration compressors are at work.

Examining Tables II-2 and II-3

A very similar approach is found in the tables prepared by Mannesmann Demag Delaval. For instance, a compressor train employed in a typical ethylene plant is being shop assembled in Fig. III-1. The three compressor cases (or casings) for charge gas (cracked gas) comprise horizontally split machines driven by a steam turbine (not shown) at 4,853 rpm and 32,350 kW. The three casings contain 4, 4, and 8 stages, respectively. The suction and discharge pressures and throughputs of the three stages are 1.5/3.0, 2.8/10.9, and 10.5/36.8 bar, or 160,618, 84,674, and 22,494 m^3/h, respectively. In Fig. III-2 a horizontally-split seven-stage propylene

Figure I-11. Two-stage integral gear expander/generator for a cryogenic process. (Source: Mafi-Trench Company, Santa Maria, California.)

Figure I-12. Two-stage integral gear expander/generator with housings removed to show expander wheels. (Source: Mafi-Trench Company, Santa Maria, California.)

Table II-1. A-C Compressor Corporation, comparison of compressors

	Compressor Type					
Application	D	DH	V/VS	VH/VHS	OFS	AXIAL
ADIP						
aMDEA						
Amine Guard FS						
Aminex/Mericat/Thiolex/Regen						
ARI LO-CAT II	X					
Bevon and others	X					
Bender	X		X			
Benfield			X			
Clintox			X			
$C0_2$ recovery			X			
$C0_2$ recovery and purification			X			
Cope	X					
CRYOMAX				X		
CrystaSulf						
D'GAASS	X					
Deoxy						
Drizo gas dehydration						
Drigas				X		
Ecotag	X		X			

Table II-1. *Continued*

Application	Compressor Type					
	D	DH	V/VS	VH/VHS	OFS	AXIAL
Ensulf	X					
Flexsorb solvents						
High conversion ratio (HCR)	X					
Hyco adsorption				X		
Hydrogen (Cryofining)						
Hydrogen (HTCR)	X					
Hydrogen (Medal)						
Hydrogen (Polybed PSA)						
Hydrogen (Polysep membrane)						
Hydrogen (steam reform)				X		
Hydrogen recovery (cryogenic)						
Hydrogen recovery (membrane)				X		
Hydrogen	X			X		
IFPEXOL						
IRON SPONGE						
KVT, Sulfox NK	X					
LNG plants			X	X		
LNG, pro				X		
Mehra NGL recovery						
Mehra Process NRU				X		
Merox	X		X			
Multipurpose gasification						
NGL recovery			X	X		
NGL, pro				X		
Nitrogen removal (reject)						
OlyClaus	X					
Paques/Thiopaq process	X					
PetroFlux			X	X		
PRICO(LNG)			X			
PURASPEC						
Purisol			X			
Rectisol			X			
Resulf						
SCOT	X					
SELEXOL	X		X			
Stretford	X					
Sulferox	X					
SRU	X					
Clinsulf-SDP	X					
Selectox						
Selexsorb						
Clauspol						
Seprex manbrane systems				X		
Sulfinol						
SULFREEN	X					
SulfaTreat—410HP						
Super Hy-Pro			X			
SUPERCLAUS	X					
Sure	X					
Syngas (adv. Reform)						
Syngas (ATR)						
Syngas (autothermal)			X			
Temperature swing adsorption (ADAPT)						
Alkylation			X			
Alkylation feed preparation						

(continued)

Table II-1. *Continued*

Application	Compressor Type					
	D	DH	V/VS	VH/VHS	OFS	AXIAL
Aromatics extraction						
Aromatics extractive distillation						
Aromatics recovery						
Benzene reduction						
Benzene saturation						
Catalytic cracking						
Catalytic dewaxing			X			
Catalytic reforming			X			
Coking			X			
Crude distillation						
Deasphalting						
Deep catalytic cracking			X			
Deep thermal conversion						
Delayed coking						
Desulfurization						
Dewaxing			X			
Electrical desalting						
Ethers			X			
Fluid catalytic cracking			X			X
Gas oil hydrotreatment				X		
Gas treating, H_2S removal						
Gasification						
Gasoline desulfurization				X		
Gasoline desulfurization, ultra deep						
Hydrocracking				X		
Hydrocracking, residue				X		
Hydrocracking/hydrotreating—VGO				X		
Hydrodearomatization				X		
Hydrodesulfurization				X		
Hydrodesulfurization—UDHDS				X		
Hydrogenation				X		
Hydrotreating				X		
Hydrotreating, catalytic dewaxing				X		
Hydrotreating, HDAr				X		
Hydrotreating, HDHDC				X		
Hydrotreating, residue						
Iso-octane						
Iso-octane/iso-octene						
Isomerization			X	X		
Lube hydroprocessing						
Lube treating						
Nox abatement	X					
Oily waste treatment						
Olefins recovery			X			
Resid catalytic cracking			X			X
Residue hydroprocessing				X		
Thermal gasoil process						
Treating			X			
Visbreaking						
Alkylbenzene, linear				X		
Alpha olefins, linear						
Amines, methyl						
Ammonia			X	X		
Ammonia, advanced			X			
Ammonia, conventional			X	X		

Table II-1. *Continued*

Application	Compressor Type					
	D	DH	V/VS	VH/VHS	OFS	AXIAL
Benzene						
Bisphenol-A						
BTX aromatics			X			
Butadiene			X		X	
Butanediol, 1,4-			X			
Butene-1						
Butyraldehyde, n and I						
Caprolactam	X		X			
Cumene						
Cyclohexane						
Dimethyl terephthalate						
EDC via oxychlorination, single stage			X			
Ethanolamines						
Ethylbenzene						
Ethylene			X	X		
Ethylene glycols		X				
Ethylene oxide		X				
Formaldehyde			X			
Maleic anhydride			X			
Methanol	X		X	X		
Mixed xylenes				X		
Octenes						
Olefins			X			
Paraxylene						
Phenol	X		X			
Phthalic anhydride			X			
Polycaproamide						
Polyethylene		X				
Polyethylene teraphthalate (PET)						
Polyethylene, HDPE						
Polyethylene, LDPE-EVA		X				
Polypropylene		X				
Polystyrene						
Propylene						
PVC (suspension)						
Styrene					X	
Teraphthalic acid						
Urea			X			
VCM removal						
Vinyl chloride monomer (VCM)			X			
Xylene isomerization						
Xylene isomers						

compressor welded-construction casing in the 8,000 kW, 6931 m^3/h size range is shown being assembled for shop tests. A steam turbine will drive this compressor at 5,616 rpm, increasing the gas pressure from a 1.45 bar suction to an outlet pressure of 14.25 bar.

A U.S. refinery operates two 11-stage axial air compressors in their fluid catalytic cracking units, or FCCs. The air enters at 1 bar and leaves at 3.42 bar. Each of the two machines (Fig. III-3) is driven by a 33,370 kW motor–steam turbine combination at 3,567 rpm. Another typical refinery compressor is shown in Fig. III-4. This 3,903 kW motor-gear driven, wet gas machine operates at 8,125 rpm with a throughput of 24,984 m^3/h and suction/discharge pressures of 1.6/16.6 bar, respectively. At 1,700 kW the four-stage recycle gas compressor train with variable speed gear unit and electric motor driver of Fig. III-5 is considerably smaller. The compressor shaft rotates at a nominal speed of 13,862 rpm; other parameters are 3,027 m^3 and gas delivery from 51.6 bar suction to 65.4 bar discharge pressure.

Returning to the petrochemical field, Mannesmann Demag Delaval provided a combined six-stage gear type compressor and two-stage expander unit for a PTA plant

Table II-2. Some typical parameters for centrifugal compressor applications in refinery processes (Source: Mannesmann Demag Delaval, Duisburg, Germany, and Trenton, NJ, USA)

Process	Licensors	Compressor service	Type of gas	Pressure ranges	Component arrangement	Driver types
Alkylation	Phillips, XOM	Refrigeration	C3, C4, IC4 mix	6–7 atm	Two-section, single body	Motor or turbine
Cat reforming	UOP; IFP	New gas to H2 users	H_2-rich, 8–12 MW	3 atm inlet, to 10 atm	Two-body, vertically split	Usually motor
Cat reforming	Others	H2 recycle	H_2-rich, N_2 regen.	14–20 atm	one-body, vertically split	Var-speed motor
Coking	All	Fractionator overhead	H-Cs, H_2S, H_2	To 15 atm	One-body, intercooled	Var-speed; motor or turbine
Crude distillation	All	Overhead gas	Mixed H-Cs	To 10 atm	One-body, one-section (usually a screw comp)	Motor
Lube dewaxing		Vaccuum	H-Cs, 25–30 MW	1 atm discharge	One-body	
		Refrigeration	Propane	To 15 atm	Two-section, one body	
Fluid cat cracking	Various	Regen air	Air	To 4 atm	Centrifugal or axial	Turbine (or expander)
		Fractionator overhead	Mixed H-Cs	To 15 atm	One-body, intercooled	
Hydrocracking	Various	H_2 Makeup	H_2-rich	From 100 to 210 atm	Reciprocal 2-4 stages	Motor
		H_2 Recycle			One-body, vertically split	Usually motor
Hydrotrating	IFP, UOP	H_2 Recycle	H_2-rich	From 75 to 110 atm	One-body, vertically split	Usually motor
	ABB	H_2 Recycle	H_2-rich	From 35 to 70 atm	One-body, vertically split	Usually motor
Hydrodesulfurization	UOP	H_2 Makeup	H_2-rich	70 to 85 atm	Two-stage reciprocal	Motor
		H_2 Recycle	H_2-rich	70 to 85 atm	One-body, vertically split	Usually motor

Table II-3. Some typical parameters for centrifugal compressor applications in petrochemical processes [Source: Mannesmann Demag Delaval, Duisburg (Germany), and Trenton, NJ (USA)]

Process	Licensors	Compressor service	Type of gas	Pressures (bars)	Component arrangement	Driver types	Power	Photo
Ammonia	KBR; Topsoe	Air	Air	Atm to 40 bars	2-body centrifugal	St. Turbine or GT	10 to 15 MW	NH_3 air
			Air	Atm to 40 bars	Integral gear	St. Turbine or GT	10 to 15 MW	IG NH_3 air
	Uhde	Syn gas/recycle	Syn Gas	To 140 bars	1–2 body w/ recycle in HP body	St Turbine	10 to 18 MW	NH_3 syn gas
		Ammonia refrigeration	NH_3	Atm to 15 bars	2-body centrifugal	St Turbine	8 to 12 MW	
Butadiene	ABB Lummus	Recycle	Butadiene	6–8 bars	Integral gear or wet screw	Motor	1 to 3 MW	
Butanadiol	BP, Lurgi	Air	Air	2–4 bars	Single-stage centrifugal	ST or Motor	2.5 to 4 MW	
		Hydrogen recycle	H_2	275 bars	Reciprocal or small high-speedcentrifugal	Motor	400 KW	
Ethylene	Various	Cracked gas	Mixed H-Cs	To 45 bars	Three-body centrifugal	St Turbine	30 to 50 MW	Cracked gas train
		Propylene refrigeration	Propylene	15 to 20 bars	One-body with side loads	St Turbine	30 to 50 MW	Propylene refrigeration
		Ethylene refrigeration	Ethylene	20 bars	One-body centrifugal	St Turbine	4 to 10 MW	
Methanol	ALL	Syn gas Makeup	Syn gas	75 to 110 bars	One-body centrifugal	St Turbine	To 20 MW	NH_3 syn gas
		Syn gas recycle	Syn gas	2 to 5 bar delta-P	Single-stage centrifugal	S. T or Motor	3 to 5 MW	
	Lurgi, Topsoe	Air, for O_2 production	Air	To 15 bars	Integral gear	S Turbine	To 20 MW	Integral gear air
Polyethylene	Various	Ethylene recycle	Ethylene	25 to 40 bars	Single-stage centrifugal	Motor	2 to 6 MW	Polyethylene recycle
Polypropylene	Various	Propylene recycle	C2–C3 mix	25 to 40 bars	Single-stage centrifugal	Motor	2 to 4 MW	
Terephthalic Acid	BP	Air	Air	10 to 20 bars	Integral gear	Expander	To 15 MW	PTA train
Urea	Snamprogetti Stamicarbon	CO_2	CO_2	150 bars	2-body centrifugal	St Turbine	8 to 15 MW	

Figure II-1. Single-stage, Model "D" manufactured by A-C Compressor Corporation.

(Fig. III-6). By opting for this arrangement, the Indian owner–purchaser probably obtained a competitive advantage. They therefore prevented the manufacturer from releasing pressure profiles and related data. The end user of the direct-driven, single-stage barrel compressor with welded casing and dry gas seals for a polyethylene plant (Fig. III-7) asked for confidentiality as well. In contrast, the U.K.-based ICI Company allowed us to know that two of the 6,720 kW electric-motor-driven, integrally-geared VK-32-6 compressors (Fig. III-8) were purchased for an ammonia plant. The three different pinions rotate at 13,366/20,048/18,712 rpm and compress 36,200 m^3 of air from 0.98 to 41 bar.

Figure II-2. Single-stage, high-pressure, type "DH" manufactured by A-C Compressor Corporation.

Figure II-3. Horizontally split, V/VS compressor by A-C Compressor Corporation (Appleton, Wisconsin.)

Figure II-4. Horizontally split, V/VS compressor train with electric motor and speed-up gear (Source: A-C Compressor Corporation, Appleton, Wisconsin.)

Figure II-5. Model VH/VS barrel compressor (A-C Compressor Corporation, Appleton, Wisconsin.)

Figure II-6. Model VH/VS barrel compressor train (A-C Compressor Corporation, Appleton, Wisconsin.)

Figure II-7. Oil-free screw compressor. (Source: A-C Compressor Corporation, Appleton, Wisconsin.)

Figure III-1. Test bed assembly of a three-casing cracked gas compressor string for an ethylene plant. (Source: Mannesmann Demag Delaval, Duisburg, Germany.)

Figure III-2. Test bed assembly of a propylene compressor with welded casing for an ethylene plant. (Source: Mannesmann Demag Delaval, Duisburg, Germany.)

Figure III-3. Assembly of an axial-flow compressor for a FCC plant in the USA, $\dot{V} = 420{,}000\ m^3/h$. (Source: Mannesmann Demag Delaval, Duisburg, Germany.)

Figure III-4. Wetgas compressor with welded casing for a refinery in Italy. (Source: Mannesmann Demag Delaval, Duisburg, Germany.)

Figure III-5. Recycle compressor with variable speed gear box for a refinery in Sweden. (Source: Mannesmann Demag Delaval, Duisburg, Germany.)

Figure III-6. Packaging of a combined six-stage, gear-type compressor and two-stage expander unit for a PTA plant in India. (Source: Mannesmann Demag Delaval, Duisburg, Germany.)

Figure III-7. Direct driven, single-stage, barrel-type compressor with welded casing and dry gas seals for a polyethylene plant. (Source: Mannesmann Demag Delaval, Duisburg, Germany.)

Figure III-8. Package of a six-stage, integrally geared turbocompressor for an ammonia plant in Great Britain. (Source: Mannesmann Demag Delaval, Duisburg, Germany.)

Segment 1

Gas Processing

PROCESS INDEX

Compressors and Modern Process Applications. By Heinz P. Bloch.

LICENSOR INDEX

Drying

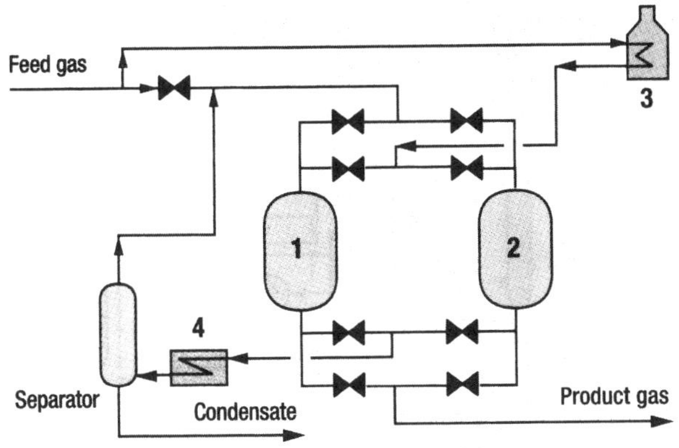

ADAPT (Gas dehydration and hydrocarbon dewpointing)

Application: Dehydration, hydrocarbon and aromatics removal, and methanol and mercaptan removal from high-pressure gases. ADAPT can be used at natural gas reception terminals, underground gas storage facilities (i.e., salt cavities, aquifers and depleted fields) and prior to LNG production. The process is suitable for pre-purification and protection facilities for gas membrane systems.

Description: Hydrocarbon and water levels in high-pressure natural gases are simultaneously reduced within a solid desiccant fixed bed (1). Tailored adsorbents selectively remove gas-phase components and control the slippage rate to the export gas, thus meeting required production specification. Once saturated, the adsorbing bed is switched to regeneration mode and a fresh bed (2) is brought online. Process flexibility enables multiple bed systems that allow very high throughputs. Using pre-heated (3) feed or product gas regenerates the saturated bed, which depends on the application requirements and economics. Regeneration temperature depends on the components being removed, but typically range from 200°C to 300°C. The rich-regeneration gas is cooled, producing a saleable hydrocarbon condensate (4). Cooler system flash gas is recycled back to the adsorbing bed for further processing. Some advantages of ADAPT over competing processes are:

- Compact process plant
- Rapid startup and shutdown
- No hot standby required
- Turndown to 10% of design flow
- High reliability and low maintenance
- Long adsorbent life.

Operating conditions: Typical operating pressures range from 30 to 120 bar, and feed gas temperatures up to 45°C. Plants can be skid mounted or modular enabling phased asset development. Typical feed gas flows from 10 MMscfd to over 1,500 MMscfd.

Economics: Costs vary with scale and application but typically range from £2 million for a 200 MMscfd plant to £9 million for a 1,500-MMscfd plant.

Installations: Plants in operation or construction across Europe, Africa and Asia. Total throughput for current ADAPT designs is approximately 9,000 MMscfd.

Reference: Metters, S. D., "An adaptable solution," *Hydrocarbon Engineering*, March 2000.

Licensor: Advantica Technologies Ltd.

Drying

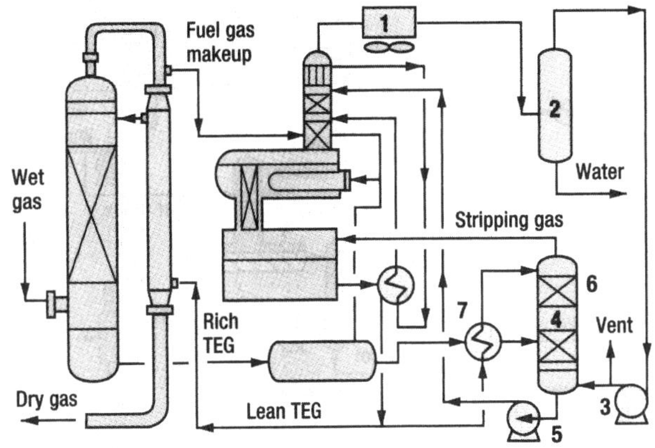

Drigas

Application: Drigas is used to dehydrate natural gas to very low dew points by glycol absorption, without using vacuum regeneration, no solvents and no consumption of stripping gas.

Description: Drigas process recycles the glycol regenerator's overhead vapors, cooled and dried, to the stripping tower.

The vent gas off of a conventional regenerator with stripping tower is cooled in overhead condenser (1) and condensed water is separated in the overhead knockout drum (2). The wet gas coming from (2) is reused as stripping gas, feeding it to the bottom of the stripping tower by means of blower (3).

Spent glycol is used to dry the recirculated gas to the stripping tower by means of a random packed atmospheric absorber (4) and a glycol pump (5) to move rich glycol from the bottom of the atmospheric absorber to the still tower.

If higher concentrations are required, a second stage of absorption (6) can be incorporated into the same atmospheric column fed by a small fraction of the regenerated, lean TEG. This arrangement gives TEG purity up to 99.99 wt%.

Main advantages of the Drigas process are:

- Very low dew points
- Stripping gas is not required
- Low operating cost
- Low pollution.

Operating conditions: Glycol flowrates up to 1,000 m^3/d, with TEG purity up to 99.99 wt%. Gas flowrate up to 15 MMscmd for each train, wet gas temperature up to 60°C and pressure up to 150 bar.

Economics: A Drigas regenerator costs marginally more than a traditional plant with stripping gas, but without any consumption of stripping gas.

Installations: One Drigas unit, with a capacity of 200 m^3/d DEG.

Reference: Franci, P. F., "New glycol regenerator adaptable to offshore use," *World Oil*, July 1993.

Licensor: SIIRTEC NIGI

Drying

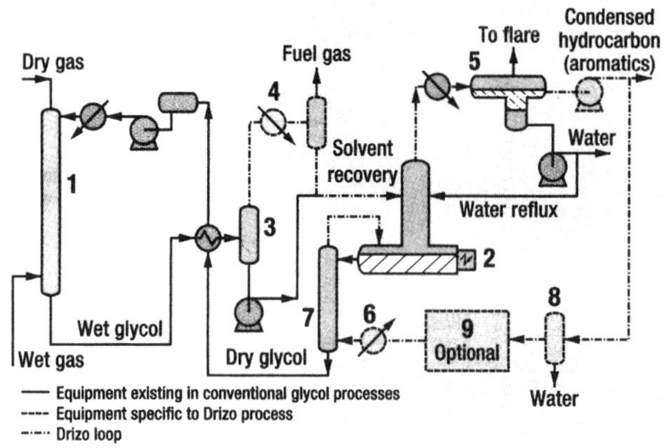

Drizo gas dehydration

Application: Low water dew points, typical water dew point depressions up to 180°F (depressions greater than 200°F achievable with Drizo HP), 95%+ recovery of BTEX vapor components.

Description: Water is absorbed (1) from natural gas by glycol (DEG, TEG or tetraethylene glycol). The glycol is then thermally regenerated in reboiler (2). The main differences with conventional glycol processes are: glycol is flashed after preheating (3) to allow high recovery of liquid hydrocarbons (4), After heteroazeotropic distillation, these liquid hydrocarbons are recovered from the still column condenser (5), vaporized (6) and used to strip the hot glycol (7). Water, still present in the liquid hydrocarbons, can be removed by a coalescer (8) and by an optional solvent dehydration package (9) (Drizo + and Drizo HP versions). Glycol purities above 99.99 wt% and up to 99.998+ wt% (Drizo HP) are obtained, thus enabling residual water content in the treated gas down to below 0.1 ppm.

Economics: Combining very low dew points with low CO_2 and BTEX emissions, Drizo is an environmentally friendly process compared with the other glycol processes. Drizo is very competitive with all dehydration processes at water dew points below –30°C. A Drizo unit would be roughly 20% cheaper than an equivalent glycol stripping unit with recompression of the stripping gas (in addition to the fact that Drizo is able to reach much lower water dew points), and can be 50% cheaper than a mol sieve unit.

Installations: More than 45 units.

References: Rigaill, C., "Solving Aromatic and CO_2 emissions with the Drizo gas/glycol dehydration process," GPA Europe Meeting, February 23rd, 2001.

Smith, R. S., "Enhancement of Drizo gas dehydration," 47th Laurance Reid Gas Conditioning Conference, March 2–5, 1997.

Smith, R. S. and S. E. Humphrey, "High Purity Glycol Design Parameters and Operating Experience," 45th Laurance Reid Gas Conditioning Conference, 1995.

Licensor: OPC Drizo, Inc., and Prosernat IFP Group Technologies.

Drying

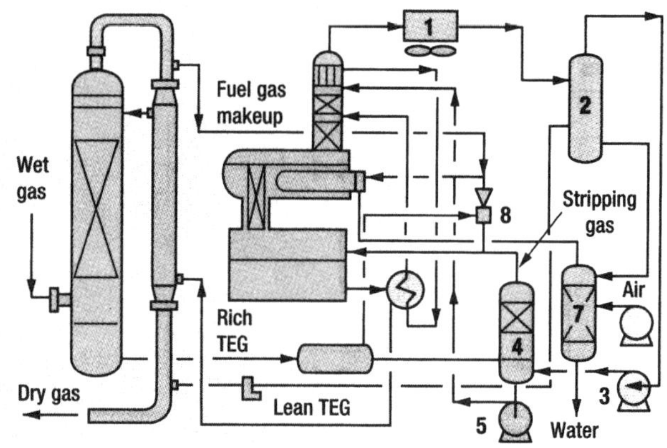

Ecoteg

Application: Ecoteg is a process that uses triethyleneglycol (TEG) to dehydrate gases rich in aromatic compounds (BTEX) where the effluent control is critical. Its emission of aromatics into the environment is nil.

Description: When TEG is used to dehydrate natural gas, it absorbs selectively not only water but also part of the BTEX that may be present. BTEX are released with all the outgoing streams of the regenerator.

The vent gas off a still column is cooled in overhead condenser (1) and condensed water and BTEX are separated in the overhead knockout drum (2). The wet gas coming from (2) is reused as stripping gas and is fed to the bottom of the stripping tower by means of blower (3).

The spent glycol is used to dry the gas recirculated to the stripping tower by means of a random-packed atmospheric absorber (4) and a glycol pump (5) to move the rich glycol from the bottom of the atmospheric absorber to the still tower.

The liquid BTEX are recovered as oil if an oil product is already present in the plant; otherwise, they are returned to the dried gas by means of pump (6) or recycled to presaturate the lean TEG.

The condensed water, before disposal, is stripped through a stripping tower (7) by means of combustion air.

Main advantages of Ecoteg process are:

- Meet more stringent regulations for disposal without additional facilities
- Low operating cost
- Low gas dew point.

Operating conditions: Gas flowrates up to 15 MMscmd for each train, wet gas temperature up to 60°C and pressure up to 150 bar.

Economics: Ecoteg is an ecological and cost-effective dehydration process that does not require additional external facilities to meet regulations for effluents. This simplifies the first installation and/or additions to an existing plant. Savings in stripping gas helps the economics and may be a determinant when low gas dew points are required.

Reference: Franci, P. F. and J. W. Clarke, "Emission free, high purity TEG regenerator," 1994 GRI Glycol Dehydrator/Gas Processing Air Toxic Conference, San Antonio.

Licensor: SIIRTEC NIGI.

Drying

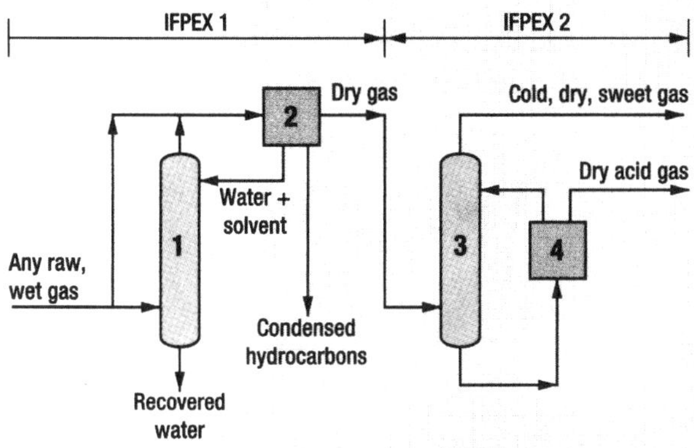

Ifpexol

Application: Treat any gas for dehydration, hydrate protection, dew-point control and acid-gas removal, using a single, low-freezing point solvent—methanol. Ifpexol is a two-step process; each step can be used independently or in combination:

- Ifpex-1—simultaneous water and hydrocarbon dew pointing (down to –100 °C)
- Ifpex-2—removal of acid gases and sulfur compounds (to sales gas specifications).

Description: *Ifpex-1:* A partial stream from the feed gas is loaded with methanol by stripping in a contactor (1). The methanol/water mixture is recycled from the cold process (2). Pure water is obtained from the bottom of the contactor (1). Overhead gas is mixed with the main gas stream and contains enough methanol to prevent freezing during the cold process (2). During this process, the gas is cooled to the required dew-point temperature by any appropriate means (J-T expansion, turbo expander or external refrigeration). The treated dry gas is recovered from the low-temperature separator along with condensed hydrocarbons and the methanol/water mixture. This methanol/water mixture is recovered as a separate liquid phase and recycled to the contactor (1).

Ifpex-2: The gas from Ifpex-1 (or any other feed gas) is contacted with refrigerated methanol-based solvent in the contactor (3). Acid gases (CO_2, H_2S) are absorbed, along with other sulfur compounds (mercaptans, COS), and a dry sweet gas is obtained on top of the contactor (3). The solvent loaded with acid gases is regenerated by a simple flash and, in some cases, by thermal regeneration. Hydrocarbon co-absorption is controlled by the solvent composition and a multi-flash regeneration recovers most of the co-absorbed hydrocarbons in a separate gas stream. Acid gases are recovered dry and under pressure (typically around 10 bar); thus, this process is particularly suitable for acid-gas reinjection applications.

Economics: For dew points below –30°C, Ifpex-1 can compete with glycol processes and offers much reduced (about 30% lower) CAPEX. Ifpex-2 offers significant savings compared to other processes for bulk acid-gas removal with acid-gas reinjection.

Installations: Fifteen Ifpex-1 units with capacities up to 350 MMscfd.

Reference: "Methanol treating schemes offers economics, versatility," *Oil & Gas Journal,* June 1, 1992, p. 65.

Minkkinen, A., et al., "Technological developments in sour gas processing," *Gas Cycling,* Les Entretiens IFP, May 14, 1998, Technip.

Licensor: Prosernat IFP Group Technologies and Titan SNC Lavalin.

Treating

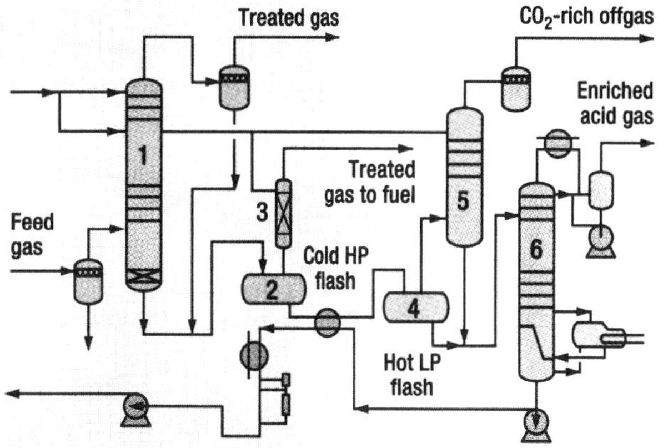

ADIP

Application: The regenerative-amine process removes hydrogen sulfide (H_2S) and carbon dioxide (CO_2) from natural gas, refinery gases and synthesis gas. Hydrogen sulfide can be reduced to low sulfur levels. The process can also be applied to remove H_2S, CO_2 and carbonyl sulfide from liquefied petroleum gas or natural gas liquids to low levels. Bulk CO_2 removal from synthesis gas with flash regeneration is another application.

The process sequence—ADIP/Claus/SCOT—can be used advantageously with an integrated ADIP system that handles selective H_2S removal upstream and the SCOT process treating the Claus offgas.

Description: The ADIP process uses aqueous solutions of the secondary amine, di-isopropanolamine or the tertiary amine, methyl di-ethananolamine. Amine concentrations up to 50%wt can be applied.

The process has low observed corrosion rates and a controllable foaming tendency of the solvent due to an optimized design. Its line-up can be very diverse, depending on the application. High integration of separate process units is possible. The simplest system resembles those of other amine processes. The ADIP liquid treating consists of an extractor followed by mixer/settlers for COS removal. Regeneration is common.

For a typical design, the co-absorbed hydrocarbons from the absorber (1) are flashed (2) from the solvent and used as fuel gas after treatment (3). The solvent is flashed again (4) to release CO_2 from the enrichment absorber (5), thereby improving the acid-gas composition from the regenerator (6) to the Claus unit.

Operating conditions: A very wide range of treating pressures and contaminant concentrations can be accommodated. Sulfur specifications below 100 ppmv H_2S in gas and 10 ppmw in liquid hydrocarbon streams can be met. In liquid hydrocarbon, COS can be removed down to 5 ppmw. Improved Claus feed gas quality can be met by improving the H_2S/CO_2 ratio in the acid gas. Bulk CO_2 removal from a high percentage to several percentage points is easily attained.

Installations: More than 400 ADIP units are in operation or under construction. Applications include: natural gas, liquefied natural gas, refinery gases and liquefied petroleum gases and synthesis gases.

References: "ADIP as the preferred solvent for amine treatment in refinery applications," Laurence Read Conference, March 1999.

"Process applications of the ADIP and Sulfinol process," Gas Processing Symposium, Dubai, United Arab Emirates, April 1999.

Licensor: Shell Global Solutions International B.V.

Treating

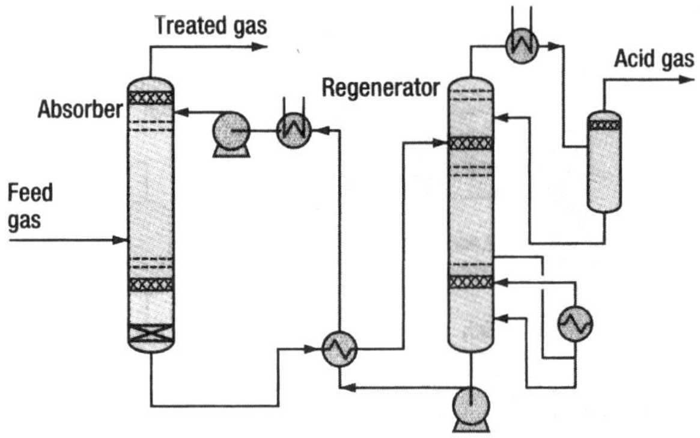

ADIP-X

Application: This regenerative-amine process is highly suitable for bulk and deep CO_2-removal from gas streams. It also removes hydrogen sulfide (H_2S), some COS and mercaptans from natural gas, refinery gas and synthesis gas. Hydrogen sulfide can be reduced to low-sulfur levels. This process achieves a higher loading capacity and reduced equipment size compared to general aqueous amine solvents.

Description: The ADIP-X process uses aqueous solutions of the tertiary amine, methyl diethananolamine and an additive. Amine concentrations up to 50%wt can be applied. The process has low observed corrosion rates and a controllable foaming tendency of the solvent due to an optimized design. Its lineup can be very diverse, depending on the application. High integration of separate process units is possible. The simplest system resembles those of other amine processes. Regeneration is common and may consist of one or more flash/steam strip regeneration steps. Solvent composition will be balanced against customer requirements.

Operating conditions: A very wide range of treating pressures and contaminant concentrations can be accommodated. Carbon dioxide specifications below 50 ppmv in gas streams can be met. H_2S specifications down to 100 mbar are also achievable. Bulk CO_2 removal from a high percentage to several percentage points is easily attained.

Installations: ADIP-X is applied in one natural gas application.

Licensor: Shell Global Solutions International B.V.

Treating

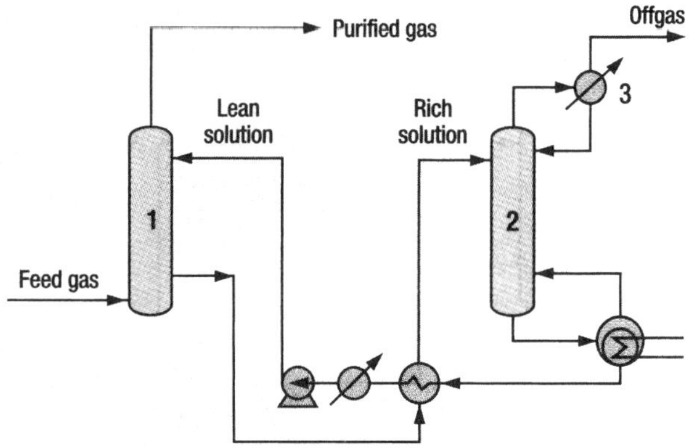

Amine Guard FS

Application: Remove CO_2, H_2S, COS and RSH from natural gas; CO_2 from ammonia syngas, etc., with a solution containing one of the Ucarsol family of formulated solvents. When desired, H_2S can be removed selectively to provide a superior Claus plant feed and reduce regeneration requirements.

Product: Purified gas to meet pipeline, LNG plant, ammonia plant or petrochemical plant specifications as appropriate.

Description: The treating solution scrubs acid gases from the feed in an absorber column (1). The rich solution is regenerated by reducing its pressure and stripping with steam in the stripper tower (2). Waste heat is commonly used to provide the steam.

Regeneration energy is minimized by choosing the optimum Ucarsol solvent for the situation, using high solvent concentrations and proper selection of process parameters.

Operating conditions: Absorption pressure is atmosphere to 2,000 psi, as available. Feed temperature is 85°F to 150°F. If the feed is available at a higher temperature, that heat will be used to supply regeneration heat. Acid gas content may be up to 50%.

Economics: For a 500-MMscfd natural gas unit having a feed gas containing 6% CO_2 and 1% H_2S, typical costs are as follows:

	Pipeline	LNG feed
Investment, $MM	14.0	17.0
Operating costs, $MM/y	6.0	7.0

Installations: More than 500 units worldwide, mostly treating natural gas, ammonia syngas and hydrogen streams.

Licensor: UOP LLC.

Treating

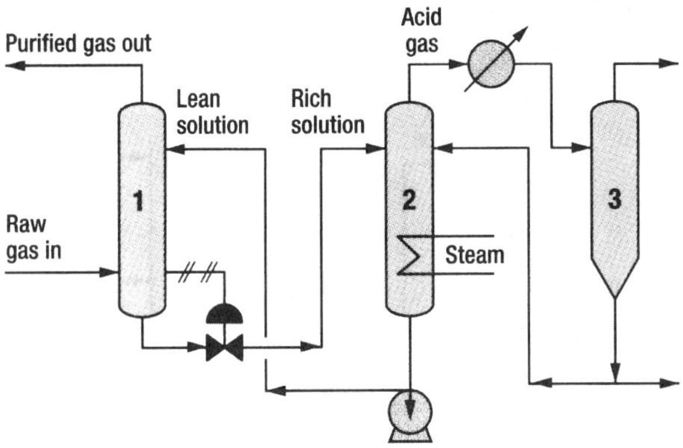

Benfield

Application: Removal of CO_2, H_2S, COS and RSH from natural gas or SNG. Removal of CO_2 from ammonia syngas, ethylene oxide recycle gas, etc.

Product: Purified gas to meet pipeline or LNG plant specifications or petrochemical specifications as appropriate.

Description: Acid gases are scrubbed from the feed in an absorber column (1) using potassium carbonate solution with Benfield additives to improve performance and avoid corrosion. The rich scrubbing solution is regenerated by reducing its pressure and stripping with steam in the stripper tower (2). Waste heat is commonly used to provide the steam. In the LoHeat version, the hot, lean solution is flashed by sending the steam through ejectors to reduce the energy requirements. In the HiPure version, acid gases are reduced to very low levels by polishing using an integrated DEA absorption loop.

Operating conditions: Absorption pressure is 50 to 2,000 psi, as available. Feed temperature is about 150°F to 250°F. If the feed is available at a higher temperature, that heat will be used to supply regeneration heat. Acid gas content may be up to 50%. Heavy hydrocarbons are easily handled. If no H_2S is present, oxygen contents of several percent are handled without difficulty or solvent degradation.

Economics: The extensive use of carbon steel, the elimination of a rich/lean solution heat exchanger, the heat recovered by LoHeat and the low cost of the solution chemical make the process attractive for a wide range of applications.

For a 500-MMscfd natural gas unit having a feed gas containing 6% CO_2 and 1% H_2S, typical costs are as follows:

	Pipeline	LNG feed
Investment, $MM	14.0	20.0
Operating costs, $MM/y	5.0	6.0

Installations: Of the more than 675 units worldwide, more than 65 treat natural gas, over 200 treat ammonia syngas and about 110 are in hydrogen plants. The remainder are in SNG, partial oxidation, coal gasification and petrochemical applications.

Licensor: UOP LLC.

Treating

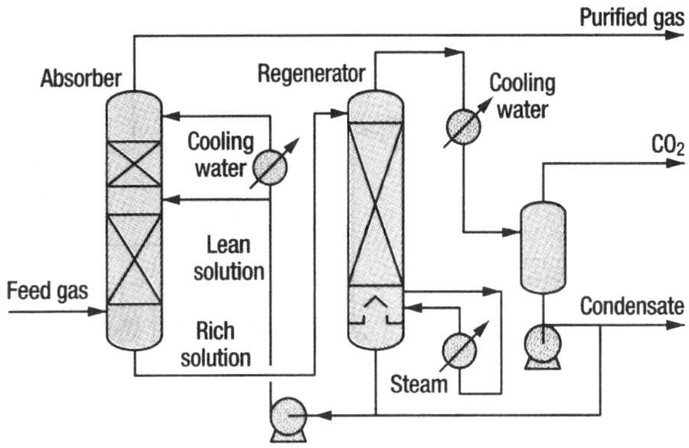

CO_2 removal—LRS 10

Application: Remove CO_2 from natural gas, SNG or ammonia syngas. The purification process is applicable in facilities for LNG plants and petrochemical applications.

Description: Rich-CO_2 feed gas is passed to the absorber column containing potassium carbonate solutions and LRS10 additives. The used solution goes through the regeneration system and purified gas exits the top of the absorber. The rich solution is regenerated either in a re-boiler facility or directly with steam. The recovered potassium carbonate/LRS10 solution is pumped back to the absorber for further reaction. Feed gas containing about 20% CO_2 has been treated successfully, typically to CO_2 levels of 1% in the processed gas, depending on process arrangement. Process configuration changes can lower CO_2 slippage levels from 500 ppm to 1,000 ppm in some designs.

Economics: A plant utilizing typically 3% LRS10 in a potassium carbonate system has been shown to offer improved performance over CO_2-removal systems such as Benfield (DEA promoted potassium carbonate) by up to 10%. This savings can normally be realized through:

- Increasing gas throughput of typically 10%
- Lower regeneration energy by about 10%
- Reduce CO_2 slippage in the outlet gas by up to 50%
- Improve column operations by moving away from column constraints.

Installations: Thirty plants worldwide, mainly retrofits, in the ammonia, hydrogen and chemicals industries.

Licensor: Advantica Technologies Ltd.

Treating

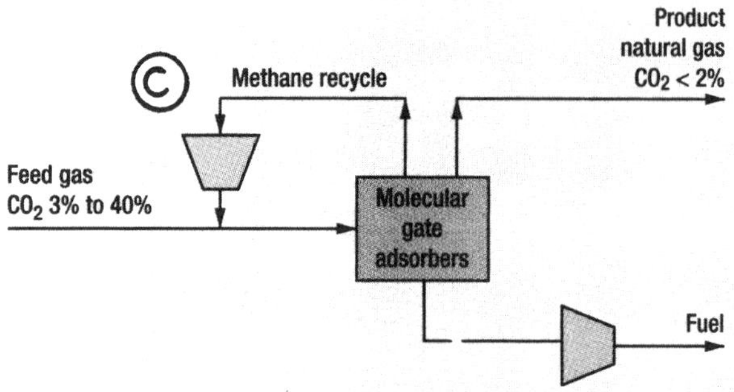

CO_2 removal—Molecular Gate

Application: Simultaneously removes carbon dioxide (CO_2) and water from contaminated natural gas. Feedstocks include coalbed methane and natural gas. Water-saturated feeds and CO_2 levels of 3 to 40% can be treated. Product is pipeline-quality natural gas with characteristic CO_2 levels of less than 2%. The process uses a specialty adsorbent for CO_2 removal in a patent-pending, proprietary, pressure swing adsorption (PSA) system.

Description: Water-saturated feed, at pressure levels between 100 and 800 psig, is routed through a series of adsorber vessels. One or more vessels are removing the water and CO_2 while pipeline-quality natural gas flows through the adsorbent bed at essentially feed pressure. Typically, between three and eight adsorber vessels are used. When the adsorbent is saturated with water and CO_2, the spent vessel is removed and replaced with a regenerated one. It is depressurized and produces a low-pressure, methane-rich stream for compression/recycle to the feed and a low-pressure fuel stream containing CO_2 to be rejected. To maximize adsorbent capacity, impurities are removed through a single-stage vacuum blower. If the feed contains C_3^+ components in large quantities a NGL recovery section can be added to produce a mixed NGL product containing these components at high recovery rates. The system is flexible for a wide range of CO_2 concentrations and has turndown capability up to 30%. Modular construction facilitates installation.

Economics: The technology can be cost-effectively applied to a wide range of flowrates. A small, 2-MMscfd unit's typical total installed cost is \$0.30/Mft^3 of feed processed. This cost decreases to less than \$0.15/Mft^3 for a 10-MMscfd design. Modular construction allows low-cost installation and equipment relocation flexibility.

Installation: The first unit started up in April 2002, removing 18% CO_2 to pipeline specifications.

Reference: Mitariten, M. J., "Eliminating carbon dioxide from coalbed methane using the Molecular Gate adsorption process," Third Annual Coalbed and Coal Mine Methane Conference, Denver, March 2002.

Licensor: Engelhard Corp.

Treating

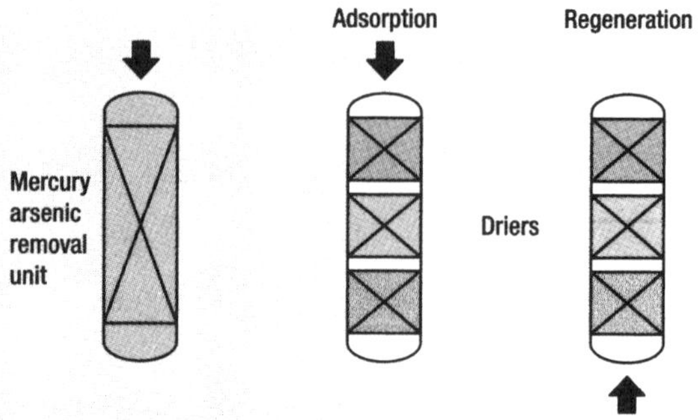

Gas contaminants removal—Multibed

Application: Remove mercury (Hg), arsenic, water, CO_2, other oxygenates, tertiary butyl catechol (TBC), NH_3 and sulfur species from natural gas, industrial gases and petrochemical streams using Axens Multibed technology.

Description: This technology uses combinations of special aluminas and zeolite molecular sieve adsorbents that are installed as beds in the same or separate vessels. (The configuration will depend on the specific application.) The aluminas function as catalysts or adsorbents for chemical or physical removal of contaminants such as HCl, Hg, AsH_3, TBC and water. The molecular sieves remove contaminants by physical adsorption.

In multibed processing, Hg can be removed either upstream or downstream of the gas dehydration step. The treated gas can contain less than 0.01μg/Nm^3 Hg and less than 1 ppmv of other contaminants. Most adsorbents are thermally regenerated with nitrogen or light-hydrocarbon streams.

Operating parameters: Operating conditions are:

Inlet temperature, °F	40–140
Inlet pressure, psig	40–1,400
Regeneration gas temperature, °F	350–600
Phase	liquid or vapor

Installations: Currently, over 60 installations worldwide are treating natural and industrial gases and other hydrocarbons streams from refining, gas and petrochemical sectors.

References: Jochem, G., "Optimization of natural gas drying and purification," *Petroleum Technology Quarterly,* Spring 1997.

Nedez, C., et al., "Optimization of the textural characteristics of an alumina to capture contaminants in natural gas," American Chemical Society, 1996.

Licensor: Axens.

Treating

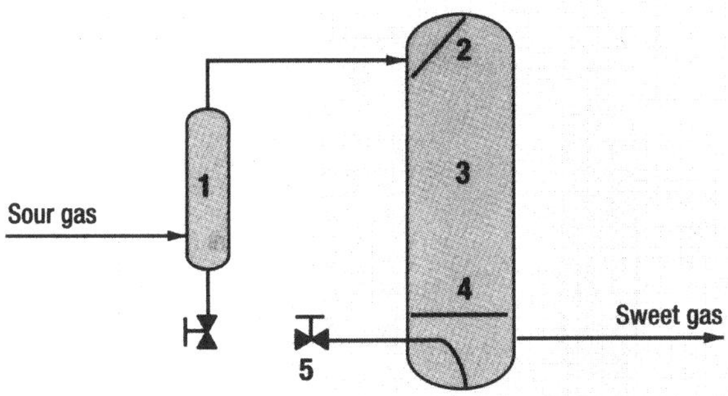

IRON SPONGE

Application: To remove hydrogen sulfide (H_2S) and mercaptans from natural gas streams at low or high-pressure conditions at or near the well head.

Process description: Iron sponge (iron oxidized onto wood shavings) uses a simple packed tower (3) on a flow-through support (4). After liquid separation (1) the deflected (2) sour gas flows down to contact with the reactive iron oxide, simply and effectively converting H_2S into a solid. The iron sulfide stays in the packed tower, effetively removing it from the gas stream. Iron sponge also removes mercaptans, the malodorous sulfur compounds found in some gas, producing a deodorized, sweet gas.

For iron sponge to effectively perform, it must be maintained within a range of water levels. This requirement is usually satisfied if the gas is saturated with water vapor, as is frequently the case. If it is not the case, a simple water spray will correct it. An excess of water is tolerated very well by iron sponge as long as the excess is drained off (5), so as not to flood the bed. Because the reaction of iron oxide with H_2S produces a small quantity of water, monitoring the drip water volume is an effective method of confirming the presence of sufficient moisture. Since the iron oxide is impregnated onto the wood, it will not wash off or migrate with the gas.

Operating conditions: The process should be located as close to the source of gas as possible, to eliminate as many corrosion problems as possible caused by the H_2S. The process should be used after a gas/liquid separator and before the dehydration process. The maximum temperature should not exceed 120°F, and the minimum of 50°F, or whatever is necessary to avoid hydrate formation for the system pressure and composition of the gas.

The gas purification is not pressure sensitive, and is not affected by other gas constituents. Carbon dioxide levels are not a problem for treatment, but liquid hydrocarbons should be removed before the IRON SPONGE process.

Since the process is so simple, minimum operator time is required. It runs unattended for days at a time.

Manufactured and supplied by: Connelly-GPM, Inc.

Treating

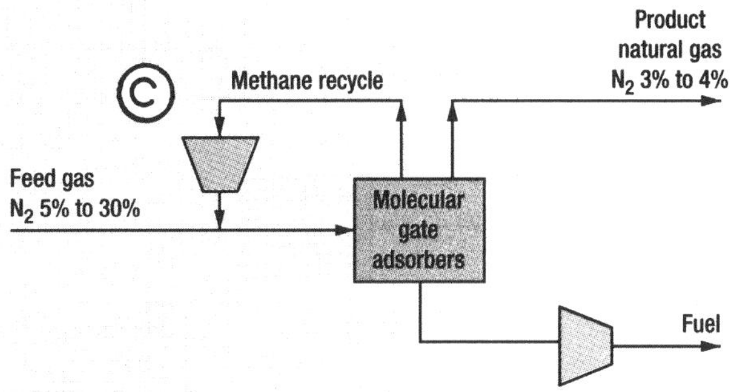

N_2 rejection—Molecular Gate

Application: Remove nitrogen (N_2) from contaminated natural gas. Feedstocks include N_2 contamination from 5% to 30%. Product is pipeline quality natural gas with typical N_2 levels of 3% to 4% using the Molecular Gate process and a patented proprietary adsorbent in a pressure swing adsorption (PSA) system.

Description: Glycol-dehydrated natural gas at pressure levels between 100-psig and 800-psig is routed through a series of adsorber vessels. One or more vessels remove N_2, while pipeline-quality natural gas flows through the adsorbent bed at essentially feed pressure. Typically, between four and nine adsorber vessels are used. When the adsorbent is saturated with N_2, the spent vessel is removed and replaced with a regenerated one. It is depressurized and produces a low-pressure, methane-rich stream for compression/recycle to the feed and a low-pressure fuel stream containing N_2 to be rejected. To maximize adsorbent capacity, N_2 is typically removed through a single-stage vacuum blower. The process is based on an adsorbent that is size selective and allows smaller N_2 molecules to fit in adsorbent pores, while the larger methane molecule is unaffected. Carbon dioxide is also completely removed with the N_2 and oxygen is removed at N_2 levels. The system is flexible for a wide range of N_2 concentrations and has turndown capability up to 30%. Modular construction facilitates installation.

Economics: The technology can be cost-effectively applied to a wide range of flowrates. A small, 2-MMscfd unit's total installed cost is \$0.50/Mft^3 of feed processed. This is decreased to less than \$0.30/Mft^3 for a 10-MMscfd design. Modular construction allows low-cost installation and equipment relocation flexibility.

Installation: The first unit has been in operation for over one year, removing 18% N_2 to pipeline specifications.

Reference: Mitariten, M. J., "New technology improves nitrogen removal economics," *Oil and Gas Journal,* April 23, 2001.

Licensor: Engelhard Corp.

Treating

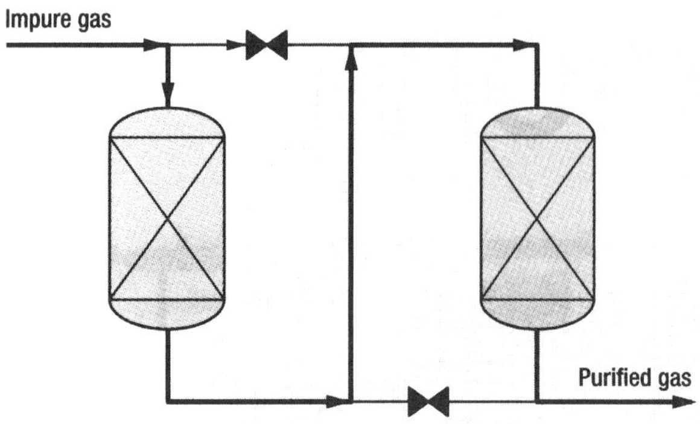

PURASPEC

Application: Removal of ppm concentrations of H_2S, COS and ppb levels of mercury from hydrocarbon gases/liquids to meet pipeline or petrochemical specifications.

Description: Fixed beds of chemical absorbents provide effectively total irreversible selective removal of impurities from wet or dry hydrocarbons without feedstock losses. Radial flow reactor designs are available if low system pressure drop is required.

Operating conditions: Temperatures: 20°F to 400°F (no heat input is required). Pressure: atmospheric to 2,000 psi.

Installations: About 100 worldwide units, treating natural gas rates up to 2 billion scfd and natural gas liquids up to 5,000 tpd for major oil/gas companies.

References: Spicer, G. W. and C. Woodward, "H_2S control keeps gas from big offshore field on spec," *Oil & Gas Journal*, May 27, 1991, p. 76.

Carnell, P. J., H. K. W. Joslin and P. R. Woodham, "Fixed bed processes provide flexibility for COS, H_2S removal," *Oil & Gas Journal,* June 5, 1995, p. 52.

Rhodes, E. F., P. J. Opensha and P. J. H. Carnell, "Fixed-bed technology purifies rich gas with H_2S, Hg," *Oil & Gas Journal,* May 31, 1999, p. 58.

Licensor: Synetix.

Treating

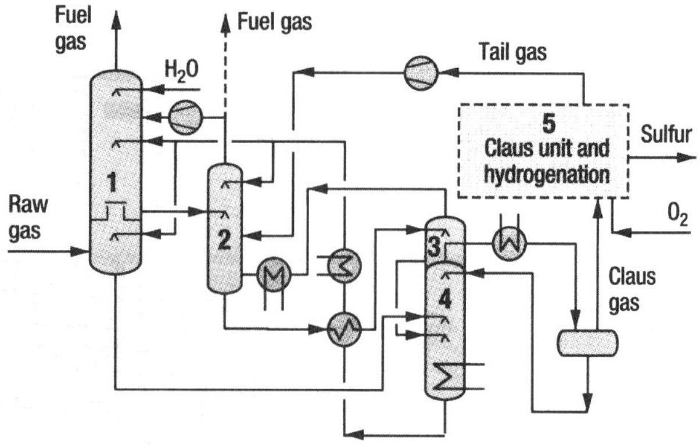

Purisol

Application: Removal of acid gases from natural gas, fuel gas and syngas by physical absorption in NMP (N-methyl-pyrrolidone). Typical cases: 1. High CO_2 contents to low residual level, 2. Bulk acid gas removal to moderate purity by simple flash regeneration, 3. Selective H_2S removal. Ideally suited for (3) in an IGCC based on POX of coal or oil, as NMP, is the most selective solvent on the market. It is a cheap, stable, noncorrosive and easily available solvent with a broad range of further industrial applications.

Description: Raw gas from a POX of heavy residue is cooled; HCN and organic sulfur compounds are removed in prewash (1). H_2S is removed in main absorber (1) by hot-regenerated, lean solvent cooled slightly below ambient temperature. NMP traces are backwashed on top of (1) with H_2O. Laden solvent from (1) is flashed at medium pressure in a reabsorber (2). H_2S traces in the flash gas are reabsorbed by a small quantity of lean NMP. The sulfur-free gas from (2) is compressed back to the produced fuel gas (1). Flashed solvent from (2) is heated by hot lean solvent and flashed again (3). Hot-flashed gas is cooled and sent back to reabsorber (2). Solvent from (3) is finally hot-regenerated in (4). The resulting, cooled acid gas, very rich in H_2S, is processed in an OxyClaus unit, the tail-gas is hydrogenated, formation water is removed by quenching, recompressed to reabsorber (2) for desulfurization and finally ending up in fuel gas.

This closed cycle is offgas free and allows for a high degree of desulfurization. It leads to a net gain of valuable gas, plus the recycled CO_2 increasing power output of the gas turbine. This increases the overall efficiency of the IGCC plant.

Material balance for a 500-MW IGCC power plant in mol%

	Raw gas	Fuel gas
H_2	43.12	43.36
N_2 + Ar	1.49	2.31
CO + CH_4	45.9	45.95
CO_2	8.27	8.38
H_2S + COS	1.20	< 50 ppm
Flow, kmol/h	18,666.3	18,610.0
Pressure, bar	52.0	49.5

Utilities:

Power (shaft) (without power recovery)	4,300 kW
Steam, medium-pressure	20.6 tph
Water, cooling (Δt = 10°C)	1,650 m^3/h
NMP vapor loss	2 kg/h
Demineralized water	2.2 tph

Installations: Seven units in operation or under construction.

Licensor: Lurgi Oel-Gas-Chemie GmbH.

Treating

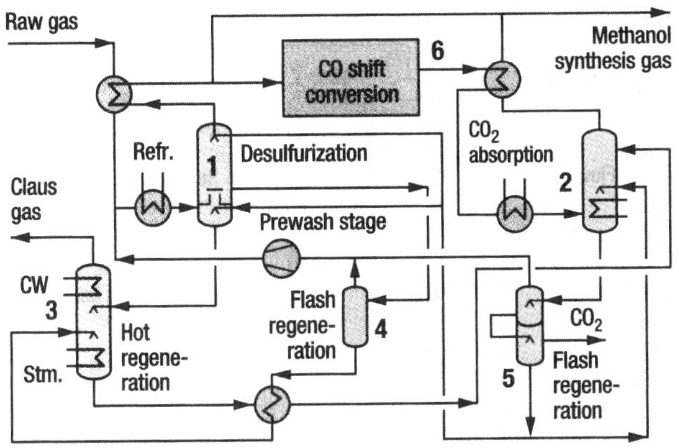

Rectisol

Application: Acid gas removal using an organic solvent at low temperatures. In general, methanol is used for H_2S, COS and CO_2 removal, whereby organic and inorganic impurities are also removed. It is possible to produce a clean gas with less than 0.1 ppm sulfur and a CO_2 content down to the ppm range. The main advantage over other processes is the use of a cheap, stable and easily available solvent, a very flexible process and low utilities.

Description: Rectisol unit for the selective desulfurization and CO_2 removal in the production of methanol synthesis gas. Raw gas (from SGP-POX) is cooled and trace components are removed in the prewash (1) with cold methanol. Prewashed gas is desulfurized (1) by using CO_2-laden solvent down to 0.1 ppm. H_2S-laden solvent is regenerated first by flashing to medium pressure (4) to recover H_2 and CO, and second, by heating to boiling temperature and stripping with methanol vapors (3). The stripped H_2S-enriched gases are sent to a Claus unit. The portion of the desulfurized gas which has been shifted in the CO-shift conversion unit (6) has a typical CO_2 content of 33%. Shifted gas re-enters the Rectisol unit, is cooled and the CO_2 is removed in a two-stage absorber (2). In the lower section, the gas CO_2 content is reduced to about 5% using flash-regenerated methanol (5). Remaining CO_2 is removed using hot regenerated (3), cold methanol in the upper section; thus, about 3% CO_2 is contained in the synthesis gas. The flashed CO_2 is free of sulfur and can be discharged to atmosphere or used further. The refrigeration balance of the system is maintained by a conventional refrigeration unit. Methanol is injected in the raw gas cooling to prevent icing. The condensed methanol/water mixture is separated in a methanol/water column (not shown).

Material balance for a 2,000-tpd methanol plant in mol%

	Raw gas	Syngas	CO_2	Claus gas
H_2	43.80	67.69	0.59	1.38
N_2 + Ar	0.25	0.25	< 0.01	0.03
CO + CH_4	52.57	29.03	0.26	8.96
CO_2	2.30	3.03	99.15	42.28
H_2S + COS	1.08	< 0.1 ppm	traces	47.35
Flow, kmol/h	8,482.5	8,415.0	1,868.7	193.5
Pressure, bar	56	48.5	1.2/Ambient	2.5

Utilities:

Power (shaft) (without power recovery)	1,640 kW
Steam, low-pressure	5.5 tph
Refrigerant at 242°K	4,200 kW
Water, cooling (Δt = 10°C)	133 m^3/h
Methanol vapor loss	40 kg/h

Installations: More than 100 units in operation or under construction.

Licensor: Lurgi Oel-Gas-Chemie GmbH and Linde AG.

Treating

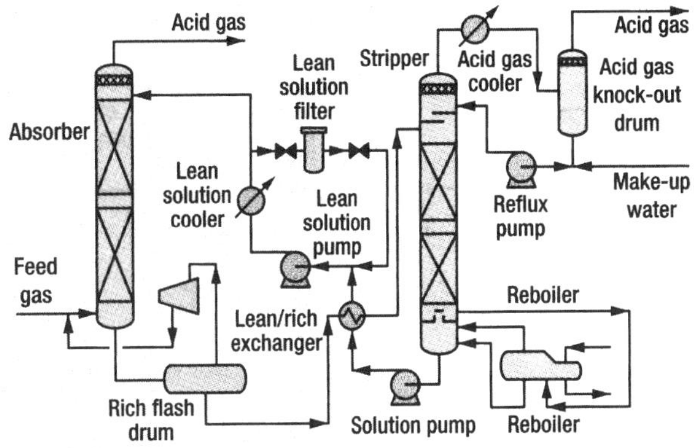

Selexol

Application: A process that can:

- Selectively remove H_2S and COS in integrated gasification combined cycle (IGCC), with high CO_2 rejection to product gas (85%+) and high sulfur (25% to 80%) feed to the Claus unit
- Selectively remove H_2S/COS plus bulk removal of CO_2 in gasification for high-purity H_2 generation for refinery or fertilizer applications, and
- Treat natural gas to achieve either LNG or pipeline specification with dew-point reduction

Description: This process uses Dow's Selexol solvent—a physical solvent made of a dimethyl ether of polyethylene glycol, which is chemically inert and not subject to degradation. The process also removes COS, mercaptans, ammonia, HCN and metal carbonyls. A variety of flow schemes permit process optimization and energy reduction. Carbon steel can be used for the materials of construction of equipment and piping due to the process's non-aqueous nature and inert chemical characteristics.

Acid gas partial pressure is the key driving force. Typical feed conditions range between 300 and 2000 psia, with acid gas composition (CO_2 + H_2S) from 5% to more than 60% by volume. The product specifications achievable depend on the application and can range from ppmv up to percent levels of acid gas.

Installations: As of January 2000, over 55 Selexol units have been put into commercial service. The Selexol process is used in many applications, ranging from natural gas to synthetic gas, and has been the dominant acid-gas removal system for gasification project awards within the past three years.

Licensor: UOP LLC.

Treating

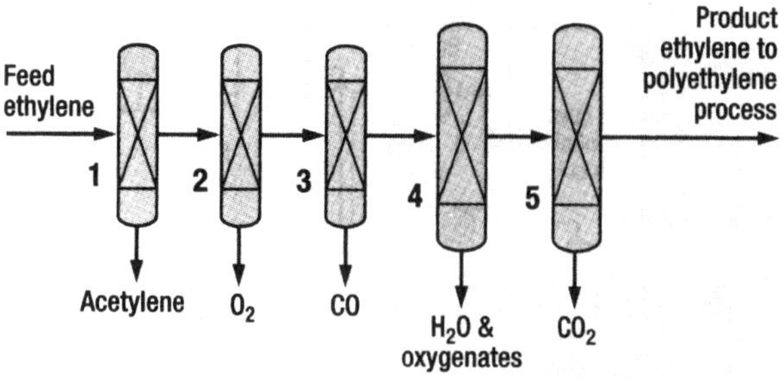

Selexsorb

Application: Purify ethylene feed stream to polyethylene production processes. Treated ethylene gas contains less than 0.2 ppmv CO_2 and H_2O and less than 1.0 ppmv oxygenates.

Description: Upstream of Selexsorb towers, the feed ethylene stream is prepurified via acetylene converter (1), tower (2) of reduced copper catalyst and bed (3) of CuO. Water and oxygenates are removed by combination bed (4) of 3A molecular sieve (H_2O adsorption) and Selexsorb CD selective adsorbent (oxygenate adsorption). Final treating step involves adsorption of CO_2 with bed (5) of Selexsorb COS selective adsorbent. Selexsorbs are thermally regenerated with nitrogen or light hydrocarbon streams. The selective adsorbents will purify: comonomer, fresh diluent, recycle diluent and reaction controlling gas (hydrogen, nitrogen) to polyethylene processes. In ethylene treating service (CO_2 adsorption), Selexsorb COS has proven its value in protecting Ziegler-Natta type catalysts in polyethylene plants worldwide.

Operating conditions: (typical)

Inlet ethylene temperature, °F	40 to 120
Inlet pressure, psig	50 to 1,500
Inlet contaminant, ppmv	
CO_2	0.5 to 50
H_2O	0.5 to saturated
Oxygenates	2.0 to 50
Regeneration gas temperature, °F	450 to 600
Regeneration gas pressure, psig	10 to 200

Installations: More than 80 installations worldwide for the treatment of ethylene and propylene feed streams to polymer plants.

Reference: Smith, D. L., "Applications for selective adsorbents in polymer production processes," *The International Journal of Hydrocarbon Engineering,* July/August 1997.

Contributor: Alcoa Inc., Alcoa World Chemicals.

Treating

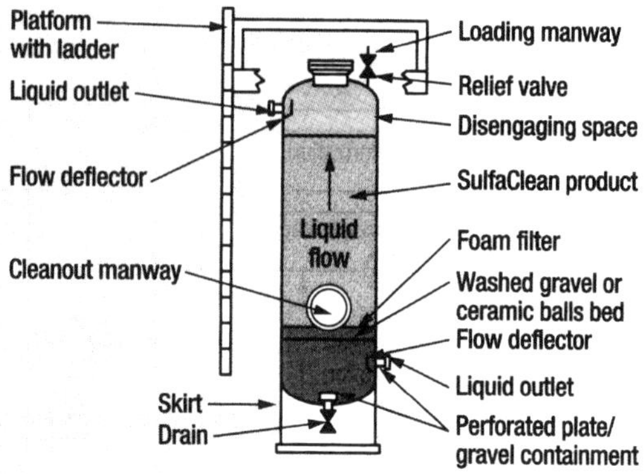

SulfaClean-HC—Liquid sweetening

Application: SulfaClean–HC is a granular material that is used to remove aggressive sulfur compounds, primarily hydrogen sulfide (H_2S), in a variety of clear liquid streams. Propane/propylene, butane, LNG, NGL, gasoline and other light liquid hydrocarbon streams treatment reduces corrosion to pass copper strip testing and meet H_2S limits. Dry or water-saturated liquid hydrocarbon can be treated, as well as water or brines for removal of dissolved H_2S prior to disposal.

Description: Liquids flow upward through the SulfaClean–HC media. Flow rates and system designs depend on fluid type and contaminants quantity. Systems range from single vessels or multiple parallel flow to lead/lag style applications. A properly designed system can adapt to variable flow rates without loss in treating efficiencies.

Economics: The SulfaClean–HC media removes up to 7.5 wt% H_2S. Its costs are often one-fourth the cost of zinc oxide products. It is competitive with caustic systems when accounting for regulatory, personal safety and disposal issues. The vessels are simple and inexpensive. Converted mole sieve, zinc oxide or iron sponge units can usually be used without modification.

Installations: This system has numerous placements worldwide. This process is currently being qualified for treatment of diverse streams such as MTBE and isooctane.

Licensor: SulfaTreat, a division of M-I L.L.C.

Treating

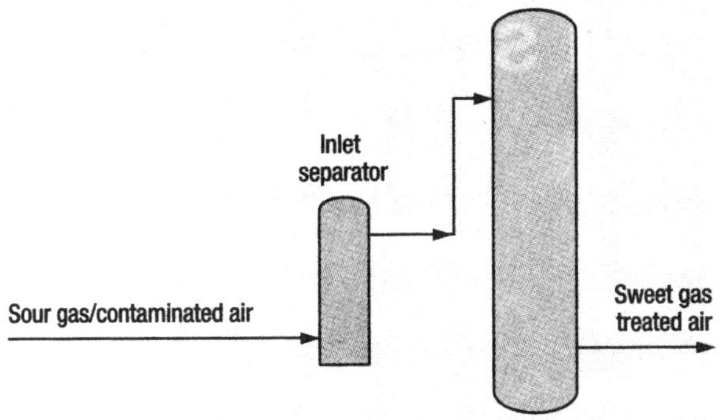

SulfaTreat—Gas or air H_2S removal

Application: Various SulfaTreat processes treat contaminated gas and air with a granular media that selectively removes hydrogen sulfide (H_2S) and some light mercaptans.

Description: Gas or vapor flows through the media, reacting with H_2S forming a stable and environmentally compliant byproduct at any point in its life cycle. Product consumption is not adversely affected by other gas components. The media prefers high humidity levels to fully water-saturate gas. Computer-assisted design matches operating conditions with desired results, such as maximum outlet allowed pressure drops (needs only a few inches W.C.) and long bed life. It works in any gas pressure or vacuum, inlet H_2S concentration and at temperatures up to 300°F. System sizes range from removing a few pounds up to over a ton H_2S daily, and to as low as 0.1-ppm maximum outlet, depending on system design.

Products:

SulfaTreat: media removes H_2S in fully water-saturated gas.

SulfaTreat–HP: media removes H_2S and light mercaptans in under-saturated gas and is faster than SulfaTreat in fully water-saturated gas.

SulfaTreat–410HP: media performs like SulfaTreat-HP but with a low pressure drop required. It is faster reacting and has twice the H_2S loading capacity in gas or air streams containing oxygen.

SulfaTreat–XLP: is a concentrated high-capacity product that lasts 2 to 4 times longer in the same vessel. It extends the operating period and reduces operating costs.

Economics: Operating costs are less than liquid reactants and have greater reliability. Operator attention is minimal. Vessels are simple and non-proprietary in design, thus lowering installation cost. Products can be used alone or may be combined with other processes that remove contaminants other than H_2S.

Installations: Over 2,000 applications globally.

Licensor: SulfaTreat, a division of M-I L.L.C.

Treating

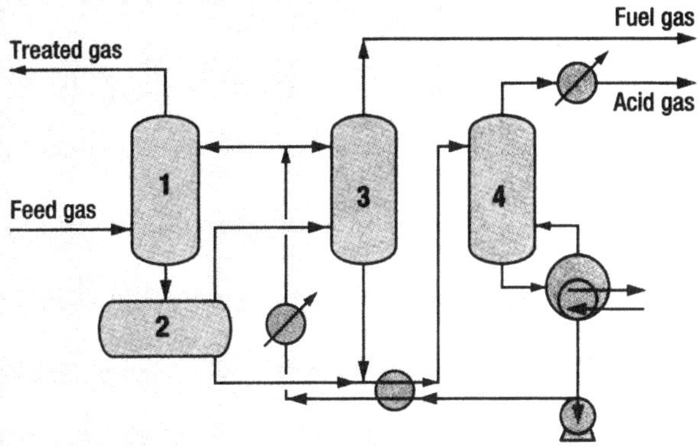

Sulfinol

Application: Removal of H_2S, COS, RSH, other organic sulfur compounds and all or part of the CO_2 from natural, synthetic and refinery gases. Total sulfur compounds in the treated gas can be reduced to ultra-low ppm levels, as required for refinery-fuel and pipeline-quality gases. An improved application is to selectively remove H_2S, COS, RSH and other organic sulfur compounds for pipeline specification, while co-absorbing only part of the CO_2.

Deep CO_2 removal for LNG plants is another application, as well as bulk CO_2 removal with flash regeneration of the solvent. The process sequence—Sulfinol/Claus/SCOT—can be used advantageously with an integrated Sulfinol system that handles selective H_2S removal upstream and the SCOT process that treats the Claus offgas.

Description: The mixed solvent consists of a chemical-reacting alkanolamine, water and physical solvent Sulfolane (tetra-hydrothiophene dioxide). The actual chemical formulation is customized for each application.

Unlike aqueous amine processes, Sulfinol removes COS, RSH and other organic sulfur compounds to stringent total sulfur specifications. The process achieves 4 ppm H_2S pipeline specification at low stream consumption. Observed corrosion rates are low and little foaming is experienced.

The system line-up resembles that of other amine processes. In most applications, co-absorbed hydrocarbons from the absorber (1) are flashed (2) from the solvent and used as fuel gas after treatment in a fuel-gas absorber (3). Loaded solvent is regenerated (4).

Operating conditions: Very wide ranges of treating pressures and contaminant concentrations can be accommodated. Refinery-fuel gas and gas pipeline specifications, such as 40 ppmv total sulfur and 100 ppmv H_2S, are readily met. Removal of organic sulfur compounds is usually done for the circulation set by H_2S and CO_2 removal. In LNG plants, a 50 ppmv CO_2 specification is easily attained.

Installations: Over 200 units have been licensed worldwide, covering natural gas treating, synthesis gas and refinery gases.

References: “Gas pretreatment and their impact on liquefaction processes,” GPAl, March 1999, Nashville.

“Process application of the ADIP and Sulfinol Process,” Gas Processing Symposium, Dubai, United Arab Emirates, April 1999.

“A mixed solvent for low total sulfur specification,” AIChE national meeting, San Diego, Aug. 21, 1990.

Licensor: Shell Global Solutions International B.V.

NGL and LNG

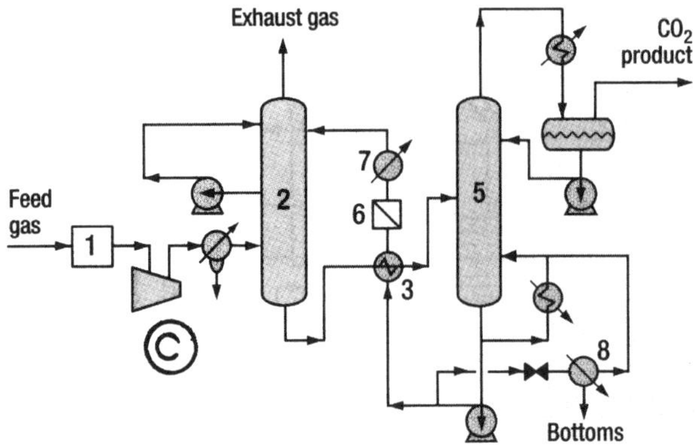

CO_2 recovery

Application: Recover high-purity including food-grade, CO_2 from oxygen-containing gases such as boiler flue gases, gas turbine exhausts and waste gases using Kerr-McGee/ABB Lummus Global absorption/stripping technology.

Description: CO_2-containing feed gases are first cooled and scrubbed (1), if necessary, to reduce SO_2 levels. The gases are boosted slightly in pressure before entering the recovery system.

The system is based on absorption/stripping using a 15% to 20% monoethanolamine (MEA) solution. Feed gases are sent to an amine absorber (2) where they are scrubbed with MEA to recover CO_2. The scrubbed gases are vented to the atmosphere after water washing in the absorber's top to minimize MEA losses. Rich solution from the MEA absorber is preheated in an exchanger (3), flashed and sent to a stripper (5) where CO_2 is recovered overhead. Condensate from the stripper overhead is returned to the system.

Lean MEA from the stripper (5) is cooled (3,7), filtered (6) and returned to the absorber. Periodically, a batch reclaiming operation (8) is conducted to purge MEA degradation products and to recover MEA by decomposing heat-stable salts. The bottoms from the reclaiming operation may be burned as boiler fuel.

CO_2 recovered from the stripper overhead may be compressed and used as a vapor product, or dried and liquefied using a standard ammonia refrigeration system to produce a liquid product.

Operating conditions: Operating units have exhibited availability factors in excess of 98%. Absorption and stripping operations take place at slightly above atmospheric pressure. Feed gases can contain up to 15 vol% oxygen, though economics are favored by high CO_2 and low oxygen concentrations in the feed. The process can recover CO_2 from flue gases containing from 3 to 15 vol% (dry basis) CO_2. Moderate levels of SO_2 and NO_x in the feed are acceptable. SO_2 prescrubbing is required only with SO_2 levels higher than 100 ppmv.

Economics: Typical capital investment for a 200-tpd CO_2 plant is $9 million. Liquefaction facilities would add roughly $4 million to the capital investment. Typical utility and chemical requirements per ton of CO_2 recovered are as follows:

Product	Gaseous	Liquid
Steam, LP, t	2.7	2.7
Water, cooling, gal	21,000	25,000
Power, kWh	100	240
Chemicals, $	2.25	3.00

Installations: Four units are operating on coal-fired boiler flue gases. Two plants produce gaseous chemical-grade CO_2 and one produces food-grade liquid. Capacities range from 200 to 800 tpd.

Licensor: ABB Lummus Global Inc.

NGL and LNG

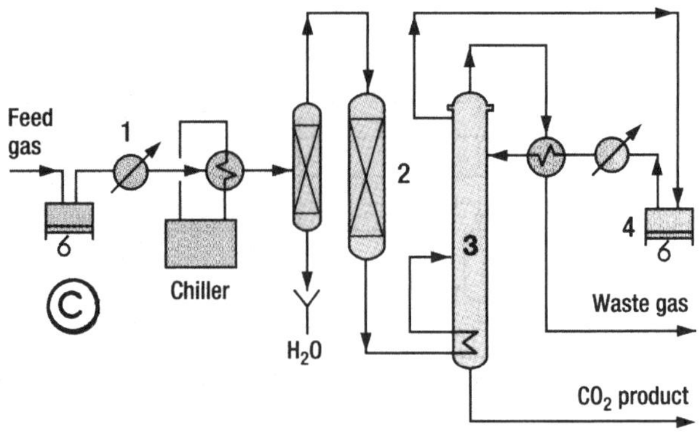

CO_2 recovery and purification

Application: Purification and liquefaction of carbon dioxide (CO_2) from process gases to remove small traces of light hydrocarbons, e.g., amine stripper off-gases. Purification of natural gas containing high levels of CO_2 and separation of ethane/CO_2, via a hybrid membrane/cryogenic process.

Products: Carbon dioxide of purities up to 99.998%. Gaseous CO_2 for use in applications such as enhanced oil recovery (EOR).

Description: Impure CO_2-rich feed (typically 90% + CO_2) is compressed and cooled to remove water (1). The gas is dried by molecular sieve driers (2) before entering the low-temperature section, where the gas is used to reboil the distillation column (3) before entering the column. The condenser is usually a reflux exchanger. Vapor rising through the exchanger passages is cooled, and the resulting liquid flows back down the same passages and has intimate contact with the rising vapor. The cold reflux to the column is provided by an external refrigeration cycle (4), typically ammonia evaporating at –28°C. The bottoms product from the column is pure liquid CO_2.

For recovery from lean-CO_2 streams, a hybrid membrane/cryogenic process can produce pure CO_2. Typically, the membrane concentrates the CO_2 content from 30% to 90%. The concentrated stream passes to a low-temperature unit for further purification. This method enables ethane/CO_2 mixtures to be separated to pure products without multiple distillation columns.

For recovery from rich-CO_2 streams, enhanced recovery can be achieved with a membrane to process the distillation column overheads and recycling of the CO_2-rich permeate into the cryogenic process.

Operating conditions: Plants from 5-tpd to 1,200-tpd pure CO_2. Virtually any feed-gas pressure can be handled. Compression is required for feed-gas pressures below 14 barg.

Economics: Typical power requirements for a cryogenic plant with a feed-gas at atmospheric pressure and producing liquid CO_2 are 0.15 kWh/kg.

Installations: Seven.

References: Limb, D. I., "The purification and liquefaction of carbon dioxide-rich vent streams containing light hydrocarbons," AIChE 1985 Spring National Meeting.

Duckett, M., "Recovery of carbon dioxide from gas mixtures," UK Patent No. 2151597.

Duckett, M. and D. I. Limb, "Process for recovering carbon dioxide," U.S. Patent No. 4639257.

Licensor: Costain Oil, Gas & Process Ltd.

NGL and LNG

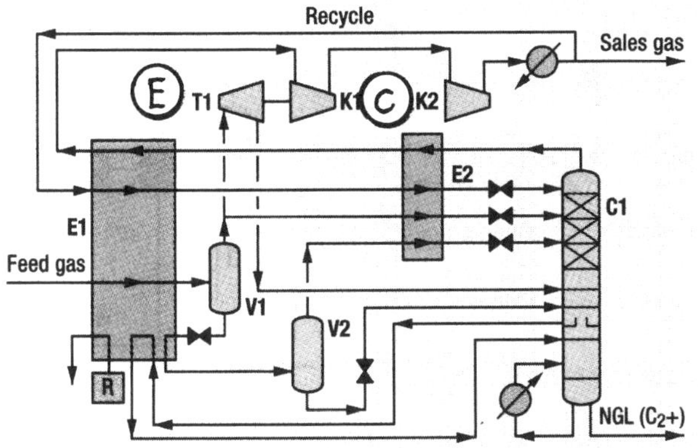

CRYOMAX MRE (multiple reflux ethane recovery)

Application: A cryogenic process for gas fractionation to recover C_2^+ hydrocarbons from natural gas. With this process, more than 95% of the ethane can be extracted from natural gas. High efficiencies are obtained through a multiple reflux concept associated with a turbo-expander. Multi-stream plate-fin heat exchangers increase efficient heat integration of the process.

Description: The high-pressure, dry feed gas at 25°C, 70 bars is cooled to −40°C in E1 and enters V1 where the liquid and gas are separated. The cold high-pressure gas is divided in two streams—the main cut (85%) is sent to the expander and to the demethanizer C1 that operates at 30 bars. The small segment (15%) is liquefied and sent as second reflux to C1.

The liquid pressure is reduced to 50 bars, and the liquid is partially vaporized in E1. Liquid and vapor are separated in V2. The vapor is liquefied in E2, used as third reflux and liquid is sent to C1.

The demethanizer overhead is reheated and compressed to sales gas pipeline specifications. A portion of the stream (10%) is recycled, cooled, liquefied and sent as first reflux. Approximately 99% ethane recovery can be reached when CO_2 content of feed gas is low.

Economics: Ethane production cost is 20% less than for a conventional process.

Installations: Locations include: France, U.S., Venezuela, Mexico and Qatar.

Reference: U.S. patents 4 689 063.5 566 554 and new U.S. patent application.

Licensor: Technip-Coflexip.

NGL and LNG

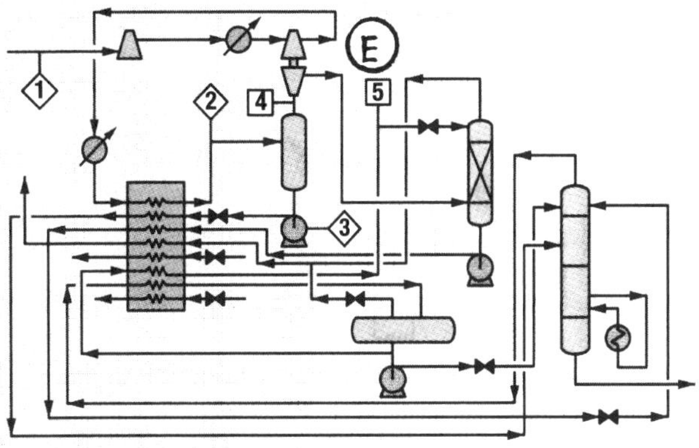

Cryo-Flex

Application: The Cryo-Flex process is designed to recover ethylene/ethane or propylene/propane from refinery/petrochemical plant offgases or natural gas streams. This operation can be quickly transformed, by the plant operations team, from an ethylene/ethane recovery to a propylene/propane recovery mode, and then returned to the original operation mode just as quickly.

Description: Dry, contaminate-free inlet gas (1) is partially condensed (2). Separated liquids (3) are heated and fractionated. Separated vapor (4) is expanded into a second fractionator where subcooled reflux liquid (5) from the first fractionator is used as reflux.

As the name Cryo-Flex implies, the plant is flexible in the recovery of the desired product and this technology spans the refining, petrochemical and natural gas processing industries. In addition to this tremendous flexibility, the methodology can offer energy savings of 5–30% in compression horsepower when compared to other state-of-the-art hydrocarbon liquid recovery technologies. Alternatively, the Cryo-Flex process can also recover more valuable product for the same compression requirement than other methods.

The technology incorporates many features that include the flexibility to handle varying feedrates and feed-gas compositions. The key benefits are:

- Higher product recovery
- Lower energy consumption
- Reliable process control
- Ease of operation
- Proven concepts and history of success.

Economics: LPP is very sensitive to client capital expenditure limits; thus, the technology offers a relatively low-cost solution and yields desirable revenue increases. It provides the opportunity to install a project with a very high rate of return and a very short (normally less than 1 to 2 years) payout. More operating gas processing plants use LPP's patented technologies than from any other manufacturer.

References: The U. S. Patent Nos. 6,311,516 B1, Process and Apparatus for C3 Recovery, and 6,278,035 B1, Process for C_2 Recovery, for Cryo-Flex.

Licensor: Linde Process Plants, Inc., a member of the Linde AG Engineering Division.

NGL and LNG

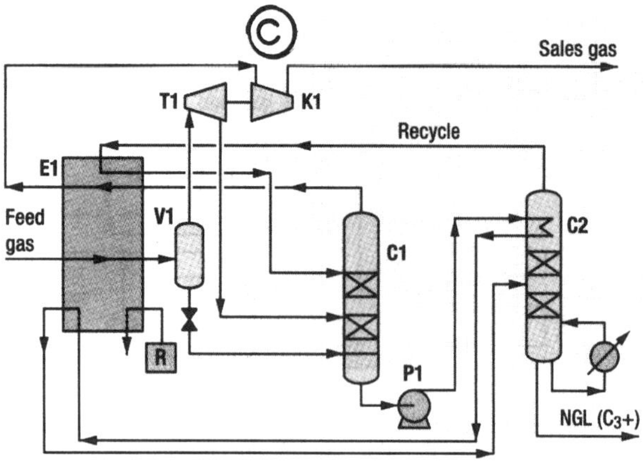

CRYOMAX DCP (dual-column propane recovery)

Application: A cryogenic process for gas fractionation to recover C_3^+ hydrocarbons from natural gas. With this process, more than 98% propane is extracted from natural gas. High efficiencies are achieved with a dual-column system associated with a turbo-expander. Multi-stream plate-fin exchangers increase efficient heat integration.

Description: The high-pressure dry feed gas at 25°C, 70 bars is cooled to –30°C in E1 and enters V1 where liquid and gas are separated. The cold high-pressure gas is expanded to 30 bars in expander T1, and the resulting stream feeds the purifier C1.

Liquid from V1 is sent to the purifier bottom. The liquid from the purifier is pumped to 33 bars and is reheated to 20°C to feed the deethanizer. The deethanizer C2 produces a vapor distillate that is ethane-rich. This stream is liquefied in E1 and sent to C1 as reflux. The treated gas at 30 bars is reheated and compressed to sales-gas pressure. Approximately 99.5% propane recovery can be reached when propane value is high.

Economics: Propane production cost is approximately 20% less than that for a conventional process.

Installations: Locations include: Russia, Qatar, Libya and UAE.

Reference: U.S. Patents 4 690 702 and 5 114 450.

Licensor: Technip-Coflexip.

NGL and LNG

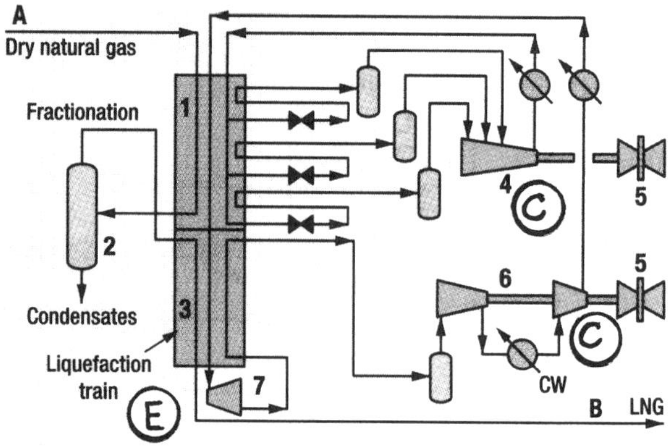

LIQUEFIN

Application: Natural gas liquefaction (LNG) process.

Description: Liquefin process—Dry natural gas (A) enters the liquefaction train's pre-cooling section (1) where it is cooled between –50°C and –80°C (–60°F and –110°F). The heat is exchanged with a mixed refrigerant in a bank of brazed aluminum plate-fin heat exchangers (PFHEs). The cooled feed stream is then sent outside the liquefaction train to fractionator (2) for condensate removal. The gas returns to the liquefaction train and enters the cryogenic section (3) where it is liquefied by heat exchange in compact, energy-efficient PFHEs with a second mixed refrigerant and leaves the train as LNG (B).

Pre-cooling refrigerant system (4)—Using mixed refrigerant reduces the feed-gas temperature to a much lower level than can be achieved with propane refrigerant. This allows the pre-cooling power requirement to be balanced with that of the cryogenic section so that two identical drivers (5) operating at optimum efficiency can be used, thus lowering investment, maintenance and operating costs.

Cryogenic refrigeration system (6)—The mixed refrigerant gas entering the pre-cooling section is completely condensed by the time it leaves the cryogenic section without using separation equipment. After leaving the cryogenic section, the refrigerant is expanded (7) and re-enters the cryogenic section where the process gas and cryogenic refrigerant are condensed.

Using two mixed-refrigerant systems and modular PFHEs in a large single train has lower investment and operating costs than systems involving single-component refrigerants or multiple cooling trains feeding a common liquefaction exchanger. Because the pre-cooling and liquefaction sections comprise several parallel modules, single trains of any size—e. g., 6 million tpy (6 MMtpy)—can be built. Proven extensively, the PFHEs are available from several vendors, which has a very positive effect on price and delivery time.

Economics: Detailed studies by international petroleum and E&C companies comparing conventional 4.5 to 6 MMtpy propane-plus-mixed refrigerant liquefaction trains with LIQUEFIN LNG installations have shown a 15-20% specific investment cost advantage for the LIQUEFIN train.

Reference: Fischer, B. and P. Boutelant, "A new LNG Process is now available," Gas Producers Association, London, February 2002.

Licensor: Axens.

NGL and LNG

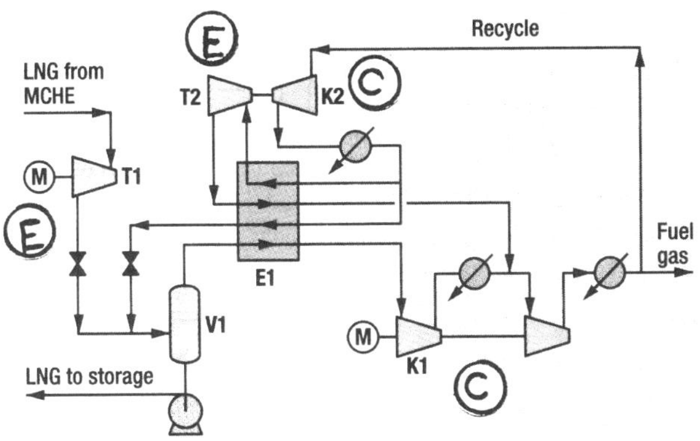

LNG end flash MLP (maxi LNG production)

Application: A process to increase capacity of LNG plants and minimize fuel gas production.

Description: LNG from the main cryogenic heat exchanger at –140°C, 40 bar is expanded in T1 turbine to produce electric power. It is mixed with liquefied fuel gas recycle before entering the LNG flash drum V1 at 1.2 bars. The LNG is pumped to storage tank through P1.

Cold flash gas is reheated to –30°C and compressed to 30 bar, thus producing fuel gas for gas turbines in K1. Part of the fuel gas is further compressed to 40 bars in K2. A portion of this stream is cooled to –80°C in E1, expanded to 8 bars at 130°C in T2. It is then reheated to 30°C and sent to the fuel gas compressor as a side stream. The second stream portion is liquefied and subcooled in E1 down to –155°C.

Economics: Additional LNG production investment of about \$60/tpy for additional LNG.

Installations: None with turbo-expander T2/K2.

Reference: U.S. patent application and LNG 14 conference, Seoul, paper PS2-1.

Licensor: Technip-Coflexip.

NGL and LNG

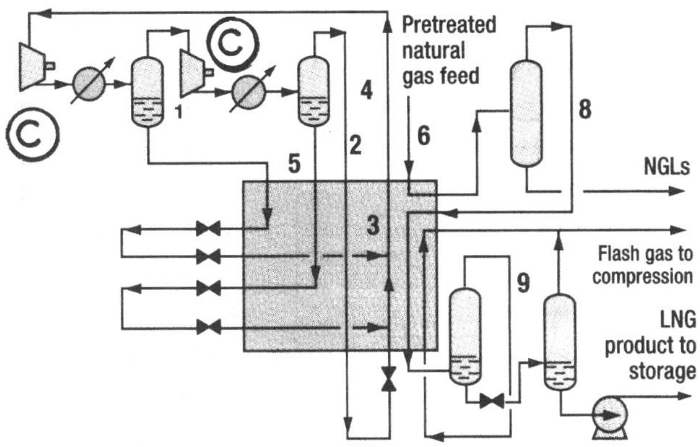

LNG plants

Application: Liquefaction of natural gas for plant capacities ranging from small peak shaving applications, up to mid-size plants (4,000 tpd, 1.4 MMtpa) using a mixed refrigerant cycle.

Products: Liquefied natural gas (LNG) at atmospheric pressure. Natural gas liquids (NGL) on larger facilities.

Description: Pretreated natural gas is cooled and condensed by a multi-stage mixed-refrigerant cycle. The refrigerant consists of a blend of nitrogen and hydrocarbons from methane through pentane.

The refrigerant is compressed by a two-stage machine (1) (normally a gas turbine driven centrifugal type depending upon plant capacity). The high-pressure mixed refrigerant is cooled (2) in the main exchanger (3), which normally consists of multiple, brazed aluminum plate-fin heat exchangers, against returning low-pressure mixed refrigerant (4). The subcooled refrigerant is then let down in pressure and evaporated to provide cooling. Liquids from refrigerant compression are cooled separately (5) in the main exchanger, let down in pressure and evaporated to provide increased process efficiency.

The natural gas is cooled (6) in the main exchanger prior to entering a hydrocarbon knockout pot (7) to remove components which would otherwise freeze in the downstream process. On large plants, the knockout pot may be replaced by a demethanizer column. NGLs recovered at this stage may be processed and used to provide refrigerant makeup. The natural gas leaving the knockout pot re-enters (8) the main exchanger and is condensed and subcooled against low-pressure refrigerant.

The subcooled LNG then enters a two-stage flash system (9) where it is let down in pressure before being pumped to storage at near atmospheric pressure. The LNG flash gas is fed to a flash gas compressor system to be used as fuel.

Economics: The mixed refrigerant cycle is often the most cost-effective process for LNG production, combining reasonable initial cost with low power requirements. The use of plate-fin heat exchangers allows the plant to be designed with high efficiency. For a 1.4 MMtpa-facility, a total installed cost of \$300 per tpa is feasible.

For small (peak-shaving) installations, expander cycles using nitrogen or methane may be cost-effective. Expander cycles can also be the technology of choice for offshore applications. For high efficiency on base-load installations, the cascade cycle, which uses multiple levels of pure refrigerants, can be the optimum choice.

Installations: Fifteen total, in partnership (six mixed refrigerant, nine expander plants).

Reference: Finn, A. J., G. L. Johnson and T. R. Tomlinson, "Developments in natural gas liquefaction," *Hydrocarbon Processing*, April 1999, pp. 47–59.
European patent No. 0090108.

Licensor: Costain Oil, Gas & Process Ltd.

NGL and LNG

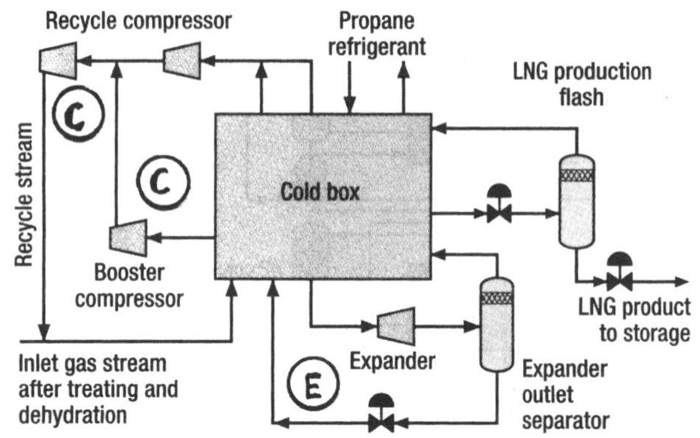

LNG-Pro

Application: To produce liquefied natural gas for transportation or storage. The process is adequate for base-load facilities, as well as for peak shaving units. The design is highly modularized, making it appropriate for remote sites or offshore applications. It is also a viable option to monetize stranded gas reserves.

Description: This process uses a hybrid technical approach for the liquefaction of natural gas. Specifically, it is a propane pre-cooled turbo-expander cycle. After being treated to remove contaminants and water that will affect the cryogenic process conditions, the feed gas is liquefied in a cold box. A side stream of the inlet gas is expanded to low pressure to generate refrigeration. The expanded gas is then sent to the cold box to supply refrigeration, and is boosted to a medium-pressure level. It is then recycled to the front end of the plant. The liquefied gas stream is expanded in a flash vessel or series of flash vessels, depending on product specifications. The liquid product is LNG, which is sent to storage. The flashed gas is sent back to provide refrigeration, and is then recycled to the front end of unit via recycle compression. This recycle stream becomes a supplementary refrigeration stream.

This process scheme achieves energy consumption comparable with the world-class base load facilities. Energy usage is within 0.19 to 0.25 Hp/lb of LNG product depending on gas sources and compositions.

Operating conditions: Ample range of pressures, temperatures and compositions. Depending on the inlet gas pressure, an inlet gas compressor could be required, but its influence is minor on energy consumption.

Performance index

	Relative HP/lb LNG
Cascade refrigeration	1.41–1.64
Mixed refrigerants	1.15–1.4
Turboexpander cycle	1
High-efficiency cycles	0.87–1.15
Utilities	
Electric power	
Recycle compr.	77%
Refrig. compr.	13%
Others	10%
Cooling duty	
Recycle compr.	60%
Booster aftercooler	11%
Refrig. condenser	29%
Fuel gas usage	
Power generation	~6%

Reference: U.S. Patent 5,755,114.

Licensor: Randall Gas Technologies, ABB Lummus Global Inc.

NGL and LNG

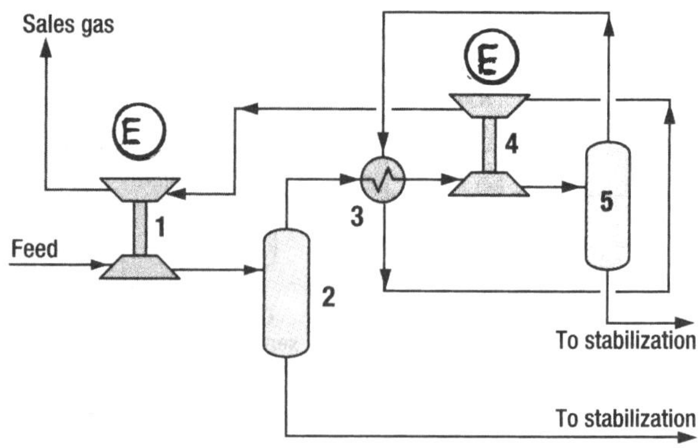

NGL recovery

Application: Deep recovery of NGL from natural gas.

Products: Sales gas and stable NGL, with a propane recovery as high as 99%. For an ethane-extraction plant, ethane recovery can be over 95%, with a propane recovery of essentially 100%.

Description: With high feed-gas pressure and particularly with dense-phase operation (above the cricondenbar), two expanders used in series enables the design pressure of the gas/gas exchanger to be reduced significantly. It also enables a plate-fin heat exchanger to be used, thus improving process efficiency as compared to using a shell and tube exchanger.

For very high-pressure ratios across the plant, two expanders in series avoid very high liquid flows in the expander exhaust, thus improving the performance and reliability of the turboexpander system.

Dehydrated feed gas is let down in pressure by the first expander (1), and the exhaust is passed to the first separator (2). Vapor from the first separator is cooled and condensed in the gas/gas exchanger (3) and let down in pressure by the second expander (4). The exhaust is passed to the second separator (5). Unstable liquids recovered from both separators are sent on to be stabilized. Cold vapor from the second separator is rewarmed and recompressed in the brake ends of the expanders to sales gas pressure.

Operating conditions: A very high propane recovery of 99% can be achieved by using a pre-absorber column and optimizing thermal integration with multistream heat exchangers.

Improved energy integration enables NGL recovery to be increased to very high levels with similar power consumption to conventional technology. Very high ethane recovery, of over 95%, can be achieved by using a portion of the sales gas for demethanizer refluxing and by utilizing refrigeration efficiently with good thermal integration.

Installations: Twelve, with capacities up to 11 MM Sm^3/h.

References: Finn, A. J., T. R. Tomlinson and G. L. Johnson, "Design equipment changes make possible high C_3 recovery," *Oil and Gas Journal,* Jan. 3, 2000, p. 37.

European patent no. 1137616 (Propane-plus)

European patent no. 1114808 (Ethane-plus).

Contributor: Costain Oil, Gas & Process Ltd.

NGL and LNG

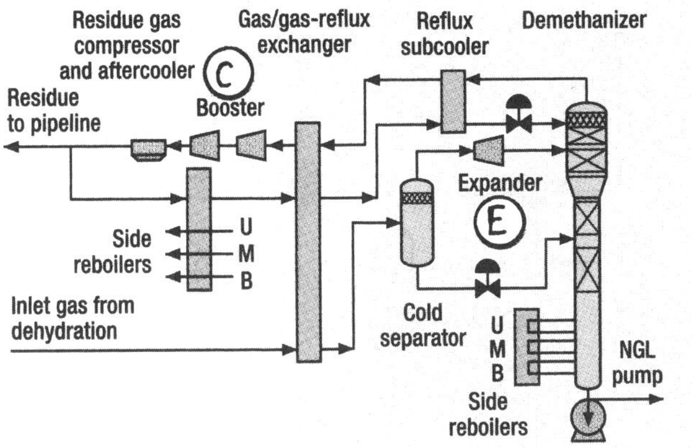

NGL-Pro

Application: To recover ethane and higher components from natural gas streams. This process is especially adequate with lean and semi-lean gas streams.

Description: The inlet gas is treated and conditioned to remove contaminants and water that cannot be processed cryogenically. The gas is partially condensed to knock out heavy hydrocarbons, and is sent to a cold separator. Removed liquids are sent to the demethanizer column, while the gas is sent to a turboexpander. The stream from the expander is sent to the demethanizer column. To increase the recovery level, a recycle/reflux stream is taken from the residue gas and is precooled, extracting refrigeration from the demethanizer column. The reflux is finally subcooled and sent to the top of the demethanizer.

This process is also adapted for the rejection of ethane and production of LPG and heavier streams. Recoveries for ethane product are 95^{+}%. Energy consumption is between 50 to 70 HP/MMscfd depending on gas conditions.

Operating conditions: Ample range of pressures and temperatures. Hydrocarbon liquid content less than 3.5 gal/Mscf.

Reference: U.S. Patent 5,890,377.

Licensor: Randall Gas Technologies, ABB Lummus Global Inc.

NGL and LNG

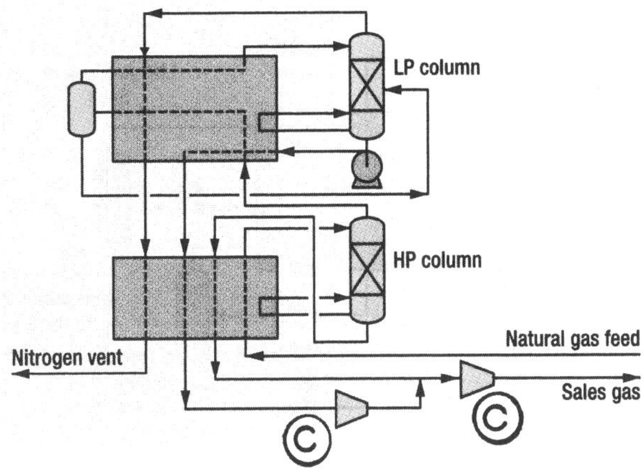

Nitrogen rejection

Application: Reject nitrogen from natural gas to increase heating value.

Description: Natural gas feed to the nitrogen rejection unit (NRU) is partially condensed and fed to the top of the high-pressure (HP) column. A reboiler driven by condensing the feed provides stripping vapors in the column. The bottoms from the first stripping column are enriched in hydrocarbons and depleted in nitrogen. This HP residue is boiled in the warm NRU exchanger and sent to residue gas compression. Enriched in nitrogen, the overhead vapor from the column is sent to the cold end of the NRU.

Vapor from the HP column is partially condensed against returning residue streams. The vapor and liquid are separated in a vertical separator drum. The vapor stream, enriched in nitrogen, from the drum is further condensed and is fed as reflux to the low-pressure (LP) column.

Liquid from the drum is the feed to the LP column. Vapor generated in the reboiler strips the descending liquid of nitrogen creating the remainder of the residue product. The liquid flowing down the column scrubs the methane from the vapor, creating a high-purity nitrogen vent stream from the top of the column.

Liquid product from the LP column is pumped to LP residue pressure. This stream is boiled against the condensing HP vapor from the overhead of the first stripping column. The nitrogen vent stream and LP residue stream are warmed in both exchangers in the NRU along with the HP residue gas stream. The residue streams are sent to recompression for export to pipeline systems. Nitrogen is vented to the atmosphere or recompressed for reinjection.

Operating conditions: The dual-column NRU process has good CO_2 tolerance and feed flexibility, handling feeds from 5 to 80 mol% nitrogen and pressures as low as 250 psig (17 barg).

Economics: The dual-column NRU has high hydrocarbon recovery (>99.9%). This process is particularly well-suited for streams with less than 20% nitrogen in the feed, and is easily adapted to recover helium (HeRU) as well. The hydrocarbons are recovered at two pressure levels, reducing recompression requirements. The amount of hydrocarbons recovered from the second column is greatly reduced requiring a smaller cryogenic pump.

Installations: Eight NRU and HERU installations, with three using the dual-column process with capacities from 30 MMscfd to 70 MMscfd.

Reference: Janzen, K. H. and S. R. Trautmann, "Innovative NRU design at Pioneer Natural Resources' Fain gas plant," 2000 GPA Convention.

Licensor: Air Products and Chemicals, Inc.

NGL and LNG

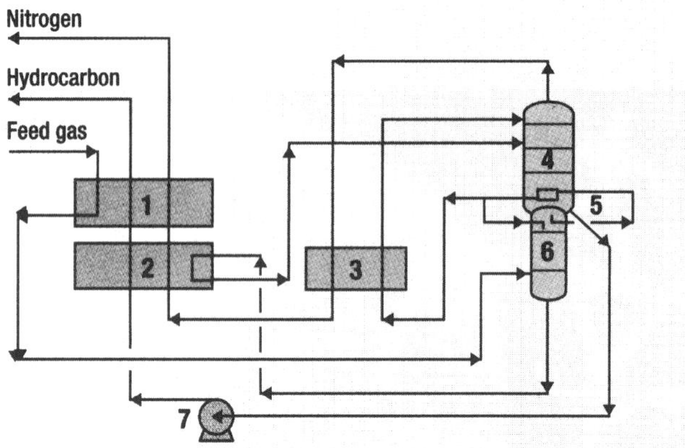

Nitrogen removal (reject)

Application: Remove nitrogen from natural gas to increase calorific value and/or to reduce gas volume for compression.

Description: Natural gas is pretreated to remove constituents that can freeze in the subsequent cryogenic process or affect product quality. After cooling (1) against hydrocarbon product and waste nitrogen, the feed is expanded into the lower (high-pressure) distillation column (6) of the linked pair. Vapor rising through the column is rectified to yield almost pure nitrogen. It condenses against boiling hydrocarbon in the condenser/reboiler (5) located in the upper (low-pressure) column (4). If helium is present in the feed gas, a purge stream containing helium can be withdrawn from the condenser/reboiler for further enrichment.

Liquid nitrogen is taken off the top of the lower column and subcooled (3) by low-pressure nitrogen. Part of the liquid nitrogen provides reflux to the upper column. Methane-rich liquid from the base of the lower column is drawn off, subcooled (2) and fed to the upper column. A waste-nitrogen stream, typically containing less than 0.5% methane, is drawn from the top of the upper column. A hydrocarbon stream is withdrawn from the base and pumped to product pressure by the hydrocarbon pump (7). Waste nitrogen and hydrocarbon product are heated to ambient against the natural gas feed to provide refrigeration to the process.

For low-nitrogen content feeds, alternative process flowsheets using a heat-pumped single-column design, or a prefractionation column upstream of one or two further columns, give improved performance.

Operating conditions: The double-column process is sufficiently flexible to handle natural gas with nitrogen concentrations varying from 5 to 80 mol%, and can be a good choice for variable content streams associated with enhanced oil recovery (EOR). Feed gas above 27 bar can be processed without any compression. For feed gas containing heavy hydrocarbons or a low-nitrogen content, a three-column process is more efficient. The third column also improves plant tolerance to CO_2, which may simplify gas pretreatment requirements.

Economics: The double-column arrangement can offer several benefits compared with conventional processes, especially for a feed gas nitrogen content above 20%. No power-consuming, heat-pump cycles are required, and machinery needs are reduced. In addition, all hydrocarbon product can leave the plant at high pressure, which reduces recompression requirements.

Installations: Five, with capacities from 80,000 Sm^3/h to 350,000 Sm^3/h.

Reference: Healy, M.J., A. J. Finn and L. Halford, "UK nitrogen removal plant starts up," *Oil and Gas Journal,* Feb. 1, 1999, p. 36.

Licensor: Costain Oil, Gas & Process Ltd.

NGL and LNG

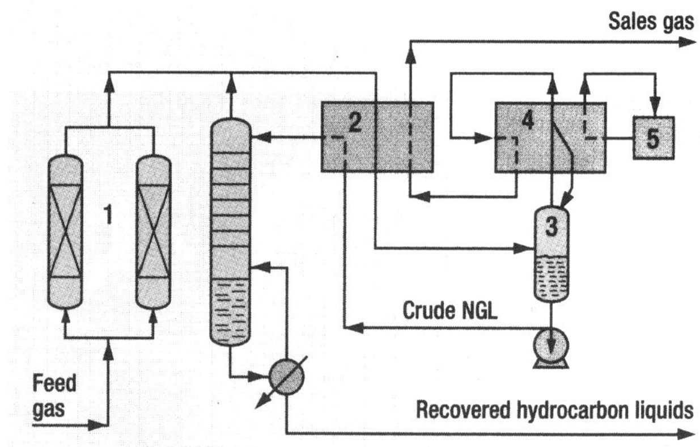

PetroFlux

Application: Recovery of light olefins—typically ethylene, propylene or butylene—from offgases using reflux exchanger technology as part of a low-temperature recovery system (LTRS). Reflux exchanger technology can also be applied to ethylene chill trains and to NGL recovery from refinery and petrochemical offgases, as well as from natural gas.

Products: Olefins with a typical recovery of 99%. For NGL recovery: 95+% recovery of propane and 100% recovery of heavier hydrocarbons.

Description: After removal of water and freezable components in the molecular sieves (1), the gas is partially liquefied in the plate-fin heat exchanger (2) and separated in the feed separator (3). The vapor formed passes upwards through the passages of a refluxing plate-fin heat exchanger (4) where it is cooled and condensed. Condensed liquids flow countercurrently to the gas down the heat exchanger passages and are fractionated before collecting in the feed separator. The resulting gas is then rewarmed and sent to product/fuel gas while the liquids collected are stabilized. Refrigeration (5) may be provided by a separate refrigeration cycle or by work expansion of the product/fuel gas. The choice of refrigeration scheme depends on both feed and product/fuel gas pressures.

Operating conditions: Typically, feed gas flows of up to 100,000 Sm^3/h can be processed. Feed gas pressure is typically between 10 and 70 barg.

Economics: The PetroFlux process offers high recoveries and significant savings in refrigeration costs, giving attractive returns on investment. It has been used widely for plant debottlenecking and to make more effective use of existing refrigeration systems.

Installations: The basic refluxing heat exchange system has been used worldwide on 28 plants.

References: Finn, A. J., "Enhance gas processing with reflux exchangers," *Chemical Engineering,* May 1994, p. 142.

Tomlinson, T. R. and D. R. Cummings, "Separation of hydrocarbon mixtures," U.S. patent no. 4622053.

Tomlinson, T. R. and B. A. Czarnecki, "Separation of hydrocarbon mixtures," U.S. patent no. 4846863.

Licensor: Costain Oil, Gas & Process Ltd.

NGL and LNG

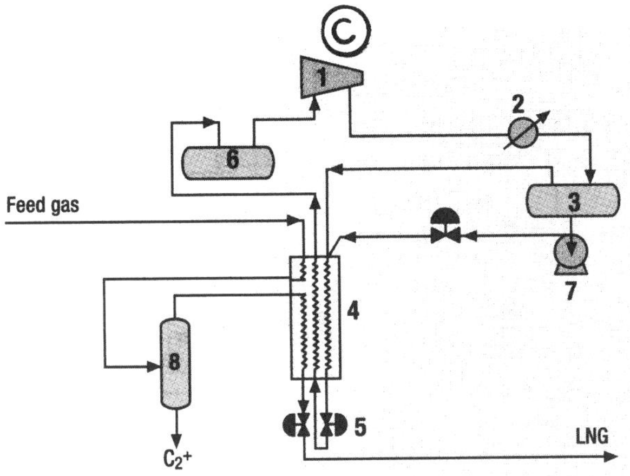

Prico (LNG)

Application: Convert natural gas to liquid form for transportation and/or storage at atmospheric pressure by using the Prico mixed refrigerant loop. Applications range from large base load LNG units to small peak-shaving units including modular applications. Also applicable to NGL extraction (high ethane recovery) from natural gas.

Product: Natural gas, essentially methane, in liquid form near –260°F at atmospheric pressure. Side products of ethane, LPG and gasoline may be recovered in liquid form, when present in the feed gas.

Description: The process is a very simple, efficient, reliable and cost-effective mixed-refrigerant cycle. A single mixed-refrigerant, composed of nitrogen and light hydrocarbons from methane through pentane, is circulated in a closed refrigeration loop. This loop contains a compressor (1), a partial condenser (2), an accumulator (3), a refrigerant heat exchanger (4), a Joule-Thomson expansion valve (5) and a refrigerant suction drum (6). Accumulator liquids are directed to the refrigerant heat exchanger with a low head centrifugal pump (7). A single case centrifugal or axial flow compressor can be used with or without intercooling.

Proportions of individual refrigerant components are adjusted to match the combined cooling duty curve.

The refrigerant heat exchanger (4) is a plate-and-fin brazed aluminum core. Multiple cores are arranged in parallel to provide desired production capacity and allow expansion easily.

Natural gas feed is pretreated for removal of CO_2 to less than 50 ppmv and dried to less than 1 ppmv water by conventional methods. Extraction of other hydrocarbons from the natural gas feed is accomplished by intermediate partial condensation (8). The heavy hydrocarbons may be separated by subsequent fractionation into ethane, propane, butanes and gasoline components. Nitrogen in the natural gas feed is dealt with according to the LNG product specification requirements by adjusting the liquefaction conditions.

Recent improvements in the system design have resulted in power savings of over 20% from previous designs while maintaining a significant cost advantage over other systems. The process matches well with gas turbine, steam turbine or electric motor driven compression systems.

Installations: Eight peak-shaving units up to 16 MMscfd each are now being successfully operated. Three 180 MMscfd base load plants with ethane and heavier hydrocarbons extraction as separate products also use the process.

Reference: Price, B. C., et al., "Deveolpment of mid-scale and floating LNG facilities," Gas Processors Association Annual Convention, San Antonio, March 1997.

Licensor: Black & Veatch Pritchard Inc.

NGL and LNG

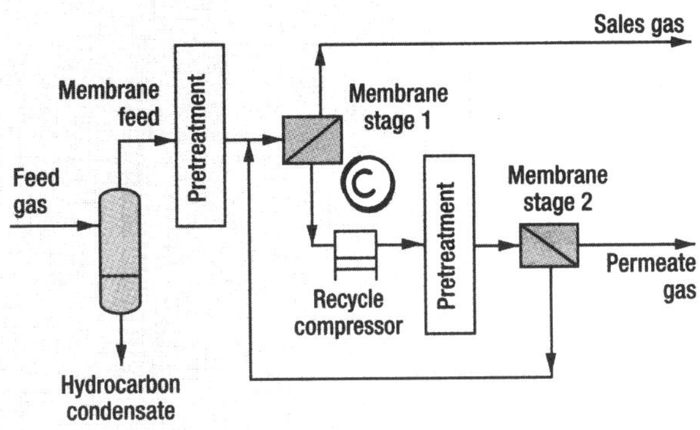

Separex membrane systems

Applications: CO_2 and water vapor removal from natural gas or associated gas to meet pipeline specifications for onshore or offshore locations. Upgrading low GHV gas for fuel. Debottlenecking existing solvent scrubbing systems or providing bulk CO_2 removal upstream of new installations. Hydrocarbon recovery from enhanced oil recovery floods for CO_2 reinjection. Landfill gas purification.

Products: Purified gas meeting pipeline specifications, high-quality fuel gas for turbine, reformer, or power generation or high-purity CO_2 gas for reinjection.

Description: Separex membrane systems are simple, dry systems requiring minimal moving parts. As the CO_2-rich feed gas mixture passes over the polymeric membrane at high pressure, it separates into two streams. Carbon dioxide, hydrogen sulfide and water vapor permeate readily through the membrane collecting on the low-pressure permeate side. Methane, ethane, other hydrocarbons and nitrogen enrich in the high-pressure residual gas.

Hydrocarbon recovery can be as high as 99% for a two-stage design, and 95% for a single stage without compression, depending upon feed composition, pressure levels, system configuration and product requirements. Feedrates vary from 1 MMscfd to 1,000 MMscfd, with CO_2 levels from 3% to 70% and feed pressures from 400 to 1,400 psig.

Designed for operational simplicity, Separex membrane systems are an excellent choice for remote locations. They require minimal rotating equipment, no chemical reagent replacement and minimal maintenance. The prefabricated units are skid mounted to minimize installation costs and plot space.

Separex membrane systems offer two membrane module configurations, spiral wound and hollow fiber, to satisfy the varied CO_2 removal requirements.

Economics: For natural gas upgrading to pipeline specification, the processing costs are lower than, or comparable to, an amine unit. However, the Separex membrane system eliminates the need for the glycol dehydration unit found in typical treating plants. The economics of smaller installations or remote operations favor membrane systems over traditional treatment options. CO_2-removal costs range between $0.05 to $0.15 per 1,000 scf of feed gas, depending on removal requirements, feed pressure, system configuration and product specifications.

Installations: Separex membrane systems have been successfully used in gas field operations since 1981. Over 50 units have been built or are in construction. The largest operating unit processes over 260 MMscfd of natural gas.

Licensor: UOP LLC.

NGL and LNG

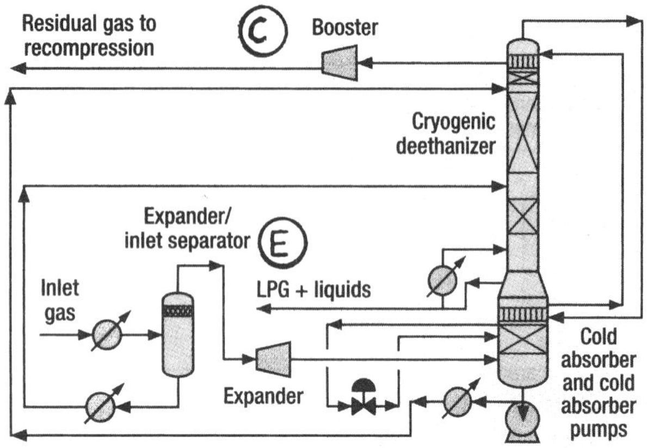

Super Hy-Pro

Application: The Super Hy-Pro process is a further development of its predecessor—the Hy-Pro process—and is designed for the high recovery of liquid hydrocarbons from natural gas, mainly LPG.

Description: The inlet gas is treated and conditioned for processing at low temperatures. After this initial step, the gas is cooled to a point to partially condense certain hydrocarbons. Gas and liquids are separated in a cold separator. The gas is routed to the turboexpander where it is expanded into a proprietary-designed cold-absorption column of the liquids-recovery section. The liquids produced in the absorption section are routed to a recovery column. The liquids-recovery column is the main process focus, where the natural gas liquids are recovered and separated. The overhead section of the column is thermally integrated with the rest of the process, following a similar arrangement as is used in the Hy-Pro process.

Depending on the gas richness, a refrigeration system may be required to aid product recovery. Product recoveries for propane exceed 95%.

Typical energy consumption, depending on the characteristics of the feed gas, is about 45 to 65 Hp/MMscfd, not considering utilities and treating.

Operating conditions: Ample range of pressures, temperatures and compositions.

Installations: More than ten facilities overseas for Hy-Pro and Super Hy-Pro.

Licensor: Randall Gas Technologies, ABB Lummus Global Inc.

Hydrogen

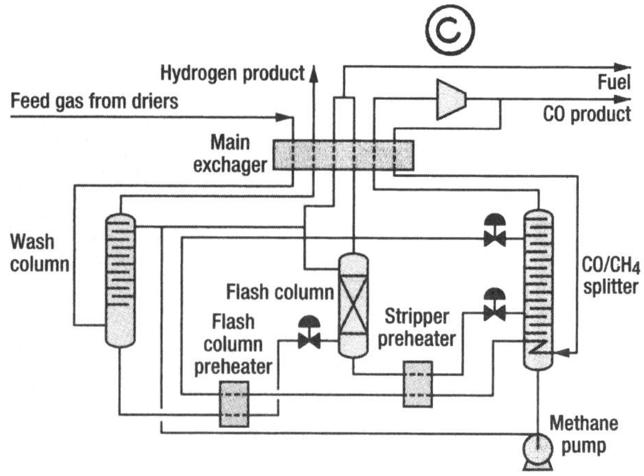

ACORN methane wash

Application: To produce a high-purity carbon monoxide (CO) stream, and a high-purity hydrogen stream, plus a ratio adjusted synthesis gas stream, if required, for use as a chemical feedstock. The synthesis gas stream is typically the product of steam methane reforming (SMR).

Description: Feed gas for CO recovery is pretreated to remove carbon dioxide and water, which will freeze at the cryogenic temperatures encountered in the process. The pretreated feed gas is cooled in the main exchanger and fed to the bottom of the wash column (1). The column is refluxed with liquid methane to produce a hydrogen wash-product free of CO, but saturated with methane (2-3%). The hydrogen is then rewarmed and recovered as a product. The liquid from the wash column is preheated, reduced in pressure and separated in the flash column (2) where hydrogen dissolved in the methane (CH_4) is rejected to fuel gas. To minimize CO losses, this column is also refluxed with liquid CH_4.

The hydrogen-free liquid from the flash column is heated and flashed to the CO/CH_4 splitter column (3). The CO from the overhead is rewarmed and compressed. Part of this stream is delivered as product; the remainder is cooled and recycled within the process. It is first used to reboil the splitter column and preheat the column feed streams. It is then flashed for refrigeration and the liquid is used as reflux for the splitter column. The CH_4 liquid from the bottom of the splitter is pumped to the wash column for use as reflux. The net CH_4 is vaporized in the main exchanger and leaves as the byproduct fuel gas.

Variations of this cycle have been developed to meet special requirements. In all cases, however, the hydrogen stream is produced at high pressure and the CO is available at low pressure. If CO is desired, a product compressor is usually required.

Installations: Six.

Licensor: Air Products and Chemicals, Inc.

Hydrogen

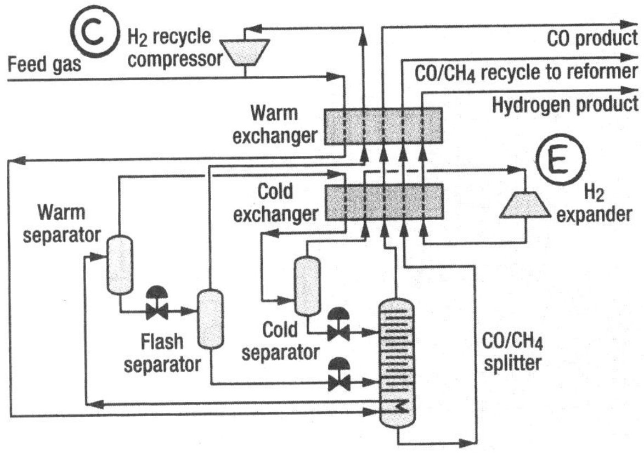

ACORN partial condensation

Application: To produce a high-purity carbon monoxide (CO) stream, and a moderate-purity hydrogen (H_2) stream, plus a ratio adjusted synthesis gas stream for use as a chemical feedstock. The synthesis gas stream is typically the product of partial oxidation reaction (POX).

Description: Feed gas for CO recovery is pretreated to remove carbon dioxide and water, which will freeze at the cryogenic temperatures encountered in the process. The feed gas is cooled against products in the warm exchanger, and is then further cooled providing heat for reboiling the CO/CH_4 splitter. Condensed CO and methane (CH_4) are removed from uncondensed vapor in the warm separator. Vapor from the warm separator is cooled in the cold exchanger where most of the remaining CO is condensed and separated in the cold separator. The liquid from this vessel is a high-purity CO stream used as reflux for the CO/CH_4 splitter.

Liquid from the warm separator is reduced in pressure and flashed in the flash separator to remove dissolved H_2. The vapor from this separator is rewarmed, compressed and recycled to the feed to recover CO. The liquid from the flash separator is sent to the CO/CH_4 splitter. The CO overhead from this tower is warmed and recovered as product. The bottoms, containing CO and CH_4, is also warmed and is available as byproduct fuel gas. The H_2 from the cold separator is warmed in the cold exchanger, expanded to provide refrigeration for the cycle, warmed in the cold and warm exchangers and leaves the process at 97–98% purity.

Variations on this basic cycle are possible depending on feed gas pressure and gas composition, and desired product purity. The H_2 product is delivered at high pressure, but the CO exits the process at low pressure. Therefore, a CO-product compressor is usually required to deliver the product to a downstream process.

Installations: Eleven.

Licensor: Air Products and Chemicals, Inc.

Hydrogen

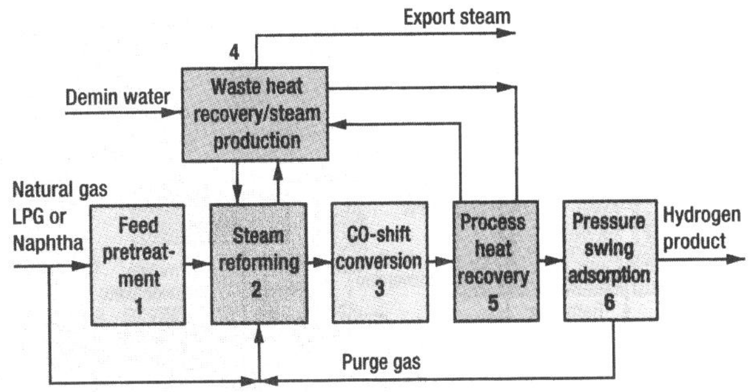

Hydrogen

Application: Produce and purify hydrogen from hydrocarbons using steam methane reforming and pressure swing adsorption (PSA).

Feedstocks: Light hydrocarbon streams ranging from natural gas to naphtha.

Description: After being raised to operating pressure and vaporized, feed gas is pre-treated (1) to remove sulfur and other components. Treatment normally consists of feed preheating, hydrogenation to convert sulfur components to H_2S and remove H_2S with ZnO-catalyst bed. Depending on the feedgas, other purification steps may be required. The desulfurized stream is then mixed with process steam and preheated against hot flue gas in a reformer convection coil.

The preheated mixed feed enters the reformer furnace (2) where the steam-reforming type reactions occur in tubes filled with a nickel-based catalyst. Because the reaction is endothermic, heat is externally supplied from multiple burners combusting PSA tail gas and supplemental fuel. Process gas flow and burner flame direction are arranged co-currently, using either a bottom fired/upflow cylindrical (can) reformer or a top-fired/downflow box reformer to optimize heat transfer along the tube length and to minimize the fuel requirement. Combustion heat is recovered in the waste heat recovery section (4) by passing the reformer flue gas over a series of coils to generate steam and preheat process streams to maximize energy efficiency.

The hot reformed gas leaving the tubes, containing H_2, CO, CO_2 and a small amount of CH_4, is collected and cooled in the process gas boiler to generate byproduct steam. The cooled gas is then fed to a shift reactor (3) where CO is reacted with excess steam present in the gas to produce additional hydrogen. Since the reaction is exothermic, the converted gas leaves the shift reactor at a higher temperature. Exothermic heat is recovered in the process heat recovery section (5) by heating feedgas, boiler feedwater and makeup demineralized water. In the process heat recovery section, process condensate is separated, collected and reused.

The cooled gas is then sent to a high-efficiency Linde PSA unit where the hydrogen is separated from the residual CO_2, CH_4 and water. The PSA unit consists of a series of vessels filled with adsorbent, which alternate between adsorption and regeneration. During adsorption, hydrogen passes through the vessel, while residual components are captured on the adsorbent material. The vessel is regenerated by alternately depressuring and purging with fresh product. The PSA purge stream is captured and sent to the reformer as primary fuel.

Installations: Over 250 plants worldwide, ranging in capacity from less than 1 MMscfd, to greater than 100 MMscfd.

Licensor: Linde AG.

Hydrogen

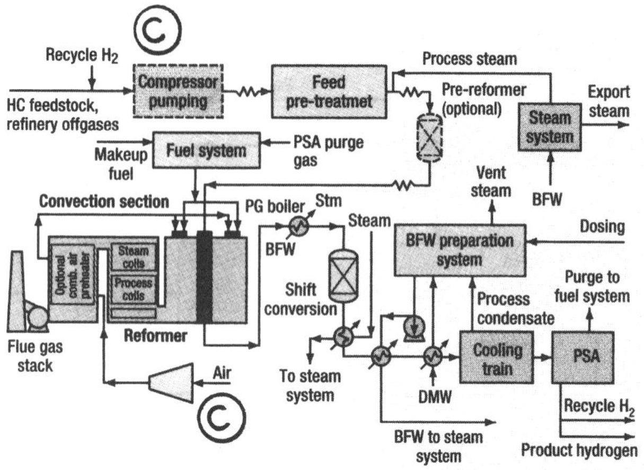

Hydrogen

Application: Production of hydrogen (H_2) from hydrocarbon (HC) feedstocks, by steam reforming.

Feedstocks: Ranging from natural gas to heavy naphtha as well as potential refinery offgases. Many recent refinery hydrogen plants have multiple feedstock flexibility, either in terms of back-up or alternative or mixed feed. Advanced automatic feedstock change-over has also successfully been applied by TECHNIP in several modern plants with multiple feedstock flexibility.

Description: The generic flowsheet consists of feed pretreatment, pre-reforming (optional), steam-HC reforming, shift conversion and hydrogen purification by pressure swing adsorption (PSA). However, it is often tailored to satisfy specific requirements.

Feed pretreatment normally involves removal of sulfur, chlorine and other catalyst poisons after preheating to 350°C to 400°C.

The treated feed gas mixed with process steam is reformed in a fired reformer (with adiadatic pre-reformer upstream, if used) after necessary superheating. The net reforming reactions are strongly endothermic. Heat is supplied by combusting PSA purge gas, supplemented by makeup fuel in multiple burners in a top-fired furnace.

Reforming severity is optimized for each specific case. Waste heat from reformed gas is recovered through steam generation before the water-gas shift conversion. Most of the carbon monoxide is further converted to hydrogen. Process condensate resulting from heat recovery and cooling is separated and generally reused in the steam system after necessary treatment. The entire steam generation is usually on natural circulation, which adds to higher reliability. The gas flows to the PSA unit that provides high-purity hydrogen product (up to < 1ppm CO) at near inlet pressures.

Typical specific energy consumption based on feed + fuel - export steam ranges between 3.0 to 3.5 Gcal/KNm^3 (330–370 Btu/ scf) LHV, depending upon feedstock, plant capacity, optimization criteria and steam-export requirements. Recent advances include integration of hydrogen recovery and generation, recuperative (post-)reforming for capacity retrofits and reduced emissions.

Installations: TECHNIP has been involved in over 220 hydrogen plants worldwide, covering a wide range of capacities. Most installations are for refinery application with basic features for high reliability and optimized cost.

Licensor: TECHNIP.

Hydrogen

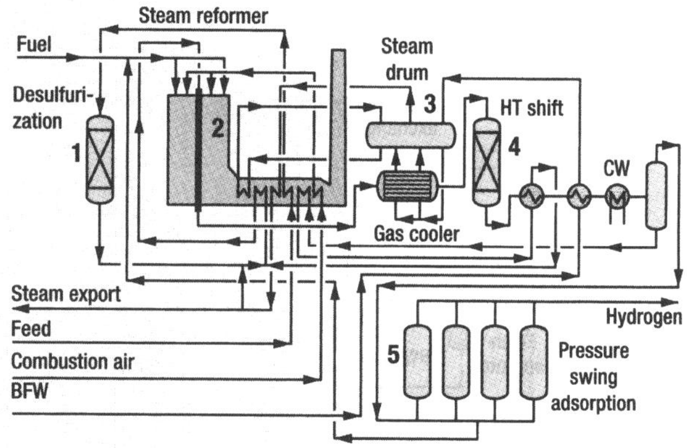

Hydrogen

Application: Produce hydrogen for refinery hydrotreating and hydrocracking or other refinery, petrochemical and industrial applications.

Feed: Natural gas, refinery offgas, LPG/butane and light naphtha.

Product: High-purity hydrogen (>99.9%). CO, CO_2 and/or electricity may also be produced separately for byproduct credit.

Description: The feedstock (natural gas, for example) is desulfurized (1), mixed with steam and converted to synthesis gas in the reformer (2) over nickel catalysts at 20 to 50 bar pressure and temperatures of 800°C to 950°C.

The Uhde steam reformer is a top-fired reformer, which has tubes made of centrifugally cast alloy steel and a proprietary "cold" outlet manifold system to enhance reliability. Subsequent high-pressure steam generation (3) and superheating permit maximum process heat exploitation to achieve an optimized energy-efficient process.

The carbon monoxide (CO) shift occurs in a single-stage, adiabatic high-temperature reactor (4). Pressure swing adsorption (5) is a well-established purification step to obtain high-purity hydrogen (99.9 % and higher).

The Uhde reformer design with the unique proprietary cold outlet manifold system enables construction and operation of world-scale reformers with hydrogen capacities to 250,000 Nm^3/h.

Economics: Typical consumption figures (feed and fuel) range between 150 and 175 GJ/metric ton of hydrogen, depending on the individual plant concept.

Installations: Uhde has recently designed and constructed two of the world's largest hydrogen plants for SINCOR C.A., Venezuela (2 × 98,000 Nm^3/h) and Shell Canada Ltd., Canada (2 × 130,000 Nm^3/h). Fifty-six Uhde reformers have been constructed worldwide.

References: Fritsch, S., "Steam reformer based hydrogen plant optimization," The International Conference HYFORUM 2000, Munich, September 2000.

Fritsch, S., "Large scale hydrogen production for SINCOR upgrader project in Venezuela," Nitrogen 99, Caracas, March 1999.

Licensor: Uhde GmbH.

Hydrogen

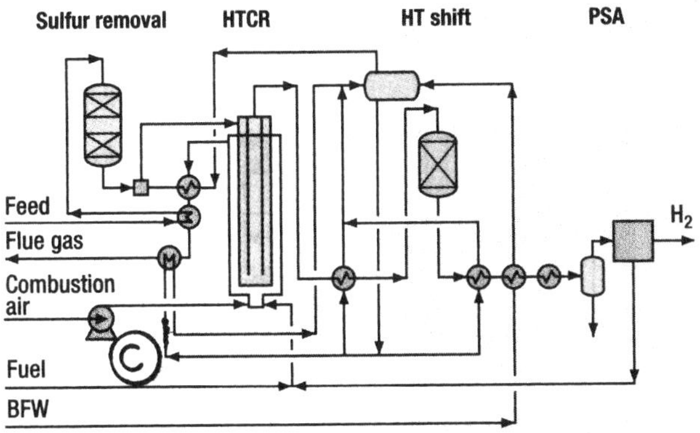

Hydrogen, HTCR based

Application: Produce hydrogen (H_2) from hydrocarbon feedstocks such as: natural gas, LPG, naphtha, refinery offgases, etc., using the Haldor Topsøe Convective Reformer (HTCR). Plant capacities range from 200 Nm^3/h to 20,000 Nm^3/h (200,000 scfd to 20 MMscfd) and hydrogen purity ranges from 99.5% to 99.999+% without steam export.

Description: The HTCR-based hydrogen plant can be tailor-made to suit the customer's needs with respect to feedstock flexibility. In a typical plant, feedstock is first desulfurized. Subsequently, process steam is added, and the mixture is fed to the HTCR. Process gases are reacted in a water-shift reactor and purified by pressure swing absorption (PSA) unit to obtain product-grade hydrogen. PSA offgases are used as fuel for the HTCR. Excess heat is efficiently used for process heating and steam generation.

A unique technology feature is high thermal efficiency. Product gas and flue gas are both cooled to about 600°C (1,100°F) to recover heat for the reforming reaction. Energy-efficient hydrogen plants based on the HTCR use high thermal efficiency and have no steam export.

Economics: HTCR-based plants provide the customer with a low-investment cost and low operating expenses. The plant can be supplied skid-mounted providing a short erection time. The plants have high flexibility, reliability and safety. Fully automated operation, startup and shutdown allow minimum operator attendance. Net energy efficiency of about 3.4 to 3.6 Gcal/1000 Nm^3 is achieved depending on size and feedstock (360 to 380 Btu/scf).

Installations: Twenty-one licensed units.

References: Dybkjær, I., et al., "Medium-size hydrogen supply using the Topsøe convection reformer," 1997 NPRA Annual Meeting, March 16–18, 1997, San Antonio.

Licensor: Haldor Topsøe A/S.

Hydrogen

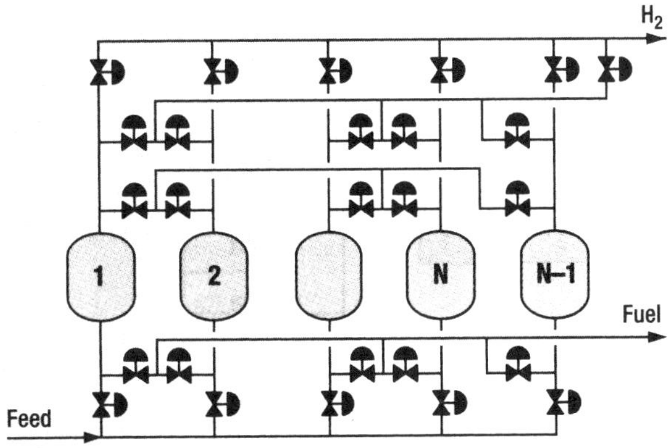

Hydrogen (Polybed PSA)

Application: Production of any purity hydrogen, typically 90% to 99.9999$^+$ mole%. Impurities efficiently removed include: N_2, CO, CH_4, CO_2, H_2O, Ar, O_2, C_2–$C_8{}^+$, CH_3OH, NH_3, H_2S and organic sulfur compounds. The technology can also be used to: purify CH_4, CO_2, He, N_2 and Cl; remove CO_2; adjust synthesis-gas stream composition ratios and separate nitrogen from hydrocarbons.

Feed: Steam reformer (at any point after the reformer), catalytic reformer net gas, other refinery purge streams, gasification offgases, ammonia plant purge gases (before or after the NH_3 waterwash), demethanizer or other ethylene plant offgases, partial oxidation gases, styrene plant offgases, methanol plant purge gases, coke-oven gas, cryogenic purification offgases or other H_2 sources. Feed pressures up to 1,000 psig have been commercially demonstrated.

Product: Recovery of H_2 varies between 60% and 90%, depending on composition, pressure levels and product requirements. Typical temperatures are 60°F to 120°F. Purity can be 99.9999$^+$ mole%.

Description: Purification is based on advanced pressure swing adsorption (PSA) technology. Purified H_2 is delivered at essentially feed pressure, and impurities are removed at a lower pressure.

Polybed PSA units contain 4 to 16+ adsorber vessels. One or more vessels are on the adsorption step, while the others are in various stages of regeneration. Single-train Polybed PSA units can have product capacities over 200 MMscfd.

All systems use advanced proprietary adsorbents and patented void-gas recovery techniques to provide maximum product recovery. Other than entrained liquid removal, no feed pretreatment is required. In addition, all impurities are removed in a single step, and purities exceeding 90% are obtained irrespective of impurities. Many units presently produce streams with less than one ppmv impurity from feed concentrations of 40$^+$ mole%.

Operation is automatic with pushbutton startup and shutdown. After startup, the unit will produce H_2 in two to four hours. Onstream factors in excess of 99.8% relative to unplanned shutdowns are typical.

Turndown capability is typically 30%. The units are built compactly with plot plans ranging from 12 × 25 ft to 60 × 120 ft. Units are skid-mounted and modular to minimize installation costs. Material for piping and vessels is carbon steel. Control can be via a local or remote-mounted control panel or by integration into the refinery's computer control system. Units are designed for outdoor, unattended operation and require no utilities other than small quantities of instrument air and power for instrumentation.

Installations: Over 700 units are in operation or under construction, including the world's first 16-bed system, and world's largest single-train system.

Licensor: UOP LLC.

Hydrogen

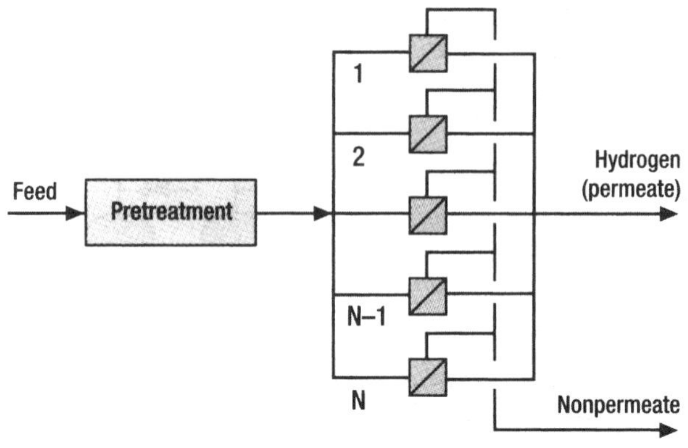

Hydrogen (Polysep membrane)

Application: Hydrogen recovery and purification or rejection from various refining, petrochemical and chemical process gas streams. Other examples are: synthesis gas ratio adjustment and carbon monoxide (CO) recovery.

Feed: Refinery streams include: catalytic reformer offgas, hydrotreater and hydrocracker purge and fluid catalytic cracking offgas. Chemical and petrochemical feed streams are: ethylene offgases, ammonia plant purges, methanol plant offgases, synthesis gas streams from steam reforming, partial oxidation or other gasification technologies.

Product: For typical hydrogen purification applications, recovery varies between 70% and 95^{+}% and purity ranges from 70 to 99 mole%, depending on feed composition, pressure levels and product requirements.

Polysep membranes systems are also designed to produce high-purity CO for petrochemical products such as polyurethanes and polycarbonates, and to ratio adjust synthesis gas streams in methanol and oxo-alcohol plants. Also, a new application is hydrogen recovery from IGCC power generation systems.

Description: The Polysep separation system is based on state-of-the-art, composite, hollow-fiber polymer membrane technology. The hollow fibers are packaged in a proprietary countercurrent-flow bundle configuration that maximizes the separation driving force and minimizes required membrane area.

The Polysep separation is a pressure-driven process. It requires a minimum of moving parts, utilities and operator attention. The systems are compact, shop-fabricated, modular units allowing reduced delivery schedules and simple inexpensive installation. Feed pretreatment equipment typically includes: a knockout drum for bulk liquid removal, a coalescing filter for particulate and entrained liquid removal, and a preheater to optimize the membranes' performance.

Operation features include: automatic startup, capacity control, product-purity control, auto depressurization and turnup/turndown. Turndown capability is typically 30% using a patented control strategy. Membrane system control is typically via integration into the refinery's control system. Once installed, a membrane system can reach steadystate operation from cold startup in a few hours with onstream factors over 99.8% relative to unplanned shutdowns.

Economics: Polysep membrane systems can be efficiently and economically scaled, from just a few modules to over 100 modules, depending on the application. Membrane-separation systems have low capital costs and plot area, and offer a rapid return on investment.

Installations: Over 50 units are in operation or under construction. Largest unit processes over 320 MMscfd of synthesis gas.

Licensor: UOP LLC.

Hydrogen

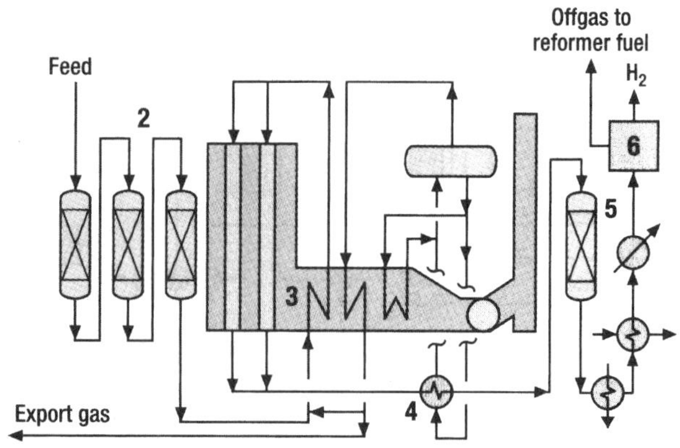

Hydrogen (steam reform)

Application: Production of hydrogen for refinery hydrotreating and hydrocracking or other refinery, petrochemical, metallurgical, and food-processing uses.

Feedstock: Light hydrocarbons such as natural gas, refinery fuel gas, LPG/butane and light naphtha.

Product: High-purity hydrogen (99.9^{+}%) at any required pressure.

Description: The feed is heated in the feed preheater and passed through the hydrotreater (1). The hydrotreater converts sulfur compounds to H_2S and saturates any unsaturated hydrocarbons in the feed. The gas is then sent to the desulfurizers (2). These adsorb the H_2S from the gas. The desulfurizers are arranged in series and designed so that the adsorbent can be changed while the plant is running.

The desulfurized feed gas is mixed with steam and superheated in the feed preheat coil. The feed mixture then passes through catalyst-filled tubes in the reformer (3). In the presence of nickel catalyst, the feed reacts with steam to produce hydrogen and carbon oxides. Heat for the endothermic reforming reaction is provided by carefully controlled external firing in the reformer.

Gas leaving the reformer is cooled by the process steam generator (4). Gas is then fed to the shift converter (5), which contains a bed of copper promoted iron-chromium catalyst. This converts CO and water vapor to additional H_2 and CO_2. Shift converter effluent gas is cooled in a feed preheater, a BFW preheater and a DA feed water preheater. Hot condensate is separated out. Process gas is then cooled in a gas air cooler and a gas trim cooler. The cooled stream flows to a cold condensate separator where the remaining condensate is separated and the gas is sent to a PSA hydrogen purification system (6).

The PSA system is automatic, thus requiring minimal operator attention. It operates on a repeated cycle having two basic steps: adsorption and regeneration. PSA offgas is sent to the reformer, where it provides most of the fuel requirement. Hydrogen from the PSA unit is sent off plot. A small hydrogen stream is then split off and recycled to the front of the plant for hydrotreating.

The thermal efficiency of the plant is optimized by recovery of heat from the reformer flue gas stream and from the reformer effluent process gas stream. This energy is utilized to preheat reformer feed gas and generate steam for reforming and export. Hot flue gas from the reformer is sent through the waste-heat recovery convection section and is discharged by an induced-draft fan to the stack.The boiler feed water deaerator and preheat circuits are integrated to maximize heat recovery. A common steam drum serves the steam generation coils and process steam generator for steam production via natural circulation.

Installations: Over 170 plants worldwide-ranging in size from less than 1 MMscfd to over 90 MMscfd capacities. Plant designs for capacities from 1 MMscfd to 280 MMscfd.

Licensor: Howe-Baker Engineers, Ltd.

Hydrogen

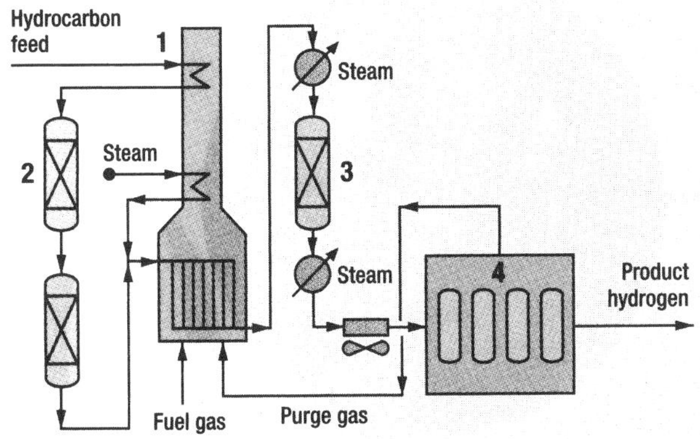

Hydrogen (steam reform)

Application: Manufacture of hydrogen for hydrotreating, hydrocracking, or other refinery or chemical use.

Feedstock: Light saturated hydrocarbons: refinery gas or natural gas, LPG or light naphtha.

Product: Typical purity for refinery use 99.9%; pressure 400 psig. Steam or CO_2 may be produced for byproduct credit.

Process description: Hydrogen is produced by steam reforming of hydrocarbons with purification by pressure swing adsorption (PSA).

Hydrocarbon feed is heated (1) and then purified (2) with zinc oxide to remove traces of sulfur compounds. Hydrogenation or chloride removal may also be required with some feedstocks.

The purified feed is mixed with steam and preheated further, then reformed at 1,400°F to 1,600°F over nickel catalyst in the tubes of the Terrace Wall reforming furnace (1). This produces a synthesis gas containing hydrogen and carbon oxides.

Because of the high temperature, heat recovery is important to project economics. This may be by steam or electricity production, by combustion air preheat, or recovery of heat for additional reforming. After cooling, the synthesis gas passes through the shift conversion section (3), where CO reacts with steam to form CO_2 and additional hydrogen.

In the PSA section (4), carbon oxides and methane are removed by beds of solid adsorbent. The adsorbent is regenerated by depressurization and purging. Purge gas from the PSA section is used as fuel for the reforming furnace.

Engineering is standardized to combine short schedules with optimized design. Smaller plants may also be fabricated in modules.

Variations include use of a gas turbine to produce electricity or as a process driver. The turbine exhaust gas is then used as reformer combustion air.

Installations: Over 100 plants, ranging from less than 1 MMscfd to 95 MMscfd in a single train, with numerous multi-train installations.

Contributor: Foster Wheeler.

Hydrogen

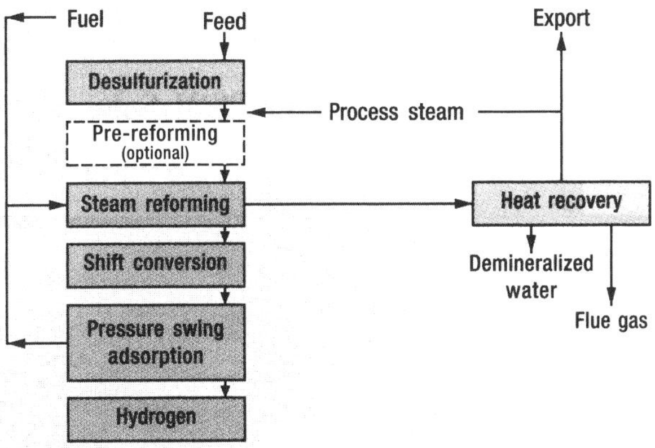

Hydrogen (steam reform)

Application: Hydrogen production from natural gas, refinery gas, associated gas, naphtha, LPG or any mixture of these. Appropriate purity product (up to 99.999%) can be used in refinery upgrade processes, chemical production and metallurgy (direct reduction). Possible byproducts are export steam or electricity, depending on cost and/or efficiency optimization targets.

Description: The hydrocarbon feedstock is admixed with some recycle hydrogen and preheated to 350–380°C. Sulfur components are totally converted to H_2S at CoMo catalyst and then adsorbed on zinc oxide by conversion to ZnS. The desulfurized feed is mixed with process steam at an optimized steam/carbon ratio.

The desulfurized feed is mixed with process steam at an optimized steam/carbon ratio, superheated to 500–650°C and fed to the Lurgi Reformer. The feed/steam mixture passing the reformer tubes is converted at 800–900°C by presence of a nickel catalyst to a reformed gas containing H_2, CO_2, CO, CH_4 and undecomposed steam. The reformed gas is cooled to approximately 33°C in a reformed gas boiler.

The Lurgi Reformer is a top-fired reformer with a low number of burners and low heat losses, almost uniform wall temperature over the entire heated tube length and low NO_x formation by very accurate fuel and combustion air equipartition to the burners.

An adiabatic pre-reformer operating at an inlet temperature of 400–500°C (dependent on feedstock) may be inserted upstream of the feed superheater as a process option. Feedgas is partly converted to H_2, CO and CO_2 with high-activity catalyst; all hydrocarbons are totally converted to methane. The pre-reformer limits steam export to maximize heat recovery from the process and increases feedstock flexibility.

The CO in the reformed gas is shift-converted with an iron-chromium catalyst, increasing hydrogen yield and reducing CO content to below 3-vol.%. The shift gas is cooled to 40°C and any process condensate is separated and recycled to the process. The gas is then routed to the PSA unit, where pure hydrogen is separated from the shift-gas stream. Offgas is used as fuel for steam reforming.

Recovered waste heat from the reformed and flue gases generates steam, which is used as process steam with the excess exported to battery limits.

Turndown rates of 30% or even less are achievable. The control concept allows fully automatic operation with load changes typically 3% of full capacity/minute.

Economics: Consumption figures based on light natural gas feedstock/1 MMscfd of H_2:

Feed + fuel, MMscfd	0.4
Demineralized water, t	1.25
Cooling water, m^3	3.0
Electricity, kWh	19
Export steam, t	0.7

Installations: More than 105 gas reforming plants, 25 being hydrogen plants, with single-train capacities ranging from 1 MMscfd to 200 MMscfd.

Licensor: Lurgi Oel-Gas-Chemie GmbH.

Hydrogen

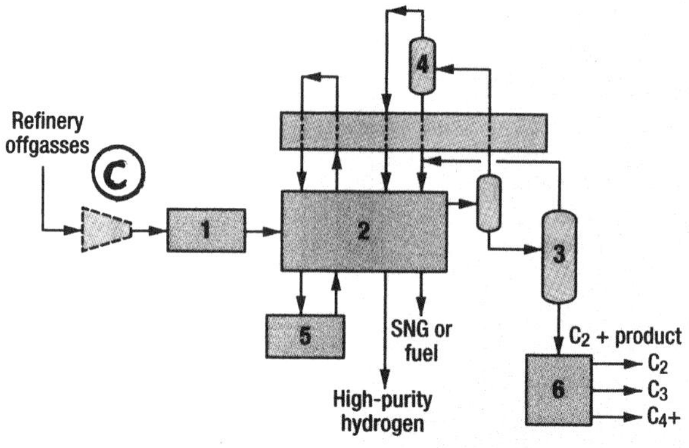

Hydrogen and liquid hydrocarbon recovery—cryogenics

Application: Recover high-purity hydrogen (H_2) and C_2+ liquid products from refinery offgases.

Description: This process is the cryogenic separation of refinery offgases and purges containing from 10% to 80% H_2 and 15% to 40% hydrocarbon liquids such as ethylene, ethane, propylene, propane and butanes. Refinery offgases are optionally compressed and then pretreated (1) to remove sulfur, CO_2, H_2O and other trace impurities. Treated feed is partially condensed in an integrated multi-passage exchanger system (2) against returning products and refrigerant.

Separated liquids are sent to a demethanizer (3) for stabilization, while hydrogen is concentrated (4) to 90% to 95+% purity by further cooling. Methane, other impurities and unrecovered products are sent to fuel or optionally split into synthetic natural gas (SNG) product and low-Btu fuel. Refrigeration is provided by a closed-loop system (5). Mixed C_2+ liquids from the demethanizer can be further fractionated (6) into finished petrochemical feeds and products such as ethane, ethylene, propane and propylene.

Operating conditions: Feed capacities from 10 to 150+ MMscfd. Feed pressures as low as 150 psig. Ethylene recoveries are greater than 95%, with higher recoveries of ethane and heavier components. Hydrogen recoveries are better than 95% recovery.

Economics: Hydrogen is economically co-produced with liquid hydrocarbon products, especially ethylene and propylene, whose high value can subsidize the capital investment. High hydrocarbon liquid products recovery is achieved without the cost for feed compression and subsequent feed expansion to fuel pressure. Power consumption is a function of hydrocarbon quantities in the feed and feed pressure. High-purity hydrogen is produced without the investment for a "back-end" PSA system. Project costs can have less than a two-year simple payback.

Installations: Five operating offgas cryogenic systems process FCC offgas, cat reformer offgas, hydrotreater purge gas, coker offgas and refinery fuel gas. Several process and refrigeration schemes have been used since 1987, with the most recent plant startup in 2001.

Reference: U.S. Patent No. 6,266,977

Trautmann, S. R. and R. A. Davis, "Refinery offgases—Alternative sources for ethylene recovery and integration," AIChE Spring Meeting, New Orleans, March 14, 2002.

Licensor: Air Products & Chemicals Inc.

Hydrogen

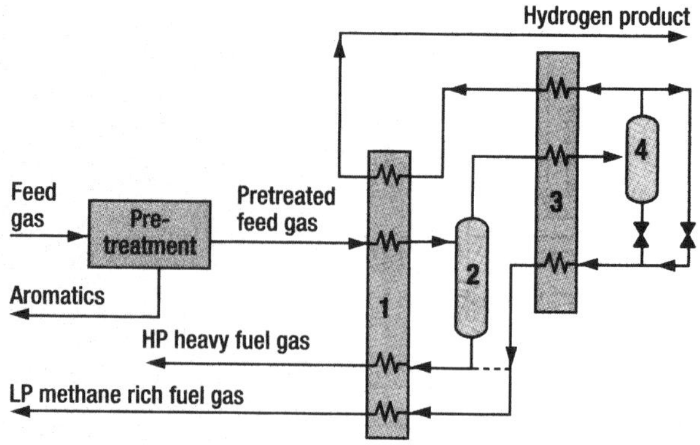

Hydrogen recovery (cryogenic)

Application: Recovery of relatively pure hydrogen from refinery and petrochemical offgas streams such as from thermal hydrodealkylation (THDA), catalytic reformers, hydrotreaters and fluid catalytic crackers. Cryogenic processing is the optimal route to produce carbon monoxide (CO) from syngas.

Products: 90% to 98% pure hydrogen. Valuable product streams, such as LPG, may also be recovered.

Description: A typical autorefrigerated cryogenic unit for recovery of hydrogen consists of two stages of cooling and partial condensation. Suitably pretreated feed gas is cooled and partially condensed against hydrogen product and fuel in the plate-fin heat exchanger (1). The hydrocarbon rich condensate is separated in the two-phase separator (2) and the vapor is further cooled and partially condensed in the second plate-fin heat exchanger (3). The methane-rich condensate is separated in the second two-phase separator (4) giving relatively pure hydrogen product, which is reheated through both exchangers.

Autorefrigerated cryogenic units use refrigeration from Joule-Thomson expansion of the condensate streams and can generate hydrogen purities up to 96%.

Pretreatment ensures that the feed gas to the cryogenic unit is dry and contains no components which would freeze in the cold section. Depending on the pretreatment scheme, additional products can be obtained.

Depending on feed gas conditions and hydrogen product requirements, one, two or three stages of separation may be optimal.

Operating conditions: Typical hydrogen recoveries are 90% to 96%.

Economics: Cryogenic recovery of hydrogen is economically favored by the ability to recover other valuable products, e.g., olefins and LPG. Compared with alternative technologies, cryogenic processing is the most efficient and has the lowest utilities cost. Cryogenic recovery has been used to treat gases with hydrogen feed concentrations as high as 80% and pressures up to 80 barg.

Installations: Thirteen.

Reference: Allen, P., "Managing hydrogen recovery," *Hydrocarbon Engineering,* April 1999, p. 71.

Licensor: Costain Oil, Gas & Process Ltd.

Hydrogen

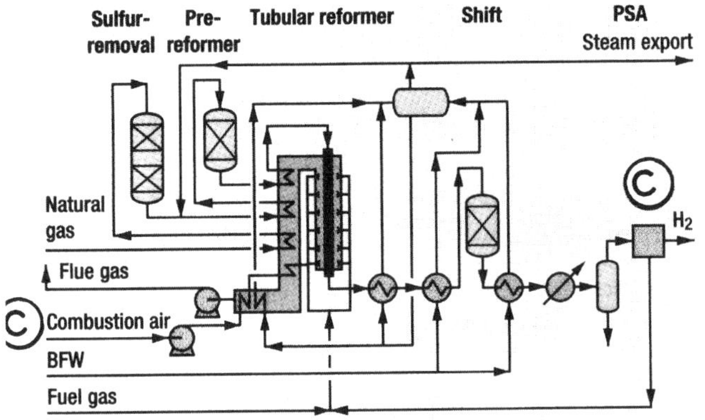

Hydrogen, steam methane reform (SMR)

Application: Produce hydrogen (H_2) from hydrocarbon feedstocks such as: natural gas, LPG, naphtha, refinery offgases, etc., using the Haldor Topsøe side-fired steam-methane reformer (SMR) process. Plant capacities range from 5,000 Nm^3/h to 200,000$^+$ Nm^3/h (4.5 MMscfd to 200$^+$ MMscfd), and hydrogen purity ranges from 99.5% to 99.999$^+$%.

Description: The Haldor Topsøe SMR-based hydrogen plant can be tailor-made to suit the customer's needs with respect to feedstock flexibility and steam export. In a typical low-steam export plant, the hydrocarbon feedstock is desulfurized. Subsequently, process steam is added, and the mixture is fed to a prereformer. Further reforming is done in side-fired SMR. Process gases are reacted in a water-gas shift reactor and purified by the pressure swing absorption (PSA) unit to obtain product-grade H_2. PSA offgases are used as fuel in the SMR. Excess heat in the plant is efficiently used for process heating and steam generation.

The SMR operates at high outlet temperatures [to about 950°C (1,740°F)] while the Topsøe reforming catalysts enables low steam-to-carbon ratios. Both conditions (advanced steam reforming) are necessary for high-energy efficiency and low hydrogen production costs. This application of Topsøe's reforming technology is in operation in several industrial plants worldwide.

Economics: The Advanced Steam Reforming conditions described can achieve a net energy efficiency as low as about 2.96 Gcal/1000 Nm^3 hydrogen using natural gas feed (315 Btu/scf).

References: Gøl, J. N., et al., "Options for hydrogen production," *HTI Quarterly,* Summer 1995.

Dybkjær, I. and S. W. Madsen, "Advanced reforming technologies for hydrogen production," *Hydrocarbon Engineering,* December/January 1997/1998.

Rostrup-Nielsen, J. R. and T. Rostrup-Nielsen, "Large scale hydrogen production," to be published in CatTech.

Licensor: Haldor Topsøe A/S.

Hydrogen

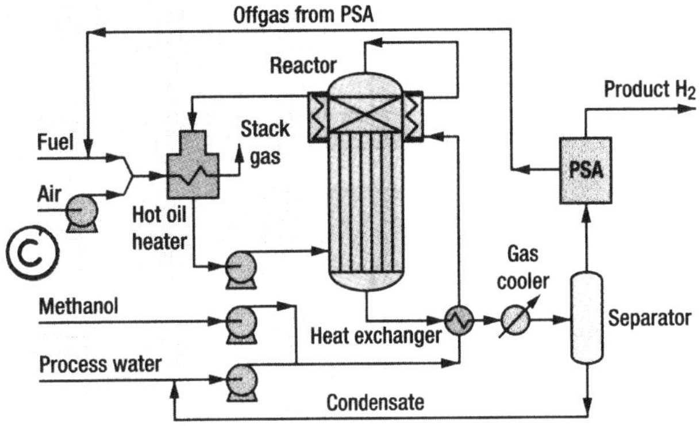

Hydrogen-methanol decomposition

Application: Produce hydrogen (H_2) in the capacity range of 100–1,000 Nm^3/h of H_2 for usage by the chemical industry and the manufacture of electronics.

Description: Feed methanol and water are mixed, evaporated and superheated before being sent to the methanol decomposition reactor. In this reactor, the methanol is reacted to form H_2, CO and CO_2. The reaction is endothermic and requires energy by heat transfer from externally heated oil or by steam.

The raw gas from the reactor is cooled, and the process condensate is separated. The separated gas is further purified by pressure swing adsorption (PSA) unit, startup, operation and shutdown are automatic.

Features: Utility requirements per Nm^3 of H_2 are:

Methanol, kg	0.63
Fuel, kcal	320
Demineralized water, kg	0.37
Electricity, kWh	0.06
Additional utilities required	Cooling water, instrument air and nitrogen

The hydrogen plant construction is compact and can be skid-mounted.

Economics: Features include:
- Low investment cost
- Low operating costs
- Low maintenance cost
- Short delivery time
- Fast installation of the skid-mounted unit.

Installations: Ten plants are in operation worldwide.

Licensor: Haldor Topsøe A/S.

Hydrogen

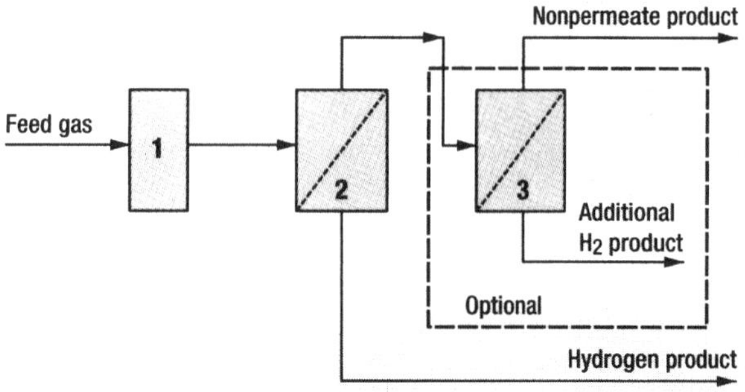

Hydrogen—PRISM membrane

Application: To recover and purify hydrogen or to reject hydrogen from refinery, petrochemical or gas processing streams. Refinery streams include hydrotreating or hydrocracking purge, catalytic reformer offgas, fluid catalytic cracker offgas or fuel gas. Petrochemical process streams include ammonia synthesis purge, methanol synthesis purge or ethylene offgas. Synthesis gas includes those generated from steam reforming or partial oxidation.

Product: Typical hydrogen (H_2) product purity is 90 to 98% and, in some cases, 99.9%. Product purity is dependent upon feed purity, available differential partial pressure and desired H_2 recovery level. Typical H_2 recovery is 80 to 95% or more.

The hydrocarbon-rich nonpermeate product is returned at nearly the same pressure as the feed gas for use as fuel gas, or in the case of synthesis gas applications, as a carbon monoxide (CO) enriched feed to oxo-alcohol, organic acid, or Fisher-Tropsch synthesis.

Description: Typical PRISM membrane systems consist of a pretreatment (1) section to remove entrained liquids and preheat feed before gas enters the membrane separators (2). Various membrane separator configurations are possible to optimize purity and recovery, and operating and capital costs such as adding a second stage membrane separator (3). Pretreatment options include water scrubbing to recover ammonia from ammonia synthesis purge stream.

Membrane separators are compact bundles of hollow fibers contained in a coded pressure vessel. The pressurized feed enters the vessel and flows on the outside of the fibers (shell side). Hydrogen selectively permeates through the membrane to the inside of the hollow fibers (tube side), which is at lower pressure. PRISM membrane separators' key benefits include resistance to water exposure, particulates and low feed to nonpermeate pressure drop.

Membrane systems consist of a pre-assembled skid unit with pressure vessels, interconnecting piping, and instrumentation and are factory tested for ease of installation and commissioning.

Economics: Economic benefits are derived from high-product recoveries and purities, from high reliability and low capital cost. Additional benefits include relative ease of operation with minimal maintenance. Also, systems are expandable and adaptable to changing requirements.

Installations: Over 270 PRISM H_2 membrane systems have been commissioned or are in design. These systems include over 54 systems in refinery applications, 124 in ammonia synthesis purge and 30 in synthesis gas applications.

Licensor: Air Products and Chemicals, Inc.

Hydrogen

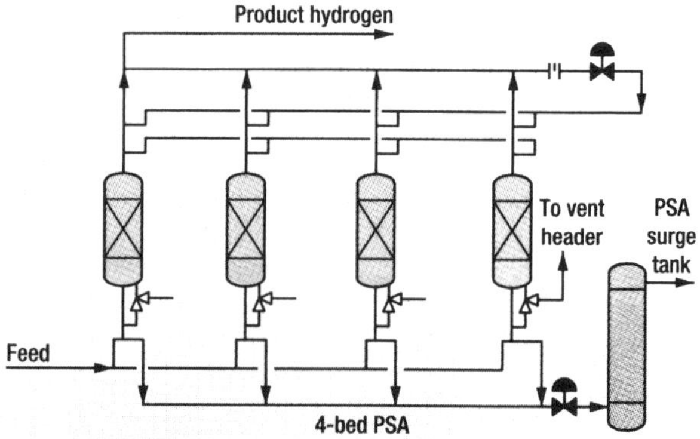

Hydrogen—PRISM PSA

Application: Purify hydrogen from dedicated H_2-production equipment and a wide range of offgas streams from refineries and hydrocarbon processing industries (HPI) to produce 95–99.999+% H_2 purity.

Feed: Wide range of H_2-containing streams, typically with feed H_2 greater than 50%, pressure from 70 psi to 600 psi (5–40 bar), and temperature from 60°F to 110°F. Examples of dedicated H_2-production units include steam-methane reformers, naphtha-fired reformers and gasifiers. Hydrogen can be recovered from ethylene, propylene, styrene, and coke oven offgases; and various refinery streams such as cat reformer, hydrotreater/hydrocracker offgases.

Products: Hydrogen recovery is 83–90% for SMR applications, with CO from 1 ppm to 50 ppm. Purity typically 95–99.9% for offgas plants, with recovery dependent on operating conditions. Integrated processes producing multiple products are practiced. For example, gasifier or refinery offgas pressure swing absorption (PSA) can be integrated with cryogenic processes for co-recovery of H_2 with CO or hydrocarbons, respectively. PSAs are typically integrated with membranes in gasifier installations to co-produce high-purity H_2 and a wide range of syngas compositions.

Description: PSA purifies H_2 by sequentially adsorbing impurities in multiple layers of adsorbents within a single PSA vessel. Adsorbent selection is critical, and is based on extensive R&D with verification from frequent plant performance tests. Hydrogen is produced from vessel(s) in their feed step, while other vessels are regenerated.

The number of vessels depends on the system capacity and the desired H_2 recovery. For example, 4-bed PSA typically produces up to 15 MMscfd H_2. As capacity increases, it is generally economical to increase the recovery and number of beds. A 10-bed PSA unit can produce over 120 MMscfd H_2 at up to 90% recovery. Single-train capacities exceed 200 MMscfd.

The systems are skid-mounted, requiring minimal fieldwork. These PSAs include all control valves, switch valves and instrumentation required. Feed, product, and tail gas monitoring, isolation, and venting systems can be integrated into the skid design to further minimize field piping associated with the PSA. The system can be controlled using any commercial PLC or DCS products. Average reliabilities exceed 99.9%.

Economics: PSA costs vary with capacity and recovery. When compared to the other hydrogen recovery technologies (cryogenics and membranes), hydrogen PSAs fall in the medium-capital cost range and have moderate scaling economics. Using a PSA, H_2 can typically be recovered for about 1.2-2X fuel value from offgas plants. When ultra-high-purity is required (e.g., ppm levels of CO, CO_2 or other contaminants), PSA is typically the best technical and economic choice.

References: Sabram, T. M,. et al., "Integrate New and Existing Hydrogen Supplies," *World Refining,* April 2001, pp. 32–34.

Installations: Over 50 operating units, including the world's largest PSA installation.

Licensor: Air Products and Chemicals, Inc.

Liquid treating

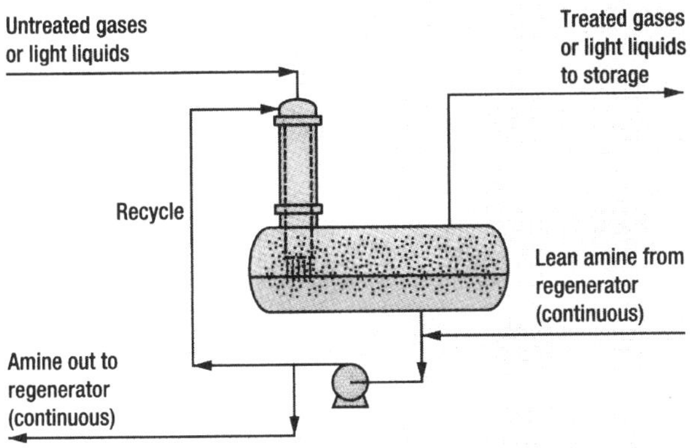

AMINEX

Application: Extract H_2S, COS and CO_2 from gases and light liquid streams with amine solution using FIBER-FILM contactor technology.

Description: The amine phase flows along the FIBER-FILM contactor fibers, which are continuously renewed, as the wet fibers are preferentially wetted by the amine phase in the AMINEX process. Hydrocarbons flow through the shroud parallel to the amine-wetted fibers where the H_2S, COS, and/or CO_2 are extracted into the amine phase. The two phases disengage in the separator vessel where the rich amine flows to the amine regeneration unit and the treated gas or light liquids goes to storage.

Economics: FIBER-FILM contactor technology requires smaller processing vessels allowing shorter separation times and less waste generation. This saves plant space and reduces capital expenditures.

Installations: Ten worldwide.

Reference: "Gas Processing Handbook," *Hydrocarbon Processing,* April 1984, p. 87.

Licensor: Merichem Chemicals & Refinery Services LLC.

Liquid treating

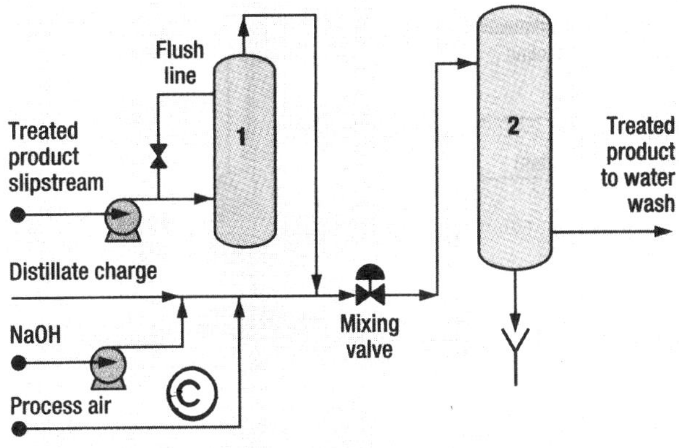

Bender

Application: Sweetens light distillates—such as LPG, gasolines and jet fuels—by converting mercaptan sulfur, in the range of 0.001 to 0.15 wt%, to disulfides.

Description: Mercaptans are converted to disulfides in the presence of a fixed-bed catalyst, air, alkali and a small amount of sulfur (1). The mixture passes over a bed (2) of solid catalyst to effect conversion. Then it is generally passed to a water wash to remove trace impurities. A diluted caustic prewash may be required if the acid number of the distillate exceeds product specifications.

Process advantages are:

- Simplicity of operation
- Low operating costs
- Uniformly sweet and noncorrosive product
- No product lost
- No continuous regeneration
- Low installation cost
- Small plant area required.

Economics:

Investment, $ per bpsd capacity	75–150
Typical requirement, unit per barrel feed	
Power, kWh	*
Steam	**
Catalyst consumption, lb	0.01–0.02
Caustic consumption, lb	0.003–0.006
Total chemical costs, approx. avg $	0.004
Maintenance, per year as % of investment	1

* Only for pumping
** For vessel steam-out and heat tracing, where required.

Licensor: Petreco International Inc.

Liquid treating

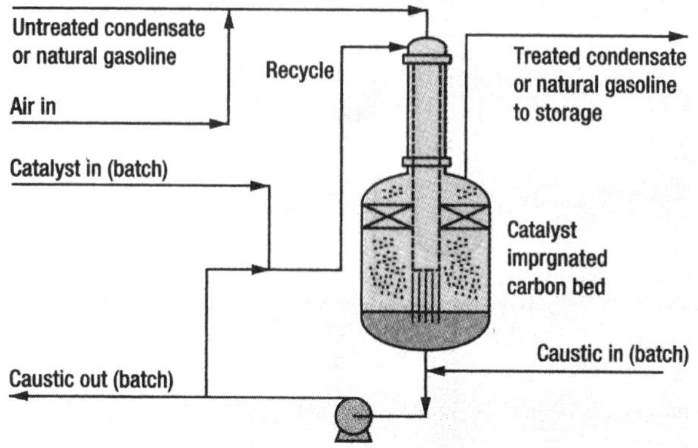

MERICAT II

Application: Oxidize mercaptans to disulfides for gas condensates and natural gasoline with air, caustic and catalyst using FIBER-FILM contactor technology. This is followed by an upflow catalyst impregnated carbon bed.

Description: The caustic phase flows along the FIBER-FILM contactor fibers, which is continuously renewed, as the fibers are preferentially wetted by the caustic phase in the MERICAT II process. Hydrocarbon flows through the shroud parallel to the caustic phase where mercaptans are extracted into the caustic phase. It is converted to disulfides by air and catalyst at the hydrocarbon-caustic interface. The two phases disengage and the hydrocarbon flows upwards through a catalyst impregnated carbon bed where the remaining mercaptans are converted to disulfides.

Economics: FIBER-FILM contactor technology requires smaller processing vessels allowing shorter separation times and less waste generation. This saves plant space and reduces capital expenditures.

Installations: Twenty-eight worldwide.

Reference: *Hydrocarbon Technology International,* 1993.

Licensor: Merichem Chemicals & Refinery Services LLC.

Liquid treating

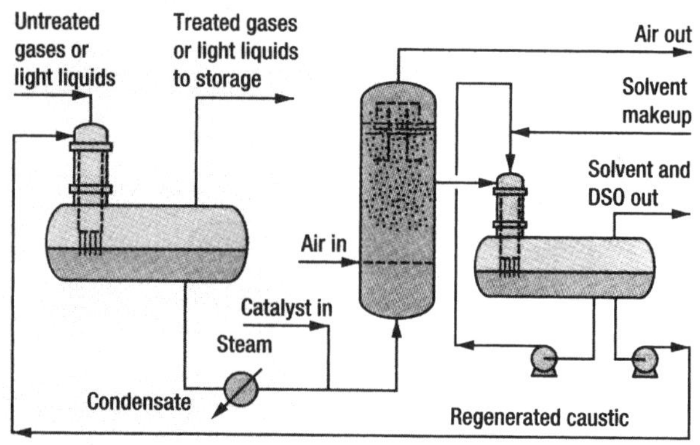

THIOLEX/REGEN

Application: Extract H_2S, COS and mercaptans from gases and light liquid streams, including gasoline, with caustic using FIBER-FILM contactor technology. It can also be used to hydrolize COS contained in LPG.

Description: The caustic phase flows along the FIBER-FILM contactor fibers, which is continuously renewed, as the wet fibers are preferentially wetted by the caustic phase in the THIOLEX process. Hydrocarbons flow through the shroud parallel to the caustic phase where the H_2S and mercaptans are extracted into the caustic phase. The two phases disengage and the caustic flows to the REGEN system. The spent caustic is regenerated using air and catalyst in the oxidizer for reuse, which converts the extracted mercaptans to disulfides. The disulfides are removed from the caustic by a FIBER-FILM contactor solvent wash system.

Economics: FIBER-FILM contactor technology requires smaller processing vessels, allowing shorter separation times and less waste generation. This saves plant space and reduces capital expenditures.

Installations: Over 200 worldwide.

Reference: *Oil & Gas Journal,* Aug. 12, 1985, p. 78.

Licensor: Merichem Chemicals & Refinery Services LLC.

Synthesis gas

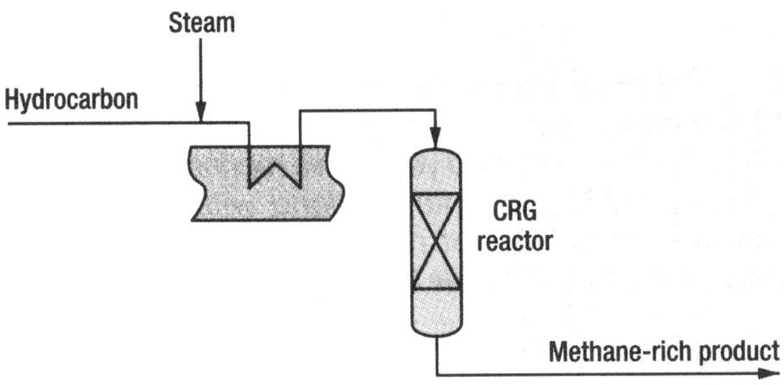

CRG processes—pre-reforming, derichment, methanation

Application: Adiabatic steam reforming of hydrocarbon from natural gas through LPG to naphtha feeds. May be used for the derichment of natural gas (LNG plants), as an adiabatic pre-reformer or in the methanation of H_2/CO-rich streams in SNG manufacture.

Description: Fixed bed of nickel-based catalyst converts hydrocarbon feeds in the presence of steam to a product stream containing only methane together with H_2, CO, CO_2 and unreacted steam. This stream may be exported as product, used as feed for further processing in a conventional fired reformer or as feed to additional CRG processing steps when the methane content of the product can be further enhanced. Using a CRG pre-reformer enables capital cost savings in primary reformer as a result of reduction in radiant box heat load and allows high-activity gas reforming catalyst to be used. The ability to increase preheat temperatures and transfer radiant duty to the convection section of the primary reformer can minimize involuntary steam production.

Operating conditions: CRG processes operate over a wide range of temperatures from 250°C to 650°C, and at pressures up to 75 bara.

Installations: CRG process technology covers 40 years of experience with over 150 plants built and operated. Ongoing development of the catalyst has lead to almost 50 such units since 1990.

References: Littlewood, S., et al., "Prereforming: Based on high activity catalyst to meet market demands," Ammonia Plant Safety & Related Facilities, Vol. 40, p. 3, AIChE.

Licensor: The CRG Process and catalyst are licensed by Davy Process Technology. The CRG Process is available through a number of process licensees worldwide, and the catalyst is manufactured and supplied under license by Synetix.

Synthesis gas

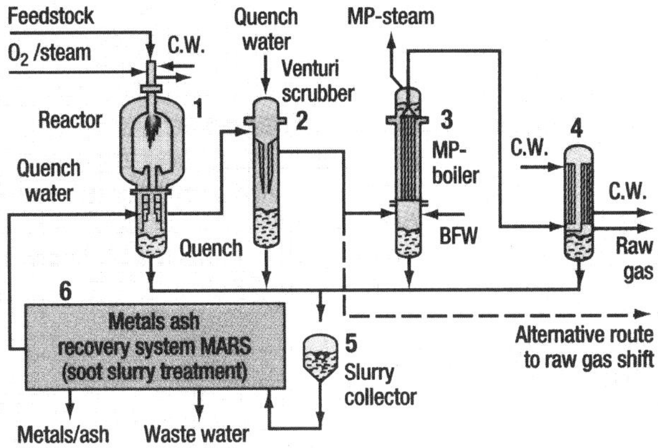

Multipurpose gasification

Application: Production of synthesis gas, essentially H_2 and CO, from a wide range of gaseous to extra heavy liquid hydrocarbons, as well as emulsions and slurries. Recent new applications are in (chemical) waste gasification. The main advantage over comparable processes is its extreme feedstock flexibility in the quench mode. A boiler mode for highest efficiency is also available.

Description: Continuous noncatalytic partial oxidation process. The quench mode is shown above: hydrocarbon feedstock, moderator (H_2O, CO_2 or N_2) and oxidant (pure or diluted O_2, air) are fed through a special burner into the reactor (1), a refractory-lined pressure vessel. Operating conditions are automatically controlled. Hot gas leaves the reactor at the bottom, passing the quench where water is injected to lower the temperature near the saturation value. Quench water washes out most particulates as unconverted carbon (soot) and ash.

Further cleaning occurs in a venturi scrubber (2) from where the gas passes to a medium-pressure steam boiler (3) for heat recovery and to the final cooler (4) before further processing. In hydrogen production, the hot, wet gas from the venturi is passed directly to a raw gas shift conversion. The soot/ash slurry from the process contains virtually all metals and ashes from the feedstock. It is withdrawn via a slurry collector (5) and processed in the metals ash recovery system (MARS) (6). There, soot/ash is filtered from the slurry and incinerated under controlled conditions yielding a saleable metal/ash product. Filtered water is returned for quenching. Excess water is stripped and sent to conventional wastewater treatment.

Operating conditions: Actual gasification temperatures of 1,200°C to 1,600°C, pressures from atmospheric to 70 bar (or higher, if economically justified). Feedstock and oxidant preheat possible in a wide range from 100°C to 600°C, depending on type of feed. Product yields and composition vary with moderator rate and type of feed. Water quench is selected for highest feedstock flexibility. At low-salt contents, the boiler mode can recover heat as high-pressure steam, raising overall efficiency.

Economics: Characteristic consumption and production rates per ton of heavy residue feedstock: 1 to 1.1 t O_2 (100%), export 0.5 t MP steam to 2.2 t HP steam, 2.2 t raw syngas (dry) equiv. to 2,600 Nm^3 H_2 + CO. Cold gas efficiency is 82% to 85%. In boiler mode, thermal efficiencies including HP steam generated are about 95% based on feedstock HHV. This makes the process attractive for syngas production and for an IGCC power plant. A highly integrated and efficient power complex will be in the range of $900/kW total invested cost.

Installations: A large-scale industrial plant operates in Germany, demonstrating full feedstock and product flexibility by feeding to a methanol and IGCC complex.

Reference: Liebner, W., and C. Erdmann, "MPG—Lurgi Multipurpose Gasification—Recent Applications and Experiences," World Petroleum Congress 2000, Calgary, Canada, June 2000.

Licensor: Lurgi Oel-Gas-Chemie GmbH.

Synthesis gas

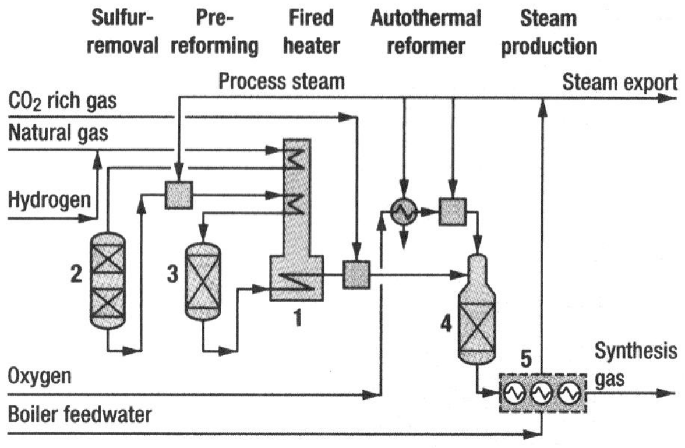

Syngas (ATR)

Application: Produce CO-rich synthesis gas.

Products: Pure carbon monoxide (CO) and hydrogen (H_2) or synthesis gas for methanol, synthetic fuels, oxo-synthesis or ammonia. Very large-scale production of synthesis gas for GTL plants.

Description: The process combines partial oxidation and adiabatic steam reforming using a fixed catalyst bed (nickel catalyst). Soot-free operation is secured by proprietary burner design and application.

The unit consists of a feed preheater (1), feed desulfurization (2), pre-reforming (optional) (3), autothermal reformer reactor with burner, combustion chamber and catalyst bed (4), heat-recovery section (5) and, when required, gas purification section, e.g., CO_2 removal (not shown).

The autothermal reformer burner features low metal temperatures and high resistance to mechanical wear, thus ensuring long burner lifetime. The burner is manufactured in commercially available high-temperature alloys without cooling circuits.

The hydrocarbon feedstock can be natural gas, LPG or naphtha. The oxygen feed can be pure oxygen, air or enriched air. CO_2 recycle or CO_2 import can be applied to adjust synthesis gas composition.

Operating conditions: Typically, CO-rich synthesis gas is obtained at feed ratios of H_2O/CH_4 = 0.2 to 1.5 and CO_2/CH_4 = 0.0 to 2.0, resulting in synthesis gas ratios H_2/CO—0.8 to 2.5 at reactor exit temperatures of 950°C to 1,050°C. Reactor pressure ranges between 20 and 70 bar.

Installations: Since 1958, 26 complete installations have been licensed. New burner technology with improved lifetime was introduced in 1992 and is in operation in 20 autothermal reforming plants and 8 secondary reforming plants. Six burners are on order for GTL plants with up to 17,000-bpd capacity per line.

References: Christensen, T. S. and I. I. Primdahl, "Improve Syngas Production using Autothermal Reforming, *Hydrocarbon Processing,* March 1994, p. 39.

Christensen, T. S., et al., "Burners for Secondary and Autothermal Reforming—Design and Industrial Performance," *Ammonia Plant Safety,* Vol. 35, 1994, p. 205.

Christensen, T. S., et al., "Syngas Preparation by Autothermal Reforming for Conversion of Natural Gas to Liquid Products (GTL)," Monetizing Stranded Gas Reserves Conference, San Francisco, 1998.

Ernst, W. S., et al., "Push Syngas Production Limits," *Hydrocarbon Processing,* March 2000, p. 100-C.

Licensor: Haldor Topsøe A/S.

Synthesis gas

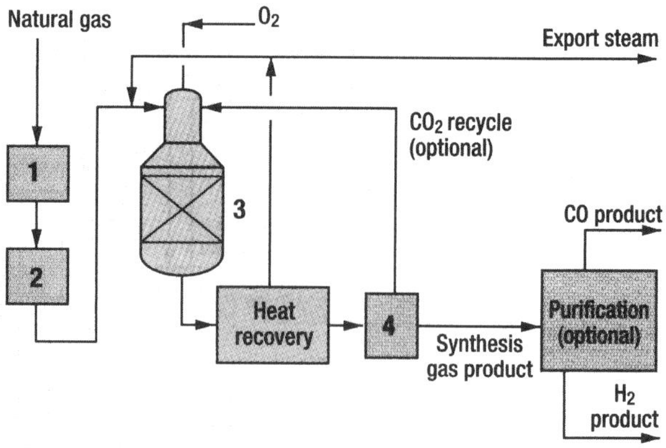

Syngas (autothermal)

Application: Production of carbon monoxide and hydrogen for petrochemical use. Typical consumers are oxo-alcohol synthesis units and methanol synthesis units.

Products: Synthesis gas containing carbon monoxide and hydrogen. The synthesis gas can be used directly for chemical production or alternatively can be further processed to yield high-purity carbon monoxide and high-purity hydrogen.

Description: The feed is preheated (1) and then desulfurized in a conventional hydrotreating-zinc oxide system (2). Steam is added to the desulfurized feed. Carbon dioxide recycle (optional) is also added. The feed mixture is sent to the autothermal reformer (3) a refractory-lined vessel containing catalyst and a burner. The feed mixture is burned with oxygen in the burner located near the top of the reformer vessel. Partial oxidation reactions occur in a combustion zone just below the burner. The mixture then passes through a catalyst bed where reforming reactions occur. The gas exits at about 1,700°F to 1,900°F, depending on the final product specifications.

The exit gas is cooled and passed through a carbon dioxide removal unit (4). The resulting process gas consists primarily of carbon monoxide and hydrogen and is available as product synthesis gas. This synthesis gas can be used to make a variety of chemicals, including methanol and oxo alcohols. Alternatively, the gas can be further processed (typically by cryogenic separation) to yield high-purity carbon monoxide and high-purity hydrogen.

Carbon dioxide can be recycled to adjust the H_2/CO product ratio. For natural gas feedstocks, the H_2/CO product ratio ranges from about 2.7 (for no CO_2 recycle) to 1.6 (for full CO_2 recycle).

Autothermal reforming technology is similar to secondary reforming for ammonia production, except that oxygen is used as feedstock instead of air. Oxygen is required since nitrogen would dilute the H_2/CO product gas purity.

Economics: The economics can be favorable for autothermal reforming when oxygen is available at relatively low cost. For natural gas feedstocks, the optimum H_2/CO product ratio is about 1.6 to 2.7.

Reference: Tindall, B. M., and M. A. Crews, "Alternative technologies to steam-methane reforming," *Hydrocarbon Processing*, November 1995.

Licensor: Howe-Baker Engineers, Ltd.

Synthesis gas

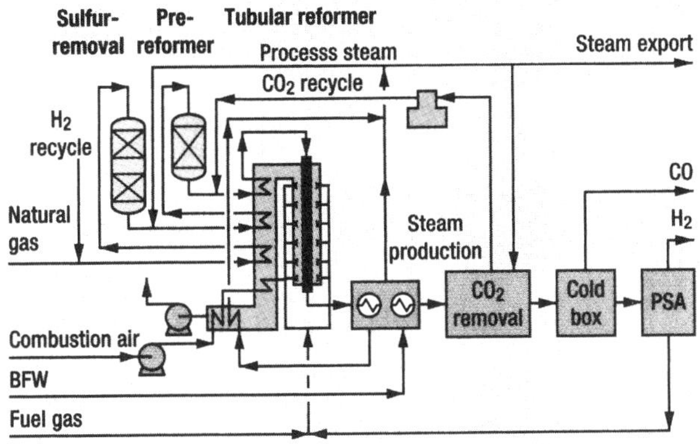

Syngas-advanced SMR

Application: Produce a CO-rich synthesis gas.

Products: Pure CO and H_2, or mixtures of CO and H_2 (synthesis gas), are used to manufacture many chemicals, e.g., acetic acid, oxo-alcohols and isocyanates.

Description: Advanced steam reforming in a fired-tubular reformer is the predominant process route for medium-sized synthesis gas plants. The Topsøe side-fired reformer and reforming catalysts enable operating at low steam-to-carbon ratios, high reformer outlet temperature and high heat flux.

The hydrocarbon feedstock can be natural gas, LPG or naphtha. For heavy feedstocks, an adiabatic pre-reformer is needed upstream of the tubular reformer. Carbon dioxide (CO_2) import and/or CO_2 recycle is applied to produce CO-rich synthesis gas.

The unit typically consists of a feed desulfurization, pre-reforming, tubular reforming, CO_2 recovery and recycle, and final purification. Purification of the synthesis gas by membrane, cold box and PSA is dependent on the required end-product specifications.

Operating conditions: Typically, CO-rich synthesis gas is obtained at feed ratios of H_2O/CH_4 = 1.5 to 2.0 with CO_2 recycle and/or CO_2 import. This results in H_2/CO ratios down to 1.8 at reformer exit temperatures of 950°C (1,740°F). Even lower H_2/CO ratios can be obtained by CO_2 reforming using the SPARG process or by applying a high-activity, noble-metal reforming catalyst.

Economics: Production CO-rich synthesis gases at advanced steam reforming conditions offers significant savings in operating and investment costs.

References: Vannby, R. and C. Stub Nielsen, "Operating experience in advanced steam reforming", Symposium on Large Chemical Plants, Antwerp, October 1992.

Winter Madsen, S. and J-H. Bak Hansen, "Industrial aspects of CO_2 reforming," AIChE spring meeting, Houston, March 1997.

Winter Madsen, S., et al., "Advanced reforming technologies for synthesis gas production," Symposium on Large Chemical Plants, Antwerp, September 1998.

Licensor: Haldor Topsøe A/S.

Synthesis gas

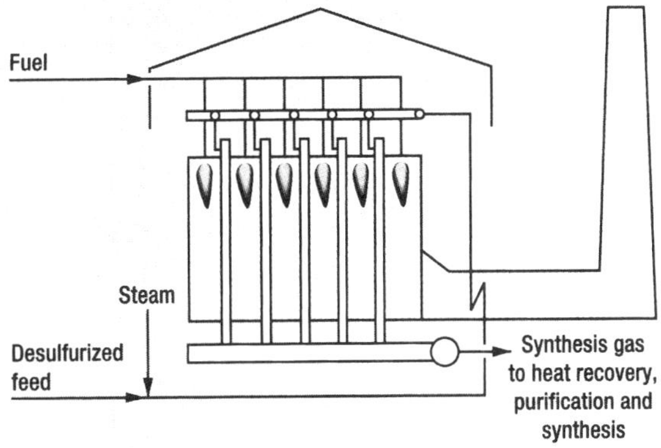

Syngas—steam reforming

Application: Produce hydrogen and/or H_2/CO-rich gas using advanced steam-reforming methods.

Feedstock: Natural gas, refinery offgas, LPG/butane and light naphtha.

Product: Synthesis gas for hydrogen, ammonia, methanol, Fischer-Tropsch-synfuel, oxo-synthesis products, etc.

Description: The steam-reforming process comprises the high-temperature reaction of methane or light hydrocarbons over a nickel catalyst that produces hydrogen and carbon monoxide (CO). The reforming occurs in tubes packed with catalyst and arranged vertically in gas-fired steam reformers. A nickel catalyst is used and applied to a supporting structure. The operating conditions of the steam reformer vary and depend on the application, with discharge temperatures ranging from 740°C to 950°C and pressures up to 50 bar. This wide range of operating conditions necessitates a versatile reformer design.

The Uhde steam reformer is a top-fired reformer which has tubes made of centrifugally cast alloy steel and a proprietary "cold" outlet manifold system. The reformer and outlet system design have proved its suitability over the past decades. It satisfies the demands of various applications and permits constructing and operating world-scale reformers with unrestricted reliability. The largest reformer, based on Uhde technology, consisted of 960 tubes.

Installations: Fifty-six Uhde reformers exist world-wide. The first Uhde reformer with a cold outlet manifold went onstream in 1966. In 1977 and 1984, two large Uhde steam reformers with 540 reformer tubes each were commissioned, and these units are still in operation today. The company has recently commissioned two large ammonia/urea complexes based on natural gas in Qatar and Egypt with ammonia syngas capacities of 96,000 Nm^3/h of CO + H_2 and 81,000 Nm^3/h of CO + H_2, respectively. A plant for oxo-chemicals supplied by Uhde to South Korea has an oxo-syngas production capacity of 9,700 Nm^3/h of CO + H_2. Uhde has recently designed and constructed two world-scale hydrogen plants for SINCOR C.A., Venezuela (2 × 98,000 Nm^3/h) and Shell Canada Ltd., Canada (2 × 130,000 Nm^3/h).

Reference: Fritsch, S., "Steam reformer based hydrogen plant optimization," The International Conference HYFORUM 2000, Munich, September 2000.

Licensor: Uhde GmbH.

Synthesis gas

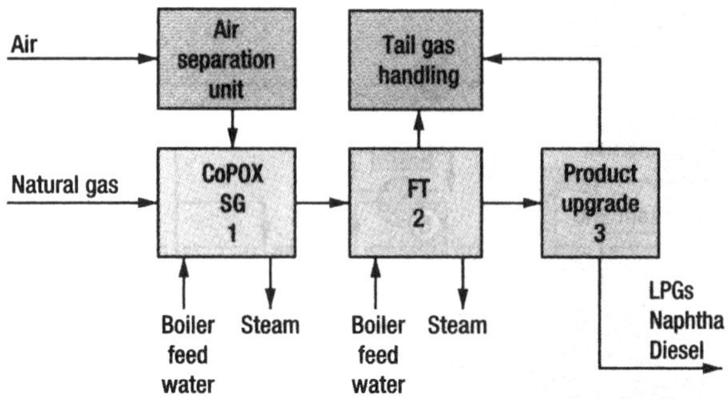

Gas-to-liquids (GTL)

Application: To produce transportable middle distillate products from natural gas. The process has a high carbon efficiency built around Conoco's proprietary CoPOX catalytic partial oxidation syngas technology and a proprietary Fischer-Tropsch (FT) gas conversion process. Conoco GTL is a viable option to monetize stranded gas reserves.

Description: The front-end process (1) is based on the unique and highly proprietary CoPOX catalytic partial oxidation syngas technology. The hydrocarbon feed conversion and selectivity to CO and H_2 is higher than conventional equilibrium syngas production methods. The reactors have high throughput and operate at mild conditions.

The middle process (2) is based on proprietary FT gas conversion technology. The FT process features a high-activity catalyst that generates a paraffinic product spectrum with a high ASF alpha distribution and low methane selectivity. A higher alpha implies the greater production of heavy hydrocarbons at the expense of lighter hydrocarbon synthesis. Different reactor configurations can be used to minimize total reactor volume and produce the desired product slate.

The FT product is finished in the back-end product upgrading Unit (3) that includes hydroprocessing using both conventional and Conoco's proprietary technology. The efficient CoPOX syngas and FT conversion technologies are configured with a total process design that delivers a high carbon efficiency while minimizing the required capital expenditures. The integrated process maximizes the use of exothermic reaction heat and minimizes recycles.

Operating conditions: The CoPOX syngas process operates within a range of conditions depending on inlet gas composition. Typical operating temperature range is 600°C to 1,000°C. The Conoco FT process operates within a typical temperature range of 200°C to 250°C. Typical GTL plant yields from a 500 MMscfd natural gas feed are 63,800 bpd of diesel, naphtha and LPGs.

Installations: Conoco is currently constructing a $75 million 400-bpd demonstration plant in Ponca City, Oklahoma. Mechanical completion is scheduled for fourth-quarter 2002. Plans are to have a first commercial plant operating in 2008.

Licensor: Conoco Gas Solutions, a division of Conoco Inc.

Synthesis gas

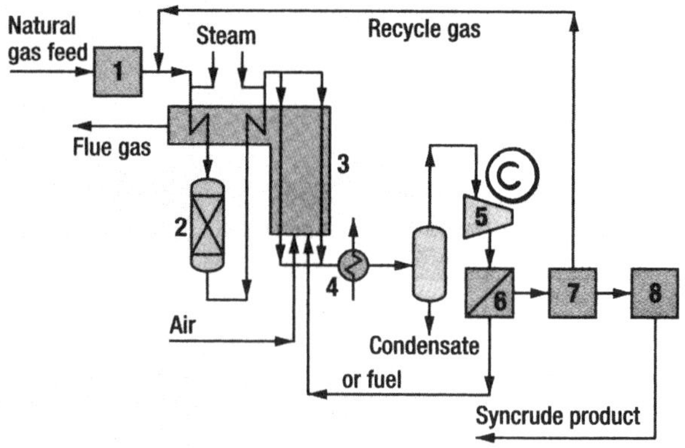

Gas-to-liquids (GTL)

Applications: To produce a zero-sulfur, pumpable syncrude from remote gas fields or from associated gas. The process can operate on a range of natural gas feedstocks including fields containing high carbon dioxide levels. The process is suited for remote or offshore locations where space and weight are of particular concern.

Description: Natural gas is pre-treated to remove sulfur using conventional desulfurization technology (1). Steam and recycle gases are added and the feed is further heated before passing to the CRG pre-reformer (2). Using a nickel catalyst, the CRG pre-reformer converts heavier hydrocarbons to methane and partially reforms the feedstock. Addition of steam and further preheating is completed before the mixed gas passes to the compact reformer (3).

The Davy/BP compact reformer is a multi-tubular, counter-current reactor, which, in the presence of a nickel catalyst, produces a mixture of carbon oxides and hydrogen. Heat for this endothermic reaction is provided by external firing of excess hydrogen produced by the process with supplementary natural gas as required. Gas leaving the reformer is cooled (4) and generates sufficient steam to satisfy process heating requirements. Excess condensate then is removed.

Dry syngas is compressed in a single-stage centrifugal compressor (5) and passes to a membrane-separation package (6) where the surplus hydrogen is recovered and reused as fuel. The non-permeate product from the membrane separation is fed to the conversion section (7) where the syngas is converted into a mixed paraffin and wax product using a cobalt catalyst. The reaction system can either be a fixed bed or slurry type depending on unit size and project needs. Unconverted syngas is recycled to the compact reformer feed.

The wax products from the conversion section can be hydrocracked to produce a pumpable syncrude using conventional hydrocracking technology (8).

Operating conditions: A wide range of reformer operating conditions are possible to optimize the process efficiency.

Installations: Operations of compact reformer and fixed-bed FT demonstration plant are scheduled for second quarter 2002.

Licensor: Davy Process Technology.

Synthesis gas

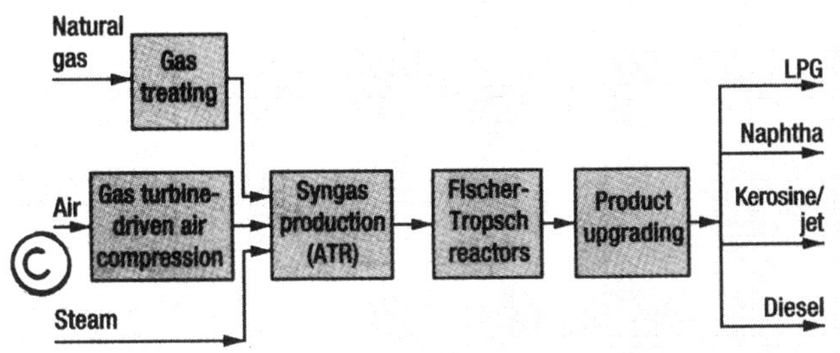

Gas-to-liquids (GTLs)

Application: To produce ultra-clean, synthetic fuels from natural gas. The fuels, diesel, kerosine, naphtha and LPG contain no sulfur, aromatics or heavy metals. This process can be designed for onshore or offshore applications. Furthermore, it is self-sufficient in utilities and can be configured to export electricity.

Description: The distinguishing characteristic of Syntroleum's Fischer-Tropsch (FT) process is using air to produce synthesis gas. Air, natural gas and steam are mixed and react in a proprietary auto-thermal reformer (ATR) to produce synthesis gas with a H_2:CO ratio of approximately 2:1. The synthesis gas is compressed and sent to the FT reactors. Using a proprietary cobalt catalyst, the carbon monoxide (CO) is hydrogenated into paraffinic, synthetic hydrocarbons. The catalysts yield an Anderson-Shultz-Fluory distribution with an alpha between 0.88 and 0.94, depending on the formulation.

Unreacted synthesis gas from the FT reactors is used as process fuel for turbines, heaters and other equipment. Both the ATR and the FT reactions generate byproduct heat and water, which are recovered and reused within the process.

The streams of synthetic crude are combined and refined into ultra-clean diesel, kerosine, naphtha and LPG. Compared to conventional crude oil, refining FT crude is less severe, i.e., lower H_2 consumption, lower temperature and pressure, and longer catalyst life, due to the absence of sulfur, aromatics and heavy metals.

Operating conditions: Approximately 10,000 scf of gas produces one barrel of product. Plants can be economically designed for gas feedrates from 25 MMscfd to 1,000 MMscfd or more.

Economics: With the Syntroleum process, air is used in place of pure oxygen from an air-separation unit. Using air combined with the high-activity catalyst technology, this process can offer considerable capital and operating cost savings as compared to other competing processes.

Reference: Patent 6,265,453.

Licensor: Syntroleum Corp.

Sulfur

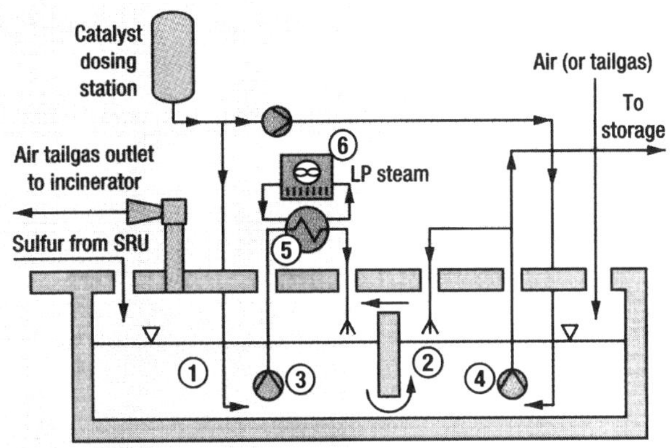

AQUISULF

Application: Decrease H_2S content in liquid sulfur condensed in SRUs and routed to sulfur degassing facilities. Maximum H_2S content is 10 ppm.

Description: A degassing unit is required to achieve 10 ppm H_2S in liquid sulfur. The degassing is done in a concrete pit or sulfur degassing vessel that is divided into two compartments. The first compartment (1) is equipped with a recirculation and cooling pump (3) and a spraying system. The second (2) is equipped with a recirculation and transfer pump (4) also with a spraying system. A partition wall separates the two areas.

Sulfur arrives continuously into the first compartment where it is sprayed. It flows to the second compartment through an opening located at the bottom of the partition wall. The sulfur is sprayed again in the second compartment where degassing is completed. The degassed liquid sulfur is then transferred under level control to sulfur storage. To quickly mix the AQUISULF catalyst with sulfur, the catalyst is injected at the suction side of each sulfur recirculation pump.

The optimal temperature for degassing is reached by cooling the sulfur (5). The heat generates LP steam that is condensed in the air cooler (6). The pit vapor phase is swept with atmospheric air sucked in by a steam ejector. Sweeping gas flows from the second compartment to the first through a hole located at the top of the partition wall. The gas and H_2S are sent to the incinerator through a steam ejector.

Economics: AQUISULF accounts for approximately 20% of the Claus unit cost.

Installations: More than 80 AQUISULF units are in operation or under design worldwide.

Reference: Nougayrede, J. and R. Voirin, "Liquid catalyst efficiently removes H_2S from liquid sulfur," *Oil and Gas Journal,* July 1989.

Licensor: Lurgi Oel-Gas-Chemie GmbH.

Sulfur

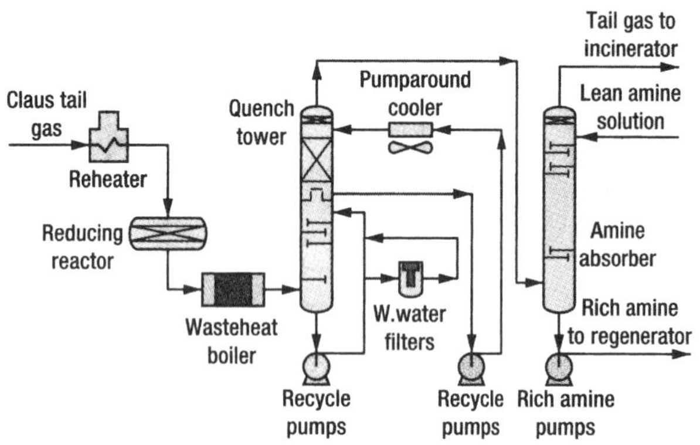

HCR—high Claus ratio

Application: Remove sulfur compounds present in tail gases from Claus plants and meet severe air pollution standards.

Description: The high Claus ratio (HCR) process consists of two sections :

- Hydrogenation and hydrolysis of sulfur compounds present in tail gases (COS, CS_2, S_x and SO_2). Tail gas is heated to approx 300°C and, without hydrogen addition and is treated with Co/Mo catalyst. Gas passes through a waste-heat boiler (WHB) and is cooled to approximately 40°C in a direct contact tower.
- H_2S removal and recycle of acid gas to Claus plant.The gas is washed in a carefully designed amine absorber, and the treated gas is incinerated. Rich amine is processed and recycled.

The process requires adjusting the operating criteria for the Claus unit by increasing the H_2S/SO_2 ratio in tail gas. The operation is very steady and has high service factors that are achieved during upset conditions from upstream units. Hydrogen or reducing gas from external sources are not required in the hydrogenation reactor.

Operating conditions: The pressure drop of the unit is 0.20–0.30 bar and the operating pressure is almost atmospheric. Treated gas contains less than 250 ppmv of H_2S.

Economics: Process uses standard equipment and carbon steel almost everywhere. No consumption of reducing gas or caustic chemicals is required. Process analysers are not mandatory.

Reduction in utilities and chemical costs are approximately $1.50/t of sulfur produced. Lower operating and maintenance labor is about ⅛ man per shift.

Installations: The first HCR commercial plant was started up in November 1988 at Agip Plas's facility (an affiliate of Agip Petroli S.p.A.) in Robassomero, Italy. Since then, more than 10 HCR plants are under construction with capacities ranging from 1.5 tpd up to 270 tpd.

Licensor: SIIRTEC NIGI.

Sulfur

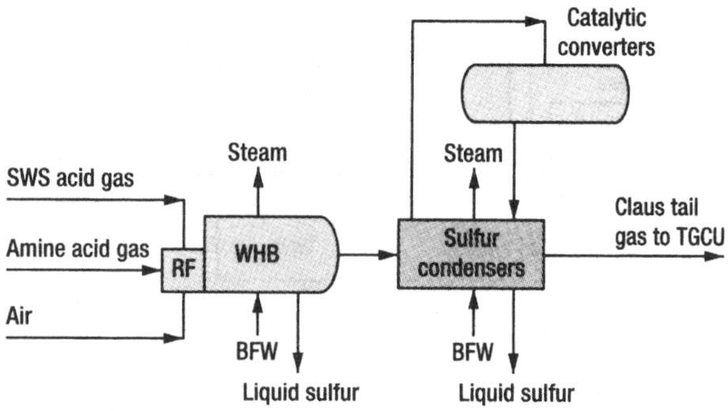

Claus, modified

Application: Recover sulfur from acid-gas streams that contain hydrogen sulfide (H_2S) and ammonia (NH_3).

Product: Bright yellow sulfur with 99.9 % purity and less than 10 ppmw of dissolved H_2S after degassing. Front-end Claus tail gas is either processed in a tail gas clean-up (TGCU) unit for further sulfur recovery, or routed to the Claus incinerator where residual H_2S is oxidized to SO_2.

Description: Acid gases from sweetening units and sour-water strippers are substoichiometrically burnt with ambient air (or air plus oxygen) in a refractory-lined furnace to convert ⅓ of H_2S to SO_2. Subsequently, elemental sulfur is produced in accordance with the Claus reaction between ⅔ of the H_2S and produced SO_2. Ammonia and hydrocarbons contained in the feed gas are also destroyed. High-pressure steam is generated in a waste-heat boiler (WHB), which cools the acid gas from the high-flame temperature to the lower catalytic reactor (converter) temperature. Further sulfur conversion is achieved in two or three stages of catalytic reaction in converters. Each converter is normally preceded by a reheater and followed by a sulfur condenser. Several methods are available for reheating process gas.

Operating conditions: The temperature inside the combustion chamber depends upon the type and quantities of species accompanying the H_2S and O_2. Generally, the operating temperature ranges between 925°C–1,200°C, but can be increased to 1,450°C if NH_3 is present in the feed gas. Total pressure drop of the process gas depends upon the numbers of converter stages. Typically, pressure drop can vary between 0.3 bar and 0.5 bar.

Sulfur recovery efficiency depends on the feedstream composition and the number of catalytic stages; typically, it ranges between 94.5% and 97.5 %.

Economics: Capital cost is approximately U.S.$10 million/100-tpd sulfur recovery unit designed to achieve 95% recovery efficiency from typical amine off gas. Operating costs can be considered negligible if credit for steam generation is taken into account.

Installations: Since the 1970s, more than 60 modified Claus units have been built worldwide by Siirtec Nigi.

Licensor: AMOCO through SIIRTEC NIGI.

Sulfur

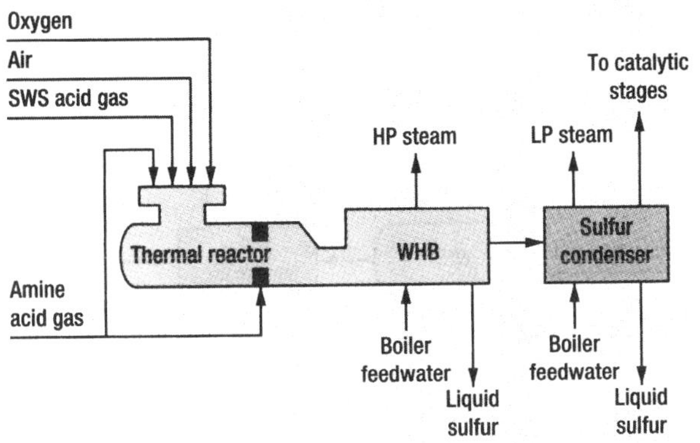

Claus, oxygen-enriched

Application: Debottleneck existing sulfur recovery units (SRU) or reduce size, capital and operating costs for new facilities by using oxygen, either to enrich or to replace combustion air.

Product: Bright yellow, high-purity sulfur. Claus tail gas is either processed in a tail gas clean-up unit (TGCU), or routed to the Claus incinerator where residual hydrogen sulfide (H_2S) is oxidized to SO_2.

Description: In an air-based Claus plant, nitrogen from the combustion air usually comprises more than half of the molar flow through the plant. By replacing air with oxygen, plant capacity can be increased significantly. The level of air enrichment with oxygen and, hence, the level of uprating depend upon the feedgas composition.

Process variations:

- Up to about 30% oxygen , only minor modifications to the plant would be expected.
- Above 30% oxygen concentration, a proprietary oxygen-compatible burner would be required using the SURE burner. Some limited modifications could be expected.
- For capacity increases in excess of about 100%, the SURE double combustion process could be used to achieve the desired expansion.

Operating conditions: The process uses a higher combustion temperature. The operating temperature is kept below 1,650°C—the normal refractory limit. Sulfur recovery efficiency for an oxygen-based Claus process is slightly better than that of air-based Claus and typically ranges between 95% and 98%, depending on process variables.

Economics: The cost of revamp is generally between 10% and 30%—only includes the cost for additional new plant capacity. The cost of a new oxygen-based Claus facility can save up to 35% of the installed cost for an air-based alternative.

Installations: Several units have been designed and revamped by SIIRTEC NIGI in cooperation with Parsons and BOC. Two new units were constructed for IGCC project.

Licensor: SIIRTEC NIGI (up to 30% O_2), SURE Parsons/BOC through SIIRTEC NIGI.

Sulfur

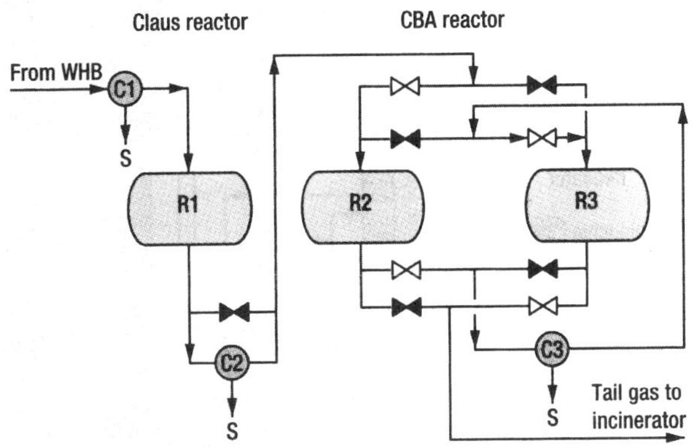

Cold Bed Adsorption (CBA)

Application: Recover elemental sulfur from acid-gas streams that contain hydrogen sulfide (H_2S), or treat Claus tail gas for additional sulfur recovery. CBA beds can be retrofitted in existing units to increase the sulfur recovery.

Product: Bright yellow, commercial grade sulfur. Tail-gas leaving the CBA reactor is normally routed to the incinerator and thermally oxidized.

Description: The CBA is a dry-bed catalytic process which extends and enhances the characteristics of the Claus reaction in two ways:

- Operating the CBA reactors near the sulfur dew point extends the Claus reaction equilibrium and can achieve higher sulfur conversion
- Using the catalyst as an in-situ capture point for the produced sulfur drives the reaction to completion.

A common flowsheet involves two CBA reactors, operating cyclically, downstream of a Claus converter. Gas leaving the sulfur condenser is fed directly to a CBA reactor on adsorption duty. The formed sulfur is adsorbed on the catalyst bed. The bed on regeneration duty is heated by diverting some hot gas from the Claus reactor to drive off and recover the sulfur. This regeneration gas stream is re-combined with the main process gas for treatment in the adsorption bed. Effluent gas from the plant may be sent to a thermal oxidizer. Other flowsheet configurations may also be used, where required, to meet specific sulfur recovery or retrofit objectives.

Operating conditions: The temperature in the CBA reactor can vary from 120°C to 150°C during the adsorption cycle and 300°C to 350°C during regeneration. The total pressure drop of the CBA section is around 0.1–0.15 bar. Sulfur recovery can be over 99%.

Materials of construction: Mainly carbon steel. CBA reactors can be either aluminized or refractory-lined carbon steel. Cyclic valves should be constructed of SS 316.

Installations: Many CBA plants have been built with capacities from 2 Mtpd to 1,300 Mtpd. The most recent CBA plant built by SIIRTEC NIGI is for Hindustan Petroleum Co. Ltd., at Visakhapatnam, India. This plant uses two units with 65-tpd capacity for each unit.

Licensor: AMOCO through SIIRTEC NIGI.

Sulfur

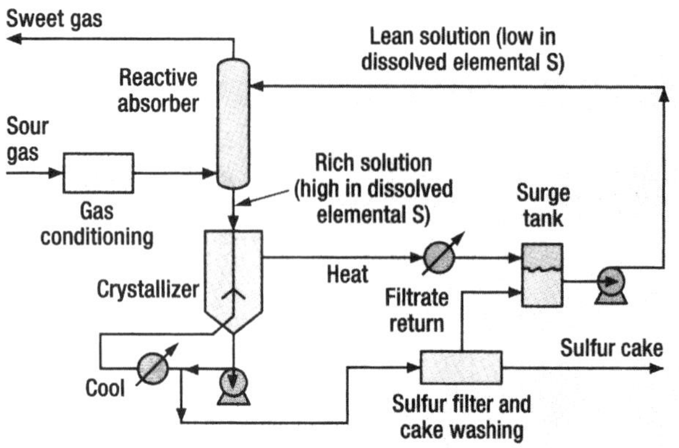

CrystaSulf

Application: Removes H_2S from gas streams in direct treat applications such as natural gas, refinery fuel gas, refinery HDS recycle stream or offgas, high-CO_2 streams in EOR plants and geothermal vent gas. Also removes H_2S, SO_2 and elemental sulfur vapor from Claus tail gas in natural gas processing plants and petroleum refineries. Can treat any high- or low-pressure gas stream, typically those containing between 0.2 and 25 tons of sulfur/day.

Description: Gaseous H_2S is converted to elemental sulfur (S) using a proprietary nonaqueous sorbent. The elemental S formed remains dissolved in the solution; thus, no solids are present in the circulating scrubbing liquor. Sulfur is removed from the system by mild cooling in a crystallizer/filter system. No surfactants, wetting agents or antifoams are needed.

For direct treat applications, SO_2 needed for the reactions is formed from a portion of the inlet H_2S or product S. Scrubbing solution circulation rates are low (e.g., 20–50 gpm/ltd sulfur), and product S purity is high (98+% sulfur, <2% moisture, only traces of hydrocarbons) and can be blended with Claus sulfur or used in agriculture.

Chemical makeup costs are < $250/ltS. Capital and operating costs are less than other options. There is no solution blowdown. For Claus tail-gas treating, the tail gas is fed directly to the CrystaSulf absorber, thus eliminating the reducing gas generator, waste-heat boiler, hydrogenation reactor or water quench. There is also no recycle back to the Claus unit.

Operating conditions: Atmospheric to 2,000+ psi; 120°F to 180°F. CO_2, COS and CS_2 in the inlet gas do not react and do not affect the system. Inlet gas hydrocarbons do not cause foaming or sulfur settling problems.

Installations: Eighteen months of pilot-plant testing was conducted on a 300-psig CO_2 stream. The first commercial installation is schedule for June 2002 and a second unit is also scheduled for fourth-quarter 2002.

References: Rueter, C. O., "CrystaSulf process fills mid-size niche for sulfur recovery in multiple applications," *Gas Tips,* Winter 2002.

Licensor: CrystaTech is GTI's exclusive licensor. (GTI holds the process patents.) The Hanover Co. and CrystaTech have an engineering, sales and marketing agreement for natural gas applications in the U.S. and Canada.

Sulfur

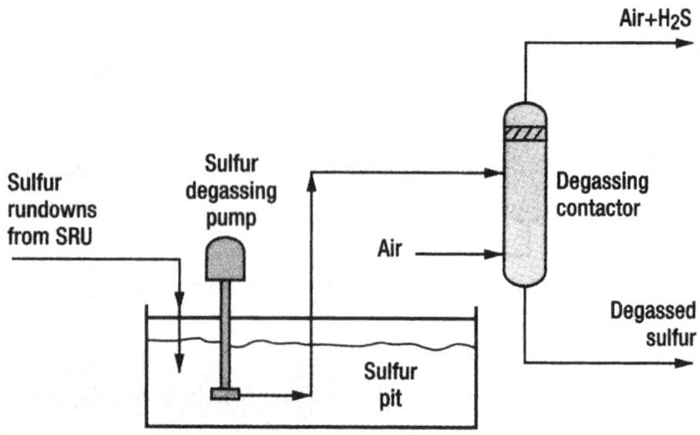

D'GAASS

Application: Removal of dissolved H_2S and H_2S_x from produced liquid sulfur. Undegassed sulfur can create odor problems and poses toxic and explosive hazards during the storage and transport of liquid sulfur.

Description: Degasification is accomplished in a pressurized vertical vessel where undegassed sulfur is efficiently contacted with pressurized process air (instrument or clean utility air). The contactor vessel may be located at any convenient location. The undegassed sulfur is pumped to the vessel and intimately contacted with air across special fixed vessel internals.

Operation at elevated pressure and a controlled temperature accelerates the oxidation of H_2S and polysulfides (H_2S_x) to sulfur.

The degassed sulfur can be sent to storage or directly to loading. Operation at elevated pressure allows the overhead vapor stream to be routed to the traditional incinerator location, or to the SRU main burner or TGTU line burner—thus eliminating the degassing unit as an SO_2 emission source.

Economics: D'GAASS achieves 10 ppmw combined H_2S/H_2S_x in product sulfur without catalyst addition. Elevated pressure results in the following benefits: low capital investment, very small footprint, low operating cost and low air requirement. Operation is simple, requiring minimal operator and maintenance time. No chemicals, catalysts, etc., are required.

Installations: Six plants in operation. Nine additional trains in engineering and construction phase with total capacity over 13,000 ltpd.

Reference: U.S. Patent 5,632,967. Nasato, E., and T. A. Allison, "Sulfur Degasification—The D'GAASS Process," 1998 Laurance Reid Gas Conditioning Conference, Norman, Oklahoma, March 1998.

Fenderson, S., "Continued development of the D'GAASS sulfur degasification process," Brimstone Sulfur Recovery Symposium, Canmore, Alberta, Canada, May 2001.

Licensor: Goar, Allison & Associates, Inc.

Sulfur

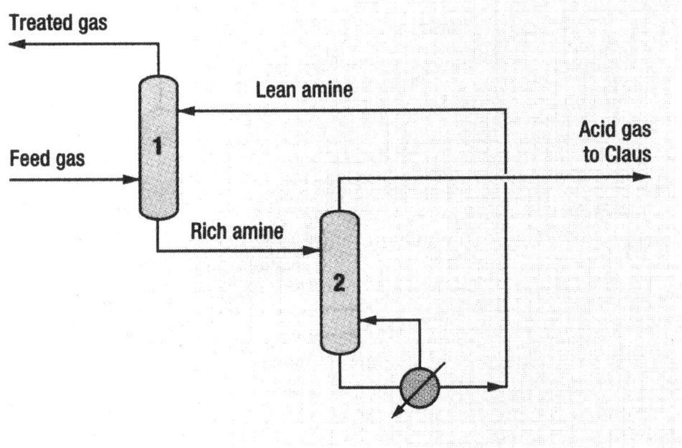

FLEXSORB solvents

Application: Remove H_2S selectively, or remove a group of acidic impurities (H_2S, CO_2, COS, CS_2 and mercaptans) from a variety of streams, depending on the solvent used. Flexsorb SE technology has been used in refineries, natural gas production facilities and petrochemical operations.

Flexsorb SE or SE Plus solvent is used on: hydrogenated Claus plant tail gas to give H_2S, ranging down to H_2S <10 ppmv; pipeline natural gas to give H_2S <0.25 gr/100 scf; or Flexicoking low Btu fuel gas. The resulting acid gas byproduct stream is rich in H_2S.

Hybrid Flexsorb SE solvent is used to selectively remove H_2S, as well as organic sulfur impurities commonly found in natural gas.

Flexsorb PS solvent yields a treated gas with: H_2S <0.25 gr/100 scf, CO_2 <50 ppmv, COS and CS_2 <1 ppmv, mercaptans removal >95%. This solvent is primarily aimed at natural gas or syngas cleanup. The byproduct stream is concentrated acid gases.

Description: A typical amine system flow scheme is used. The feed gas contacts the treating solvent in the absorber (1). The resulting rich solvent bottom stream is heated and sent to the regenerator (2). Regenerator heat is supplied by any suitable heat source. Lean solvent from the regenerator is sent through rich/lean solvent exchangers and coolers before returning to the absorber.

Flexsorb SE solvent is an aqueous solution of a hindered amine. Flexsorb SE Plus solvent is an enhanced aqueous solution, which has improved H_2S regenerability yielding <10 vppm H_2S in the treated gas. Hybrid Flexsorb SE solvent is a hybrid solution containing Flexsorb SE amine, a physical solvent and water. Flexsorb PS solvent is a hybrid consisting of a different hindered amine, a physical solvent and water.

Economics: Lower investment and energy requirements based primarily on requiring 30% to 50% lower solution circulation rates.

Installations: Total gases treated by Flexsorb solvents are about 2 billion scfd and the total sulfur recovery is about 900 long tpd.

Flexsorb SE—26 plants operating, two startups in 2002, one in design
Flexsorb SE Plus—14 plants operating, one startup in 2002, one in design
Hybrid Flexsorb SE—two plants operating
Flexsorb PS—four plants operating and one startup scheduled in 2000.

Reference: Garrison, J., et al, "Keyspan Energy Canada Rimbey acid gas enrichment with FLEXSORB SE Plus technology," 2002 Laurance Reid Gas Conditioning Conference, Norman, Oklahoma.

Adams-Smith, J., et al, Chevron USA Production Company Carter Creek Gas Plant FLEXSORB tail gas treating unit," 2002 GPA Annual Meeting, Dallas.

Licensor: ExxonMobil Research and Engineering Co.

Sulfur

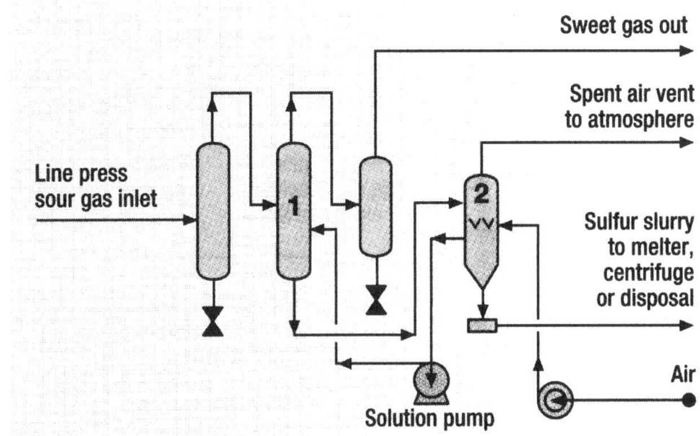

LO-CAT

Application: Removal of H_2S and the production of high-purity sulfur from both anaerobic and aerobic gas streams including wellhead gas, fuel gas, acid gas, natural gas, carbon dioxide, Claus tail gas, synthesis gas and ventilating-air streams. Infinite turndown with respect to H_2S concentration, sulfur loading and gas rate with sulfur capacities ranging from a few pounds per day to greater than 25 ltpd. Recovery of the sulfur product as a slurry, a filter cake or as high-purity, molten sulfur. In most cases, sulfur cake can be deposited in a non-hazardous landfill. Autocirculation configuration when treating amine acid gas allows for remote or unattended operation typical for gas field operation.

Description: Three processing configurations are available depending on the type of gas and final use of the sweet gas. The conventional scheme, shown above, is used to process both combustible gas streams and product gas streams. The sour gas contacts a dilute, proprietary, iron chelate catalyst solution in an absorber (1), where the H_2S is absorbed and oxidized to solid sulfur. The sweet gas leaves the absorber for use by the client. The reduced catalyst solution returns to the oxidizer (2), where sparged air reoxidizes the catalyst solution. The catalyst solution is then returned to the absorber. The continuous regeneration of the catalyst solution allows for very low chemical operating costs.

In the patented autocirculation scheme, the absorber (1) and the oxidizer (2) are combined in one vessel, but separated internally by baffles. Sparging of the sour gas and regeneration air into the specially designed baffle system creates a series of "gas lift" pumps eliminating the external circulation pumps. This configuration is ideally suited for treating amine acid gas streams.

The third processing scheme sweetens air streams contaminated with H_2S. The absorption and oxidation of H_2S to sulfur, as well as the regeneration of the catalyst solution, occur in one vessel. The air in the gas stream is used to regenerate the catalyst, eliminating the oxidizer air blowers.

Operating conditions: Operating pressures from vacuum conditions to several hundred psi. Operating temperatures range from 40°F to 140°F. Hydrogen sulfide concentrations from a few ppm to 100%. Sulfur loadings from a few lb/d to 25^+ ltpd. No restrictions on type of gas to be treated; however, some contaminants such as SO_2 may increase operating costs.

Installations: 119 operating, 3 under construction and 151 under license.

Reference: Hardison, L. C. and D. E. Ramshaw, "H_2S to S: Process improvement," *Hydrocarbon Processing,* Vol. 71, Jan. 1992, pp. 89–90.

Licensor: Gas Technology Products LLC, a Merichem Co.

Sulfur

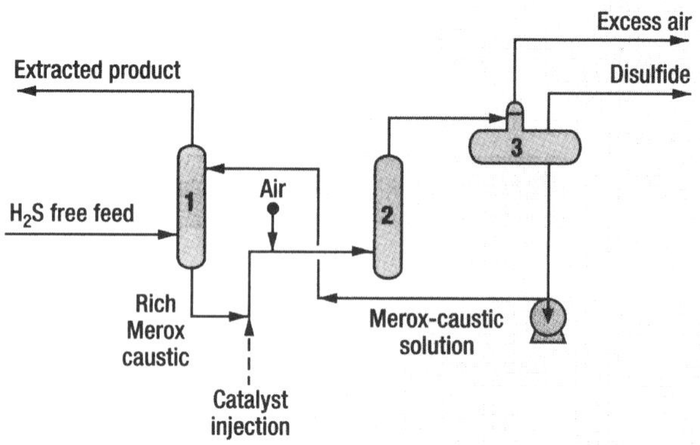

Merox

Application: Extraction of mercaptans from gases, LPG, lower boiling fractions and gasolines, or sweetening of gasoline and heavier stocks by in situ conversion of mercaptans into disulfides.

Products: Essentially mercaptan sulfur-free, i.e., less than 5 ppmw, and concomitant reduced total sulfur content when treated by Merox extraction technique.

Description: Merox units are designed in several flow configurations, depending on feedstock type and processing objectives. All are characterized by low capital and operating costs, ease of operation and minimal operator attention. Three types are described.

Gases, LPG and light naphtha are countercurrently extracted (1) with caustic containing Merox catalyst. Mercaptans in the rich caustic are oxidized (2) with air to disulfides that are decanted (3) before the regenerated caustic is recycled.

Minalk is now the most prevalent Merox gasoline and condensate sweetening scheme. Conversion of mercaptans into disulfides is accomplished with a fixed bed of Merox catalyst that uses air and a continuous injection of only minute amounts of alkali. Sweetened gasoline from the reactor typically contains less than one ppm sodium. A new additive, Merox Plus reagent, can be used to greatly extend catalyst life.

Heavy gasoline and condensate may be sweetened in a fixed-bed unit that closely resembles Minalk, except that a larger amount of more concentrated caustic is recirculated intermittently over the catalyst bed.

Economics: Typical capital investment and operating costs of some Merox process schemes are given based on 2002 dollars for a 10,000-bpsd capacity liquid unit and 10 million scfd gas unit with modular design and construction.

Product	Gas	LPG	Gasoline	Jet fuel/kerosine fixed bed
Scheme	Ext.	Ext.	Minalk	
Est. plant capital, (modular, \$10^3)	2,200	1,600	900	2,100
Direct operating cost, ¢/bbl (¢/10^6 scf)	(1.5)	0.4	0.2	1.5

Installations: Capacity installed and under construction exceeds 13 MMbpsd. Almost 1,600 units have been commissioned as of January 2002, with capacities between 40 and 140,000 bpsd. UOP has licensed gas Merox extraction units with capacities as high as 2.9-billion-scfd for mercaptan control.

Licensor: UOP LLC.

Sulfur

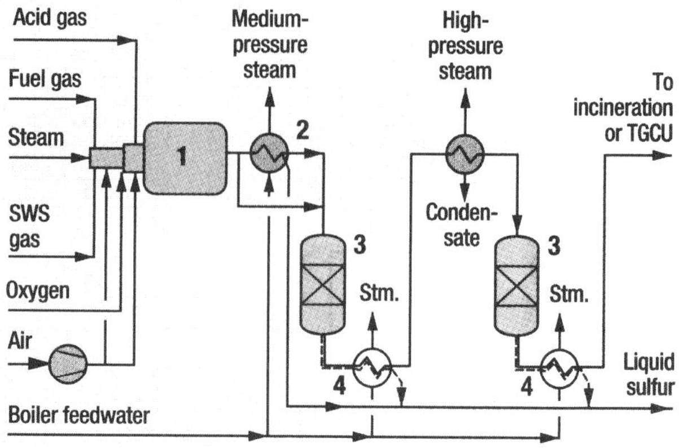

OxyClaus

Application: Increase capacity up to 200% in existing Claus sulfur recovery units, or for a more economical design of grassroots Claus sulfur recovery units.

Description: The modified Claus reaction is carried out with direct oxygen combustion. By using a proprietary thermal reactor burner (1), levels of 80–90% net oxygen can be utilized. Combustion temperature moderation is achieved without the need for any type of gas recycle. Oxygen is combusted with the acid gas in the center of an extremely hot flame core, while air is introduced around the outside of this flame, combusting the balance of the acid gas. Considerable cracking of H_2S to hydrogen and sulfur occurs in the hot flame core as thermodynamic equilibrium is approached. Carbon dioxide is also reduced to carbon monoxide. These endothermic reactions provide proven temperature moderation consistent with conventional refractory/insulating brick materials. The level of produced hydrogen then decreases in the waste-heat boiler as the hot gas is cooled, since equilibrium of the H_2S-cracking reaction is favored by high temperatures. Heat generated by the exothermic reverse reaction is removed in the waste-heat boiler (2). Downstream recovery of elemental sulfur is accomplished by the conventional modified Claus process using a series of catalytic reactors (3) and sulfur condensers (4). No specialized equipment or changes to conventional design practices are required.

Ammonia-containing sour water stripper offgas can also be processed. The ammonia is combusted with air in a separate central burner muffle at near-oxidizing conditions.

Units may be operated in a base-load mode with air only. Peak shaving, as well as operation at full design capacity, is accomplished with air and oxygen.

Economics: For a reference 200-tpd sulfur recovery unit (Claus and tail gas unit) requiring 99.9% overall sulfur recovery, capital cost savings of \$1.6 MM to \$2.5 MM are achievable with oxygen enrichment as compared to an air-only design.

Based on typical pipeline oxygen costs of \$35 per ton, even if oxygen enrichment were used 100% of the time, it would take over eight years for oxygen costs to equal the incremental capital savings.

Installations: More than 30 Claus sulfur recovery units with OxyClaus are in operation or under design worldwide.

Reference: Stevens, D. K., L. H. Stern and W. Nehb, "OxyClaus technology for sulfur recovery," Laurance Reid Gas Conditioning Conference, Norman, Oklahoma, 1996.

Licensors: Lurgi Oel-Gas-Chemie GmbH and The Pritchard Corporation (U.S. only).

Sulfur

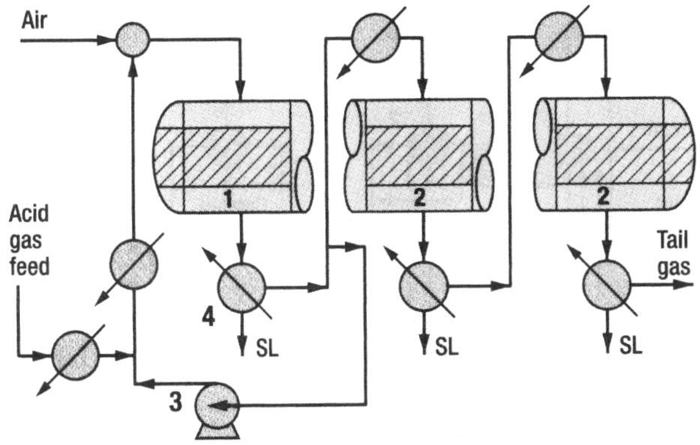

Selectox

Application: Recover sulfur from low H_2S concentration acid gas streams (typ. <30 vol% H_2S). Products are bright yellow sulfur (99.9% pure) and clean tail gas for further sulfur recovery or incineration.

Description: Selectox and Recycle Selectox units resemble Claus units, except the burner and reaction furnace are replaced by a fixed bed of Selectox catalyst (1). Once-through Selectox processes acid gases with 0.3% to 5% H_2S. Over the catalyst, air oxidizes H_2S to SO_2 which reacts with additional H_2S to produce elemental sulfur. About 80% of the inlet H_2S is recovered as liquid sulfur. Exit gas containing H_2S, SO_2 and sulfur vapor passes through one or more catalytic Claus reactor stages using alumina catalyst (2) which boosts sulfur recovery to 90% to 95%.

Recycle Selectox processes acid gases with 5% to 100% H_2S. To control the temperature rise in the Selectox reactor, a recycle blower (3) dilutes the acid gas feed with gas from the Selectox reactor condenser (4). In other respects, Recycle Selectox is the same as Once-through Selectox. Average Recycle Selectox recovery is about 97%.

Tail gas from either process can be incinerated (thermal or catalytic) or fed to a tail gas cleanup process. In two units, the incinerated tail gas is scrubbed with caustic to achieve total sulfur removal.

Operating conditions: Temperatures: 350°F (inlet) to 700°F (outlet). Sulfur recovery: 90% to 98%. Incineration of tail gases followed by caustic scrubbing gives > 99.9% sulfur removal.

Economics: The estimated cost for a 15 to 20-ltpd Recycle Selectox sulfur recovery unit designed to achieve 96% to 98% recovery from an acid gas containing 15% to 25% H_2S is $3–4 million.

Installations: Twenty-one total Selectox licensed plants with sulfur capacities from 0.5 ltpd to 30 ltpd.

Reference: Gowdy, H. W., R. V. Bertram and R. K. Sahlin, "Selectox processes for sulfur management of hydrogen sulfide streams," Proceedings of the Gas Research Institute's Seventh Sulfur Recovery Conference, Austin, Texas, September 1995.

Licensors: UOP LLC and Parsons; Sub-licensors: Goar, Allison & Associates, Inc. and KTI.

Sulfur

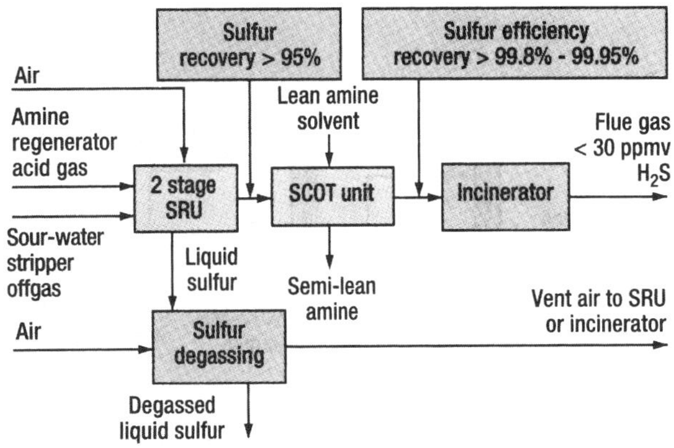

SRU

Application: To recover liquid sulfur from H_2S-containing gases such as acid gas from gas-sweetening units, acid gas from sour-water strippers or offgas from sulfur degassing facilities.

Description: The main reactions are well-known Claus reactions. Ammonia, present in the sour-water stripper offgas, is processed in the main burner (1) directly with other feed gas streams. Practical complete ammonia destruction is ensured by choosing the proper high-intensity burner and preheating feed gas. The Claus unit is designed to suppress SO_3 formation at the reaction furnace and subsequent line burners. The sulfur complexes, even with ammonia in the feed, can typically run for 3 to 4 years without a maintenance shutdown.

The Claus tail gas is routed to a SCOT unit, where the sulfur components are recovered (see SCOT). Upsets, such as those based on partial oxidation, usually pass almost unnoticed in the SCOT unit. For example, a hydrocarbon upset in the feed to the Claus will not have any other effect on the SCOT, other than a slight increase in H_2S content from the SCOT absorber. A temporary ammonia breakthrough from the SRU has no other effect than a slight pH increase in the quench water.

Liquid sulfur produced in the Claus unit is degassed in the Shell sulfur degassing facilities (see "sulfur degassing" for more details). In this process, the H_2S level in liquid sulfur is reduced to below 10 ppmw, without using catalyst, which improves safety during liquid sulfur handling. Offgas of the sulfur stripper can be routed to the Claus unit or to an incinerator.

The incinerator downstream the Claus/SCOT unit is designed such that a maximum of 10 ppmv. H_2S can slip through, the remainder of the H_2S and any other sulfur component are oxidized to SO_2. A thermal incinerator can be designed with a heat-recovery section for superheating steam produced in the Claus unit or gas preheating. A catalytic incinerator, using propriety catalyst, more selectively oxidizers sulfur compounds, thus significantly reducing the risk from temperature runaways.

Installations: Units with sulfur capacities up to 4,000 tsd for refineries and gas plants.

References: Hoksberg, et al., "Sulfur recovery for the New Millennium," *Hydrocarbon Engineering,* November 1999.

Verhulst, "Safeguarding of the sulfur recovery process," Bovar/Western Research Sulfur Seminar, Budapest, 1993.

Licensor: Shell Global Solutions International B.V.

Sulfur

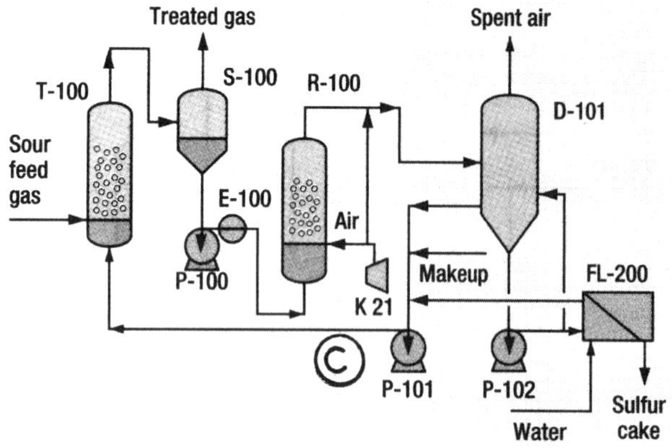

SulFerox

Application: Removal of H_2S from a large range of gas streams giving a sulfur production of 0.1–20 tpd. Gas applications include: natural gas, amine tail gas, enhanced oil recovery CO_2 recycle, refinery gases, geothermal, syngas, offshore production gas, digester offgas and offgas from wastewater treatment plants. The full range of H_2S concentrations (from a few ppmv to almost 100%V of H_2S) can be treated to 1 ppmv H_2S. Turndown properties—both on H_2S concentration and total gas flow—are very good, and the process shows excellent flexibility. If gas flows become too large (in excess of 10 $MMNm^3/d$) the combination of an amine unit coupled to a SulFerox unit becomes more economical. The sulfur product can be obtained as a filter cake or alternatively as molten sulfur of high quality. The sulfur cake can be land filled as a nonhazardous waste or—depending on local regulations—directly used as a fertilizer.

Description: SulFerox is a redox-based process that converts H_2S in sour gas to elemental sulfur (S) solution in the contactor through reaction with a proprietary aqueous ferric iron chelate solution. Various contactor types are available, such as sparged towers, spray towers and pipeline contactors. The sparged tower is the most versatile contactor and will be selected whenever pressure drop allows. After contacting the sour gas and SulFerox solution in the (sparged tower) contactor, the gas/liquid mixture is separated. In the separator, the solution and treated gas is separated yielding the treated sweet gas—leaving the unit via an optional knockout vessel—and the depleted SulFerox solution. After optional degassing, the reduced iron chelate solution is regenerated via reaction of Fe^{2+} back to Fe^{3+} with oxygen from an air source in the regenerator vessel. Via a thickener vessel, part of the solution is sent for filtration and sulfur recovery; the major portion of the stream is returned to the contactor. In addition, the resulting filtrate is returned to the process for maximum solution recovery, thus, optimally using the chelate solution. Depending on the feed gas conditions, the contactor and separator can be combined in one vessel as can be done with the regenerator and surge vessel. This gives a two-vessel configuration suited for amine regenerator offgas.

Operating conditions: Operating pressures vary from just over atmospheric to 500 psig. The SulFerox process itself operates at temperatures of 110°F–140°F. However, the feed gas temperature can be between 75°F–130°F. The feed gas must be free of hydrocarbon liquids. Gas at other conditions may need pretreatment first (cooling, dew pointing). Although the applicability of the SulFerox process is very wide, some feed gas contaminants such as high levels of NH_3, HCN and SO_2 might affect the economics of the process.

Installations: Twenty-nine units operating, one under construction.

References: Smit, C. J. and E. C. Heyman, "Present status SulFerox process," Proc. GRI Sulfur Recovery Conf. 9th meeting, 1999.

Oostwouder, S. P., "SulFerox Process Update," Proc. GRI Sulfur Recovery Conf. 8th meeting, 1997.

Licensor: Shell Global Solutions International B.V.

Sulfur

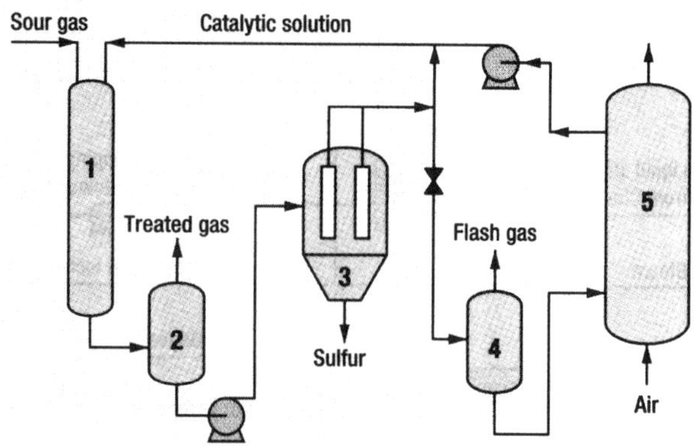

Sulfint HP

Application: Selective hydrogen sulfide (H_2S) removal from high-pressure gases with direct conversion of H_2S into elemental sulfur (S). Residual H_2S levels in the treated gas can be lower than 1 ppm vol.

Description: The Sulfint HP uses the well-known concept of redox desulfurization. However, this process has been specifically designed to treat high-pressure gases. The feed gas is contacted with the redox catalytic solution (aqueous iron chelate based solution) in a co-current absorber (1). The H_2S is absorbed by the catalytic solution, reacts with the catalyst and is converted into elemental S. Desulfurized gas is recovered from the top of the separator vessel (2). The sulfur-loaded solution is then pumped through a multi-cartridge, high-pressure filter (3). The filtered solution can be partly recycled to the absorber (1) and partially expanded for regeneration.

The expanded solution, after separation of the dissolved gases (4) is regenerated with air in the oxidizer vessel (5). The whole process operates at near-ambient temperature, and no thermal regeneration nor chilling is needed. High-pressure filtration mitigates foaming/plugging problems, especially after the expansion of the solution. Due to direct recycle of the filtered solution to the absorber, pumping costs are minimized. This process is highly selective; only H_2S is removed (no CO_2 co-absorption and very little hydrocarbons co-absorption).

Economics: For units treating 100–300 MMscfd of gas with a sulfur production 0.1–10 tpd, the CAPEX will be 20–50% lower than for conventional processes.

Installations: An industrial pilot plant is treating up to 1.7 MMSCFD of gas under 1,160 psia and has been operated successfully for more than 6,000 hours.

Reference: Le Strat, P. Y., et al., "New redox process successful in high pressure gas streams," *Oil & Gas Journal,* Nov. 26, 2001, pp. 46–54.

Licensors: Prosernat IFP Group Technologies and Le Gaz Integral.

Sulfur

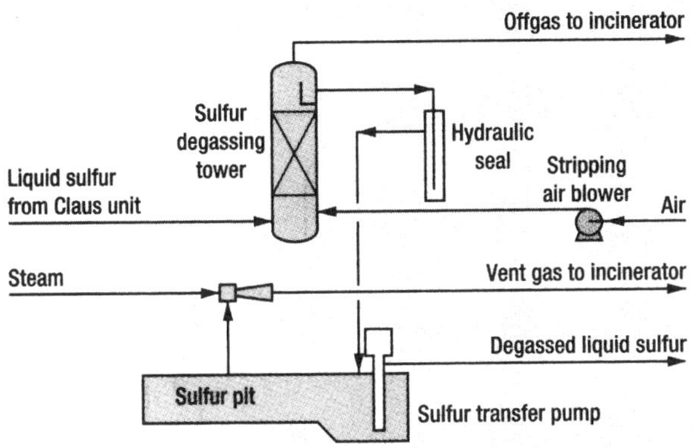

Sulfur degassing

Application: Remove hydrogen sulfide (H_2S) and hydrogen poly-sulfides (H_2S_x) dissolved in liquid sulfur.

Safety: Hydrogen sulfide is a highly toxic and potentially explosive gas. A concentration of 600 ppmv is lethal and is explosive at approximately 3.5% vol. Dissolved H_2S_x is decomposed, and H_2S is released in storage, during loading and unloading and also during transportation, thus leading to potentially dangerous conditions.

Product: Liquid sulfur containing 10 ppmw of H_2S or less.

Description: Liquid sulfur flowing from the Claus plant to the sulfur pit contains typically 250–350 ppmw of $H_2S + H_2S_x$. Sulfur is degassed using an active gas-liquid contacting system to release dissolved gas. Adding chemicals is not required. Sulfur from the pit is pumped into the degassing tower where it is contacted counter-currently with hot compressed air over a fixed catalyst bed. Degassed sulfur is returned to the product section of the sulfur pit.

Operating conditions: The operating temperature in the sulfur degassing tower can vary from 125°C to 150°C depending on the temperature in the sulfur pit. The operating pressure drop is around 0.3–0.5 bar. Treated sulfur has a residual H_2S level in the range of 5–10 ppmw.

Materials of construction: Mainly carbon steel. Aluminized steel may be used for the degassing tower.

Installations: Several units are in operation, design or under construction with unit capacities up to 300 tpd.

Licensor: AMOCO through SIIRTEC NIGI.

Sulfur

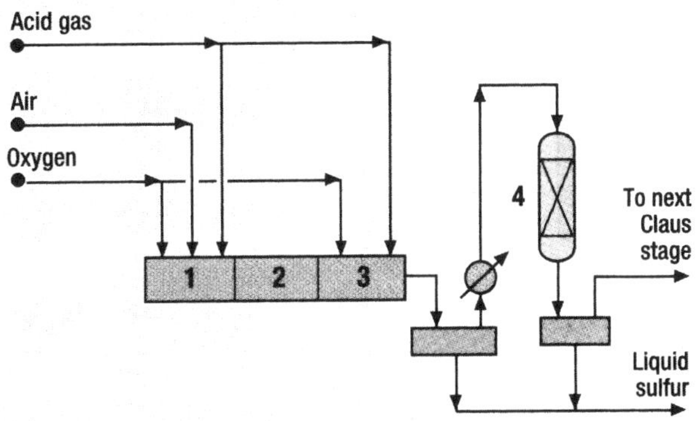

Sure

Application: Oxygen-enhanced recovery of sulfur from H_2S-containing gas streams. Products are bright yellow sulfur (99.9% pure) and clean tail gas for further recovery or incineration.

Description: The capacity of a Claus-type sulfur recovery unit can be increased by thermal combustion of H_2S in two or more stages with an oxidant which is an oxygen-rich gas stream comprised of pure oxygen or a mixture of air and oxygen. A portion of the oxidant is fed to a first combustion zone (1) with all or a portion of the acid gas or ammonia-containing acid gas. The reacted mixture is cooled (2) and the remaining gas streams are fed to a second combustion zone (3). After condensing sulfur, the remaining gas stream is then fed to one or more Claus converters (4).

Operating conditions: Pressures are near atmospheric. Oxygen concentration in oxidant is 21% to 100%. Sulfur recovery is 90% to 98%.

Economics: Replacing air with oxygen will more than double the capacity of a Claus-type sulfur plant and subsequent tail gas treating unit. Use of the Sure process is particularly attractive for revamping existing Claus and tail gas units to substantially increase the acid gas removal and for intermittent operation of such units at higher thruputs. Using the Sure process in new installations significantly reduces capital investment.

Installations: One in Japan, one in the UK, eight in Italy and six in the U.S. Twelve more plants in various engineering and design phases.

Licensors: BOC Gases and Parsons Energy & Chemicals Group, Inc.

Effluent cleanup

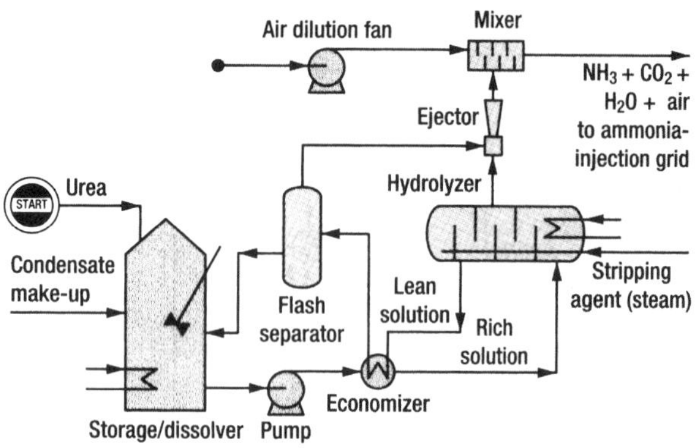

AMMOGEN

Application: AMMOGEN provides gaseous ammonia to fossil-fuel-burning plants to operate pollution-control systems such as $DeNO_x$/$DeSO_x$ SCR, SNCR and flue-gas treatment units. This process produces gaseous ammonia onsite and on demand with harmless and easy-to-handle feedstocks—urea and water. It eliminates the hazard of transporting and storing toxic compounds such as anhydrous or aqueous ammonia.

Products: The gaseous stream of ammonia (15–35%v), carbon dioxide and water vapor.

Description: Dry urea enters the dissolver/storage tank where it is dissolved using lean-recycle solution and condensate. The stirrer reduces mixing time, and the rich-urea solution (20–60%w) is pumped to the hydrolyzer through the economizer where the lean solution sensible heat is recovered. A multistage hydrolysis, at 180–250°C and 15–30 bar, is done in the baffled hydrolyzer, while the reaction products, i.e., ammonia and carbon dioxide, are removed by a stripping fluid (typically steam). The heat of reaction is supplied by an internal heater and stripping fluid. After the urea decomposition, the lean solution is flashed at atmospheric pressure in the flash separator, from where almost pure water is recycled to the dissolver. The flashed vapor, rich in NH_3, joins the main stream produced by the hydrolyzer and both are sent to the static mixer and diluted with air before delivery to the ammonia-injection grid of the flue-gas treatment system.

Main advantages include:

- Gaseous ammonia produced onsite and on demand
- Utmost safety (no governmental reportable amounts or contingency plans)
- Urea feedstock is harmless, easy-to-handle and widely available
- Simple and safe noncatalytic process
- No carryover of compounds that can damage SCR systems
- Very rapid response time and maximum turndown availability
- Quick startup, shutdown and standby
- Automatic operation and low maintenance
- Limited plant footprint for easier installation; also can be skid-mounted or module
- Low capital and operating cost
- Capacity from several to thousands kg/h of ammonia.

Installations: Four plants in the U.S. with design capacity from 270 kg/h to 1,600 kg/h.

Licensors: SIIRTEC NIGI and HERA.

Effluent cleanup

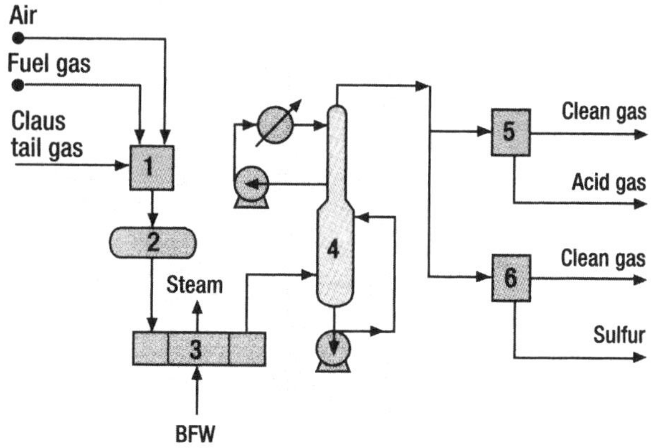

Beavon-others

Application: Purify tail gas from sulfur recovery units (such as Claus units) and other gas streams containing low concentrations of SO_2. The type of process to be combined with the Beavon treatment depends on the intended disposition of the treated product gas (e.g., additional sulfur recovery, other component recovery, incineration or exhaust, while meeting stringent air pollution standards).

Beavon processing converts sulfur compounds to H_2S.
Beavon-MDEA processing adds H_2S separation.
Beavon-Selectox processing converts H_2S to elemental sulfur.
Beavon-Hi-Activity processing converts H_2S to elemental sulfur.

Description: In the Beavon step, essentially all sulfur compounds in the feed gas (SO_2, S_x, COS, CS_2) are converted to H_2S. The feed gas is heated (1) to reaction temperature by mixing with the hot combustion products of fuel gas and air. This combustion is carried out with a deficiency of air to provide sufficient H_2 and CO to convert all of the sulfur and sulfur compounds to H_2S. The heated gas mixture is then passed through a catalyst bed (2) where all sulfur compounds are converted to H_2S by hydrogenation and hydrolysis. The hydrogenated gas stream is cooled in a steam generator (3), and then by direct contact (4) with a buffer solution before entering the selected H_2S removal process.

Beavon-MDEA. One of several processes used to remove H_2S is by absorption (5) in a solution of MDEA (methyl diethanolamine) or one of the recently developed highly selective amine type solvents. The clean tail gas contains less than 10 ppm H_2S when using the newer solvents. When this combination is operating on a Claus tail gas, the separated H_2S can be recycled to the Claus unit.

Beavon-Selectox. An alternative (6) for removing the H_2S is to convert it to elemental sulfur by the Selectox process.

Beavon-Hi-Activity. Another alternate for removing the H_2S is to oxidize it directly to elemental sulfur by the Hi-Activity process.

Operating conditions: All pressures are near atmospheric. The Beavon hydrogenation/hydrolysis reactor operating temperature is in the range of 550°F to 750°F. Equipment is essentially all carbon steel. Sulfur recovery of Claus plus Beavon-Selectox or Beavon-MDEA is more than 99% or 99.9%, respectively.

Installations: There are more than 100 Beavon-type tail gas treating units worldwide. Among these are more than 30 Beavon-MDEA plants operating worldwide. Two Beavon-Selectox plants are operating in the U.S. and Germany; one Beavon-Hi-Activity plant operating in Germany. Two Beavon-Hi-Activity plants to be installed in China and Venezuela.

Licensors: Parsons Energy & Chemicals Group, Inc. and UOP LLC.

Effluent cleanup

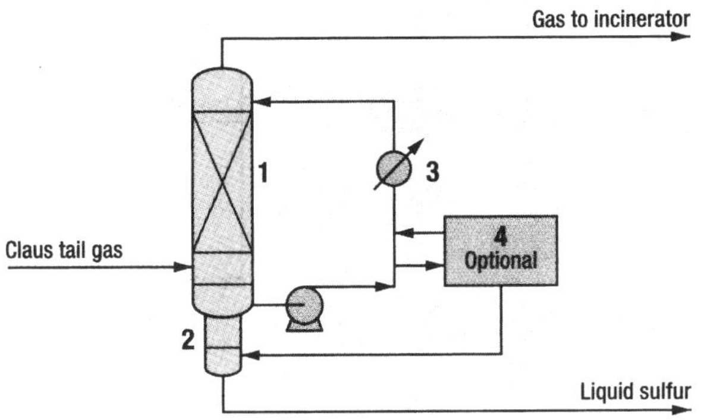

Clauspol

Application: Claus tail-gas treatment with total sulfur recoveries up to 99.9+%. Liquid-phase conversion of H_2S and SO_2 into liquid elemental sulfur (S).

Description: Claus tail gas is contacted counter-currently with an organic solvent in a low-pressure drop packed column (1). Hydrogen sulfide (H_2S) and SO_2 are absorbed in the solvent and react to form liquid elemental S according to the Claus reaction, which is promoted by an inexpensive dissolved catalyst. The solvent is pumped around the contactor (1), and the heat of reaction is removed through a heat exchanger (3) to maintain a constant temperature slightly above the sulfur melting point. Due to the limited solubility of S in the solvent, pure liquid S separates from the solvent and is recovered from a settling section (2) at the bottom of the contactor (1). This standard Clauspol II flow scheme allows S recovery up to 99.8% (Claus + Clauspol). The recovery level can be customized by adapting the size of the contactor (1).

The latest development is the optional solvent desaturation section (4). By removing the dissolved sulfur from the circulating solvent, the overall sulfur recovery can be raised up to 99.9+%.

Economics: For a Clauspol unit treating a typical Claus tail gas, in the 99.7–99.9% recovery, the CAPEX are typically 60–80% and OPEX 40–60% of those for a conventional hydrogenation/amine plant. Contrarily to the hydrogenation/amine process, Clauspol does not recycle any H_2S to the Claus unit, thus saving Claus plant capacity.

Installations: More than 40 units.

Reference: Barrère-Tricca, C., et al., "Thirty Years of Operating Experience with a wet subdewpoint TGT process," GPA Europe Annual Conference, Amsterdam, Sept. 26–28, 2001.

Licensor: Prosernat IFP Group Technologies.

Effluent cleanup

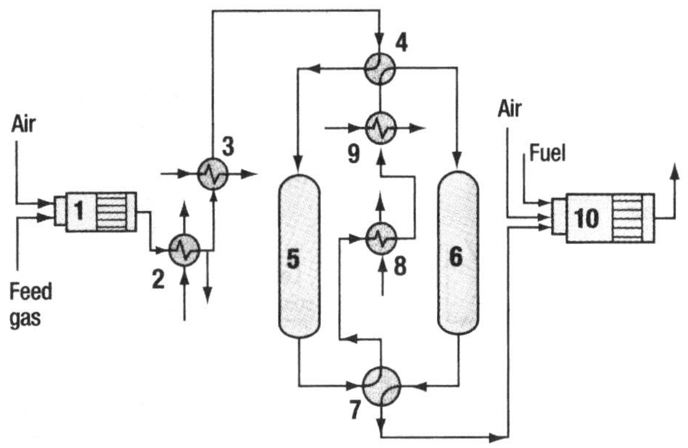

Clinsulf-SDP

Application: Sulfur recovery from typical Claus gas rich in H_2S, from industrial gases, sour water stripper gas.

Products: Elemental sulfur of Claus-grade quality, purified offgas, steam.

Description: Clinsulf-SDP is an improved Claus process. Downstream of the Claus furnace (1), sulfur condensor (2) and reheater (3) are two catalytic reactors integrating the Claus step and a SubDewPoint tail-gas treatment. Its main feature is the use of internally cooled reactors, which allow the reaction temperatures to be controlled by cooling coils in the catalyst beds. The first bed (5) operates at high temperature, the second bed (6) below the sulfur dew point. This reactor adsorbs sulfur and must be regenerated. That is done by switching the cool second reactor to the position of the hot first reactor and vice versa. The use of four-way valves (4, 7) makes the switchover between adsorption and regeneration simple and reliable. The desulfurized gas is incinerated (10).

Conventional Claus catalysts are applied. They ensure up to 99.8% sulfur recovery.

In no operating mode are there any dead pipes. This avoids the related problems of corrosion and sulfur plugging.

Operating conditions: Typical feed gas is Claus gas rich in H_2S. Clinsulf-SDP is capable of handling wide fluctuations of feed gas flows and composition. Turndown ratios up to 6/1 have been realized. Sulfur recovery rates up to 99.8% were observed. The process operates fully automatically, including the switchover.

Economics: Capital cost depends on the amount of feed gas and on the required sulfur recovery rate. Products are bright yellow sulfur and steam. Utility consumptions are very low, as in other Claus processes.

Installations: Five plants have been built. Two are under construction. Five of the plants are 1-reactor plants, two are 2-reactor plants.

Reference: Kunkel, J. and M. Heisel, "Clinsulf-SDP makes its debut," *SULPHUR,* no. 243.

Licensor: Linde AG.

Effluent cleanup

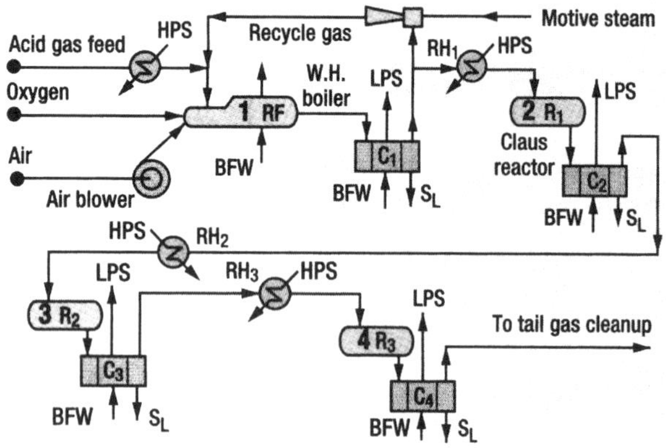

COPE

Application: Increase capacity and recovery of existing Claus sulfur recovery/tail gas cleanup units, provide redundant sulfur processing capacity and improve combustion performance of units processing lean acid gas through O_2 enrichment.

Description: The sulfur processing capacity of typical Claus sulfur recovery units can be increased to more than 200% of the base capacity through partial to complete replacement of combustion air with pure O_2. SRU capacity is typically limited by hydraulic pressure drop. As O_2 replaces combustion air, the quantity of inert nitrogen is reduced allowing additional acid gas to be processed. The process can be implemented in two stages. As the O_2 enrichment level increases, the combustion temperature (1) increases. COPE Phase I, which does not use a recycle stream, can often achieve 50% capacity increase through O_2 enrichment to the maximum reaction furnace refractory temperature limit of about 2,700°F to 2,800°F. Higher O_2 enrichment levels are possible with COPE Phase II, which uses an internal process recycle stream to moderate the combustion temperature allowing enrichment up to 100% O_2. Flow through the remainder of the SRU (2, 3, & 4) and the tail gas cleanup unit is greatly reduced. Ammonia and hydrocarbon acid gas impurity destruction and thermal stage conversion are improved at the higher O_2-enriched combustion temperatures. Overall SRU sulfur recovery is increased by 0.5% to 1%. A single proprietary COPE burner handles acid gas, recycle gas, air and oxygen.

Operating conditions: Combustion pressure from 6 to 12 psig; combustion temperature up to 2,800°F. Oxygen concentration from 21% to 100%. SRU sulfur recovery is 95% to 98%.

Economics: Expanded SRU and tail gas unit retrofit sulfur processing capacity at capital cost of 15% to 25% of new plant cost. New plant savings of up to 25% and redundant capacity at 15% of base capital cost. Operating costs are a function of oxygen cost. Reduced incinerator fuel and reduced operating and maintenance labor costs.

Installations: Seventeen COPE trains are in operation at 11 locations. Five additional trains are in engineering and/or the construction phase.

Reference: U.S. Patent 4,552,747.

Sala, L., W. P. Ferrell and P. Morris, "The COPE process—increase sulfur recovery capacity to meet changing needs," European Fuels Week Conference, Giardini Naxos, Taormina, Italy, April 2000.

Licensors: Goar, Allison & Associates, Inc. and Air Products and Chemicals, Inc.

Effluent cleanup

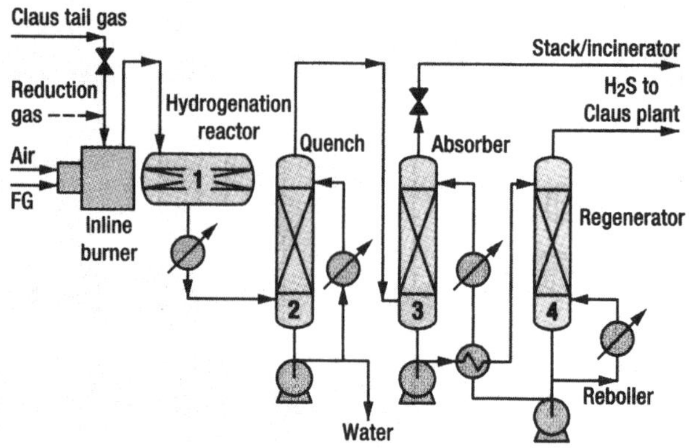

LTGT (Lurgi tail-gas treatment process)

Application: Wet-scrubbing process purifies Claus tail gas for total sulfur recovery ranging from 99.8% to 99.9+%.

Description: The optimum Claus tail-gas treatment converts sulfur species to H_2S and recovers it in a wet-scrubbing process. The Lurgi tail-gas treatment process (LTGT) is an amine treating system with generic MDEA solvent, structured packing and plate heat exchangers when possible. This process enables using smaller diameter columns, plot size, worldwide treating solution availability, proven technology and high selectivity. Installing smaller equipment lowers total investment costs.

Different sulfur species from the incoming Claus tail gas are converted (1) to H_2S. Process water formed by the Claus reaction is removed in a direct cooler (2), and H_2S is expunged by MDEA solution in an absorber column (3). The amine solution is regenerated in a steam-heated stripper column (4) and produces a H_2S gas stream that is recycled back to the Claus section. Due to hydrogenation in the tail-gas treatment, both acid gases—H_2S and CO_2—are produced and routed to the absorber column.

Tertiary amines like MDEA, which are used in the LTGT, have the ability to selectively absorb H_2S due to chemical structure and do not co-absorb CO_2. Primary (MEA) and secondary amines (DEA) will absorb H_2S along with most of the CO_2. In tertiary amine solutions, CO_2 can only be absorbed by an indirect acid/base reaction forming bicarbonates; this is a very slow reaction.

Economics: Investment amounts are approximately 90–100% of the Claus unit cost. Using a common regeneration along with the upstream amine unit, investments are approximately 65–75% of the Claus unit cost.

Installations: Six LTGT units for processing Claus tail gases are in operation or under design.

Reference: Connock, L., "Emerging sulphur recovery technologies," *Sulphur,* March/April 2001.

Licensor: Lurgi Oel-Gas-Chemie GmbH.

Effluent cleanup

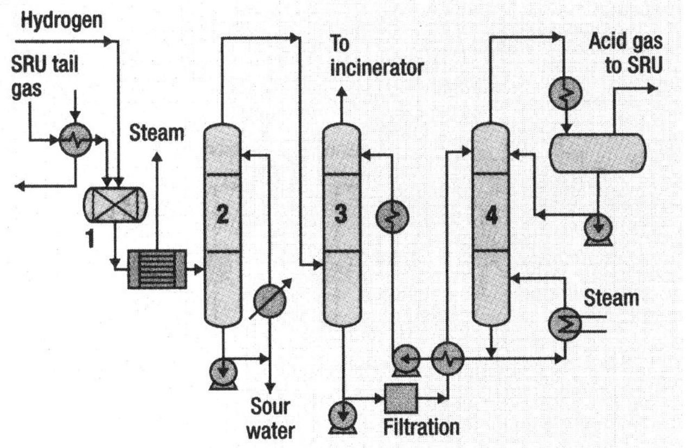

Resulf

Application: Purification of sulfur recovery unit (SRU) tail gas for incineration. Resulf, Resulf-10 and Resulf-MM units are easily retrofitted to existing SRU complexes. They feature a low unit pressure drop and can use the latest specialty solvents to lower energy consumption and maximize flexibility.

Products: Treated vent gas from a Resulf-MM unit typically contains 1,000 ppm H_2S and must be incinerated. Treated vent gas from a Resulf unit typically contains less than 150 ppmv H_2S and is oxidized in an incinerator before venting to the atmosphere. Vent gas from a Resulf-10 unit has a maximum of 10 ppmv H_2S and may not require incineration.

Description: SRU tail gas is heated in the feed heater, then mixed with a reducing gas containing H_2. The heated stream passes through the reactor (1), where the SO_2, elemental sulfur and other sulfur-containing compounds, such as COS and CS_2, are converted to H_2S. Hot gas leaving the reactor is cooled in a waste-heat steam generator. The gas is further cooled in a direct contact water cooler (2). The overhead gas stream is fed to the absorber (3).

Lean solvent is also fed to the absorber. The downward flowing solvent contacts the upward flowing gas and absorbs nearly all the H_2S and only part of the CO_2. Rich solvent is sent to the regenerator (4) where the H_2S and CO_2 are removed by steam stripping. Acid gas from the regenerator is recycled to the SRU. Lean solvent from the regenerator is cooled and returned to the absorber.

Operating conditions: Resulf units use MDEA or formulated MDEA as a solvent. Resulf-10 units are designed using specialty amines such as formulated MDEA. Resulf units use generic MDEA solvents. Resulf-MM units use amine from the primary amine unit (MEA, DEA or MDEA).

Economics: Plate and frame heat exchangers have been used to reduce capital costs. Modular designs can also be used to reduce capital costs while maintaining critical project schedules. The cost for Resulf-MM is significantly lower than for Resulf or Resulf-10 due to the lower recovery. Key features of the Resulf-MM and Resulf technologies are that they can be inexpensively upgraded.

Installations: TPA has licensed and designed worldwide:

Resulf units: 39
Resulf-10 units: 3
Resulf-MM units: 2.

Licensor: TPA Howe-Baker, Ltd.

Effluent cleanup

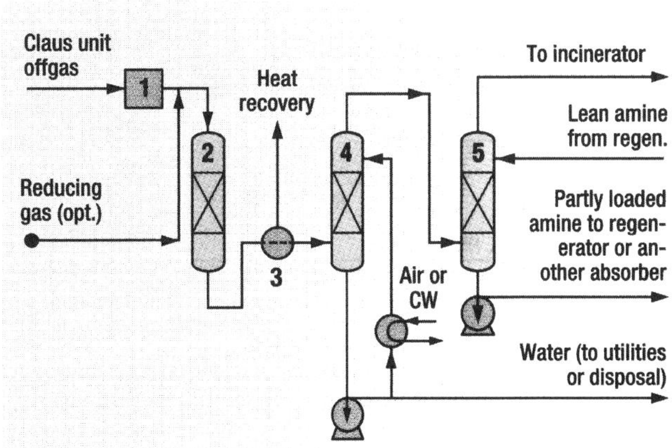

SCOT

Application: A low-pressure drop, high-sulfur recovery efficiency process to recover sulfur components from tail gas of sulfur plants. The SCOT process is insensitive to variations in the upstream SRU, such as in the H_2S/SO_2 ratio, hydrocarbon or ammonia breakthrough.

Products: Offgas from a SCOT unit contains a *total* sulfur content less than 120 ppmv; offgas from a Super SCOT unit contains maximum 50 ppmv total sulfur content.

Description: The Claus tail gas feed to the SCOT unit is heated to 250°C to 300°C using an inline burner or heat exchanger (1) with optionally added H_2 or a mixture of H_2/CO. If reducing gas, H_2 or CO is not available, an inline burner (1) is operated in an air-deficient mode to produce reducing gas. The heated gas then flows through a catalyst bed (2) where sulfur components, SO_2, elemental sulfur (S), COS and CS_2 are practically completely converted to H_2S. The gas is cooled to 40°C in a heat-recovery system (3) and a water-quench tower (4), followed by selective H_2S removal in an amine absorber (5) to typically 30 to 100 ppmv H_2S. The semi-loaded amine is often further loaded in another absorber. The H_2S absorbed in the SCOT or Super SCOT process is recycled to the Claus unit via the amine regenerator. The absorber offgas is incinerated.

The process is continuous, has a pressure drop of 4 psi or lower, provides excellent sulfur recovery and can be operated at high reliability with less than 1% unscheduled downtime.

Economics: The total sulfur recovery efficiency of the SCOT process combined with the upstream Claus unit typically guarantees a sulfur recovery of at least 99.8%. In case of a Super SCOT unit, an overall sulfur recovery efficiency of 99.95% can be guaranteed. An integrated and cascaded SCOT unit has capital and operational expenditures comparable to other tail-gas treating technologies, but has higher total sulfur recovery efficiency.

References: Hoksberg, et al., "Sulfur recovery for the new millennium," *Hydrocarbon Engineering,* November 1999.

Verhulst, "Recent developments in SCOT tail gas technology," Bovar/Western Research Sulfur Seminar, Istanbul, 1994.

Kuypers, *Ten years SCOT experience,* Erdol and Kohle, January 1985.

Licensor: Shell Global Solutions International B.V.

Effluent cleanup

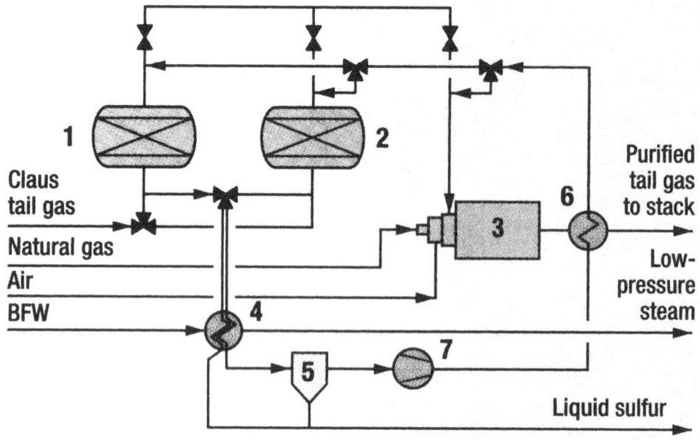

SULFREEN

Application: Catalytic purification of Claus tail gas or lean H_2S waste gas for an overall sulfur recovery ranging from 99% to 99.9%. Different versions of the SULFREEN process are available.

Description: The SULFREEN process is based on the well-known Claus reaction in which the components—H_2S and SO_2 in tail gas—are catalytically converted into elemental sulfur. The process occurs in the gas phase; the operating conditions being those at which the tail gas leaves the upstream Claus plant. The catalyst, which is arranged in fixed beds, consists of impregnated activated alumina, the properties of which are similar to those of Claus catalysts.

Tail gas leaving the Claus plant at temperatures of 120°C to 140°C passes through one of the two reactors (1) (2), where most of the H_2S and SO_2 are converted into elemental sulfur and adsorbed on the catalyst.

The sulfur-laden catalyst is regenerated by using part of the Claus plant tail gas. The regeneration gas is heated in a gas/gas-heat exchanger (6), using the heat in the hot offgases from the incineration unit (3). A gas-fired heat exchanger, a direct-fired heater or an electrical heater can also be used for heating purposes.

The desorbed sulfur contained in the hot regeneration gas is recovered in the sulfur condenser (4) (5). The regeneration-gas blower (7) serves for overcoming the pressure drop of the closed regeneration loop. After subsequent cooling of the catalyst bed with purified tailgas, the reactor (1) (2), is again ready to be switched to adsorption.

If high COS and CS_2 concentrations are expected to be present in the tailgas, a modified version of the process, named HYDROSULFREEN, is available. The HYDROSULFREEN process includes a pretreatment of the tail gas in an hydrolysis and oxidation reactor, located upstream of the SULFREEN reactors. For sulfur recoveries up to 99.9%, the DOXOSULFREEN process can be applied. This process includes an additional direct oxidation step downstream of the SULFREEN reactors.

Economics: SULFREEN investment amounts to 30–45% of the Claus unit cost for the conventional version, and 40–65% for the improved versions. Operating costs are much lower than for solvent-based processes.

Installations: More than 75 SULFREEN units—for processing tail gases of Claus plants, ranging from 5 tpd to 2,200 tpd of sulfur—are in operation or under design.

Reference: Willing, W., and T. Lindner, "Lurgi's TGT processes and new operational results from SULFREEN plants," presented at the SULPHUR '94 Conference, Tampa, Florida, Nov. 6–9, 1994.

Licensor: Lurgi Oel-Gas-Chemie GmbH.

Flue gas treatment

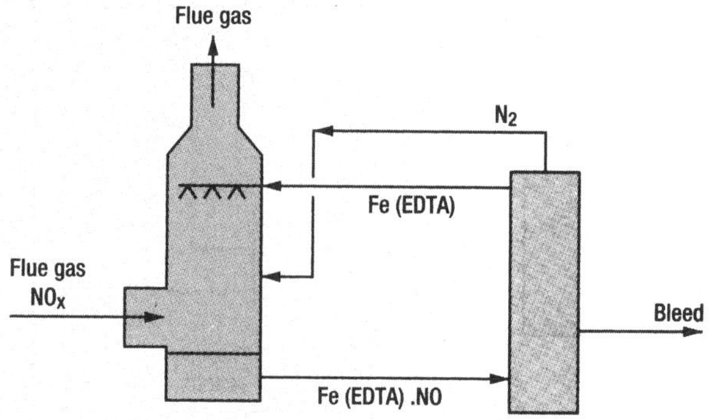

BioDeNO$_x$

Application: The BioDeNO$_x$ biological process removes NO_x from flue gases. Iron chelate selectively absorbs NO_x; the NO_x is reduced to nitrogen with ethanol in the presence of microorganisms.

Description: The BioDeNO$_x$ system uses a wet gas scrubber to contact the circulation liquid with the flue-gas feed and absorb the NO_x. In the sump underneath the scrubber, the absorbed NO_x is biologically reduce to nitrogen and ethanol is also consumed. Thus, the iron chelate solution is regenerated. The presence of oxygen and acidic compounds in the flue gas, such as HCl and HF, oxidizes a part of iron chelate to ferric state. Therefore, a purge and makeup of iron chelate is necessary to eliminate this oxidized ferric material. To minimize iron chelate consumption, a nano-filtration is installed. Bleed from the unit is passed through the filter and the chelate is recovered.

Operating conditions: The scrubber and the sump operate close to the atmospheric pressure and 60–70°C.

Economics: The costs associated with using BioDeNO$_x$ to treat a typical refinery FCC flue gas for 670,000 Nm^3/hr and 200 ppmv of NO_x with the removal efficiency greater than 90%:

Total installation cost	$4,800,000
Daily operating cost	$3,076

Installations: The process has been developed over the past seven years. During 2002, two projects are planned: A full-scale demonstration plant at a power station and a demonstration scale plant at a refinery FCC unit are currently operating. Results are expected to be published in the last quarter 2002.

Licensors: UOP LLC and BioStar B.V.

Flue gas treatment

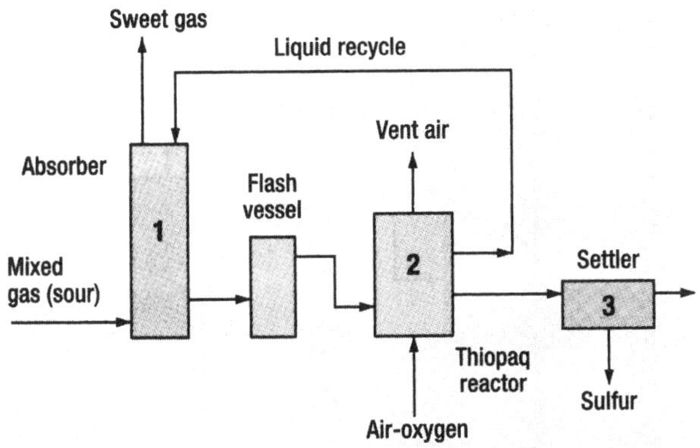

Shell-Paques/THIOPAQ process

Application: Biological desulfurization of high-pressure natural gas, synthesis gas, fuel gas streams, acid gas from amine regenerators and treatment of spent caustic. The two processes are based on the same principle, but marketed under the name Shell Paques when high-pressure natural gas, synthesis gas and Claus tail gas are treated; for all other applications, the name Thiopaq is used.

Products: The Paques or Thiopaq unit can be designed such that the treated gas stream contains less than 5 ppmv H_2S; the resulting sulfur recovery is consequently 99.999%, based on gas streams. The bio-sulfur produced can be used directly as fertilizer, since it has a hydrophilic character. Thus, the sulfur is more accessible in soil for oxidation and subsequent uptake by plants. Alternatively, the bio-sulfur can be washed and remelted to produce a final liquid sulfur product that will meet industrial specifications. The hydrophilic character of the bio-sulfur is lost after remelting.

Description: In the Thiopaq process, H_2S is directly oxidized to elemental sulfur (S) using colorless sulfur bacteria (Thiobacilli). These naturally occurring bacteria are not genetically modified. Feed gas is sent to a caustic scrubber (1) in which the H_2S reacts to sulfide. The sulfide is converted to elemental S and caustic by the bacteria when air is supplied in the bioreactor (2). Sulfur particles are covered with a (bio-) macropolymer layer, which keeps the sulfur in a milk-like suspension that does not cause fouling or plugging.

In this process, a sulfur slurry is produced, which can be concentrated to a cake containing 60% dry matter. This cake can be used directly for agricultural purposes, or as feedstock for sulfuric acid manufacturing. Alternatively, the biological sulfur slurry can be purified further by melting to high-quality sulfur to meet international Claus sulfur specifications.

Economics: The Shell Paques/Thiopaq process achieves a very low H_2S content in the treated gas; a very high-sulfur recovery efficiency of 99.999% is achievable. This process can thus replace the combination of an amine/Claus/TGTU or, for smaller applications, liquid redox processes.

It compares favorably in terms of capital expenditure with practically all liquid redox applications, and with the traditional amine/Claus/TGTU for sulfur capacities, up to around 50 tpd. The capital and operating costs for this biological process decrease with decreasing CO_2/H_2S ratios.

Reference: Janssen, A. J. H., et al, "Biological process for H_2S removal from high-pressure gas: The Shell-Paques/THIOPAQ gas desulfurization process," *Sulphur,* 2001.

Licensors: Shell Paques: Paques B.V. and Shell Global Solutions International B.V.

Flue gas treatment

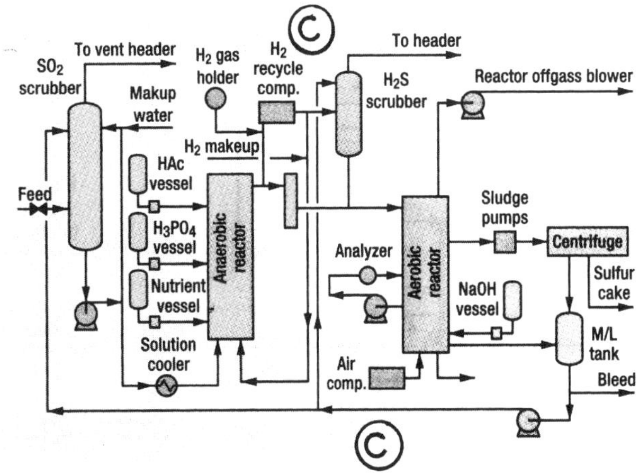

Thiopaq $DeSO_x$

Application: The Thiopaq $DeSO_x$ biological process selectively removes and converts SO_x in the flue gases to elemental sulfur or H_2S.

Product: The sulfur produced is hydrophilic; thus, it prevents equipment from fouling or blocking. Moreover, this characteristic makes the product sulfur suitable for agricultural use as fertilizer or as an insecticide. Alternatively, the sulfur can be molten to a high-purity product and meet international Claus sulfur specifications. Alternatively, H_2S can be produced and the gas can be sent to the normal sulfur handling process.

Description: The Thiopaq $DeSO_x$ is a commercial, two-step biological process. It can convert sulfite and sulfate to elemental sulfur. A sodium biphosphate solution quenches and removes particulates and absorbs SO_x from the flue gas. The scrubbing liquid is regenerated and recycled to the scrubber.

The heart of the Thiopaq process is the proprietary anaerobic and aerobic bioreactors. In these air-lift-loop reactors, the absorbed sulfite is reduced to sulfide (HS^-) inside the anaerobic reactor in the presence of microorganism with hydrogen. The sulfide is then oxidized under controlled conditions to elemental sulfur via the microorganisms. The produced elemental sulfur has a hydrophilic nature and is separated from the aqueous effluent in a proprietary three-phase separator.

Operating conditions: The scrubber operates close to the atmospheric pressure and at saturation temperature of the scrubbing liquid. The anaerobic bioreactor can be operated at pressures of up to 6 bar and 60–70°C. The aerobic bioreactor operates at atmospheric pressure and 30–35°C.

Economics: The costs associated with using the Thiopaq process to treat a typical refinery FCC flue gas for 150,000 Nm^3/hr at 284°C, particulate loading of 250 mg/Nm^3 and 2400 wppm of SO_x with the removal efficiency greater than 99%:

Total installation cost	$7,000,000
Annual operating cost	$1,125,000

Installation: Twenty Thiopaq units operating worldwide to remove sulfur from gas and liquid streams for a variety of industries. Currently, two 13-mtpy units are under construction for refinery service in Egypt and Cyprus.

Licensors: UOP LLC, Monsanto Environchem System Inc., and Paques Bio System B.V.

Flue Gas Treatment

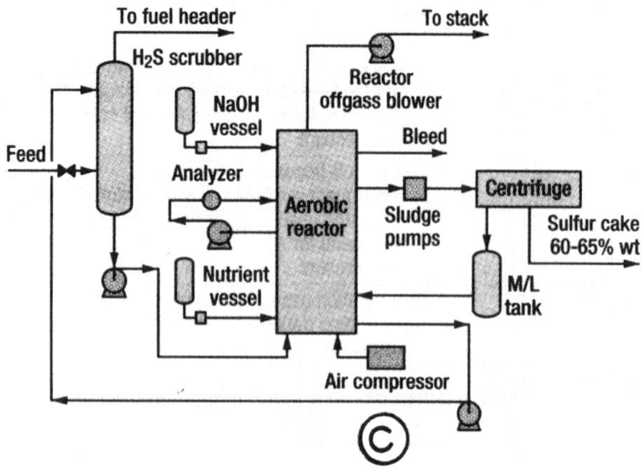

Thiopaq—H_2S Removal

Application: The biological Thiopaq process selectively removes and converts H_2S and light mercaptans from gas streams, aqueous streams and/or light hydrocarbons to elemental sulfur or sulfate.

Product: The sulfur produced is hydrophilic; thus, it prevents equipment from fouling and blocking. Moreover, this characteristic makes the product suitable for agricultural use as fertilizer or as an insecticide. Alternatively, the sulfur can be molten to a high-purity product meeting international Claus sulfur specifications.

Description: The Thiopaq process consists of three integrated process sections: an absorption section to remove the H_2S from the gas stream, bioreactor(s) and a sulfur-separation section.

The heart of this process is the proprietary bioreactor. In this air-lift-loop reactor, sulfide (HS^-) is oxidized under controlled conditions to elemental sulfur in the presence of microorganisms. These aerobic (oxygen consuming) organisms use the released energy from the sulfide oxidation for metabolic processes. The elemental sulfur produced has a hydrophilic nature and is separated from the aqueous effluent in a proprietary three-phase separator.

The scrubbing step to remove H_2S from the gas streams is integrated into the Thiopaq process and regenerates the scrubbing solution, rather than its disposal. Regeneration is possible because the alkalinity consumption due to the absorption of H_2S is compensated by the oxidation of H_2S to elemental sulfur.

Operating conditions: The absorber operates at the pressure of feed and at bioreactor temperature. The bioreactor operates at atmospheric pressure and 30–35°C. If the feed is available at a higher temperature, then it requires cooling before entering the absorber.

Economics: The costs associated with using Thiopaq to treat a typical refinery spent caustic of a flowrate of 9,510 gpd containing 2,970 lbs/day of sulfide:

Capital cost, budget	$1,500,000
Annual operating cost	$158,000

The cost of an LPG flow of 1,500 BSD and recovery of 1,500 lbs/d of sulfur and achieving < 5ppm H_2S in the treated LPG:

Capital cost, budget	$1,330,600
Annual operating cost	$180,000

The costs for an offgas flowrate of 24,200 lbs/hr and recovery of 7.8 tpd of sulfur:

Capital cost, budget	$2,600,000
Annual operating cost	$300,000

Installations: Thirty-one units are operating worldwide to remove sulfur from gas and liquid streams in a variety of industries. Currently, two 13-mtpd units under construction for refinery service in Egypt and Cyprus.

Licensors: UOP LLC, Shell International Oil Products B.V., and Paques Bio System B.V.

Segment 2

Refining Processes

PROCESS INDEX

Compressors and Modern Process Applications. By Heinz P. Bloch.

LICENSOR INDEX

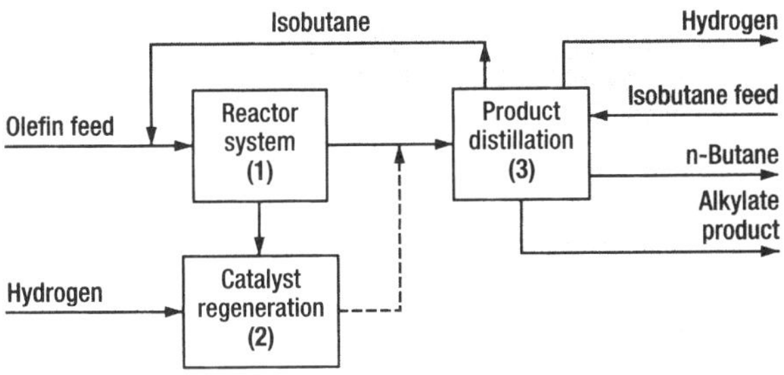

Alkylation

Application: The AlkyClean process converts light olefins into alkylate by reacting the olefins with isobutane over a true solid acid catalyst. AlkyClean's unique catalyst, reactor design and process scheme allows operation at low external isobutene to olefin ratios while maintaining excellent product quality.

Products: Alkylate is a high-octane, low-Rvp gasoline component used for blending in all grades of gasoline.

Description: The light olefin feed is combined with the isobutene make-up and recycle and sent to the alkylation reactors which convert the olefins into alkylate using a solid acid catalyst (1). The AlkyClean process uses a true solid acid catalyst to produce alkylate eliminating the safety and environmental hazards associated with liquid acid technologies. Simultaneously, reactors are undergoing a mild liquid-phase regeneration using isobutene and hydrogen and, periodically, a reactor undergoes a higher temperature vapor phase hydrogen strip (2). The reactor and mild regeneration effluent is sent to the product-fractionation section, which produces propane, n-butane and alkylate, while also recycling isobutene and recovering hydrogen used in regeneration for reuse in other refinery hydroprocessing units (3). AlkyClean does not produce any acid soluble oils (ASO) or require post treatment of the reactor effluent or final products.

Product: The C_5^+ alkylate has a RON of 93–98 depending on processing conditions and feed composition.

Economics:

Investment (basis 10,000-bpsd Unit) $/bpsd	3,100
Operating cost, $/gal	0.47

Installation: Demonstration unit at Fortum's Porvoo, Finland Refinery.

Reference: "The Process: A new solid acid catalyst gasoline alkylation technology," NPRA 2002 Annual Meeting, March 17–19, 2002.

Licensor: ABB Lummus Global Inc., Akzo Nobel Catalysts and Fortum Oil and Gas.

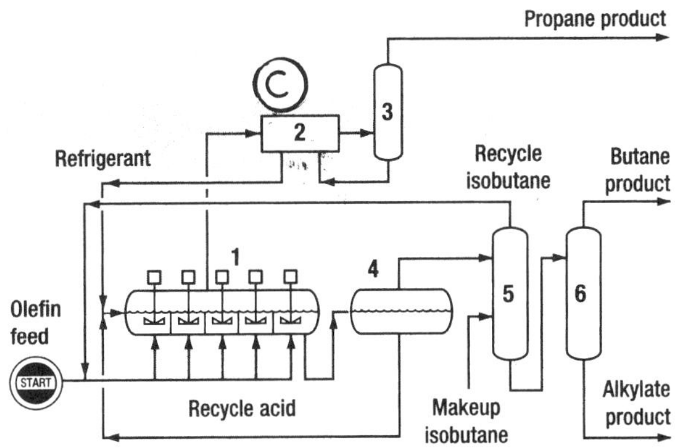

Alkylation

Application: Combines propylene, butylene and pentylene with isobutane, in the presence of sulfuric acid catalyst, to form a high-octane, mogas component.

Products: A highly isoparaffinic, low Rvp, high-octane gasoline blendstock is produced from the alkylation process.

Description: Olefin feed and recycled isobutane are introduced into the stirred, autorefrigerated reactor (1). Mixers provide intimate contact between the reactants and the acid catalyst. Reaction heat is removed from the reactor by the highly efficient autorefrigeration method. The hydrocarbons that are vaporized from the reactor, and that provide cooling to the 40°F level, are routed to the refrigeration compressor (2) where they are compressed, condensed and returned to the reactor. A depropanizer (3), which is fed by a slipstream from the refrigeration section, is designed to remove any propane introduced to the plant with the feeds. The reactor product is sent to the settler (4), where the hydrocarbons are separated from the acid that is recycled. The hydrocarbons are then sent to the deisobutanizer (5) along with makeup isobutane. The isobutane-rich overhead is recycled to the reactor. The bottoms are then sent to a debutanizer (6) to produce a low Rvp alkylate product with an FBP less than 400°F.

Major features of the reactor are:

- Use of the autorefrigeration method of cooling is thermodynamically efficient. It also allows lower temperatures, which are favorable for producing high product quality with low power requirements.
- Use of a staged reactor system results in a high average isobutane concentration, which favors high product quality.
- Use of low space velocity in the reactor design results in high product quality and eliminates any corrosion problems in the fractionation section associated with the formation of esters.
- Use of low reactor operating pressure means high reliability for the mechanical seals for the mixers.
- Use of simple reactor internals translates to low cost.

Yields:

Alkylate yield	1.78 bbl C_5^+/bbl butylene feed
Isobutane (pure) required	1.17 bbl/bbl butylene feed
Alkylate quality	96 RON/94 MON

Economics:

Utilities, typical per barrel of alkylate produced:

Water, cooling (20°F rise), 1,000 gal	2.1
Power, kWh	10.5
Steam, 60 psig, lb	200
H_2SO_4, lb	19
NaOH, 100%, lb	0.1

Installation: 115,000-bpd capacity at 11 locations with the sizes ranging from 2,000 to 30,000 bpd. Single reactor/settle trains with capacities up to 9,500 bpsd.

Reference: Lerner, H., "Exxon sulfuric acid alkylation technology," *Handbook of Petroleum Refining Processes,* 2nd ed., R. A. Meyers, Ed., pp. 1.3–1.14.

Licensor: ExxonMobil Research & Engineering Co.

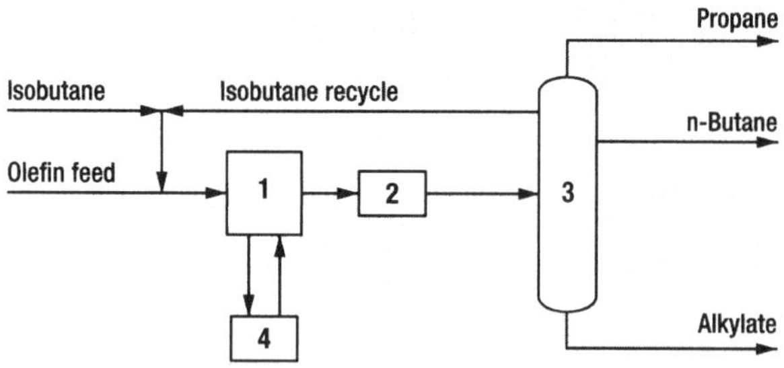

Alkylation

Application: The Topsøe fixed-bed alkylation (FBA) technology applies a unique fixed-bed reactor system with a liquid superacid catalyst absorbed on a solid support. FBA converts isobutane with propylene, butylene and amylenes to produce branched chain hydrocarbons. As an alternative, FBA can conveniently be used to alkylate isopentane as a means of disposing isopentane for RVP control purpose.

Products: A high-octane, low-RVP and ultra-low-sulfur blending stock for motor and aviation gasoline.

Description: The FBA process combines the benefits of a liquid catalyst with the advantages of a fixed-bed reactor system. Olefin and isobutane feedstocks are mixed with a recycle stream of isobutane and charged to the reactor section (1). The olefins are fully converted over a supported-liquid-phase catalyst confined within a mobile, well-defined catalyst zone. The simple fixed-bed reactor system allows easy monitoring and maintenance of the catalyst zone with no handling of solids.

Traces of dissolved acid in the net reactor effluent are removed quantitatively in a compact and simple-to-operate effluent treatment unit (2). In the fractionation section (3), the acid-free net reactor effluent is split into propane, isobutane, n-butane and alkylate. The unique reactor concept allows an easy and selective withdrawal of small amounts of passivated acid. The acid catalyst is fully recovered in a compact catalyst activity maintenance unit (4). The integrated, inexpensive, on-site catalyst activity maintenance is a distinct feature of the FBA process. Other significant features of FBA include:

- High flexibility (feedstock, operation temperature)
- Low operating costs
- Low catalyst consumption.

Process performance:

	Olefin feed type		
	MTBE raffinate	FCC C_4 cut	C_3–C_5 cut
Alkylate product			
RON (C_5+)	98	95	93
MON (C_5+)	95	92	91

Economics: (Basis: MTBE raffinate, inclusive feed pretreatment and on-site catalyst activity maintenance)

Investment (basis: 6,000 bpsd unit), $ per bpsd	5,600
Utilities, typical per bbl alkylate:	
Electricity, kWh	10
Steam, MP (150 psig), lb	60
Steam, LP (50 psig), lb	200
Water, cooling (20°F rise), gal10^3	2.2

Licensor: Haldor Topsøe A/S.

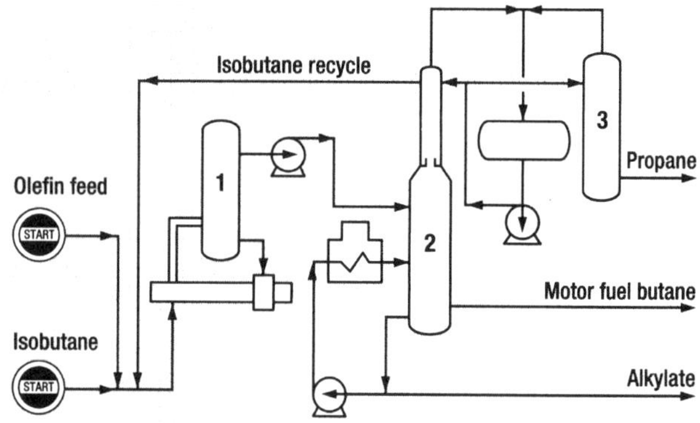

Alkylation

Application: Convert propylene, amylenes, butylenes and isobutane to the highest quality motor fuel using ReVAP alkylation.

Products: An ultra-low-sulfur, high-octane and low-Rvp blending stock for motor and aviation fuels.

Description: Dry liquid feed containing olefins and isobutane is charged to a combined reactor-settler (1). The reactor uses the principle of differential gravity head to effect catalyst circulation through a cooler prior to contacting highly dispersed hydrocarbon in the reactor pipe. The hydrocarbon phase that is produced in the settler is fed to the main fractionator (2), which separates LPG-quality propane, isobutane recycle, n-butane and alkylate products. Small amount of dissolved catalyst is removed from the propane product by a small stripper tower (3). Major process features are:

- Gravity catalyst circulation (no catalyst circulation pumps required)
- Low catalyst consumption
- Low operating cost
- Superior alkylate qualities from propylene, isobutylene and amylene feedstocks
- Onsite catalyst regeneration
- Environmentally responsible (very low emissions/waste)
- Between 60% and 90% reduction in airborne catalyst release over traditional catalysts
- Can be installed in all licensors' HF alkylation units.

With the proposed reduction of MTBE in gasoline, ReVAP offers significant advantages over sending the isobutylene to a sulfuric-acid-alkylation unit or a dimerization plant. ReVAP alkylation produces higher octane, lower RVP and endpoint product than a sulfuric-acid-alkylation unit and nearly twice as many octane barrels as can be produced from a dimerization unit.

Yields:

	Feed type	
	Butylene	Propylene-butylene mix
Composition (lv%)		
Propylene	0.8	24.6
Propane	1.5	12.5
Butylene	47.0	30.3
i-Butane	33.8	21.8
n-Butane	14.7	9.5
i-Pentane	2.2	1.3
Alkylate product		
Gravity, API	70.1	71.1
RVP, psi	6–7	6–7
ASTM 10%, °F	185	170
ASTM 90%, °F	236	253
RONC	96.0	93.5
Per bbl olefin converted		
i-Butane consumed, bbl	1.139	1.175
Alkylate produced, bbl	1.780	1.755

Installation: 107 alkylation units licensed worldwide.

Licensor: Fuels Technology Division of ConocoPhillips Co.

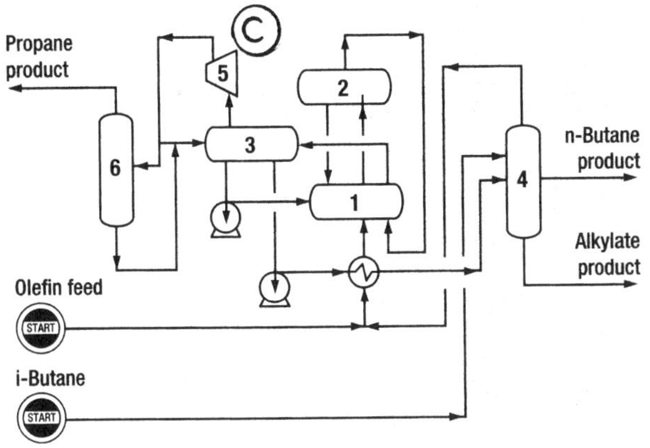

Alkylation

Application: To combine propylene, butylenes and amylenes with isobutane in the presence of strong sulfuric acid to produce high-octane branched chain hydrocarbons using the Effluent Refrigeration Alkylation process.

Products: Branched chain hydrocarbons for use in high-octane motor fuel and aviation gasoline.

Description: Plants are designed to process a mixture of propylene, butylenes and amylenes. Olefins and isobutane-rich streams along with a recycle stream of H_2SO_4 are charged to the STRATCO Contactor reactor (1). The liquid contents of the Contactor reactor are circulated at high velocities and an extremely large amount of interfacial area is exposed between the reacting hydrocarbons and the acid catalyst from the acid settler (2). The entire volume of the liquid in the Contactor reactor is maintained at a uniform temperature, less than 1°F between any two points within the reaction mass. Contractor reactor products pass through a flash drum (3) and deisobutanizer (4). The refrigeration section consists of a compressor (5) and depropanizer (6).

The overhead from the deisobutanizer (4) and effluent refrigerant recycle (6) constitutes the total isobutane recycle to the reaction zone. This total quantity of isobutane and all other hydrocarbons is maintained in the liquid phase throughout the Contactor reactor, thereby serving to promote the alkylation reaction. Onsite acid regeneration technology is also available.

Product quality: The total debutanized alkylate has RON of 92 to 96 clear and MON of 90 to 94 clear. When processing straight butylenes, the debutanized total alkylate has RON as high as 98 clear. Endpoint of the total alkylate from straight butylene feeds is less than 390°F, and less than 420°F for mixed feeds containing amylenes in most cases.

Economics (basis: butylene feed):

Investment (basis: 10,000-bpsd unit), $ per bpsd	3,500
Utilities, typical per bbl alkylate:	
Electricity, kWh	13.5
Steam, 150 psig, lb	180
Water, cooling (20°F rise), 10^3 gal	1.85
Acid, lb	15
Caustic, lb	0.1

Installation: Nearly 600,000 bpsd installed capacity.

Reference: *Hydrocarbon Processing,* Vol. 64, No. 9, September 1985, pp. 67–71.

Licensor: Stratco, Inc.

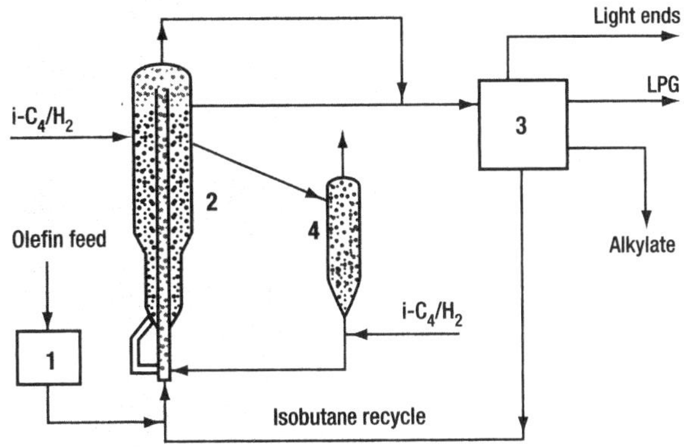

Alkylation

Application: The Alkylene process uses a solid catalyst to react isobutane with light olefins (C_3 to C_5) to produce a branched-chain paraffinic fuel. The performance characteristics of this catalyst and novel process design have yielded a technology that is competitive with traditional liquid-acid-alkylation processes. Unlike liquid-acid-catalyzed technologies, significant opportunities to continually advance the catalytic activity and selectivity of this exciting new technology are possible. This process meets today's demand for both improved gasoline formulations and a more "environmentally friendly" light olefin upgrading technology.

Description: Olefin charge is first treated to remove impurities such as diolefins and oxygenates (1). The olefin feed and isobutane recycle are mixed with reactivated catalyst at the bottom of the reactor vessel riser (2). The reactants and catalyst flow up the riser in a cocurrent manner where the alkylation reaction occurs. Upon exiting the riser, the catalyst separates easily from the hydrocarbon effluent liquid by gravity and flows downward into the cold reactivation zone of the reactor. The hydrocarbon effluent flows to the fractionation section (3), where the alkylate product is separated from the LPG product. There is no acid soluble oil (ASO) or heavy polymer to dispose of as with liquid acid technology.

The catalyst flows slowly down the annulus section of the reactor around the riser as a packed bed. Isobutane saturated with hydrogen is injected to reactivate the catalyst. The reactivated catalyst then flows through standpipes back into the bottom of the riser. The reactivation in this section is nearly complete, but some strongly adsorbed material remains on the catalyst surface. This is removed by processing a small portion of the circulating catalyst in the reactivation vessel (4), where the temperature is elevated for complete reactivation. The reactivated catalyst then flows back to the bottom of the riser.

Product quality: Alkylate has ideal gasoline properties such as: high research and motor octane numbers, low Reid vapor pressure (Rvp), and no aromatics, olefins or sulfur. The alkylate from an Alkylene unit has the particular advantage of lower 50% and 90% distillation temperatures, which is important for new reformulated gasoline specifications.

Economics: (basis: FCC source C_4 olefin feed)

Investment (basis: 6,000-bpsd unit), $ per bpsd	6,100
Operating cost ($/gal)	0.45

Licensor: UOP LLC.

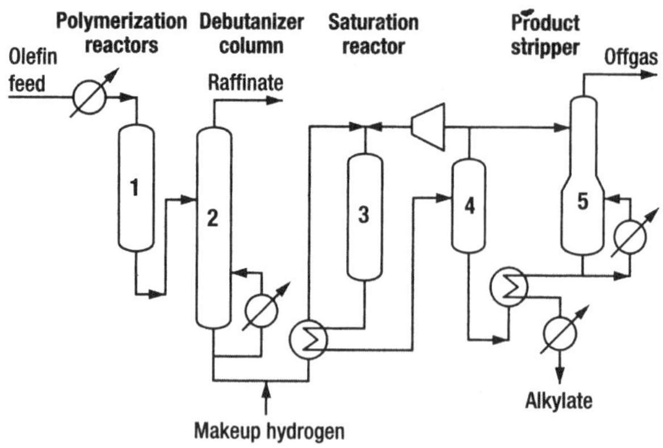

Alkylation

Application: The UOP Indirect Alkylation (InAlk) process uses solid catalysts to react isobutylene with light olefins (C_3 to C_5) to produce a high-octane, low-vapor pressure, paraffinic gasoline component similar in quality to traditional motor alkylate.

Description: The InAlk process combines two, commercially proven technologies: polymerization and olefin saturation. Isobutylene is reacted with light olefins (C_3 to C_5) in the polymerization reactor (1), the resulting mixture is stabilized (2) and the isooctane-rich stream is saturated in the saturation reactor (3). Recycle hydrogen is removed (4) and the product is stripped (5) to remove light-ends.

The InAlk process is more flexible than the traditional alkylation processes. Using a direct alkylation process, refiners must match the isobutane requirement with olefin availability. The InAlk process does not require isobutane to produce a high-quality product. Additional flexibility comes from being able to revamp existing catalytic condensation and MTBE units easily to the InAlk process.

The flexibility of the InAlk process is in both the polymerization and saturation sections. Both sections have different catalyst options based on specific operating objectives and site conditions. This flexibility allows existing catalytic condensation units to revamp to the InAlk process with the addition of the saturation section and optimized processing conditions. Existing MTBE units can be converted to the InAlk process with only minor modifications.

Product quality: High-octane, low Rvp, mid-boiling-range paraffinic gasoline blending component with no aromatic content, low-sulfur content and adjustable olefin content.

Economics: (basis: C_4 feed from FCC unit)

Investment (basis: 2,800-bpsd unit), $/bpsd	
Grassroots	3,000
Revamp of MTBE unit	1,580
Utilities (per bbl alkylate)	
Hydrogen, lb	5.2
Power, kW	7.5
Steam, HP, lb	385
Steam, LP, lb	50

Licensor: UOP LLC.

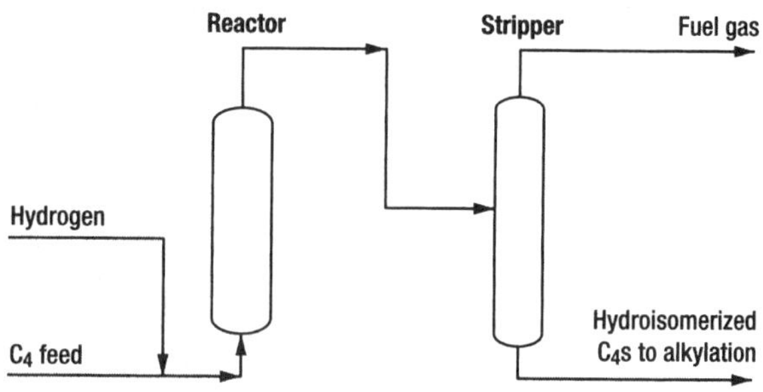

Alkylation—feed preparation

Application: Upgrades alkylation plant feeds with Alkyfining process.

Description: Diolefins and acetylenes in the C_4 (or C_3–C_4) feed react selectively with hydrogen in the liquid-phase, fixed-bed reactor under mild temperature and pressure conditions. Butadiene and, if C_3s are present, methylacetylene and propadiene are converted to olefins.

The high isomerization activity of the catalyst transforms 1-butene into *cis*- and *trans*-2-butenes, which affords higher octane-barrel production.

Good hydrogen distribution and reactor design eliminate channeling while enabling high turndown ratios. Butene yields are maximized, hydrogen is completely consumed, and essentially, no gaseous byproducts or heavier compounds are formed. Additional savings are possible when pure hydrogen is available eliminating the need for a stabilizer. The process integrates easily with the C_3/C_4 splitter.

Alkyfining performance and impact on HF alkylation product:

The results of an Alkyfining unit treating an FCC C_4 HF alkylation unit feed containing 0.8% 1,3-butadiene are:

Butadiene in alkylate, ppm	< 10
1-butene isomerization, %	70
Butenes yield, %	100.5
RON increase in alkylate	2
MON increase in alkylate	1
Alkylate end point reduction, °C	–20

The increases in MON, RON and butenes yield are reflected in a substantial octane-barrel increase while the lower alkylate end point reduces ASO production and HF consumption.

Economics:

Investment:	
Grassroots ISBL cost:	
For an HF unit, $/bpsd	430
For an H_2SO_4 unit, $/bpsd	210

Annual savings for a 10,000-bpsd alkylation unit:

HF unit	4.1 million U.S.$
H_2SO_4 unit	5.5 million U.S.$

Installation: Over 80 units are operating with a total installed capacity of 700,000 bpsd

Licensor: Axens, Axens NA.

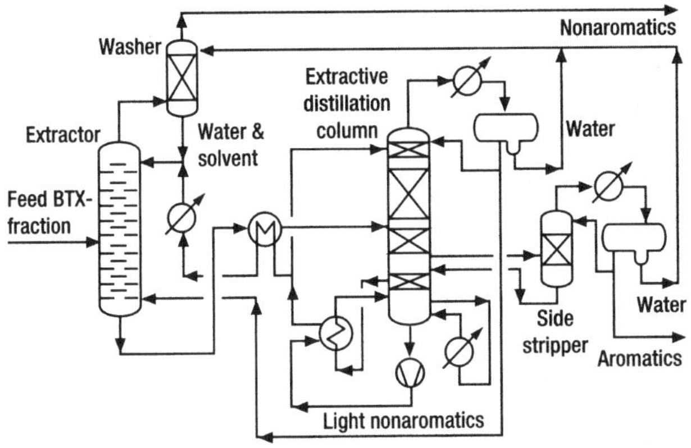

Aromatics extraction

Application: Simultaneous recovery of benzene, toluene and xylenes (BTX) from reformate or pyrolysis gasoline (pygas) using liquid-liquid extraction.

Description: At the top of extractor operating at 30°C to 50°C and 1 to 3 bar, the solvent, N-Formylmorpholin with 4% to 6% water, is fed as a continuous phase. The feedstock—reformate or pygas—enters several stages above the base of the column. Due to density differences, the feedstock bubbles upwards, countercurrent to the solvent. Aromatics pass into the solvent, while the nonaromatics move to the top, remaining in the light phase. Low-boiling nonaromatics from the top of the extractive distillation (ED) column enter the base of the extractor as countersolvent.

Aromatics and solvent from the bottom of the extractor enter the ED, which is operated at reduced pressure due to the boiling-temperature threshold. Additional solvent is fed above the aromatics feed containing small amounts of nonaromatics that move to the top of the column. In the bottom section, as well as in the side rectifier, aromatics and practically water-free solvent are separated.

The water is produced as a second subphase in the reflux drum after azeotropic distillation in the top section of the ED. This water is then fed to the solvent-recovery stage of the extraction process.

Economics:

Consumption per ton of feedstock	
Steam (20 bar), t/t	0.46
Water, cooling (T=10°C), m³/t	12
Electric power, kWh/t	18
Production yield	
Benzene, %	~100
Toluene, %	99.7
EB, Xylenes,%	94.0
Purity	
Benzene, wt%	99.999
Toluene, wt%	>99.99
EB, Xylenes, wt%	>99.99

Installation: One Morphylex plant was erected.

Reference: Emmrich, G., F. Ennenbach and U. Ranke, "Krupp Uhde Processes for Aromatics Recovery," European Petrochemical Technology Conference, June 21–22, 1999, London.

Licensor: Uhde GmbH.

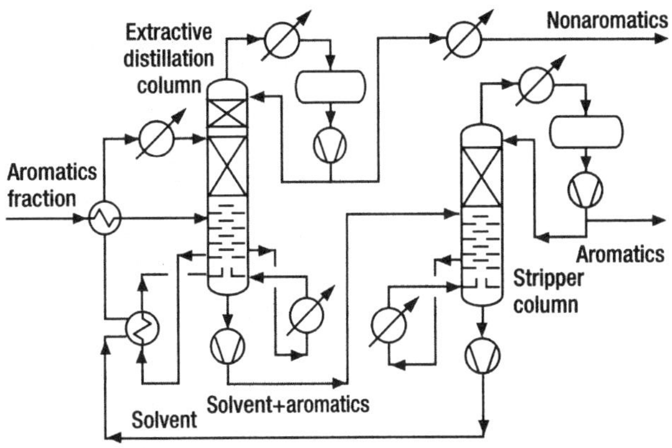

Aromatics extractive distillation

Application: Recovery of high-purity aromatics from reformate, pyrolysis gasoline or coke-oven light oil using extractive distillation.

Description: In the extractive distillation (ED) process, a single-compound solvent, N-Formylmorpholin (NFM) alters the vapor pressure of the components being separated. The vapor pressure of the aromatics is lowered more than that of the less soluble nonaromatics.

Nonaromatics vapors leave the top of the ED column with some solvent, which is recovered in a small column that can either be mounted on the main column or installed separately.

Bottom product of the ED column is fed to the stripper to separate pure aromatics from the solvent. After intensive heat exchange, the lean solvent is recycled to the ED column. NFM perfectly satisfies the necessary solvent properties needed for this process including high selectivity, thermal stability and a suitable boiling point.

Economics:

Pygas feedstock:

	Benzene	Benzene/toluene
Production yield		
Benzene	99.95%	99.95%
Toluene	–	99.98%
Quality		
Benzene	30 wt ppm NA*	80 wt ppm NA*
Toluene	–	600 wt ppm NA*
Consumption		
Steam	475 kg/t ED feed	680 kg/t ED feed**

Reformate feedstock with low aromatics content (20wt%):

	Benzene
Quality	
Benzene	10 wt ppm NA*
Consumption	
Steam	320 kg/t ED feed

*Maximum content of nonaromatics.
**Including benzene/toluene splitter.

Installation: 45 Morphylane plants (total capacity of more than 6 MMtpa).

Reference: Emmrich, G., F. Ennenbach and U. Ranke, "Krupp Uhde Processes for Aromatics Recovery," European Petrochemical Technology Conference, June 21–22, 1999, London.

Licensor: Uhde GmbH.

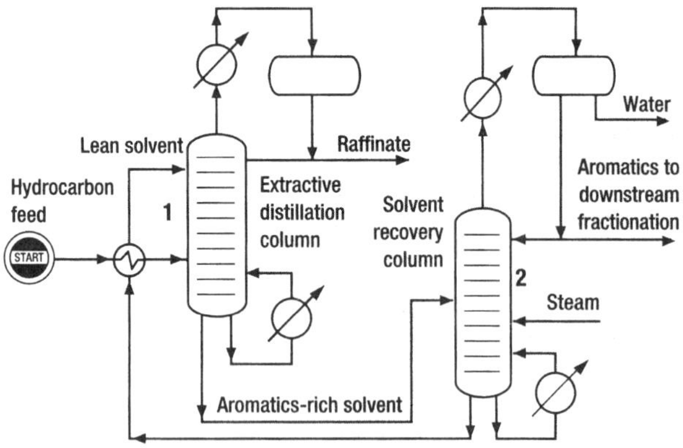

Aromatics recovery

Application: GT-BTX is an aromatics recovery process. The technology uses extractive distillation to remove benzene, toluene and xylene (BTX) from refinery or petrochemical aromatics streams such as catalytic reformate or pyrolysis gasoline. The process is superior to conventional liquid-liquid and other extraction processes in terms of lower capital and operating costs, simplicity of operation, range of feedstock and solvent performance. Flexibility of design allows its use for grassroots aromatics recovery units, debottlenecking or expansion of conventional extraction systems.

Description: The technology has several advantages:

- Less equipment required, thus, significantly lower capital cost compared to conventional liquid-liquid extraction systems
- Energy integration reduces operating costs
- Higher product purity and aromatic recovery
- Recovers aromatics from full-range BTX feedstock without pre-fractionation
- Distillation-based operation provides better control and simplified operation
- Proprietary formulation of commercially available solvents exhibits high selectivity and capacity
- Low solvent circulation rates
- Insignificant fouling due to elimination of liquid-liquid contactors
- Fewer hydrocarbon emission sources for environmental benefits
- Flexibility of design options for grassroots plants or expansion of existing liquid-liquid extraction units.

Hydrocarbon feed is preheated with hot circulating solvent and fed at a midpoint into the extractive distillation column (EDC). Lean solvent is fed at an upper point to selectively extract the aromatics into the column bottoms in a vapor/liquid distillation operation. The nonaromatic hydrocarbons exit the top of the column and pass through a condenser. A portion of the overhead stream is returned to the top of the column as reflux to wash out any entrained solvent. The balance of the overhead stream is the raffinate product, requiring no further treatment.

Rich solvent from the bottom of the EDC is routed to the solvent-recovery column (SRC), where the aromatics are stripped overhead. Stripping steam from a closed-loop water circuit facilitates hydrocarbon removal. The SRC is operated under a vacuum to reduce the boiling point at the base of the column. Lean solvent from the bottom of the SRC is passed through heat exchange before returning to the EDC. A small portion of the lean circulating solvent is processed in a solvent-regeneration step to remove heavy decomposition products.

The SRC overhead mixed aromatics product is routed to the purification section, where it is fractionated to produce chemical-grade benzene, toluene and xylenes.

Economics: Estimated installed cost for a 15,000-bpd GT-BTX extraction unit processing BT-Reformate feedstock is $12 million (U.S. Gulf Coast 2002 basis).

Installations: Three grassroots applications.

Licensor: GTC Technology Inc.

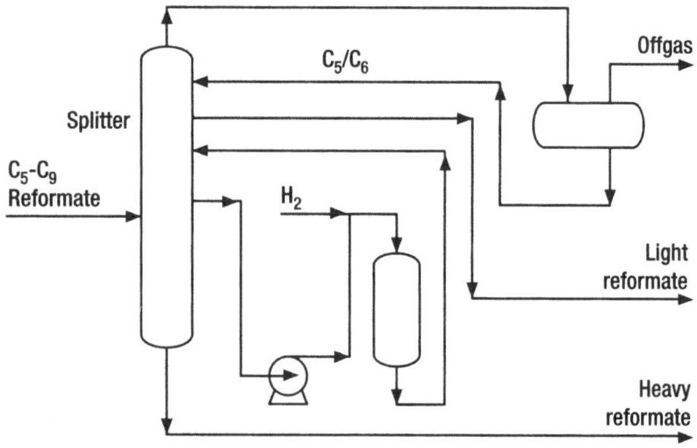

Benzene reduction

Application: Benzene reduction from reformate, with the Benfree process, using integrated reactive distillation.

Description: Full-range reformate from either a semiregenerative or CCR reformer is fed to the reformate splitter column, shown above. The splitter operates as a dehexanizer lifting C_6 and lower-boiling components to the overhead section of the column. Benzene is lifted with the light ends, but toluene is not. Since benzene forms azeotropic mixtures with some C_7 paraffin isomers, these fractions are also entrained with the light fraction.

Above the feed injection tray, a benzene-rich light fraction is withdrawn and pumped to the hydrogenation reactor outside the column. A pump enables the reactor to operate at higher pressure than the column, thus ensuring increased solubility of hydrogen in the feed.

A slightly higher-than-chemical stoichiometric ratio of hydrogen to benzene is added to the feed to ensure that the benzene content of the resulting gasoline pool is below mandated levels, i.e., below 1.0 vol% for many major markets. The low hydrogen flow minimizes losses of gasoline product in the offgas of the column. Benzene conversion to cyclohexane can easily be increased if even lower benzene content is desired. The reactor effluent, essentially benzene-free, is returned to the column.

The absence of benzene disrupts the benzene-iso-C_7 azeotropes, thereby ensuring that the latter components leave with the bottoms fraction of the column. This is particularly advantageous when the light reformate is destined to be isomerized, because iso-C_7 paraffins tend to be cracked to C_3 and C_4 components, thus leading to a loss of gasoline production.

Economics:

Investment, Grassroots ISBL cost, $/bpsd:	300
Combined utilities, $/bbl	0.17
Hydrogen	Stoichiometric to benzene
Catalyst, $/bbl	0.01

Installation: Eighteen benzene reduction units have been licensed.

Licensor: Axens, Axens NA.

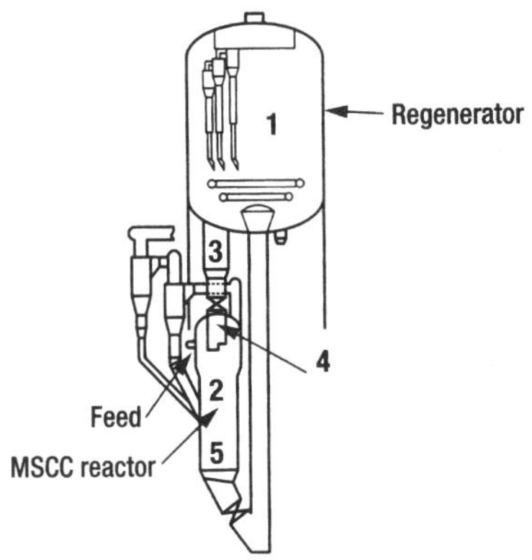

Catalytic cracking

Application: To selectively convert gas oils and residual feedstocks to higher-value cracked products such as light olefins, gasoline and distillates.

Description: The Milli-Second Catalytic Cracking (MSCC) process uses a fluid catalyst and a novel contacting arrangement to crack heavier materials into a highly selective yield of light olefins, gasoline and distillates. A distinguishing feature of the process is that the initial contact of oil and catalyst occurs without a riser in a very short residence time followed by a rapid separation of initial reaction products. Because there is no riser and the catalyst is downflowing, startup and operability are outstanding.

The configuration of an MSCC unit has the regenerator (1) at a higher elevation than the reactor (2). Regenerated catalyst falls down a standpipe (3), through a shaped opening (4) that creates a falling curtain of catalyst, and across a well-distributed feed stream. The products from this initial reaction are quickly separated from the catalyst. The catalyst then passes into a second reaction zone (5), where further reaction and stripping occurs. This second zone can be operated at a higher temperature, which is achieved through contact with regenerated catalyst.

Since a large portion of the reaction product is produced under very short time conditions, the reaction mixture maintains good product olefinicity and retains hydrogen content in the heavier liquid products. Additional reaction time is available for the more-difficult-to-crack species in the second reaction zone/stripper.

Stripped catalyst is airlifted back to the regenerator where coke deposits are burned, creating clean, hot catalyst to begin the sequence again.

Installations: A new MSCC unit began operation earlier this year. Four MSCC units are currently in operation.

Reference: "Short-Contact-Time FCC," AIChE 1998 Spring Meeting, New Orleans.

Licensor: UOP LLC (in cooperation with BARCO).

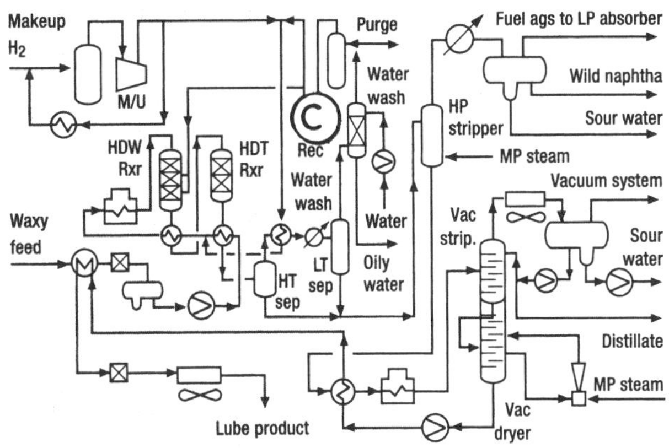

Catalytic dewaxing

Application: Use the ExxonMobil Selective Catalytic Dewaxing (MSDW) process to make high VI lube base stock.

Products: High VI/low-aromatics lube base oils (light neutral through bright stocks). Byproducts include fuel gas, naphtha and low-pour diesel.

Description: MSDW is targeted for hydrocracked or severely hydrotreated stocks. The improved selectivity of MSDW for the highly isoparaffinic-lube components, which results in higher lube yields and VI's. The process uses multiple catalyst systems with multiple reactors. Internals are proprietary (the Spider Vortex Quench Zone technology is used). Feed and recycle gases are preheated and contact the catalyst in a down-flow-fixed-bed reactor. Reactor effluent is cooled, and the remaining aromatics are saturated in a post-treat reactor. The process can be integrated into a lube hydrocracker or lube hydrotreater. Postfractionation is targeted for client needs.

Operating conditions:

Temperatures, °F	550 to 800
Hydrogen partial pressures, psig	500 to 2,500
LHSV	0.4 to 3.0

Conversion depends on feed wax content

Pour point reduction as needed.

Yields:

	Light neutral	Heavy neutral
Lube yield, wt%	94.5	96.5
C_1 to C_4, wt%	1.5	1.0
C_5–400°F, wt%	2.7	1.8
400°F–Lube, wt%	1.5	1.0
H_2 cons, scf/bbl	100–300	100–300

Economics: \$3,000–5,500 per bpsd installed cost (U. S. Gulf Coast).

Installation: Three units are operating, one under construction and one being converted.

Licensor: ExxonMobil Research & Engineering Co.

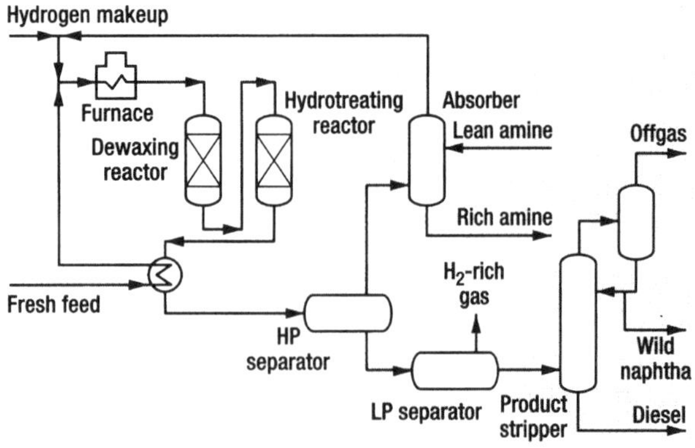

Catalytic dewaxing

Application: Catalytic dewaxing process improves the cold flow properties (pour point, CFPP) of distillate fuels so that deeper cuts can be made at the crude unit. Thus, middle-distillate fuel production can be increased. The waxy n-paraffins are selectively cracked to produce a very high yield of distillate with some fuel gas, LPG and naphtha.

Description: The heart of the dewaxing process is the zeolitic catalyst, which operates at typical distillate hydrotreating conditions. This feature allows low-cost revamp for existing hydrotreaters into a HDS/DW unit by adding reactor volume. The dewaxing step requires a very small increase in hydrogen consumption; thus, the incremental operating cost is low. Since the dewaxing catalyst is tolerant of sulfur and nitrogen components in the feed, it can be located upstream of the HDS catalyst. The run length for the dewaxing catalyst can be designed to match the HDS catalyst.

Economics: The cost of a new HDS/DW is estimated at 1,000-2,000 $/bbl depending primarily on hydrotreating requirements.

Installation: One unit is operating, and one ultra-low-sulfur/dewaxing unit is under design.

Licensor: Haldor Topsøe A/S.

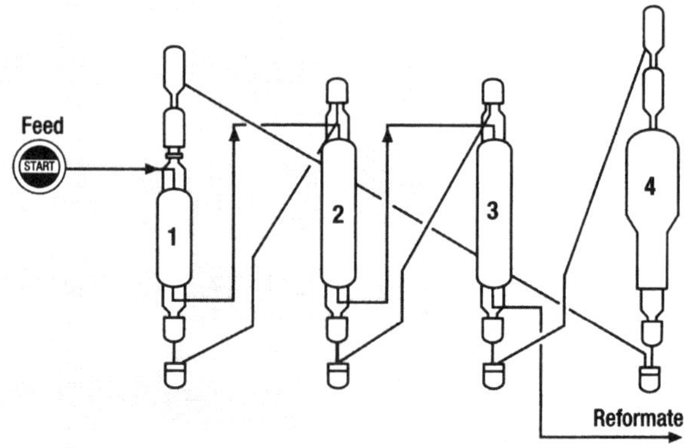

Catalytic reforming

Application: Upgrade various types of naphtha to produce high-octane reformate, BTX and LPG.

Description: Two different designs are offered. One design is conventional where the catalyst is regenerated in place at the end of each cycle. Operating normally in a pressure range of 12 to 25 kg/cm2 (170 to 350 psig) and with low pressure drop in the hydrogen loop, the product is 90 to 100 RONC. With its higher selectivity, trimetallic catalyst RG582 and RG682 make an excellent catalyst replacement for semi-regenerative reformers.

The second, the advanced Octanizing process, uses continuous catalyst regeneration allowing operating pressures as low as 3.5 kg/cm^2 (50 psig). This is made possible by smooth-flowing moving bed reactors (1–3) which use a highly stable and selective catalyst suitable for continuous regeneration (4). Main features of Axens's regenerative technology are:

- Side-by-side reactor arrangement, which is very easy to erect and consequently leads to low investment cost.
- The Regen C catalyst regeneration system featuring the dry burn loop, completely restores the catalyst activity while maintaining its specific area for more than 600 cycles.

Finally, with the new CR401 (gasoline mode) and AR501 (aromatics production) catalysts specifically developed for ultra-low operating pressure and the very effective catalyst regeneration system, refiners operating Octanizing or Aromizing processes can obtain the highest hydrogen, C_5^+ and aromatics yields over the entire catalyst life.

Yields: Typical for a 90°C to 170°C (176°F to 338°F) cut from light Arabian feedstock:

	Conventional	Octanizing
Oper. press., kg/cm^2	10–15	<5
Yield, wt% of feed		
Hydrogen	2.8	3.8
C_5^+	83	88
RONC	100	102
MONC	89	90.5

Economics:

Investment (basis 25,000 bpsd continuous octanizing unit, battery limits, erected cost, mid-2002 Gulf Coast), U.S.$ per bpsd 1,800

Utilities: typical per bbl feed:

Fuel, 10^3 kcal	65
Electricity, kWh	0.96
Steam, net, HP, kg	12.5
Water, boiler feed, m^3	0.03

Installation: Of 110 units licensed, 60 units are designed with continuous regeneration technology capability.

Reference: "Continuing Innovation In Cat Reforming," NPRA Annual Meeting, March 15–17, 1998, San Antonio.

"Fixed Bed Reformer Revamp Solutions for Gasoline Pool Improvement," *Petroleum Technology Quarterly*, Summer 2000.

"Increase reformer performance through catalytic solutions," ERTC 2002, Paris.

Licensor: Axens, Axens NA.

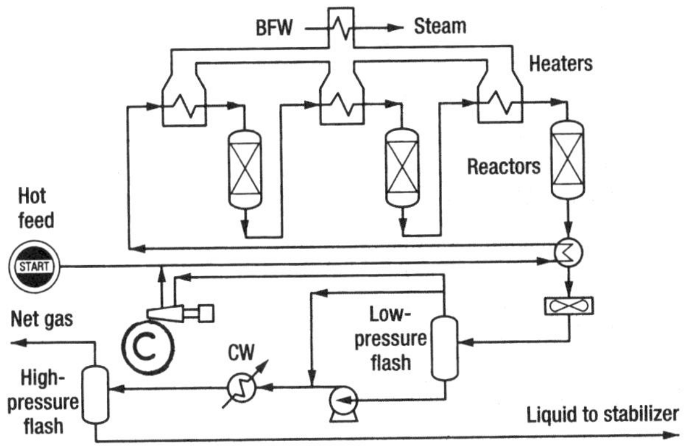

Catalytic reforming

Application: Increase the octane of straight-run or cracked naphthas for gasoline production.

Products: High-octane gasoline and hydrogen-rich gas. Byproducts may be LPG, fuel gas and steam.

Description: Semi-regenerative multibed reforming over platinum or bimetallic catalysts. Hydrogen recycled to reactors at the rate of 3 to 7 mols/mol of feed. Straight-run and/or cracked feeds are typically hydrotreated, but low-sulfur feeds (<10 ppm) may be reformed without hydrotreatment.

Operating conditions: 875°F to 1,000°F and 150 to 400 psig reactor conditions.

Yields: Depend on feed characteristics, product octane and reactor pressure. The following yields are one example. The feed contains 51.4% paraffins, 41.5% naphthenes and 7.1% aromatics, and boils from 208°F to 375°F (ASTM D86). Product octane is 99.7 RONC and average reactor pressure is 200 psig.

Component	wt%	vol%
H_2	2.3	1,150 scf/bbl
C_1	1.1	—
C_2	1.8	—
C_3	3.2	—
iC_4	1.6	—
nC_4	2.3	—
C_5+	87.1	—
LPG	—	3.7
Reformate	—	83.2

Economics:

Utilities, (per bbl feed)	
Fuel, 10^3 Btu release	275
Electricity, kWh	7.2
Water, cooling (20°F rise), gal	216
Steam produced (175 psig sat), lb	100

Licensor: Howe-Baker Engineers, Ltd., a subsidiary of Chicago Bridge & Iron Co.

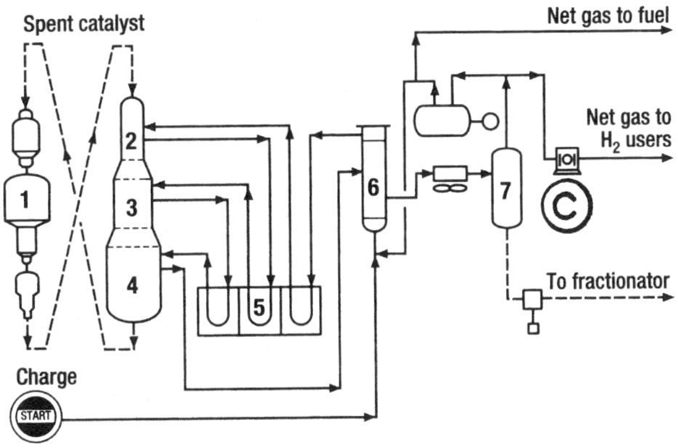

Catalytic reforming

Application: Upgrade naphtha for use as a gasoline blendstock or feed to a petrochemical complex with the UOP CCR Platforming process. The unit is also a reliable, continuous source of high-purity hydrogen.

Description: Constant product yields and onstream availability distinguish the CCR Platforming process featuring catalyst transfer with minimum lifts, no valves closing on catalyst and gravity flow from reactor to reactor (2,3,4). The CycleMax regenerator (1) provides simplified operation and enhanced performance at a lower cost than other designs. The product recovery section downstream of the separator (7) is customized to meet site-specific requirements. The R-270 series catalysts offer the highest C_5^+ and hydrogen yields while also providing the R-230 series attributes of CCR Platforming process unit flexibility through reduced coke make.

Semiregenerative reforming units also benefit from the latest UOP catalysts. R-86 catalyst provides the high stability with excellent yields at low cost. Refiners use UOP engineering and technical service experience to tune operations, plan the most cost-effective revamps, and implement a stepwise approach for conversion of semiregenerative units to obtain the full benefits of CCR Platforming technology.

Yields:

Operating mode	Semiregen.	Continuous
Onstream availability, days/yr	330	360
Feedstock, P/N/A LV%	63/25/12	63/25/12
IBP/EP,°F	200/360	200/360
Operating conditions		
Reactor pressure, psig	200	50
C_5^+ octane, RONC	100	100
Catalyst	R-86	R-274
Yield information		
Hydrogen, scfb	1,270	1,690
C_5^+, wt%	84.8	91.6

Economics:

Investment (basis: 20,000 bpsd CCR Platforming unit, 50 psig reactor pressure, 100 C_5^+ RONC, 2002, U.S. Gulf Coast ISBL):
$ per bpsd 2,100

Installation: UOP has licensed more than 800 platforming units; 37 customers have selected CCR platforming for two or more catalytic reforming units. Twenty-nine refiners operate 100 of the 173 operating units. Twenty units are designed for initial semiregenerative operation with the future installation of a CCR regeneration section.

	Operating	Design & const.
Total CCR Platforming units	173	47
Ultra-low 50 psig units	44	31
Units at 35,000+ bpsd	29	5
Semiregenerative units with a stacked reactor	14	5

Licensor: UOP LLC.

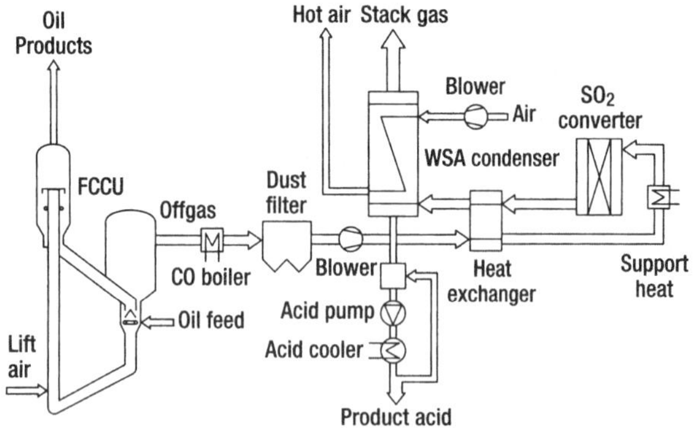

Catalytic SO_x removal

Application: The Wet gas Sulfuric Acid (WSA) process catalytically removes more than 99% + of sulfurous compounds from moist acid gases without prior drying and recovers concentrated sulfuric acid.

WSA combined with selective catalytic reduction, the SNOX process, efficiently removes nitrogen oxides (up to 95%) and sulfur oxides from flue gases and offgases. The main applications in refineries are H_2S gases, onsite regeneration of alkylation acid (spent acid recovery (SAR)), FCC regenerator offgases (example below), and boiler offgases, especially flue gases from petroleum coke and heavy residual oil fired boilers.

Description: Flue gas from the FCC regenerator is cooled to 430°F (220°C) in the waste-heat boiler. By means of an electrostatic precipitator, catalyst and coke particulates are reduced to less than 0.18 lb/MMscf. The flue gas is heated to approximately 770°F (410°C) before entering the SO_2 reactor. In the SO_2 reactor, SO_2 is oxided to SO_3, and all remaining particulates are deposited in the catalyst panels. The reactor consists of several parallel catalyst panels, which can be individually cleaned and reloaded without interrupting plant operation.

After the SO_2 converter, the gas is cooled to near the acid dew point. In the last step, concentrated sulfuric acid is condensed, and flue gas is cooled in the WSA condenser. Hot cooling air from the condenser may be used for preheating of boiler feedwater or as preheated air for the FCC regenerator/CO boiler.

All equipment, except the condenser, is made of carbon steel or low alloy steel. The WSA process has few moving parts, low maintenance costs and high onstream availability. The process can be applied to new or revamp installations.

The WSA process is characterized by:

- 99% or more of the flue gas sulfur is recovered as commercial grade concentrated sulfuric acid
- Particulates are essentially completely removed
- No waste solids or wastewater is produced. No absorbents or auxiliary chemicals are used
- Operating costs decrease with increasing sulfur content in flue gas
- Process is fully automated, contains few moving parts and does not use a circulation of slurries or solids
- Simple operation allows wide flexibility in operating loads.

Installation: More than 40 units worldwide.

Licensor: Haldor Topsøe A/S.

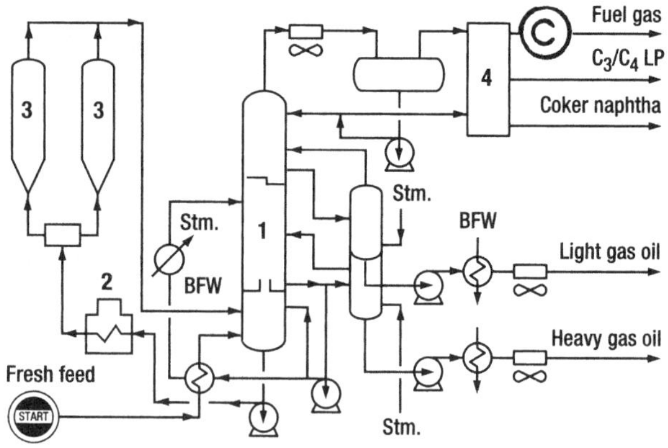

Coking

Application: Conversion of vacuum residues (virgin and hydrotreated), various petroleum tars and coal tar pitch through delayed coking.

Products: Fuel gas, LPG, naphtha, gas oils and fuel, anode or needle grade coke (depending on feedstock and operating conditions).

Description: Feedstock is introduced (after heat exchange) to the bottom of the coker fractionator (1) where it mixes with condensed recycle. The mixture is pumped through the coker heater (2) where the desired coking temperature is achieved, to one of two coke drums (3). Steam or boiler feedwater is injected into the heater tubes to prevent coking in the furnace tubes. Coke drum overhead vapors flow to the fractionator (1) where they are separated into an overhead stream containing the wet gas, LPG and naphtha; two gas oil sidestreams; and the recycle that rejoins the feed.

The overhead stream is sent to a vapor recovery unit (4) where the individual product streams are separated. The coke that forms in one of at least two (parallel connected) drums is then removed using high-pressure water. The plant also includes a blow-down system, coke handling and a water recovery system.

Operating conditions:

Heater outlet temperature, °F	900–950
Coke drum pressure, psig	15–90
Recycle ratio, vol/vol feed, %	0–100

Yields:

Feedstock	Middle East vac. residue	Vacuum residue of hydrotreated bottoms	Coal tar pitch
Gravity, °API	7.4	1.3	211.0
Sulfur, wt%	4.2	2.3	0.5
Conradson carbon, wt%	20.0	27.6	—
Products, wt%			
Gas + LPG	7.9	9.0	3.9
Naphtha	12.6	11.1	—
Gas oils	50.8	44.0	31.0
Coke	28.7	35.9	65.1

Economics:

Investment (basis: 20,000 bpsd straight-run vacuum residue feed, U.S. Gulf Coast 2002, fuel-grade coke, includes vapor recovery), U.S. $ per bpsd (typical)	4,000
Utilities, typical/bbl of feed:	
Fuel, 10^3 Btu	145
Electricity, kWh	3.9
Steam (exported), lb	20
Water, cooling, gal	180

Installation: More than 55 units.

Reference: Mallik, Ram, Gary and Hamilton, "Delayed coker design considerations and project execution," NPRA 2002 Annual Meeting, March 17–19, 2002.

Licensor: ABB Lummus Global Inc.

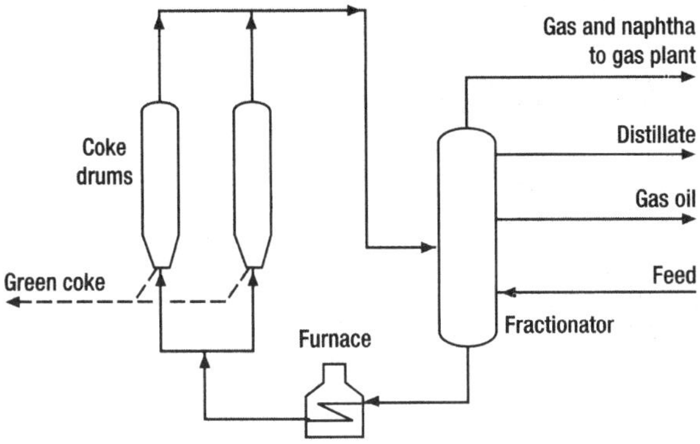

Coking

Application: Upgrading of petroleum residues (vacuum residue, bitumen, solvent-deasphalter pitch and fuel oil) to more valuable liquid products (LPG, naphtha, distillate and gas oil). Fuel gas and petroleum coke are also produced.

Description: The delayed coking process is a thermal process and consists of fired heater(s), coke drums and main fractionator. The cracking and coking reactions are initiated in the fired heater under controlled time-temperature-pressure conditions. The reactions continue as the process stream moves to the coke drums. Being highly endothermic, the coking-reaction rate drops dramatically as coke-drum temperature decreases. Coke is deposited in the coke drums. The vapor is routed to the fractionator, where it is condensed and fractionated into product streams—typically fuel gas, LPG, naphtha, distillate and gas oil.

When one of the pair of coke drums is full of coke, the heater outlet stream is directed to the other coke drum. The full drum is taken offline, cooled with steam and water and opened. The coke is removed by hydraulic cutting. The empty drum is then closed, warmed-up and made ready to receive feed while the other drum becomes full.

Benefits of Conoco-Bechtel's delayed coking technology are:

- Maximum liquid-product yields and minimum coke yield through low-pressure operation, patented distillate recycle technology and zero (patented) or minimum natural recycle operation
- Maximum flexibility; distillate recycle operation can be used to adjust the liquid-product slate or can be withdrawn to maximize unit capacity
- Extended furnace runlengths between decokings
- Ultra-low-cycle-time operation maximizes capacity and asset utilization
- Higher reliability and maintainability enables higher onstream time and lowers maintenance costs
- Lower investment cost.

Economics: For a delayed coker processing 35,000 bpsd of heavy, high-sulfur vacuum residue, the U.S. Gulf Coast investment cost is approximately U.S.$145–160 million.

Installation: Low investment cost and attractive yield structure has made delayed coking the technology of choice for bottom-of-the-barrel upgrading. Numerous delayed coking units are operating in petroleum refineries worldwide.

Licensor: Bechtel Corp. and Conoco Inc.

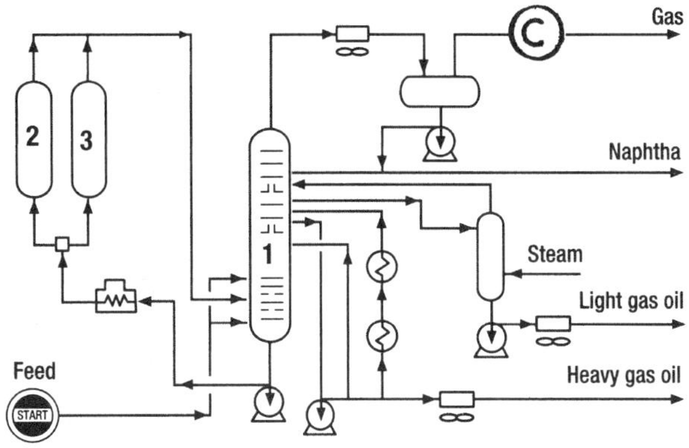

Coking

Application: Manufacture petroleum coke and upgrade residues to lighter hydrocarbon fractions using the Selective Yield Delayed Coking (SYDEC) process.

Products: Coke, gas, LPG, naphtha and gas oils.

Description: Charge is fed directly to the fractionator (1) where it combines with recycle and is pumped to the coker heater where it is heated to coking temperature, causing partial vaporization and mild cracking. The vapor-liquid mix enters a coke drum (2 or 3) for further cracking. Drum overhead enters the fractionator (1) to be separated into gas, naphtha, and light and heavy gas oils. There are at least two coking drums, one coking while the other is decoked using high-pressure water jets.

Operating conditions: Typical ranges are:

Heater outlet temperature, °F	900–950
Coke drum pressure, psig	15–100
Recycle ratio, equiv. fresh feed	0.05–1.0

Increased coking temperature decreases coke production; increases liquid yield and gas oil end point. Increasing pressure and/or recycle ratio increases gas and coke make, decreases liquid yield and gas oil end point.

Yields:

Feed, source	Venezuela	N. Africa	—
Type	Vac. resid	Vac. resid	Decant oil
Gravity, °API	2.6	15.2	–0.7
Sulfur, wt%	4.4	0.7	0.5
Concarbon, wt%	23.3	16.7	—
Operation	Max dist.	Anode coke	Needle coke
products, wt%			
Gas	8.7	7.7	9.8
Naphtha	10.0	19.9	8.4
Gas oil	50.3	46.0	41.6
Coke	31.0	26.4	40.2

Economics:

Investment (basis: 65,000–10,000 bpsd, 4Q 1999, U.S. Gulf), $ per bpsd	2,500–4,000
Utilities, typical per bbl feed:	
Fuel, 10^3 Btu	120
Electricity, kWh	3.6
Steam (exported), lb	(40)
Water, cooling, gal	36

Installation: More than 58,000 tpd of fuel, anode and needle coke.

Reference: *Handbook of Petroleum Refining Processes,* 2nd Ed., pp. 12.25–12.82; *Oil & Gas Journal,* Feb. 4, 1991, pp. 41–44; *Hydrocarbon Processing,* Vol. 71, No. 1, January 1992, pp. 75–84.

Licensor: Foster Wheeler/UOP LLC

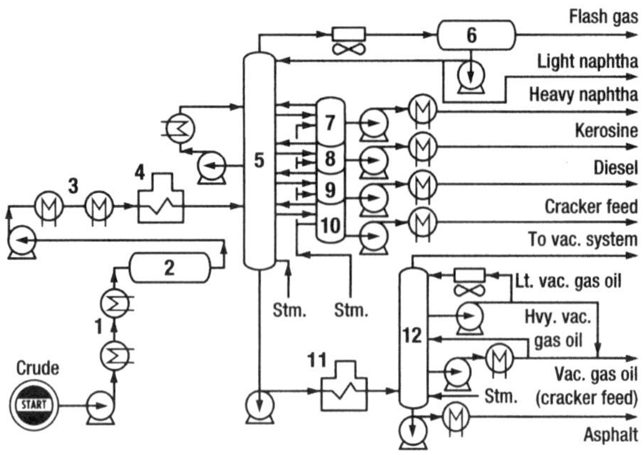

Crude distillation

Application: Separates and recovers the relatively lighter fractions from a fresh crude oil charge (e.g., naphtha, kerosine, diesel and cracking stock). The vacuum flasher processes the crude distillation bottoms to produce an increased yield of liquid distillates and a heavy residual material.

Description: The charge is preheated (1), desalted (2) and directed to a preheat train (3) where it recovers heat from product and reflux streams. The typical crude fired heater (4) inlet temperature is on the order of 550°F, while the outlet temperature is on the order of 675°F to 725°F. Heater effluent then enters a crude distillation column (5) where light naphtha is drawn off the tower overhead (6); heavy naphtha, kerosine, diesel and cracking stock are sidestream drawoffs. External reflux for the tower is provided by pumparound streams (7–10). The atmospheric residue is charged to a fired heater (11) where the typical outlet temperature is on the order of 750°F to 775°F.

From the heater outlet, the stream is fed into a vacuum tower (12), where the distillate is condensed in two sections and withdrawn as two sidestreams. The two sidestreams are combined to form cracking feedstock. An asphalt base stock is pumped from the bottom of the tower. Two circulating reflux streams serve as heat removal media for the tower.

Yields: Typical for Merey crude oil:

Crude unit products	wt%	°API	Pour, °F
Overhead & naphtha	6.2	58.0	—
Kerosine	4.5	41.4	-85
Diesel	18.0	30.0	-10
Gas oil	3.9	24.0	20
Lt. vac. gas oil	2.6	23.4	35
Hvy. vac. gas oil	10.9	19.5	85
Vac. bottoms	53.9	5.8	(120)*
Total	100.0	8.7	85

*Softening point, °F

Note: Crude unit feed is 2.19 wt% sulfur. Vacuum unit feed is 2.91 wt% sulfur.

Economics:

Investment (basis: 100,000–50,000 bpsd, 2nd Q, 2002, U.S. Gulf), $ per bpsd 890–1,100

Utility requirements, typical per bbl fresh feed

Steam, lb	24
Fuel (liberated), 10^3 Btu	(80–120)
Power, kWh	0.6
Water, cooling, gal	300–400

Installation: Foster Wheeler has designed and constructed crude units having a total crude capacity in excess of 10 MMbpsd.

Reference: *Encyclopedia of Chemical Processing and Design,* Marcel-Dekker, 1997, pp. 230–249.

Licensor: Foster Wheeler.

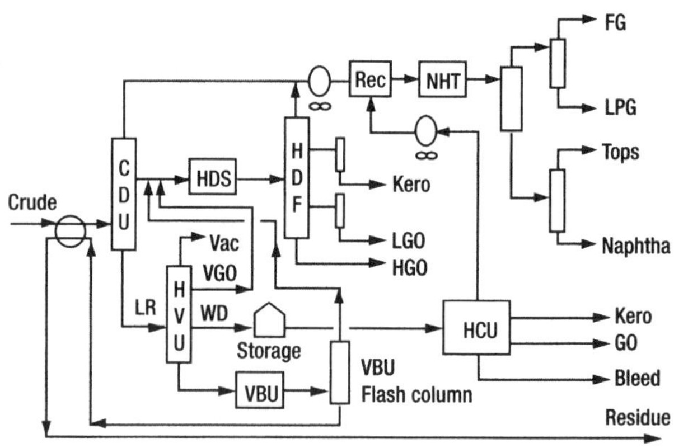

Crude distillation

Application: The Shell Bulk CDU is a highly integrated concept. It separates the crude in long residue, waxy distillate, middle distillates and a naphtha minus fraction. Compared with stand-alone units, the overall integration of a crude distillation unit (CDU), hydrodesulfurization unit (HDS), high vacuum unit (HVU) and a visbreaker (VBU) results in a 50% reduction in equipment count and significantly reduced operating costs. A prominent feature embedded in this design is the Shell deep-flash HVU technology. This technology can also be provided in cost-effective process designs for both feedprep and lube oil HVU's as stand-alone units. For each application, tailor-made designs can be produced.

Description: The basic concept of the bulk CDU is the separation of the naphtha minus and the long residue from the middle distillate fraction which is routed to the HDS. After desulfurization in the HDS unit, final product separation of the bulk middle distillate stream from the CDU takes place in the HDS fractionator (HDF), which consists of a main atmospheric fractionator with side strippers.

The long residue is routed hot to a feedprep HVU, which recovers the waxy distillate fraction from long residue as the feedstock for a cat cracker or hydrocracker unit (HCU). Typical flashzone conditions are 415°C and 24 mbara. The Shell design features a deentrainment section, spray sections to obtain a lower flashzone pressure, and a VGO recovery section to recover up to 10 wt% of as automotive diesel. The Shell furnace design prevents excessive cracking and enables a 5-year run length between decoke.

Yields: Typical for Arabian light crude

Products		% wt
Gas	C_1–C_4	0.7
Gasoline	C_5–150°C	15.2
Kerosine	150–250°C	17.4
Gasoil (GO)	250–350°C	18.3
VGO	350–370°C	3.6
Waxy distillate (WD)	370–575°C	28.8
Residue	575°C+	16.0

Economics: Due to the incorporation of Shell high capacity internals and the deeply integrated designs, an attractive CAPEX reduction can be achieved. Investment costs are dependent on the required configuration and process objectives.

Installation: Over 100 Shell CDU's have been designed and operated since the early 1900s. Additionally, a total of some 50 HVU units have been built while a similar number has been debottlenecked, including many third-party designs of feedprep and lube oil HVU's.

Licensor: Shell Global Solutions International B.V.

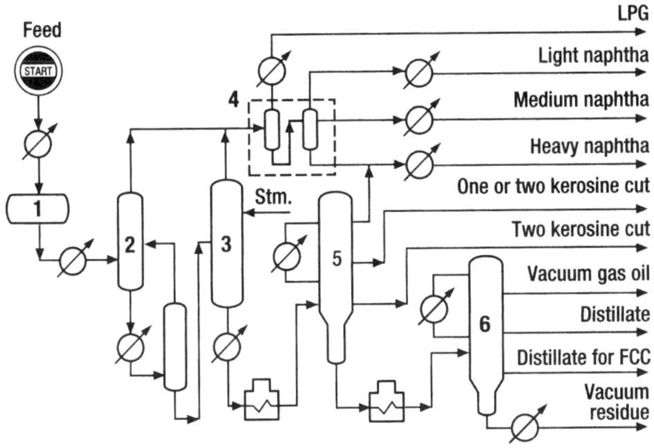

Crude distillation

Application: The D2000 process is progressive distillation to minimize the total energy consumption required to separate crude oils or condensates into hydrocarbon cuts, which number and properties are optimized to fit with sophisticated refining schemes and future regulations. This process is applied normally for new topping units or new integrated topping/vacuum units but the concept can be used for debottlenecking purpose.

Products: This process is particularly suitable when more than two naphtha cuts are to be produced. Typically the process is optimized to produce three naphtha cuts or more, one or two kerosine cuts, two atmospheric gas oil cuts, one vacuum gas oil cut, two vacuum distillates cuts, and one vacuum residue.

Description: The crude is preheated and desalted (1). It is fed to a first dry reboiled pre-flash tower (2) and then to a wet pre-flash tower (3). The overhead products of the two pre-flash towers are then fractionated as required in a gas plant and rectification towers (4).

The topped crude typically reduced by ⅔ of the total naphtha cut is then heated in a conventional heater and conventional topping column (5). If necessary the reduced crude is fractionated in one deep vacuum column designed for a sharp fractionation between vacuum gas oil, two vacuum distillates (6) and a vacuum residue, which could be also a road bitumen.

Extensive use of pinch technology minimizes heat supplied by heaters and heat removed by air and water coolers.

This process is particularly suitable for large crude capacity from 150,000 to 250,000 bpsd.

It is also available for condensates and light crudes progressive distillation with a slightly adapted scheme.

Economics:

Investment (basis 230,000 bpsd including atmospheric and vacuum distillation, gas plant and rectification tower) 750 to 950 $ per bpsd (U.S. Gulf Coast 2000).

Utility requirements, typical per bbl of crude feed:

Fuel fired, 10^3 btu	50–65
Power, kWh	0.9–1.2
Steam 65 psig, lb	0–5
Water cooling, (15°C rise) gal	50–100

Total primary energy consumption:

for Arabian Light or Russian Export Blend:	1.25 tons of fuel per 100 tons of Crude
for Arabian Heavy	1.15 tons of fuel per 100 tons of Crude

Installation: Technip has designed and constructed one crude unit and one condensate unit with the D2000 concept. The latest revamp project currently in operation shows an increase of capacity of the existing crude unit of 30% without heater addition.

Licensor: TOTALFINAELF, Technip-Coflexip.

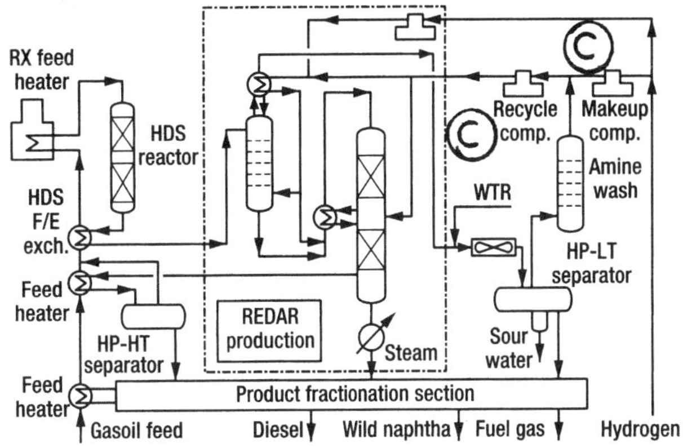

Dearomatization—middle distillate

Application: Deep dearomatization of middle distillates and upgrading of light cycle oil (LCO).

Description: The process uses the REDAR catalyst, developed by Engelhard Corp. It is capable of aromatics hydrogenation in presence of sulfur and nitrogen at low-operating pressure and temperature. The process operates in conjunction with a conventional hydrotreating step to remove sulfur. Therefore, it is ideal as an add-on to existing hydrotreaters with sulfur levels reaching 250 ppm-wt and nitrogen levels up to 100 ppm-wt.

The process also offers an excellent cost-effective opportunity to upgrade LCOs from FCC units. Depending on the LCO blend ratio in the feed, the first stage reactor may require a highly active NiMo catalyst followed by the REDAR catalyst in the second stage. A hot-hydrogen stripper is recommended for both applications. An important feature of this process is cascading of treat gas from the REDAR stage to the hydrotreating stage, thereby taking advantage of higher hydrogen partial pressure in the REDAR stage.

Key features of the process are: low gas and naphtha make, sulfur in product meets all proposed fuel regulations, significant aromatics reduction and boiling point shift, which allows higher boiling point LCO in the feed.

Typical product properties: (LCO upgrading)

Properties	Feed	HDS stage	REDAR stage
Sulfur, ppm-wt	8,800	83	4
Density, @60°F, °API	18.0	24.5	32.7
5% BP, D-86, °F	361	345	310
95% BP, D-86, °F	675	650	630
Cetane index, D-4737	24	31	40
Mono aromatics, IP 391/95 wt%	16.7	53.1	7.5
Di aromatics, IP 391/95, wt%	32.2	8.2	0.2
Tri and higher, IP 391/95, wt%	13.0	4.0	0.0
Total aromatics	61.9	65.3	7.7
Gas yield, Wt% on feed		<0.5	<0.1
C_5^- 392°F, Wt% on feed		<1.0	<3.0

Economics: Investment cost for a grassroots 15,000-bpsd LCO upgrading unit is approximately $35 million on U.S.GC, 2Q 2002 basis. Cost of retrofitting an existing 40,000-bpsd hydrotreater will be significantly less and is estimated at $15 million for U.S.GC, 2Q, 2002.

Licensor: Badger Technology Center of Washington Group International, in association with Engelhard Corp.

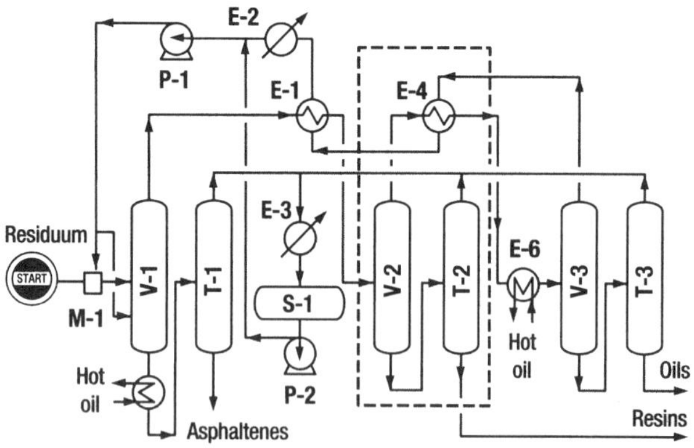

Deasphalting

Application: Extract lubricating oil blend stocks and FCCU or hydrocracker feedstocks with low metal and Conradson carbon contents from atmospheric and vacuum resid using ROSE Supercritical Fluid Technology. Can be used to upgrade existing solvent deasphalters. ROSE may also be used to economically upgrade heavy crude oil.

Products: Lube blend stocks, FCCU feed, hydrocracker feed, resins and asphaltenes.

Description: Resid is charged through a mixer (M-1), where it is mixed with solvent before entering the asphaltene separator (V-1). Countercurrent solvent flow extracts lighter components from the resid while rejecting asphaltenes with a small amount of solvent. Asphaltenes are then heated and stripped of solvent (T-1). Remaining solvent solution goes overhead (V-1) through heat exchange (E-1) and a second separation (V-2), yielding an intermediate product (resins) that is stripped of solvent (T-2). The overhead is heated (E-4, E-6) so the solvent exists as a supercritical fluid in which the oil is virtually insoluble. Recovered solvent leaves the separator top (V-3) to be cooled by heat exchange (E-4, E-1) and a cooler (E-2). Deasphalted oil from the oil separator (V-3) is stripped (T-3) of dissolved solvent. The only solvent vaporized is a small amount dissolved in fractions withdrawn in the separators. This solvent is recovered in the product strippers. V-1, V-2 and V-3 are equipped with high-performance ROSEMAX internals. These high-efficiency, high-capacity internals offer superior product yield and quality while minimizing vessel size and capital investment. They can also debottleneck and improve operations of existing solvent deasphalting units.

The system can be simplified by removing equipment in the outlined box to make two products. The intermediate fraction can be shifted into the final oil fraction by adjusting operating conditions. Only one exchanger (E-6) provides heat to warm the resid charge and the small amount of extraction solvent recovered in the product strippers.

Yields: The extraction solvent composition and operating conditions are adjusted to provide the product quality and yields required for downstream processing or to meet finished product specifications. Solvents range from propane through hexane and include blends normally produced in refineries.

Economics:

Investment (basis: 30,000 bpsd, U.S. Gulf Coast), $ per bpsd	1,250
Utilities, typical per bbl feed:	
Fuel absorbed, 10^3 Btu	80–110
Electricity, kWh	2.0
Steam, 150-psig, lb	12

Installation: Thirty-three licensed units with a combined capacity of over 600,000 bpd.

Reference: Northup, A. H., and H. D. Sloan, "Advances in solvent deasphalting technology," 1996 NPRA Annual Meeting, San Antonio.

Licensor: Kellogg Brown & Root, Inc.

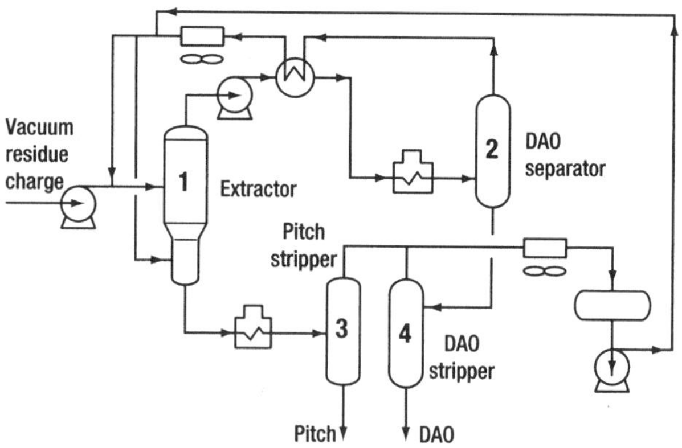

Deasphalting

Application: Prepare quality feed for FCC units and hydrocrackers from vacuum residue and blending stocks for lube oil and asphalt manufacturing.

Products: Deasphalted oil (DAO) for catalytic cracking and hydrocracking feedstocks, resins for specification asphalts, and pitch for specification asphalts and residue fuels.

Description: Feed and light paraffinic solvent are mixed then charged to the extractor (1). The DAO and pitch phases, both containing solvents, exit the extractor. The DAO and solvent mixture is separated under supercritical conditions (2). Both the pitch and DAO products are stripped of entrained solvent (3,4). A second extraction stage is utilized if resins are to be produced.

Operating conditions: Typical ranges are: solvent, various blends of C_3–C_7 hydrocarbons including light naphthas. Pressure: 300 to 600 psig. Temp.: 120°F to 450°F. Solvent-to-oil ratio: 4/1 to 13/1.

Yields:

Feed, type	**Lube oil**		**Cracking stock**
Gravity, °API	6.6		6.5
Sulfur, wt.%	4.9		3.0
CCR, wt%	20.1		21.8
Visc, SSU@ 210°F	7,300		8,720
NI/V, wppm	29/100		46/125
DAO			
Yield, vol.% of Feed	30	53	65
Gravity, °API	20.3	17.6	15.1
Sulfur, wt%	2.7	1.9	2.2
CCR, wt%	1.4	3.5	6.2
Visc., SSU@ 210°F	165	307	540
Ni/V, wppm	0.25/0.37	1.8/3.4	4.5/10.3
Pitch			
Softening point, R&B,°F	149	226	240
Penetration@77°F	12	0	0

Economics:

Investment (basis: 2,000–40,000 bpsd 4Q 2000, U.S. Gulf), $/bpsd 800–3,000

Utilities, typical per bbl feed:	**Lube oil**	**Cracking stock**
Fuel, 10^3 Btu	81	56
Electricity, kWh	1.5	1.8
Steam, 150-psig, lb	116	11
Water, cooling (25°F rise), gal	15	nil

Installations: 50+. This also includes both UOP and Foster Wheeler units originally licensed separately before the merging the technologies in 1996.

Reference: *Handbook of Petroleum Refining Processes,* 2nd Ed., McGraw Hill, 1997, pp.10.15–10.60.

Licensor: UOP LLC/Foster Wheeler.

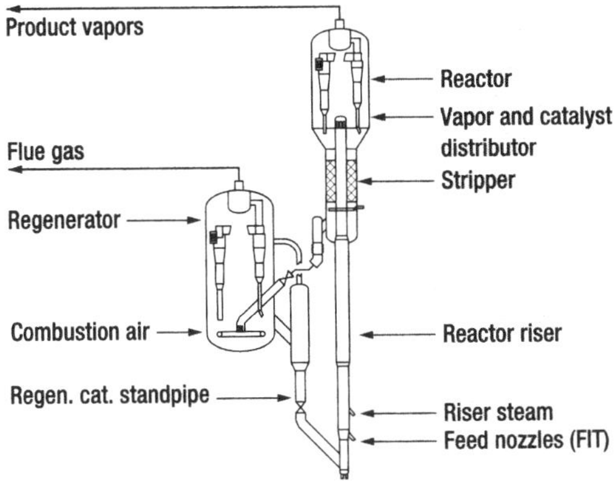

Deep catalytic cracking

Application: Selective conversion of gasoil and paraffinic residual feedstocks.

Products: C_2–C_5 olefins, aromatic-rich, high-octane gasoline and distillate.

Description: DCC is a fluidized process for selectively cracking a wide variety of feedstocks to light olefins. Propylene yields over 24 wt% are achievable with paraffinic feeds. A traditional reactor/regenerator unit design uses a catalyst with physical properties similar to traditional FCC catalyst. The DCC unit may be operated in two operational modes: maximum propylene (Type I) or maximum iso-olefins (Type II).

Each operational mode utilizes unique catalyst as well as reaction conditions. Maximum propylene DCC uses both riser and bed cracking at severe reactor conditions while Type II DDC uses only riser cracking like a modern FCC unit at milder conditions.

The overall flow scheme of DCC is very similar to that of a conventional FCC. However, innovations in the areas of catalyst development, process variable selection and severity and gas plant design enables the DCC to produce significantly more olefins than FCC in a maximum olefins mode of operation.

This technology is quite suitable for revamps as well as grassroot applications. Feed enters the unit through nozzles proprietary feed as shown in the schematic. Integrating DCC technology into existing refineries as either a grassroots or revamp application can offer an attractive opportunity to produce large quantities of light olefins.

In a market requiring both proplylene and ethylene, use of both thermal and catalytic processes is essential, due to the fundamental differences in the reaction mechanisms involved. The combination of thermal and catalytic cracking mechanisms is the only way to increase total olefins from heavier feeds while meeting the need for an increased propylene to ethylene ratio. The integrated DCC/steam cracking complex offers significant capital savings over a conventional standalone refinery for propylene production.

Products (wt%)	DCC Type I	DCC Type II	FCC
Ethylene	6.1	2.3	0.9
Propylene	20.5	14.3	6.8
Butylene	14.3	14.6	11.0
in which $IC_4^=$	5.4	6.1	3.3
Amylene	—	9.8	8.5
in which $IC_5^=$	—	6.5	4.3

Installation: Five units are curently operating in China and one in Thailand. Several more units are under design in China.

Reference: Chapin, Letzsch and Zaiting, "Petrochemical options from deep catalytic cracking and the FCCU," NPRA Annual Meeting , March 1998.

Licensor: Stone & Webster Inc., a Shaw Group Co., and Research Institute of Petroleum Processing, Sinopec

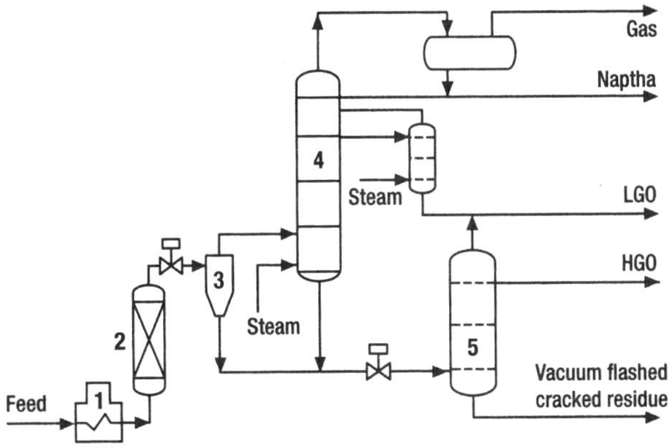

Deep thermal conversion

Application: The Shell Deep Thermal Conversion process closes the gap between visbreaking and coking. The process yields a maximum of distillates by applying deep thermal conversion of the vacuum residue feed and by vacuum flashing the cracked residue. High-distillate yields are obtained, while still producing a stable liquid residual product, referred to as liquid coke. The liquid coke, not suitable for blending to commercial fuel, is used for speciality products, gasification and/or combustion, e.g., to generate power and/or hydrogen.

Description: The preheated short residue is charged to the heater (1) and from there to the soaker (2), where the deep conversion takes place. The conversion is maximized by controlling the operating temperature and pressure. The soaker effluent is routed to a cyclone (3). The cyclone overheads are charged to an atmospheric fractionator (4) to produce the desired products like gas, LPG, naphtha, kero and gasoil. The cyclone and fractionator bottoms are subsequently routed to a vacuum flasher (5), which recovers additional gasoil and waxy distillate. The residual liquid coke is routed for further processing depending on the outlet.

Yields: Depend on feed type and product specifications.

Feed, vacuum residue	Middle East
Viscosity, cSt @ 100°C	770
Products in % wt. on feed	
Gas	4.0
Gasoline ECP 165°C	8.0
Gas oil ECP 350°C	18.1
Waxy distillate ECP 520°C	22.5
Residue ECP 520°C+	47.4

Economics: The investment ranges from 1,300 to 1,600 U.S.$/bbl installed excl. treating facilities and depending on the capacity and configuration (basis: 1998)

Utilities, typical per bbl @ 180°C	
Fuel, Mcal	26
Electricity, kWh	0.5
Net steam production, kg	20
Water, cooling, m^3	0.15

Installation: To date, four Deep Thermal Conversion units have been licensed. In two cases this involved a revamp of an existing Shell Soaker Visbreaker unit. In addition, two units are planned for revamp, while one grassroots unit is currently under construction. Post startup services and technical services on existing units are available from Shell.

Reference: Visbreaking Technology, Erdöl and Kohle, January 1986.

Licensor: Shell Global Solutions International B.V. and ABB Lummus Global B.V.

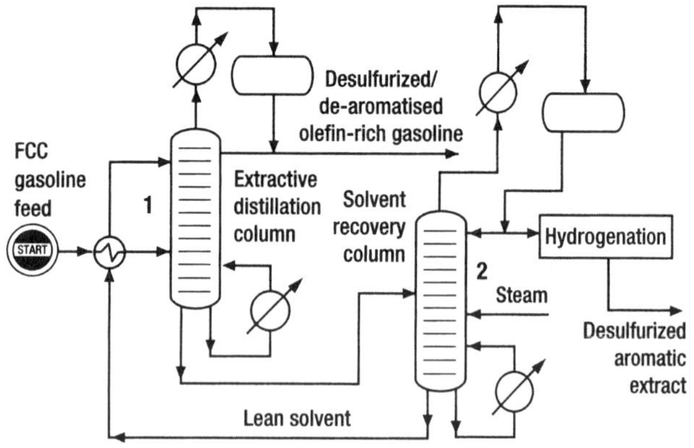

Desulfurization

Application: GT-DeSulf addresses overall plant profitability by desulfurizing the FCC stream with no octane loss and decreased hydrogen consumption by using a proprietary solvent in an extractive distillation system. This process also recovers valuable aromatics compounds.

Description: FCC gasoline, with endpoint up to 210°C, is fed to the GT-DeSulf unit, which extracts sulfur and aromatics from the hydrocarbon stream. The sulfur and aromatic components are processed in a conventional hydrotreater to convert the sulfur into H_2S. Because the portion of gasoline being hydrotreated is reduced in volume and free of olefins, hydrogen consumption and operating costs are greatly reduced. In contrast, conventional desulfurization schemes process the majority of the gasoline through hydrotreating and caustic-washing units to eliminate the sulfur. That method inevitably results in olefin saturation, octane downgrade and yield loss.

GT-DeSulf has these advantages:

- Segregates and eliminates FCC-gasoline sulfur species to meet a pool gasoline target of 20 ppm
- Preserves more than 90% of the olefins from being hydrotreated in the HDS unit; and thus, prevents significant octane loss and reduces hydrogen consumption
- Fewer components (only those boiling higher than 210°C and the aromatic concentrate from ED unit) are sent to the HDS unit; consequently, a smaller HDS unit is needed and there is less yield loss
- High-purity BTX products can be produced from the aromatic-rich extract stream after hydrotreating
- Olefin-rich raffinate stream (from the ED unit) can be recycled to the FCC unit to increase the light olefin production.

FCC gasoline is fed to the extractive distillation column (EDC). In a vapor-liquid operation, the solvent extracts the sulfur compounds into the bottoms of the column along with the aromatic components, while rejecting the olefins and nonaromatics into the overhead as raffinate. Nearly all of the nonaromatics, including olefins, are effectively separated into the raffinate stream. The raffinate stream can be optionally caustic washed before routing to the gasoline pool, or to a $C_3^=$ producing unit.

Rich solvent, containing aromatics and sulfur compounds, is routed to the solvent recovery column, (SRC), where the hydrocarbons and sulfur species are separated, and lean solvent is recovered in columns bottoms. The SRC overhead is hydrotreated by conventional means and used as desulfurized gasoline, or processed through an aromatics recovery unit. Lean solvent from the SRC bottoms are treated and recycled back to the EDC.

Economics: Estimated installed cost of $1,000/bpd of feed and production cost of $0.50/bbl of feed for desulfurization and dearomatization.

Licensor: GTC Technology Inc.

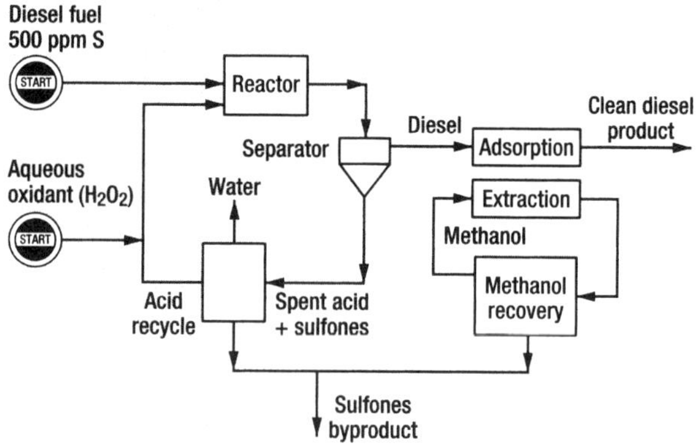

Desulfurization

Application: To produce ultra-low sulfur fuels, having less than 10 ppm sulfur, from distillate feeds containing 20 to 3,000 ppm sulfur. The UniPure ASR-2 process is based on oxidation chemistry. It requires no hydrogen and uses no fired heaters. The process also reduces nitrogen compounds to ultra-low levels.

Applications are anticipated in refineries and stand-alone plants. Skid-mounted or truck-mounted units can remediate off spec products at distribution terminals. Extensions are under development for other refinery hydrocarbon streams and for lube oils.

Description: Diesel fuel, or other distillate feed, is introduced at about 200°F into the oxidation reactor operating at about 1 bar pressure. An aqueous oxidizing solution comprised primarily of recycled formic acid containing a small amount of hydrogen peroxide and water is also introduced into the reactor. After a short residence time, the sulfur species are completely oxidized to the corresponding sulfones. The acid extracts about half of the oxidized sulfur compounds and is separated from the hydrocarbon in a gravity separator. Spent acid and sulfones are processed further to reject the sulfones and regenerate the acid by removing water introduced in the process.

The oxidized diesel from the gravity separator, which contains no residual peroxide, is water-washed, dried and then passed over a solid alumina adsorbing bed to extract the remaining sulfones. The product stream typically has less than 5 ppm sulfur. Two alumina columns are operated in cycles. While one is being used for adsorption of oxidized sulfur, the other is regenerated with methanol. The methanol extract containing sulfones is then flash distilled to separate the methanol from the mixed sulfones. The sulfones recovered from the alumina extraction are combined with those recovered from the spent acid to form a small byproduct stream.

Product quality: Product properties other than sulfur and nitrogen are virtually unchanged. Initial indications are that diesel lubricity is not reduced. The process can achieve sulfur and nitrogen levels below 1 ppm. The diesel product is usually water white.

Economics: (basis: 25,000-bpsd unit, 500 ppm S feed)

Investment, $/bpsd	1,000
Operating cost (utilities and reagents), $/bbl	0.70–0.90

Commercialization status: A demonstration plant with a 50- bpd capacity will start up at a Gulf Coast refinery by early 2003. Commercial readiness for diesel is anticipated before mid-2003.

Reference: *Hydrocarbon Engineering,* Vol. 7, No. 7, July 2002, pp. 25–28.

Licensor: UniPure Corp.

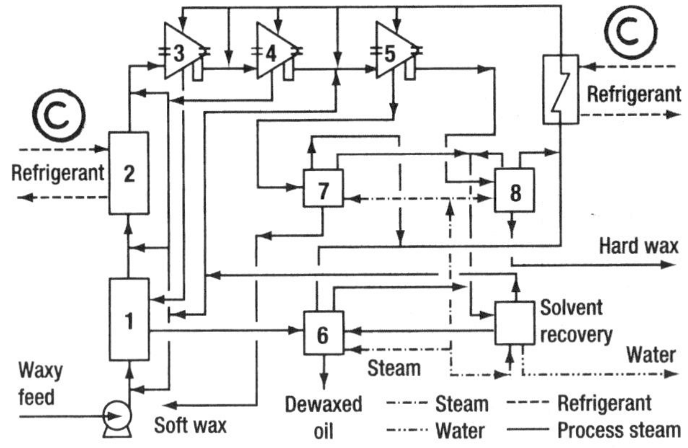

Dewaxing/wax deoiling

Application: Bechtel's dewaxing/wax fractionation processes remove waxy components from lubrication base-oil streams to simultaneously meet desired low-temperture properties for dewaxed oils and produce hard wax as a premium byproduct.

Description: The two-stage, solvent-dewaxing process can be expanded to simultaneously produce hard wax by adding a third deoiling stage using the wax fractionation process.

Waxy feedstock (raffinate, distillate or deasphalted oil) is mixed with a binary-solvent system and chilled in a very closely controlled manner in scraped-surface double-pipe exchangers (1) and refrigerated chillers (2) to form a wax/oil/solvent slurry. This slurry is filtered through the primary-filter stage (3) and dewaxed-oil mixture is routed to the dewaxed-oil-recovery section (6) for separation of solvent from the oil. Prior to solvent recovery, the primary filtrate is used to cool the feed/solvent mixture (1). Wax from the primary stage is slurried with cold solvent and filtered again in the repulp filter (4) to reduce the oil content to approximately 10%. The repulp filtrate is reused as dilution solvent in the feed-chilling train.

The low-oil-content slack wax is then warmed by mixing with warm solvent to melt the low-melting-point waxes (soft wax) and is filtered in a third stage of filtration (5) to separate the hard wax from the soft wax. The hard and soft wax mixtures are each routed to solvent-recovery sections (7,8) to strip solvent from the product streams (hard wax and soft wax). The recovered solvent is collected, dried and recycled back to the chilling and filtration sections.

Economics:

Investment (basis: 7,000-bpsd feedrate capacity, 2002 U.S. Gulf Coast), \$/bpsd	10,500
Utilities, typical per bbl feed:	
Fuel, 10^3 Btu (absorbed)	280
Electricity, kWh	46
Steam, lb	60
Water, cooling (25°F rise), gal	1,500

Installation: Seven in service.

Licensor: Bechtel Corp.

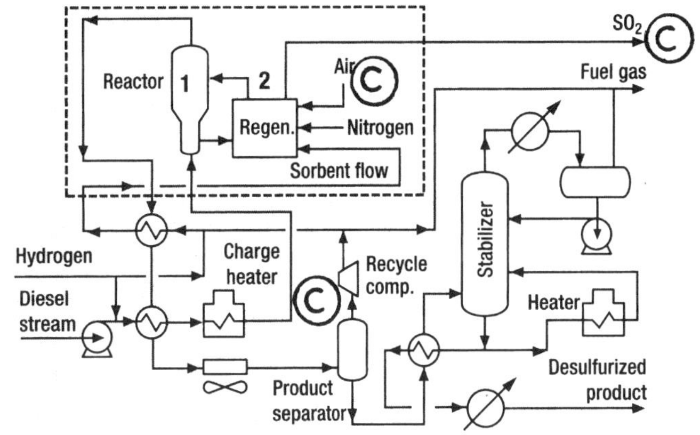

Diesel desulfurization

Application: Convert high-sulfur diesel streams into a very-low sulfur diesel product using S Zorb sulfur removal technology.

Products: A "zero" sulfur product for diesel motor fuels.

Description: Diesel-weight streams from a variety of refinery sources is combined with a small hydrogen stream and heated. Stream enters the expanded fluid-bed reactor (1), where the proprietary sorbent removes sulfur from the feed. The process can be designed to operate with no net chemical hydrogen consumption.

Regeneration: The sorbent is continuously withdrawn from the reactor and transferred to the regenerator section (2), where the sulfur is removed as SO_2. The cleansed sorbent is reconditioned and returned to the reactor. The rate of sorbent circulation is controlled to help maintain the desired sulfur concentration in the product.

General operating conditions:

Reactor temperature, °F	700–800
Reactor pressure, psig	275–500

The following example case studies show the performance of S Zorb technology for processing various distillate streams.

Feed properties:

Sulfur ppm	API gravity	D86 IBP, °F	10%	50%	90%
523	33.20	385	440	513	604
460	36.05	346	402	492	573
2,000	41.27	291	318	401	496
2,400	20.38	409	480	537	611

Product properties:

Sulfur, ppm	API gravity	D86 IBP, °F	10%	50%	90%
6	33.22	380	438	513	603
<1	36.23	347	403	491	574
<1	41.51	290	317	400	495
10	21.99	409	480	537	611

Operating conditions:

LHSV	H_2 consumption, scf/bbl
2	–5
2	–15
6	42
1	186

Installation: Licensed for use at 28 sites, as of 2Q 2002.

Licensor: Fuels Technology Division of ConocoPhillips Co.

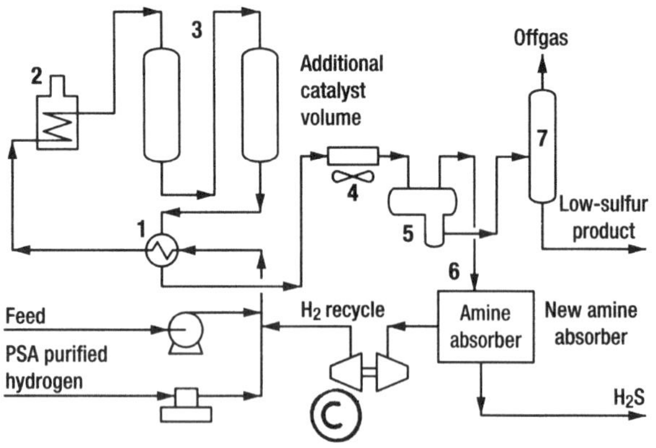

Diesel hydrotreatment

Application: Produce ultra-low sulfur diesel and high quality diesel fuel (low aromatics, high cetane) via Prime-D toolbox of proven state-of-the art technology, catalysts and services.

Description: In the basic process as shown above, feed and hydrogen are heated in the feed-reactor effluent exchanger (1) and furnace (2) and enter the reaction section (3), with possible added volume for revamp cases. The reaction effluent is cooled by exchanger (1) and air cooler (4) and separated in the separator (5). The hydrogen-rich gas phase is treated in an existing or new amine absorber for H_2S removal (6) and recycled to the reactor. The liquid phase is sent to the stripper (7) where small amounts of gas and naphtha are removed and high-quality product diesel is recovered.

Whether the need is for a new unit or for maximum reuse of existing diesel HDS units, the Prime-D hydrotreating toolbox of solutions meets the challenge. Process objectives ranging from low-sulfur, ultra-low sulfur, low-aromatics, and/or high cetane number are met with minimum cost by:

- Selection of the proper catalyst from the HR 400 series, based on the feed analysis and processing objectives. HR 400 catalysts cover the range of ULSD requirements with highly active and stable catalysts. HR 426 CoMo exhibits high desulfurization rates at low-to-medium pressures, HR 448 NiMo has higher hydrogenation activity at higher pressures, and HR 468 NiCoMo is very effective for ULSD in the case of moderate pressures.
- Use of proven, efficient reactor internals, EquiFlow, that allow near-perfect gas and liquid distribution and outstanding radial temperature profiles.
- Loading catalyst in the reactor(s) with the Catapac dense loading technique for up to 20% more reactor capacity. Over 8,000 tons of catalyst have been loaded quickly and safely in recent years using the Catapac technique.
- Application of Advanced Process Control for dependable operation and longer catalyst life.
- Sound engineering design based on years of R&D, process design and technical service feed-back to ensure the right application of the right technology for new and revamp projects.

Whatever the diesel quality goals—ULSD, high cetane or low aromatics—Prime-D's Hydrotreating Toolbox approach will attain your goals in a cost-effective manner.

Installation: Over 100 middle distillate hydrotreaters have been licensed or revamped. They include 23 low- and ultra-low sulfur diesel units (< 50 ppm), as well as a number of cetane boosting units. Most of those units are equipped with Equiflow internals.

References: "Getting Total Performance with Hydrotreating," *Petroleum Technology Quarterly*, Spring 2002.

"Premium Performance Hydrotreating with Axens HR 400 Series Hydrotreating Catalysts," NPRA Annual Meeting, March 2002, San Antonio.

"The Hydrotreating Toolbox Approach", *Hart's European Fuel News*, May 29, 2002.

Licensor: Axens, Axens NA.

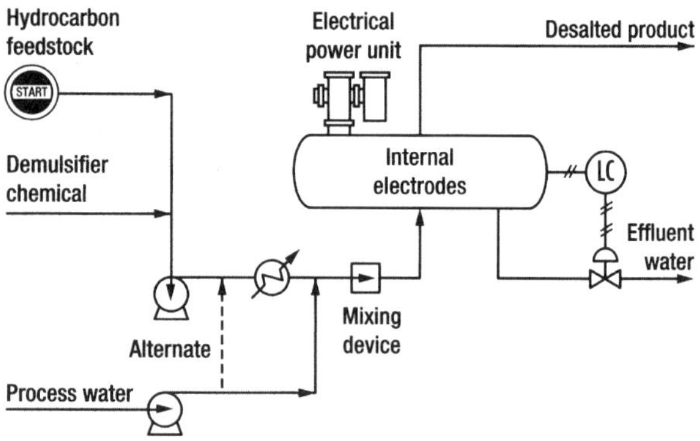

Electrical desalting

Application: For removal of undesirable impurities such as salt, water, suspended solids and metallic contaminants from unrefined crude oil, residuums and FCC feedstocks.

Description: Salts such as sodium, calcium and magnesium chlorides are generally contained in the residual water suspended in the oil phase of hydrocarbon feedstocks. All feedstocks also contain, as mechanical suspensions, such impurities as silt, iron oxides, sand and crystalline salt. These undesirable components can be removed from hydrocarbon feedstocks by dissolving them in washwater or causing them to be water-wetted. Emulsion formation is the best way to produce highly intimate contact between the oil and washwater phases.

The electrical desalting process consists of adding process (wash) water to the feedstock, generating an emulsion to assure maximum contact and then utilizing a highly efficient AC electrical field to resolve the emulsion. The impurity-laden water phase can then be easily withdrawn as underflow.

Depending on the characteristics of the hydrocarbon feedstock being processed, optimum desalting temperatures will be in the range of 150°F to 300°F. For unrefined crude feedstocks, the desalter is located in the crude unit preheat train such that the desired temperature is achieved by heat exchange with the crude unit products or pumparound reflux. Washwater, usually 3 to 6 vol%, is added upstream and/or downstream of the heat exchanger(s). The combined streams pass through a mixing device thereby creating a stable water-in-oil emulsion. Properties of the emulsion are controlled by adjusting the pressure drop across the mixing device.

The emulsion enters the desalter vessel where it is subjected to a high voltage electrostatic field. The electrostatic field causes the dispersed water droplets to coalesce, agglomerate and settle to the lower portion of the vessel. The water phase, containing the various impurities removed from the hydrocarbon feedstock, is continuously discharged to the effluent system. A portion of the water stream may be recycled back to the desalter to assist in water conservation efforts. Clean, desalted hydrocarbon product flows from the top of the desalter vessel to subsequent processing facilities.

Desalting and dehydration efficiency of the oil phase is enhanced by using EDGE (Enhanced Deep-Grid Electrode) technology which creates both high and low intensity AC electrical fields inside the vessel. Demulsifying chemicals may be used in small quantities to assist in oil/water separation and to assure low oil contents in the effluent water.

Licensor: Howe-Baker Engineers, Ltd., a subsidiary of Chicago Bridge & Iron Co.

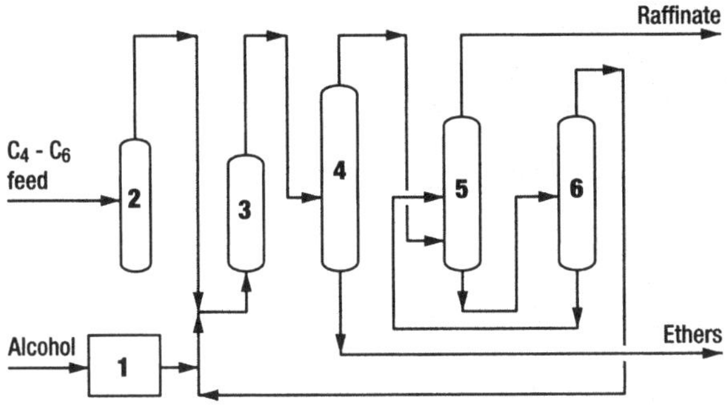

Ethers

Application: Production of high-octane reformulated gasoline components (MTBE, ETBE, TAME and/or higher molecular-weight ethers) from C_1 to C_2 alcohols and reactive hydrocarbons in C_4 to C_6 cuts.

Description: Different arrangements have been demonstrated depending on the nature of the feeds. All use acid resins in the reaction section. The process includes alcohol purification (1), hydrocarbon purification (2), followed by the main reaction section. This main reactor (3) operates under adiabatic upflow conditions using an expanded-bed technology and cooled recycle. Reactants are converted at moderate well-controlled temperatures and moderate pressures, maximizing yield and catalyst life. The main effluents are purified for further applications or recycle.

More than 90% of the total per pass conversion occurs in the expanded-bed reactor. The effluent then flows to a reactive distillation system (4), Catacol. This system, operated like a conventional distillation column, combines catalysis and distillation. The catalytic zones of the Catacol use fixed-bed arrangements of an inexpensive acidic resin catalyst that is available in bulk quantities and easy to load and unload.

The last part of the unit removes alcohol from the crude raffinate using a conventional waterwash system (5) and a standard distillation column (6).

Yields: Ether yields are not only highly dependent on the reactive olefins' content and the alcohol's chemical structure, but also on operating goals: maximum ether production and/or high final raffinate purity (for instance, for downstream 1-butene extraction) are achieved.

Economics: Plants and their operations are simple. The same inexpensive (purchased in bulk quantities) and long lived, non-sophisticated catalysts are used in the main reactor section catalytic region of the Catacol column, if any.

Installation: Over 25 units, including ETBE and TAME, have been licensed. Twenty-four units, including four Catacol units, are in operation.

Licensor: Axens, Axens NA.

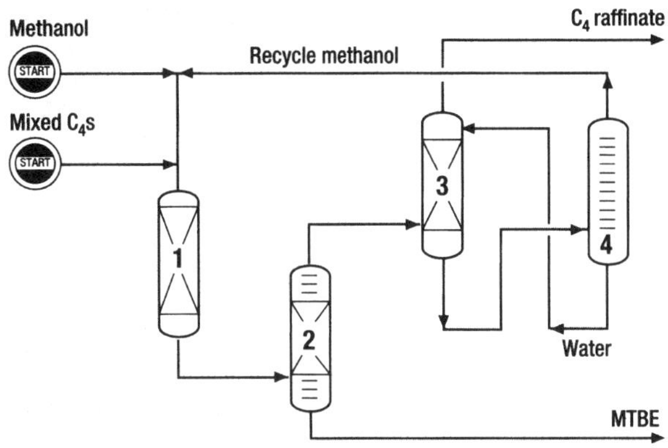

Ethers

Application: To produce high-octane, low-vapor-pressure oxygenates such as methyl tertiary butyl ether (MTBE), tertiary amyl methyl ether (TAME) or heavier tertiary ethers for gasoline blending to reduce olefin content and/or meet oxygen/octane/vapor pressure specifications. The processes use boiling-point/tubular reactor and catalytic distillation (CD) technologies to react methanol (MeOH) or ethanol with tertiary isoolefins to produce respective ethers.

Description: For an MTBE unit, the process can be described as follows. Process description is similar for production of heavier ethers. The C_4s and methanol are fed to the boiling-point reactor (1)—a fixed-bed, downflow adiabatic reactor. In the reactor, the liquid is heated to its boiling point by the heat of reaction, and limited vaporization occurs. System pressure is controlled to set the boiling point of the reactor contents and hence, the maximum temperatures. An isothermal tubular reactor is used, when optimum, to allow maximum temperature control. The equilibrium-converted reactor effluent flows to the CD column (2) where the reaction continues. Concurrently, MTBE is separated from unreacted C_4s as the bottom product.

This scheme can provide overall isobutylene conversions up to 99.99%. Heat input to the column is reduced due to the heat produced in the boiling-point reactor and reaction zone. Over time, the boiling-reactor catalyst loses activity. As the boiling-point reactor conversion decreases, the CD reaction column recovers lost conversion, so that high overall conversion is sustained. CD column overhead is washed in an extraction column (3) with a countercurrent water stream to extract methanol. The water extract stream is sent to a methanol recovery column (4) to recycle both methanol and water.

C_4s ex-FCCU require a well-designed feed waterwash to remove catalyst poisons for economic catalyst life and MTBE production.

Conversion: The information below is for 98% isobutylene conversion, typical for refinery feedstocks. Conversion is slightly less for ETBE than for MTBE. For TAME and TAEE, isoamylene conversions of 95%+ are achievable. For heavier ethers, conversion to equilibrium limits are achieved.

Economics: Based on a 1,500-bpsd MTBE unit, (6,460-bpsd C_4s ex-FCCU, 19% vol. isobutylene, 520-bpsd MeOH feeds) located on the U.S. Gulf Coast, the inside battery limits investment is:

Investment, $ per bpsd of MTBE product	3,500
Typical utility requirements, per bbl of product	
Electricity, kWh	0.5
Steam, 150-psig, lb	210
Steam, 50-psig, lb	35
Water, cooling (30° F rise), gal	1,050

Installation: Over 60 units are in operation using catalytic distillation to produce MTBE, TAME and ETBE. More than 100 ether projects have been awarded to CDTECH since the first unit came onstream in 1981. Snamprogetti has over 20 operating ether units using tubular reactors.

Licensor: CDTECH (CDTECH and Snamprogetti are cooperating to further develop and license their ether technologies.)

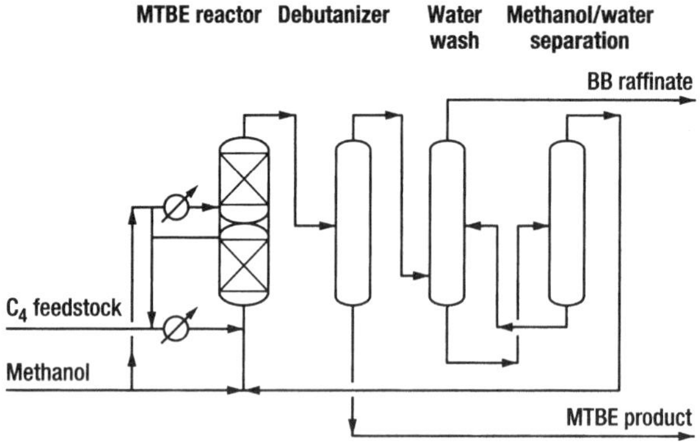

Ethers—MTBE

Application: The Uhde EDELEANU MTBE process combines methanol and isobutene to produce the high-octane oxygenate—methyl tertiary butyl ether (MTBE).

Feeds: C_4-cuts from steam cracker and FCC units with isobutene contents range from 12% to 30%.

Products: MTBE and other tertiary alkyl ethers are primarily used in gasoline blending as an octane enhancer to improve hydrocarbon combustion efficiency.

Description: The Uhde Edeleanu technology features a two-stage reactor system of which the first reactor is operated in the recycle mode. With this method, a slight expansion of the catalyst bed is achieved which ensures very uniform concentration profiles within the reactor and, most important, avoids hot spot formation. Undesired side reactions such as the formation of dimethyl ether (DME) is minimized.

The reactor inlet temperature ranges from 45°C at start-of-run to about 60°C at end-of-run conditions. One important factor of the two-stage system is that the catalyst may be replaced in each reactor separately, without shutting down the MTBE unit.

The catalyst used in this process is a cation-exchange resin and is available from several catalyst manufacturers. Isobutene conversions of 97% are typical for FCC feedstocks. Higher conversions are attainable when processing steam-cracker C_4-cuts that contain isobutene concentrations of 25%.

MTBE is recovered as the bottoms product of the distillation unit. The methanol-rich C_4-distillate is sent to the methanol-recovery section. Water is used to extract excess methanol and recycle it back to process. The isobutene-depleted C_4-stream may be sent to a raffinate stripper or to a molsieve-based unit to remove other oxygenates such as DME, MTBE, methanol and tert-butanol.

Very high isobutene conversion, in excess of 99%, can be achieved through a debutanizer column with structured packings containing additional catalyst. This reactive distillation technique is particularly suited when the raffinate-stream from the MTBE unit will be used to produce a high-purity butene-1 product.

For a C_4-cut containing 22% isobutene, the isobutene conversion may exceed 98% at a selectivity for MTBE of 99.5%.

Utility requirements, (C_4-feed containing 21% isobutene; per ton of MTBE):

Steam, MP, kg	100
Electricity, kWh	35
Water, cooling, m^3	15
Steam, LP, kg	900

Installation: Uhde Edeleanu's proprietary MTBE process has been successfully applied in five refineries. The accumulated licensed capacity exceeds 1 MMtpy.

Licensor: Uhde Edeleanu GmbH.

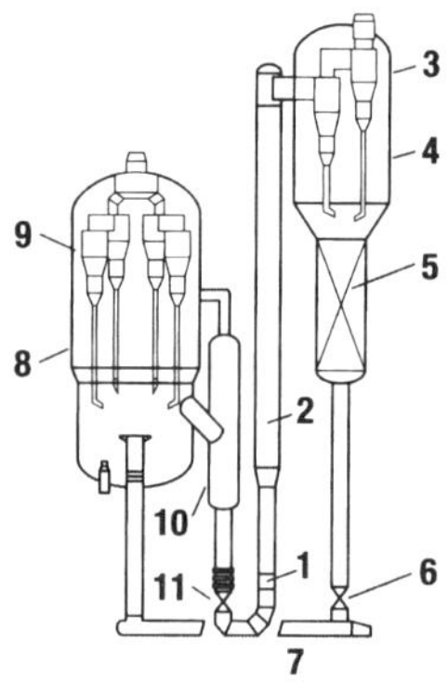

Fluid catalytic cracking

Application: Selective conversion of a wide range of gas oils into high-value products. Typical feedstocks are virgin or hydrotreated gas oils but may also include lube oil extract, coker gas oil and resid.

Products: High-octane gasoline, light olefins and distillate. Flexibillity of mode of operation allows for maximizing the most desirable product. The new Selective Component Cracking (SCC) technology maximizes propylene production.

Description: The Lummus process incorporates an advanced reaction system, high-efficiency catalyst stripper and a mechanically robust, single-stage fast fluidized bed regenerator. Oil is injected into the base of the riser via proprietary Micro-Jet feed injection nozzles (1). Catalyst and oil vapor flow upwards through a short-contact time, all-vertical riser (2) where raw oil feedstock is cracked under optimum conditions.

Reaction products exiting the riser are separated from the spent catalyst in a patented, direct-coupled cyclone system (3). Product vapors are routed directly to fractionation, thereby eliminating nonselective, post-riser cracking and maintaining the optimum product yield slate. Spent catalyst containing only minute quantities of hydrocarbon is discharged from the diplegs of the direct-coupled cyclones into the cyclone containment vessel (4). The catalyst flows down into the stripper (5).

Trace hydrocarbons entrained with spent catalyst are removed in the stripper using stripping steam. The net stripper vapors are routed to the fractionator via specially designed vents in the direct-coupled cyclones. Catalyst from the stripper flows down the spent-catalyst standpipe and through the slide valve (6). The spent catalyst is then transported in dilute phase to the center of the regenerator (8) through a unique square-bend-spent catalyst transfer line (7). This arrangement provides the lowest overall unit elevation. Catalyst is regenerated by efficient contacting with air for complete combustion of coke. For resid-containing feeds, the optional catalyst cooler is integrated with the regenerator. The resulting flue gas exists via cyclones (9) to energy recovery/flue gas treating. The hot regenerated catalyst is withdrawn via an external withdrawal well (10). The well allows independent optimization of catalyst density in the regenerated catalyst standpipe, maximizes slide valve (11) pressure drop and ensures stable catalyst flow back to the riser feed injection zone.

Economics:

Investment (basis: 30,000 bpsd including reaction/regeneration system and product recovery. Excluding offsites, power recovery and flue gas scrubbing U.S. Gulf Coast 2001.)

$/bpsd (typical)	2,200–3,000

Utilities, typical per bbl fresh feed:

Electricity, kWh	0.8–1.0
Steam, 600 psig (produced)	50–200
Maintenance, % of investment per year	2–3

Installation: Fourteen grassroots units in operation and one in design stage. Fifteen units revamped and two in design stage.

Licensor: ABB Lummus Global Inc.

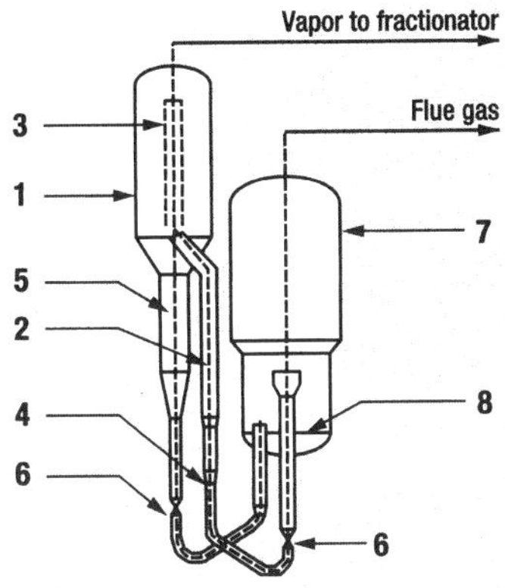

Fluid catalytic cracking

Application: FLEXICRACKING IIIR converts high-boiling hydrocarbons including residues, gasoils, lube extracts, and/or deasphalted oils to higher value products.

Products: Light olefins for gasoline processes and petrochemicals, LPG, blend stocks for high-octane gasoline, distillates, and fuel oils.

Description: The FLEXICRACKING IIIR technology includes process design, hardware details, special mechanical and safety features, control systems, flue gas processing options, and a full range of technical services and support. The reactor (1) incorporates many features to enhance performance, reliability, and flexibility, including a riser (2) with patented high efficiency close-coupled riser termination (3), enhanced feed injection system (4), and efficient stripper design (5). The reactor design and operation maximizes the selectivity of desired products, such as naphtha and propylene.

The technology uses an improved catalyst circulation system with advanced control features, including cold-walled slide valves (6). The single vessel regenerator (7) has proprietary process and mechanical features for maximum reliability and efficient air/catalyst distribution and contacting (8). Either full or partial combustion is used. With increasing residue processing and the need for additional heat balance control, partial burn operation with outboard CO combustion is possible, or KBR dense phase catalyst cooler technology may be applied. The ExxonMobil wet gas scrubbing or the ExxonMobil-KBR Cyclofines TSS technologies can meet flue gas emission requirements.

Yields: Typical examples:

	Resid feed mogas operation	VGO+lube extracts distillate operation	VGO feed mogas operation
Feed			
Gravity, °API	22.9	22.2	25.4
Con carbon, wt%	3.9	0.7	0.4
Quality	80% Atm. Resid (Hydrotreated)	20% Lube Extracts	50% TBP – 794°F
Product yields			
Naphtha, lv% ff	78.2	40.6	77.6
(C_4/FBP)	(C_4/430°F)	(C_4/260°F)	(C_4/430°F)
Mid Dist., lv% ff	13.7	49.5	19.2
(IBP/FBP)	(430/645°F)	(260/745°F)	(430/629°F)

Installation: More than 70 units with a design capacity of over 2.5-million bpd fresh feed.

References: Ladwig, P. K., "Exxon FLEXICRACKING IIIR fluid catalytic cracking technology," *Handbook of Petroleum Refining Processes,* Second Ed., R. A. Meyers, Ed., pp. 3.3–3.28.

Licensor: ExxonMobil Research & Engineering Co. and Kellogg Brown & Root, Inc.

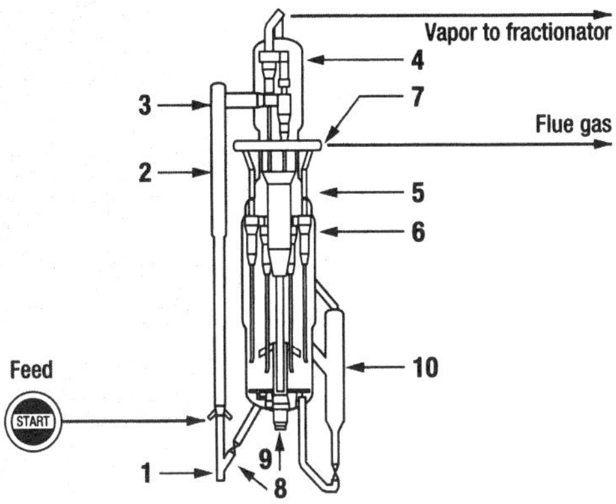

Fluid catalytic cracking

Application: Conversion of gas oils and residues to high-value products using the efficient and flexible Orthoflow catalytic cracking process.

Products: Light olefins, high-octane gasoline, and distillate.

Description: The converter is a one-piece modularized unit that efficiently combines KBR's proven Orthoflow features with ExxonMobil's advanced design features. Regenerated catalyst flows through a wye (1) to the base of the external vertical riser (2). Feed enters through the proprietary ATOMAX-2 feed injection system. Reaction vapors pass through a patented right-angle turn (3) and are quickly separated from the catalyst in a patented closed-cyclone system (4). Spent catalyst flows through a two-stage stripper equipped with DynaFlux baffles (5) to the regenerator (6) where advanced catalyst distribution and air distribution are used. Either partial or complete CO combustion may be used in the regenerator, depending on the coke-forming tendency of the feedstock. The system uses a patented external flue gas plenum (7) to improve mechanical reliability. Catalyst flow is controlled by one slide valve (8) and one plug valve (9). An advanced dense-phase catalyst cooler (10) is used to optimize profitability when heavier feeds are processed.

Economics:

Investment (basis: 50,000-bpsd fresh feed including converter, fractionator, vapor recovery and amine treating, but not power recovery; battery limit, direct material and labor, 2002 Gulf Coast)	
$ per bpsd	1,950–2,150
Utilities, typical per bbl fresh feed	
Electricity, kWh	0.7–1.0
Steam, 600 psig (produced) lb	40–200
Catalyst, makeup, lb/bbl	0.10–0.15
Maintenance, % of plant replacement cost/yr	3

Installation: More than 150, resulting in a total of over 4 million bpd fresh feed, with 20 designed in the past 12 years.

References: "New developments in FCC feed injection and stripping technologies," NPRA 2000 Annual Meeting, March 2000.

"RegenMax technology: staged combustion in a single regenerator," NPRA 1999 Annual Meeting, March 1999.

Licensor: Kellogg Brown & Root, Inc.

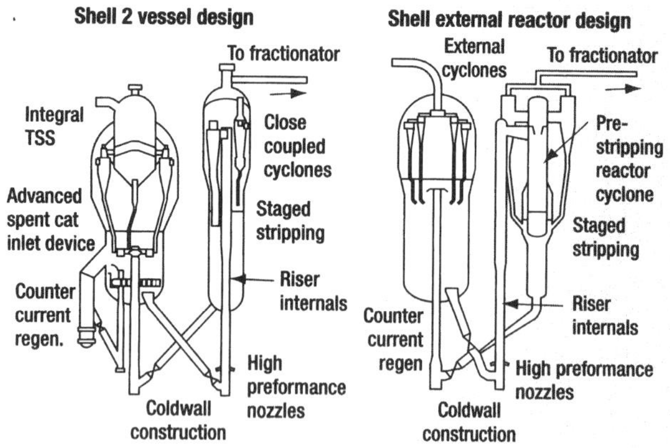

Fluid catalytic cracking

Application: The Shell FCC process converts heavy petroleum distillates and residues to high-value products. Profitability is increased by a reliable and robust process, which has flexibility to process heavy feeds and maximize product upgrading including propylene production where required.

Products: Light olefins, LPG, high-octane gasoline, distillate and propylene.

Description: Hydrocarbon is fed to a short contact-time-riser by Shell's high performance feed nozzle system, ensuring good mixing and rapid vaporisation. Proprietary riser internals lower pressure drop and reduce back mixing. The riser termination design provides rapid catalyst/hydrocarbon separation to maximise desired product yields and a staged stripper achieves low hydrogen in coke without excessive gas or coke formation. A single stage partial burn regenerator delivers excellent performance at low cost (full burn can also be applied). Cat coolers can be added for feedstock flexibility. Flue gas cleanup is by Shell's third stage separator and power recovery can be incorporated if justified.

There are currently two FCC design configurations. The Shell 2 Vessel design is recommended for feeds (including residue) with mild coking tendencies, the incorporation of reactor and regenerator elements within the vessels leads to low capital expenditure. The Shell External Reactor design is the preferred option for heavy feeds with high coking tendencies, delivering improved robustness.The pre-stripping cyclone positioned inside the rough-cut cyclones prevents post riser coke make and the external reactor design eliminates stagnant areas for coke growth. Cost effectiveness is achieved through a simple, low-elevation design. Proprietary catalyst circulation enhancement techniques are vital in achieving that. The designs have proven to be reliable due to incorporation of Shell's extensive operating experience.

Shell can also provide advanced distillation designs, advanced process control and optimizers as part of an integrated FCC design solution.

Installation: Shell has designed and licensed over 30 grassroots units, including seven for residue feed. Shell has revamped over 30 units, including the designs of other licensors. Shell has converted seven existing distillate units to residue operation. A Shell close-coupled riser termination system has been designed for 14 units, Shell's high performance feed nozzles for 15 units, catalyst circulation enhancement for 8 units and third-stage separators for 58 units. Many licenses are for non-Shell customers. Shell has over 1,000+ years of own FCC operational experience.

Reference: "Chapter FCC," *Handbook Fluidization,* Wen-Cing Yang, Ed., 2002.

"FCC cyclones—a vital elment in profitability," *Petroleum Technology Quarterly,* Spring, 2001.

"New advances in third-stage separators," *World Refining,* October 2000.

"Update on Shell Residue FCC Process and Operation," AIChE 1998 Spring Meeting."Design and Operation of Shell's Residue Catalytic Crackers in East Asia," ARTC 1998 Conference.

Licensor: Shell Global Solutions International B.V.

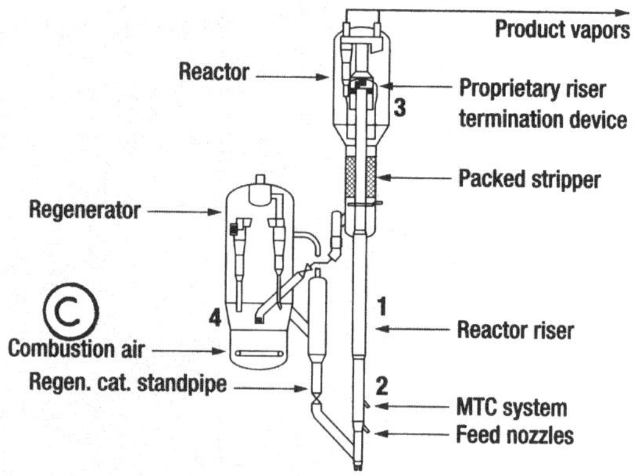

Fluid catalytic cracking

Application: Selective conversion of gas oil feedstocks.

Products: High-octane gasoline, distillate and C_3–C_4 olefins.

Description: Catalytic and selective cracking in a short-contact-time riser (1) where oil feed is effectively dispersed and vaporized through a proprietary feed-injection system. Operation is carried out at a temperature consistent with targeted yields. The riser temperature profile can be optimized with the proprietary mixed temperature control (MTC) system (2). Reaction products exit the riser-reactor through a high-efficiency, close-coupled, proprietary riser termination device RS^2 (Riser Separator Stripper) (3). Spent catalyst is pre-stripped followed by an advanced high-efficiency packed stripper prior to regeneration. The reaction product vapor may be quenched using Amoco's proprietary technology to give the lowest possible dry gas and maximum gasoline yield. Final recovery of catalyst particles occurs in cyclones before the product vapor is transferred to the fractionation section.

Catalyst regeneration is carried out in a single regenerator (4) equipped with proprietary air and catalyst distribution systems, and may be operated for either full or partial CO combustion. Heat removal for heavier feedstocks may be accomplished by using reliable dense-phase catalyst cooler, which has been commercially proven in over 24 units and is licensed exclusively by Stone & Webster/Axens. As an alternative to catalyst cooling, this unit can easily be retrofitted to a two-regenerator system in the event that a future resid operation is desired.

The converter vessels use a cold-wall design that results in minimum capital investment and maximum mechanical reliability and safety. Reliable operation is ensured through the use of advanced fluidization technology combined with a proprietary reaction system. Unit design is tailored to the refiner's needs and can include wide turndown flexibility. Available options include power recovery, wasteheat recovery, flue gas treatment and slurry filtration. Revamps incorporating proprietary feed injection and riser termination devices and vapor quench result in substantial improvements in capacity, yields and feedstock flexibility within the mechanical limits of the existing unit.

Installation: Stone & Webster and Axens have licensed 26 full-technology units and performed more than 100 revamp projects.

Reference: Letzsch, W. S., "1999 FCC Technology Advances," 1999 Stone & Webster Eleventh Annual Refining Seminar at NPRA Q&A, Dallas, Oct. 5, 1999.

Licensor: Stone & Webster Inc., a Shaw Group Co./Axens, IFP Group Technologies

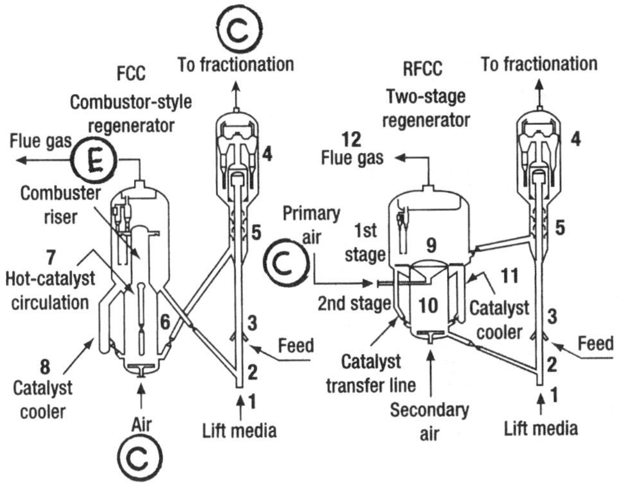

Fluid catalytic cracking

Application: Selectively convert gas oils and resid feedstocks into higher-value products using the FCC/RFCC/PETROFCC process.

Products: Light olefins (for alkylation, polymerization, etherification or petrochemicals), LPG, high-octane gasoline, distillates and fuel oils.

Description: The combustor-style unit is used to process gas oils and moderately contaminated resids, while the two-stage unit is used for more contaminated resids.

In either unit style, the reactor section is similar. A lift media of light hydrocarbons, steam or a mixture of both contacts regenerated catalyst at the base of the riser (1). This patented acceleration zone (2), with elevated Optimix feed distributors (3), enhances the yield structure by effectively contacting catalyst with finely atomized oil droplets.

The reactor zone features a short-contact-time riser and a state-of-the-art riser termination device (4) for quick separation of catalyst and vapor, with high hydrocarbon containment (VSS/VDS technology). This design offers high gasoline yields and selectivity with low dry gas yields. Steam is used in an annular stripper (5) to displace and remove entrained hydrocarbons from the catalyst. Existing units can be revamped to include these features (1–5).

The combustor-style regenerator (6) burns coke, in a fast-fluidized environment, completely to CO_2 with very low levels of CO. The circulation of hot catalyst (7) from the upper section to the combustor provides added control over the burn-zone temperature and kinetics and enhances radial mixing. Catalyst coolers (8) can be added to new and existing units to reduce catalyst temperature and increase unit flexibility for commercial operations of feeds up to 6 wt% Conradson carbon.

For heavier resid feeds, the two-stage regenerator is used. In the first stage, upper zone (9), the bulk of the carbon is burned from the catalyst, forming a mixture of CO and CO_2. Catalyst is transferred to the second stage, lower zone (10), where the remaining coke is burned in complete combustion, producing low levels of carbon on regenerated catalyst. A catalyst cooler (11) is located between the stages. This configuration maximizes oxygen use, requires only one train of cyclones and one flue gas stream (12), avoids costly multiple flue gas systems and creates a hydraulically-simple and well-circulating layout. The two-stage regenerator system has processed feeds up to 10 wt% Conradson carbon.

PETROFCC is a customized application using mechanical features such as RxCAT technology for recontacting carbonized catalyst, high-severity processing conditions and selected catalyst and additives to produe high yields of propylene, light olefins and aromatics for petrochemical applications.

Installations: All of UOP's technology and equipment are commercially proven for both process performance and mechanical reliability. UOP has been an active designer and licensor of FCC technology since the early 1940s and has licensed more than 210 FCC, Resid FCC, and MSCC process units. Today, more than 150 of these units are operating worldwide. In addition to applying our technology and skills to new units, UOP is also extensively involved in the revamping of existing units. During the past 15 years, UOP's FCC Engineering department has undertaken 40 to 60 revamp projects or studies per year.

Licensor: UOP LLC.

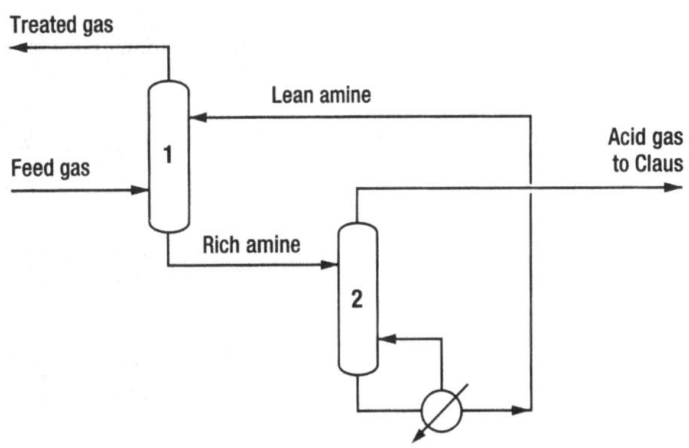

Gas treating—H_2S removal

Application: Remove H_2S selectively, or remove a group of acidic impurities (H_2S, CO_2, COS, CS_2 and mercaptans) from a variety of streams, depending on the solvent used. FLEXSORB SE technology has been used in refineries, natural gas production facilities and petrochemical operations.

FLEXSORB SE or SE Plus solvent is used on: hydrogenated Claus plant tail gas to give H_2S, ranging down to H_2S <10 ppmv; pipeline natural gas to give H_2S <0.25 gr/100 scf; or Flexicoking low Btu fuel gas. The resulting acid gas byproduct stream is rich in H_2S.

Hybrid FLEXSORB SE solvent is used to selectively remove H_2S, as well as organic sulfur impurities commonly found in natural gas.

FLEXSORB PS solvent yields a treated gas with: H_2S <0.25 gr/100 scf, CO_2 <50 ppmv, COS and CS_2 <1 ppmv, mercaptans removal >95%. This solvent is primarily aimed at natural gas or syngas cleanup. The byproduct stream is concentrated acid gases.

Description: A typical amine system flow scheme is used. The feed gas contacts the treating solvent in the absorber (1). The resulting rich solvent bottom stream is heated and sent to the regenerator (2). Regenerator heat is supplied by any suitable heat source. Lean solvent from the regenerator is sent through rich/lean solvent exchangers and coolers before returning to the absorber.

FLEXSORB SE solvent is an aqueous solution of a hindered amine. FLEXSORB SE Plus solvent is an enhanced aqueous solution, which has improved H_2S regenerability yielding <10 vppm H_2S in the treated gas. Hybrid FLEXSORB SE solvent is a hybrid solution containing FLEXSORB SE amine, a physical solvent and water. FLEXSORB PS solvent is a hybrid consisting of a different hindered amine, a physical solvent and water.

Economics: Lower investment and energy requirements based primarily on requiring 30% to 50% lower solution circulation rates.

Installations: Total gases treated by FLEXSORB solvents are about 2 billion scfd and the total sulfur recovery is about 900 long tpd.

FLEXSORB SE—26 plants operating, two startups in 2002, one in design

FLEXSORB SE Plus—14 plants operating, one startup in 2002, one in design

Hybrid FLEXSORB SE—two plants operating

FLEXSORB PS—four plants operating and one startup scheduled in 2000.

Reference: Garrison, J., et al, "Keyspan Energy Canada Rimbey acid gas enrichment with FLEXSORB SE Plus technology," 2002 Laurance Reid Gas Conditioning Conference, Norman, Oklahoma.

Adams-Smith, J., et al, Chevron USA Production Company Carter Creek Gas Plant FLEXSORB tail gas treating unit," 2002 GPA Annual Meeting, Dallas.

Licensor: ExxonMobil Research and Engineering Co.

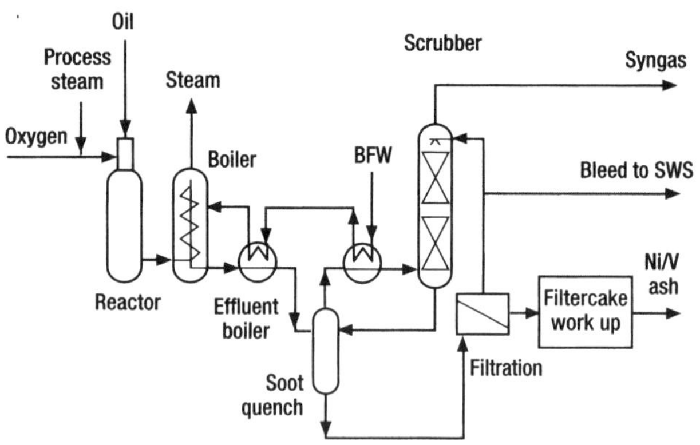

Gasification

Application: The Shell Gasification Process (SGP) converts the heaviest residual liquid hydrocarbon streams with high-sulfur and metals content into a clean synthesis gas and valuable metal oxides. Sulfur (S) is removed by normal gas treating processes and sold as elemental S.

The process converts residual streams with virtually zero value as fuel-blending components into valuable, clean gas and byproducts. This gas can be used to generate power in gas turbines and for making H_2 by the well-known shift and PSA technology. It is one of the few ultimate, environmentally acceptable solutions for residual hydrocarbon streams.

Products: Synthesis gas ($CO+H_2$), sulfur and metal oxides.

Process description: Liquid hydrocarbon feedstock (from very light such as natural gas to very heavy such as vacuum flashed cracked residue, VFCR and ashphalt) is fed into a reactor, and gasified with pure O_2 and steam. The net reaction is exothermic and produces a gas primarily containing CO and H_2. Depending on the final syngas application, operating pressures, ranging from atmospheric up to 65 bar, can easily be accommodated. SGP uses refractory-lined reactors that are fitted with both burners and a heat-recovery-steam generator, designed to produce high-pressure steam—over 100 bar (about 2.5 tons per ton feedstock). Gases leaving the steam generator are at a temperature approaching the steam temperature; thus further heat recovery occurrs in an economizer.

Soot (unconverted carbon) and ash are removed from the raw gas by a two-stage waterwash. After the final scrubbing, the gas is virtually particulate-free; it is then routed to a selective-acid-gas-removal system. Net water from the scrubber section is routed to the soot ash removal unit (SARU) to filter out soot and ash from the slurry. By controlled oxidation of the filtercake, the ash components are recovered as valuable oxides—principally vanadium pentoxide. The (clean) filtrate is returned to the scrubber.

A related process—the Shell Coal Gasification Process (SCGP)—gasifies solids such as coal or petroleum coke. The reactor is different, but main process layout and work-up are similar.

Installation: Over the past 40 years, more than 150 SGP units have been installed, that convert residue feedstock into synthesis gas for chemical applications. The latest, flagship installation is in the Shell Pernis refinery near Rotterdam, The Netherlands. This highly complex refinery depends on the SGP process for its H_2 supply. Similar projects are underway in India and Italy.

The Demkolec Power plant at Buggenum, The Netherlands produces 250 Mwe based on the SCGP process. The Shell middle distillate synthesis plant in Bintulu, Malaysia, uses SGP to convert 100 million scfd of natural gas into synthesis gas used for petrochemical applications.

Reference: "Shell Gasification Process," AIChE Spring National Meeting, March 5–9, 2000. "Shell Pernis Netherlands Refinery Residue Gasification Project," 1999 Gasification Technologies Conference, San Francisco, Oct. 17–20, 1999.

Licensor: Shell Global Solutions International B.V.

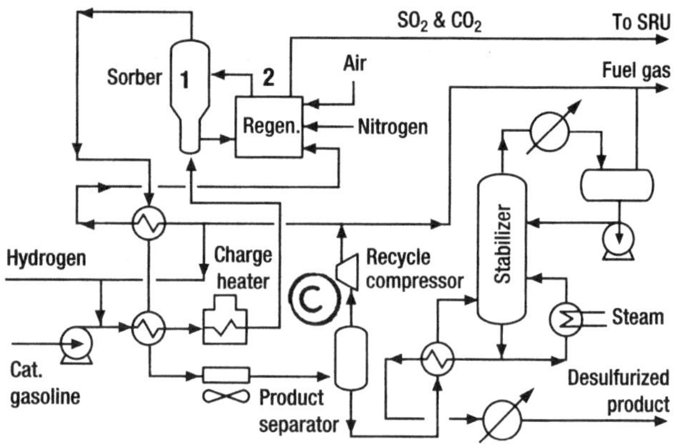

Gasoline desulfurization

Application: Convert high-sulfur gasoline streams into a low-sulfur gasoline blendstock using S Zorb sulfur-removal technology.

Products: A zero sulfur blending stock for gasoline motor fuels.

Description: Gasoline from the fluid catalytic cracker unit is combined with a small hydrogen stream and heated. Vaporized gasoline is injected into the expanded fluid-bed reactor (1), where the proprietary sorbent removes sulfur from the feed. A disengaging zone in the reactor removes suspended sorbent from the vapor, which exits the reactor to be cooled.

Regeneration: The sorbent (catalyst) is continuously withdrawn from the reactor and transferred to the regenerator section (2), where the sulfur is removed as SO_2 and sent to a sulfur-recovery unit. The cleansed sorbent is reconditioned and returned to the reactor. The rate of sorbent circulation is controlled to help maintain the desired sulfur concentration in the product.

Economics:

General operating conditions:	
Temperature, °F	650–775
Pressure, psig	100–300
Space velocity, whsv	4–10
Hydrogen purity, %	70–99
Total H_2 usage, scf/bbl	40–60
Case study premises:	
25,000-bpd feed	
775-ppm feed sulfur	
25-ppm product sulfur (97% removal)	
No cat gasoline splitter	
Results:	
C_5+ yield, vol% of feed	~ 100
Lights yield, wt% of feed	< 0.2
(R+M) Loss	
2	0.6 (or <0.6)
Capital cost, $/bbl	900
Operating cost, ¢/gal *	0.9

* Includes utilities, 4% per year maintenance and sorbent costs.

Installation: Forty-three sites licensed as of 2Q 2002.

Licensor: Fuels Technology Division of ConocoPhillips Co.

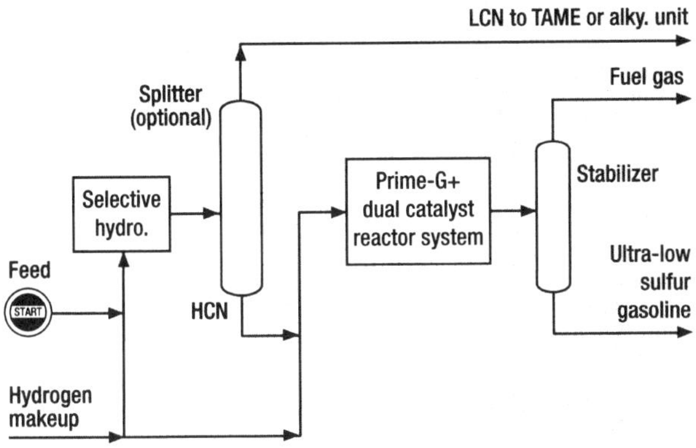

Gasoline desulfurization, ultra-deep

Application: Ultra-deep desulfurization of FCC gasoline with minimal octane penalty using Prime-G+ process.

Description: FCC debutanizer bottoms are fed directly to a first reactor wherein, under mild conditions diolefins are selectively hydrogenated and mercaptans are converted to heavier sulfur species. The selective hydrogenation reactor effluent is then usually split to produce a LCN (light cat naphtha) cut and a HCN (heavy cat naphtha).

The LCN stream is mercaptans free with a low sulfur and diolefin concentration enabling further processing in an etherification or alkylation unit. The HCN then enters the main Prime-G+ section where it undergoes in a dual catalyst reactor system; a deep HDS with very limited olefins saturation and no aromatics losses produces an ultra-low sulfur gasoline.

The process provides flexibility to advantageously co-process other sulfur containing naphthas such as light coker naphtha, steam cracker naphtha or light straight-run naphtha.

Industrial results:

Full-range FCC Gasoline, 40°C–220°C	Feed	Prime-G+ Product
Sulfur, ppm	2100	50*
(RON + MON)/2	87.5	86.5
Δ (RON + MON)/2		1.0
% HDS		97.6

≤ 30 ppm pool sulfur after blending

Pool sulfur specifications as low as less than 10 ppm are attained with the Prime-G+ process in two units in Germany.

Economics:

Investment:

Grassroots ISBL cost, $/bpsd	600–800

Installation: Fifty-three units have been licensed for a total capacity of 1.4 million bpsd. Four Prime-G+ units are already in operation, producing ultra-low sulfur gasoline. Four other units will come onstream at the end of 2002.

OATS process: In addition to the Prime-G+ technology, the OATS (olefins alkylation of thiophenic sulfur), initially developed by BP, is also exclusively offered for license by Axens for ultra-low sulfur gasoline production.

Reference: "Prime-G+: From pilot to start-up of world's first commercial 10 ppm FCC gasoline desulfurization process," NPRA Annual Meeting, March 17–19, 2002, San Antonio.

Licensor: Axens, Axens NA.

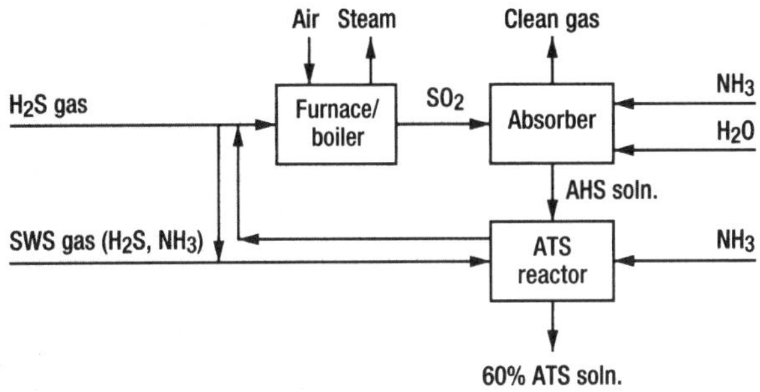

Absorber: $SO_2 + NH_3 + H_2O \rightarrow NH_4HSO_3$

ATS reactor: $4\ NH_4HSO_3 + 2\ H_2S + 2\ NH_3 \rightarrow 3\ (NH_4)_2S_2O_3 + 3\ H_2O$

H_2S and SWS gas conversion

Application: The ATS process recovers H_2S and NH_3 in amine regenerator offgas and sour water stripper gas (SWS gas) as a 60% aqueous solution of ATS—ammonium thiosulfate, $(NH_3)_2S_2O_3$, which is the standard commercial specification. The ATS process can be combined with a Claus unit; thus increasing processing capacity while obtaining a total sulfur recovery of >99.95%.

ATS is increasingly used as a fertilizer (12-0-0-26S) for direct application and as component in liquid fertilizer formulations.

Description: Amine regenerator off gas is combusted in a burner/waste-heat boiler. The resulting SO_2 with ammonia is absorbed in a two-stage absorber to form ammonium hydrogen sulfate (AHS). NH_3 and H_2S contained in the SWS gas plus imported ammonia (if required) is reacted with the AHS solution in the ATS reactor. The ATS product is withdrawn as a 60% aqueous solution that meets all commercial specifications for usage as a fertilizer. Unreacted H_2S is vented to the H_2S burner.

Except for the H_2S burner/waste-heat boiler, all process steps occurs in the water phase at moderate temperatures and neutral pressure. The AHS absorber and ATS reactor systems are chilled with cooling water.

More than 99.95% of the sulfur and practically 100% of the ammonia contained in the feed-gas streams are recovered. Typical emission values are:

SO_x	<100 ppmv
NO_x	<50 ppmv
H_2S	<1 ppmv
NH_3	<20 ppmv

Installation: One Topsøe 30,000-mtpy ATS plant is operating in Northern Europe.

Licensor: Haldor Topsøe A/S.

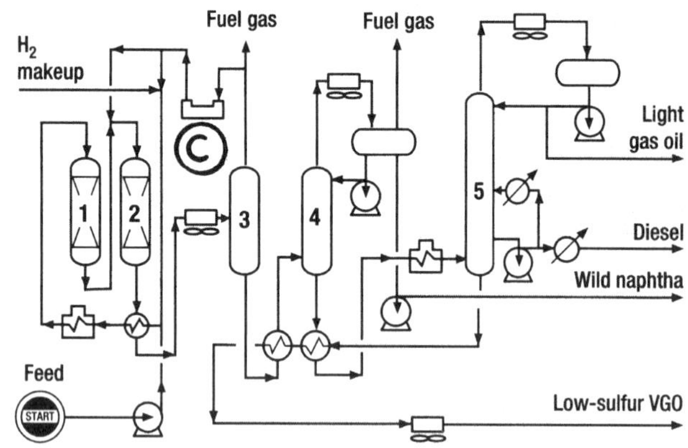

Hydrocracking

Application: Upgrade vacuum gas oil alone or blended with various feedstocks (light-cycle oil, deasphalted oil, visbreaker or coker-gas oil).

Products: Middle distillates, very-low-sulfur fuel oil, extra-quality FCC feed with limited or no FCC gasoline post-treatment or high VI lube base stocks.

Description: This process uses a refining catalyst usually followed by a zeolite-type hydrocracking catalyst. Main features of this process are:

- High tolerance toward feedstock nitrogen
- High selectivity toward middle distillates
- High activity of the zeolite, allowing for 3–4 year cycle lengths and products with low aromatics content until end of cycle.

Three different process arrangements are available: single-step/once-through; single-step/total conversion with liquid recycle; and two-step hydrocracking. The process consists of: reaction section (1, 2), gas separator (3), stripper (4) and product fractionator (5).

Product quality: Typical for HVGO (50/50 Arabian light/heavy):

	Feed, HVGO	Jet fuel	Diesel
Sp. gr.	0.932	0.800	0.826
TBP cut point, °C	405–565	140–225	225–360
Sulfur, ppm	31,700	<10	<10
Nitrogen, ppm	853	<5	<5
Metals, ppm	<2	–	–
Cetane index	–	–	62
Flash pt., °C	–	≥ 40	125
Smoke pt., mm, EOR	–	26–28	–
Aromatics, vol%, EOR	–	< 12	< 8
Viscosity @ 38°C, cSt	110	–	5.3
PAH, wt%, EOR			<2

Economics:

Investment (basis: 40,000-bpsd unit, once-through, 90% conversion, battery limits, erected, engineering fees included, 2000 Gulf Coast), $ per bpsd	2,000–2,500
Utilities, typical per bbl feed:	
Fuel oil, kg	5.3
Electricity, kWh	6.9
Water, cooling, m^3	0.64
Steam, MP	balance

Installation: Fifty references, cumulative capacity exceeding 1 million bpsd, conversion ranging from 20% to 99%.

Licensor: Axens, Axens NA.

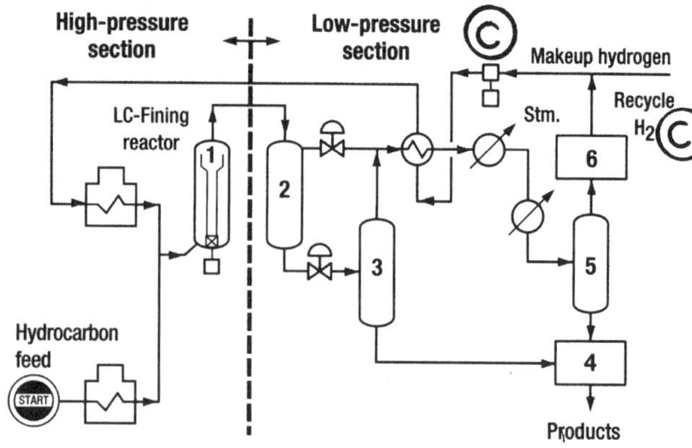

Hydrocracking

Application: Desulfurization, demetallization, CCR reduction, and hydrocracking of atmospheric and vacuum resids using the LC-Fining process.

Products: Full range of high quality distillates. Residual products can be used as fuel oil, synthetic crude or feedstock for a resid FCC, coker, visbreaker or solvent deasphalter.

Description: Fresh hydrocarbon liquid feed is mixed with hydrogen and reacted within an expanded catalyst bed (1) maintained in turbulence by liquid upflow to achieve efficient isothermal operation. Product quality is maintained constant and at a high level by intermittent catalyst addition and withdrawal. Reactor products flow to a high-pressure separator (2), low-pressure separator (3) and product fractionator (4). Recycle hydrogen is separated (5) and purified (6).

Process features include on stream catalyst addition and withdrawal. Recovering and purifying the recycled H_2 at low pressure rather than at high pressure can result in lower capital cost and allows design at lower gas rates.

Operating conditions: Typical reactor temp., 725°F to 840°F; reactor press., 1,400 to 3,500 psig; H_2 part. press., 1,000 to 2,700 psig; LHSV, 0.1 to 0.6; conversion, 40% to 97+%; desulfurization, 60% to 90%; demetallization, 50% to 98%; CCR reduction, 35% to 80%.

Yields: For Arabian Heavy/Arabian Light blends:

Feed	Atm. resid		Vac. resid	
Gravity, °API	12.40	4.73	4.73	4.73
Sulfur, wt%	3.90	4.97	4.97	4.97
Ni/V, ppmw	18/65	39/142	39/142	39/142
Conversion vol% (1,022°F+)	45	60	75	95
Products, vol%				
C_4	1.11	2.35	3.57	5.53
C_5–350°F	6.89	12.60	18.25	23.86
350–700°F (650°F)	(15.24)	30.62	42.65	64.81
700 (650°F)–1,022°F	(55.27)	21.46	19.32	11.92
1,022°F+	25.33	40.00	25.00	5.0
C_5^+°API/wt%S	23.7/0.54	22.5/0.71	26.6/0.66	33.3/0.33

Economics: Investment, estimated (U.S. Gulf Coast, 2000)

Size, bpsd fresh feed	92,000		49,000	
$ per bpsd (typical) fresh feed	2,200	3,500	4,200	5,200
Utilities, per bbl fresh feed				
Fuel fired, 10^3 Btu	56.1	62.8	69.8	88.6
Electricity, kWh	8.4	13.9	16.5	22.9
Steam (export), lb	35.5	69.2	97.0	97.7
Water, cooling, gal	64.2	163	164	248

Installation: Four LC-Fining units in operation, one LC-Fining unit in construction and two LC-Fining units in engineering.

Licensor: Chevron Lummus Global LLC.

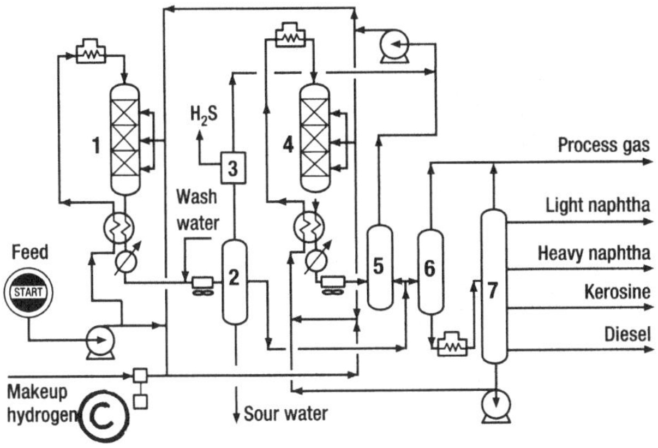

Hydrocracking

Application: Convert naphthas, AGO, VGO and cracked oils from FCCs, cokers, hydroprocessing plants and SDA plants using the Chevron Isocracking process.

Products: Lighter, high-quality, more valuable products: LPG, gasoline, catalytic reformer feed, jet fuel, kerosine, diesel, lube oils and feeds for FCC or ethylene plants.

Description: A broad range of both amorphous/zeolite and zeolite catalysts are used to obtain an exact match with the refiner's process objective. An extensive range of proprietary amorphous/zeolitic catalysts and various process configurations are used to match the refiner's process objectives. Generally, a staged reactor system consists of one reactor (1, 4) and one HP separator (2, 5) per stage optional recycle scrubber (3), LP separator (6) and fractionator (7) to provide flexibility to vary the light-to-heavy product ratio and obtain maximum catalytic efficiency. Single-stage options (both once-through and recycle) are also used when economical.

Yields: Typical from various feeds:

Feed	**Naphtha**	**LCO**	**VGO**	**VGO**
Catalyst stages	1	2	2	2
Gravity, °API	72.5	24.6	25.8	21.6
ASTM 10%/EP, °F	154/290	478/632	740/1,050	740/1,100
Sulfur, wt%	0.005	0.6	1.0	2.5
Nitrogen, ppm	0.1	500	1,000	900
Yields, vol%				
Propane	55	3.4	—	—
iso-Butane	29	9.1	3.0	2.5
n-Butane	19	4.5	3.0	2.5
Light naphtha	23	30.0	11.9	7.0
Heavy naphtha	—	78.7	14.2	7.0
Kerosine	—	—	86.8	48.0
Diesel	—	—	—	50.0
Product quality				
Kerosine smoke pt., mm	—	—	28	28
Diesel Cetane index	—	—	—	58
Kerosine freeze pt., °F	—	—	–65	–75
Diesel pour pt., °F	—	—	—	–10

Economics:

Investment (basis: 30,000 bpsd maximum conversion unit, Mid-East VGO feed, includes only on-plot facilities and first catalyst charge, 2002 U.S. Gulf Coast), $ per bpsd	3,000
Utilities, typical per bbl feed:	
Fuel, equiv. fuel oil, gal	1
Electricity, kWh	7
Steam, 150 psig (net produced), lb	(50)
Water, cooling, gal	330

Installation: More than 50 units exceeding 750,000 bpsd total capacity.

Licensor: Chevron Lummus Global LLC.

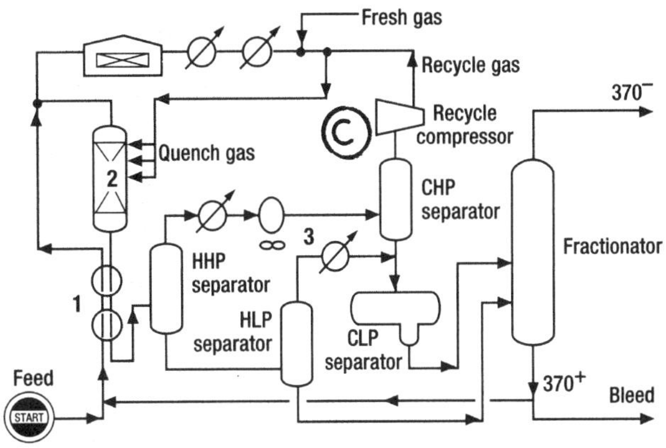

Hydrocracking

Application: The Shell hydrocracking process converts heavy VGO and other low-cost cracked and extracted feedstocks to high-value, high-quality products. Profitability is maximized by careful choice of process configuration, conditions and catalyst system to match refiners' product quality and selectivity requirements.

Products: Low-sulfur diesel and jet fuel with excellent combustion properties, high-octane light gasoline, and high-quality reformer, cat cracker, ethylene cracker or lube oil feedstocks.

Description: Heavy hydrocarbons are discharged to the reactor circuit and preheated with reactor effluent (1). Fresh hydrogen is discharged to the reactor circuit and combined with recycle gas from the cold high-pressure separator. The mixed gas is supplied as quench for reactor interbed cooling with the balance first preheated with reactor effluent followed by further heating in a single phase furnace. After mixing with the liquid feed, the reactants pass in trickle flow through the multi-bed reactor(s) which contains proprietary pre-treat, cracking and post-treat catalysts (2). Interbed ultra-flat quench internals and high dispersion nozzle trays combine excellent quench, mixing and liquid flow distribution at the top of each catalyst bed while maximizing reactor volume utilization. After cooling by incoming feed streams, reactor effluent enters a four-separator system from which hot effluent is routed to the fractionator (3). Wash water is applied via the cold separators in a novel countercurrent configuration to scrub the effluent of corrosive salts and avoid equipment corrosion.

Two-stage, series flow and single-stage unit design configurations are all available including the single reactor stacked catalyst bed which is suitable for capacities up to 10,000 tpd in partial or full conversion modes. The catalyst system is carefully tailored for the desired product slate and cycle run length.

Installation: Over 20 new distillate and lube oil units including two recent single-reactor high-capacity stacked bed units. Over a dozen revamps have been carried out on own and other licensor designs usually to debottleneck and increase feed heaviness.

References: "Performance optimisation of trickle bed processes," European Refining Technology Conference, Berlin, November 1998; "Design and operation of Shell single reactor hydrocrackers," 3rd International Petroleum Conference, New Delhi, January 1999.

Licensor: Shell Global Solutions International B.V.

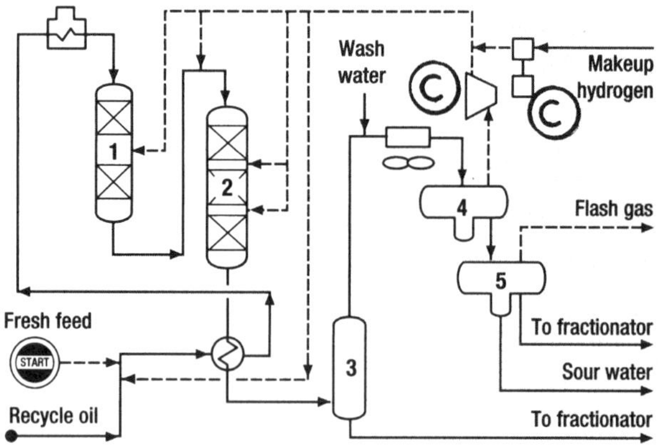

Hydrocracking

Application: Convert a wide variety of feedstocks into lower-molecular-weight products using the Unicracking and HyCycle Unicracking process.

Feed: Feedstocks include atmospheric gas oil, vacuum gas oil, FCC/RCC cycle oil, coker gas oil, deasphalted oil and naphtha for production of LPG.

Products: Processing objectives include production of gasoline, jet fuel, diesel fuel, lube stocks, ethylene-plant feedstock, high-quality FCC feedstock and LPG.

Description: Feed and hydrogen are contacted with catalysts, which induce desulfurization, denitrogenation and hydrocracking. Catalysts are based upon both amorphous and molecular-sieve containing supports. Process objectives determine catalyst selection for a specific unit. Product from the reactor section is condensed, separated from hydrogen-rich gas and fractionated into desired products. Unconverted oil is recycled or used as lube stock, FCC feedstock or ethylene-plant feedstock.

Yields: Example:

Feed type	FCC cycle oil blend	Vacuum gas oil	Fluid coker gas oil
Gravity, °API	27.8	22.7	8.4
Boiling, 10%, °F	481	690	640
End pt., °F	674	1,015	1,100
Sulfur, wt%	0.54	2.4	4.57
Nitrogen, wt%	0.024	0.08	0.269

Principal products	**Gasoline**	**Jet**	**Diesel**	**FCC feed**
Yields, vol% of feed				
Butanes	16.0	6.3	3.8	5.2
Light gasoline	33.0	12.9	7.9	8.8
Heavy naphtha	75.0	11.0	9.4	31.8
Jet fuel		89.0		
Diesel fuel			94.1	33.8
600°F + gas oil				35.0
H_2 consump., scf/bbl	2,150	1,860	1,550	2,500

Economics: Example:

Investment, $ per bpsd capacity	2,000–4,000
Utilities, typical per bbl feed:	
Fuel, 10^3 Btu	70–120
Electricity, kWh	7–10

Installation: Selected for 151 commercial units, including several converted from competing technologies. Total capacity exceeds 3.4 million bpsd.

Licensor: UOP LLC.

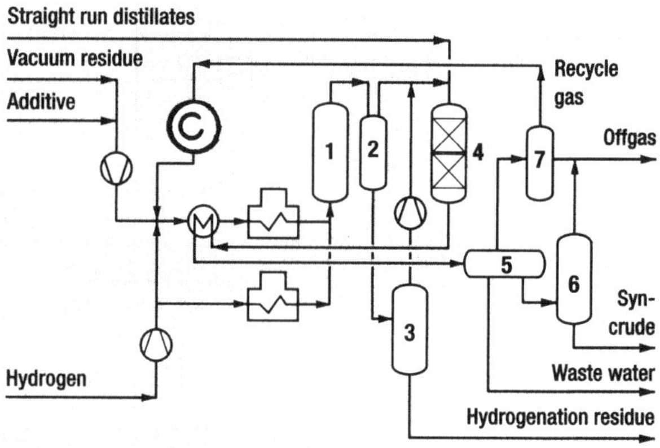

Hydrocracking

Application: Upgrading of heavy and extra heavy crudes as well as residual oils.

Products: Full-range high-quality syncrude.

Description: A hydrogen addition process applying the principles of thermal hydrocracking in liquid-phase hydrogenation reactors (LPH) (1) for primary conversion directly coupled with an integrated catalytical hydrofinishing step (GPH). In the LPH slurry phase reactors, residue is converted up to 95% at temperatures between 440°C and 500°C. In the hot separator (HS) (2), light distillates are separated from the unconverted material. By vacuum-flash distillation (3), the HS bottoms distillates are recovered. For further hydrotreating, the HS overheads, together with the recovered HS bottoms distillates and straight-run distillates (optional), are routed to the catalytical fixed-bed reactors of the GPH (4), which operates at the same pressure as the LPH. GPH pressure is typically above hydrocracking conditions, therefore, GPH mild hydrocracking can also be applied to allow a shift in yield structure to lighter products. Separation of syncrude and associated gases is performed in a cold separator system (5). The syncrude after separation is sent to a stabilizer (6) and a fractionation unit. After being washed in a lean oil scrubber (7), the gases are recycled to the LPH section.

Feed: Feedstock quality ranges covered are:

Gravity, °API	–3 to 14
Sulfur, wt%	0.7 to 7
Metals (Ni,V), ppm	up to 2,180
Asphaltenes, wt%	2 to 80

Yields:

Naphtha <180°C, wt%	15–30
Middle distillates, wt%	35–40
Vac. gasoil >350°C, wt%	15–30

Product qualities:

Naphtha: Sulfur < 5 ppm, Nitrogen < 5 ppm
Kerosine: Smoke point >20 mm, Cloud point <-50°C
Diesel: Sulfur <50 ppm, Cetane no. > 45
Vac. gasoil: Sulfur <150 ppm, CCR <0.1wt%, Metals <1 ppm

Economics: Plant capacity 23,000 bpsd.
Investment U.S. MM$190 (U.S. Gulf Coast, 1st Q. 1994)
Utilities:

Fuel oil, MW	12
Power, MW	17
Steam, tph	–34
Water, cooling, m^3/h	2,000

Installation: Two licenses have been granted.

Licensor: VEBA OEL Technologie und Automatisierung GmbH.

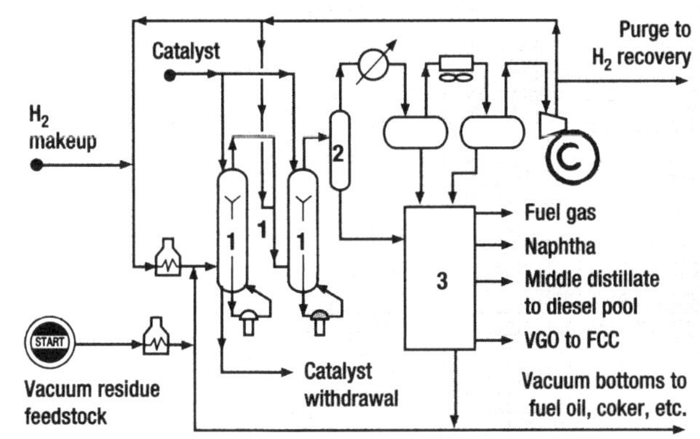

Hydrocracking, residue

Application: Catalytic hydrocracking and desulfurization of residua and heavy oils in an ebullated-bed reactor using the H-Oil process.

Products: Full-range distillates and upgraded residue, transportation fuels, FCC or coker feed, low-sulfur-fuel oil.

Description: A one, two or three stage ebullated-bed (1) reactor system. Feed consists of atmospheric or vacuum residue, recycle from downstream fractionation (3), hydrogen-rich recycle gas and makeup hydrogen. Combined feed is fed to the bottom of the reactor and expands the catalyst bed resulting in good mixing, near isothermal operation and allows for onstream catalyst replacement to maintain catalyst activity and 3 to 4 year run lengths between turnarounds. Two-phase reactor effluent is sent to high-pressure separator (2); liquid is sent to fractionation (3) to recover light liquid products and vacuum bottoms for recycle.

Operating conditions: Temperature, 770°F–840°F; hydrogen partial pressure, psi 1,000–2,500; LHSV, 0.1–1.0 hr^{-1}; conversion 40%–90%.

Process performance and yields: From commercial two-stage processing:

Vacuum residue conversion	**52 W%**	**70 W%**
Processing objective	LSFO	Conversion
Feed	Ural VR	Arab H VR+FCC slurry
Gravity, °API	13	3.6
1,000°F+, vol%	85	85
Sulfur, wt%	2.8	5.2
Performance, yields and product qualities		
HDS, wt%	85	83
HDN, wt%	40	38
Chem. H_2 Cons, scf/bbl	920	1,540
Naphtha, vol%	7	8
Diesel, vol%	25	33
VGO, vol%	31	38
Residue, vol%	41	25
Diesel sulfur, wppm	400	2,000
VGO sulfur, wt%	0.18	0.9
Residue sulfur, wt%	0.8	2.0
Residue gravity, °API	14	4.0

Economics: Basis 2002 U.S. Gulf Coast

Investment—$ per bpsd	3,500–5,500
Utilities—per bbl of feed	
Fuel, 10^3 Btu	40–100
Power, kWh	9–15
Water, cooling (20°F rise), gal	100–200
Catalyst makeup, $	0.2–1.5

Installation: Seven units currently in operation.

Licensor: Axens, Axens NA.

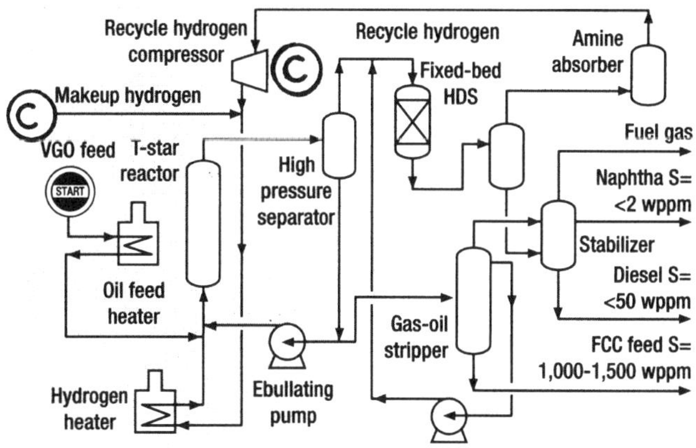

Hydrocracking/hydrotreating—VGO

Application: The T-Star Process is an ebullated-bed process for the hydrotreatment/hydrocracking of vacuum gas oils. The T-Star Process is best suited for difficult feedstocks (coker VGO, DAO), high-severity operations and applications requiring a long run length.

Description: A T-Star process flow diagram, which includes integrated mid-distillate hydrotreating, is shown above. The typical T-Star battery limits include oil/hydrogen fired heaters, an advanced hot high-pressure design for product separation and for providing recycle to the ebullating pump, recycle gas scrubbing and product separation. Catalyst is replaced periodically from the reactor, without shutdown. This ensures the maintenance of constant, optimal catalyst activity and consistent product slate and quality. After high-pressure recovery of the effluent and recycle gas, the products are separated and stabilized through fractionation. A T-Star unit can operate for four-year run lengths with conversion in the 20%–60% range and hydrodesulfurization in the 93%–99% range.

Operating conditions:

Temperature, °F	750–820
Hydrogen partial pressure, psi	600–1,500
LHSV, hr^{-1}	0.5–3.0
VGO conversion, %	20–60

Examples: In Case 1, a deep-cut Arab heavy VGO is processed at 40 wt% conversion with objectives of mild conversion and preparing specification diesel and FCC unit feed. In Case 2, a VGO blend containing 20% coker material is processed at lower conversion to also obtain specification FCC unit feedstock and high-quality diesel.

Economics: Basis 2002 U.S. Gulf Coast

Investment in $ per bpsd	1,200–2,500
Utilities, per bbl of feed	
Fuel, 10^3 Btu	60
Power, kWh	3
Catalyst makeup, $	0.05–0.20

Installation: The T-Star process is commercially demonstrated based on the ebullated-bed reactor. Axens has licensed more than 1.3 MMbpsd of capacity in gas oil, VGO and residue. Axens has seven commercially operating ebullated-bed units and one is under construction that will process a variety of VGO feedstocks.

Reference: "A novel approach to attain new fuel specifications," *Petroleum Technology Quarterly*, Winter 1999/2000.

Licensor: Axens, Axens NA.

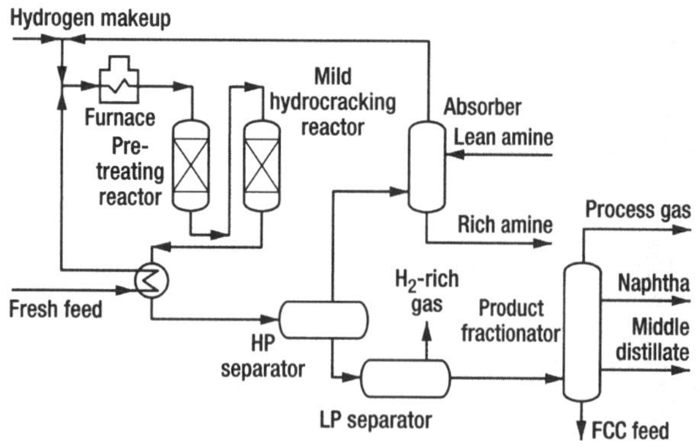

Hydrocracking (mild)/ VGO hydrotreating

Application: The Topsøe mild hydrocracking / VGO hydrotreating process upgrades, and if required, converts a variety of vacuum gasoil feedstocks including straight run (SR) and cracked components from an FCC, Coker, and visbreaker as well as deasphalting units.

Products: Low-sulfur naphtha, diesel and vacuum gasoil. The vacuum gasoil sulfur is adjusted such that when processed by the FCC, it will produce low-sulfur gasoline that does not require post treatment and can be blended directly in the gasoline pool.

Description: This process uses a combination of process conditions and hydrotreating and hydrocracking catalyst to meet the required conversion as well as required product quality specifications. Topsøe has developed amorphous and zeolitic cracking catalysts specifically designed for mild hydrocracking applications. Topsøe 's engineers utilize their process design experience to adjust the unit flow configuration to meet product quality requirements and provide a cost-effective design.

The unit design also features an industry leader graded-bed design for reactor pressure-drop control as well as state of the art reactor internals design. For FCC pretreater revamps, Topsøe has further developed the Aroshift process that considerably improves FCC profitability at little investment.

Operating conditions: Typical operating pressures range from 55 to 110 barg (800 to 1,600 psig). Typical operating temperatures range from 340°C to 410°C (644°F to 770°F).

Installation: Four units designed by Topsøe for Mild Hydrocracking/ VGO Hydrotreating are in service.

Licensor: Haldor Topsøe A/S.

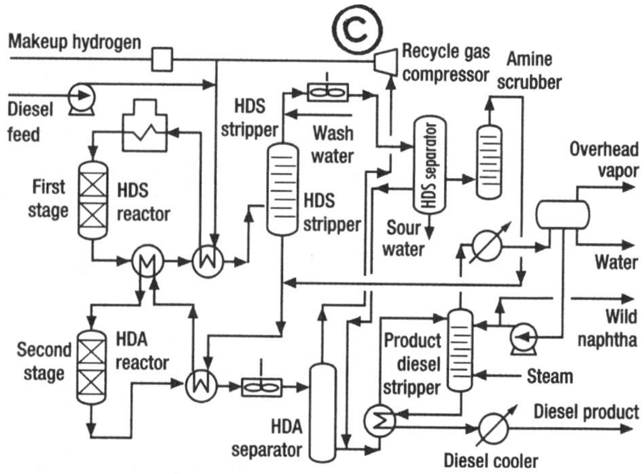

Hydrodearomatization

Application: Topsøe's two-stage hydrodesulfurization hydrodearomatization (HDS/HDA) process is designed to produce low-aromatics distillate products. This process enables refiners to meet the new, stringent standards for environmentally friendly fuels.

Products: Ultra-low sulfur, ultra-low nitrogen, low-aromatics diesel, kerosine and solvents (ultra-low aromatics).

Description: The process consists of four sections: initial hydrotreating, intermediate stripping, final hydrotreating and product stripping. The initial hydrotreating step, or the "first stage" of the two-stage reaction process, is similar to conventional Topsøe hydrotreating, using a Topsøe high-activity base metal catalyst such as TK-573 to perform deep desulfurization and deep denitrification of the distillate feed. Liquid effluent from this first stage is sent to an intermediate stripping section, in which H_2S and ammonia are removed using steam or recycle hydrogen. Stripped distillate is sent to the final hydrotreating reactor, or the "second stage." In this reactor, distillate feed undergoes saturation of aromatics using a Topsøe noble metal catalyst, either TK-907/TK-908 or TK-915, a high-activity dearomatization catalyst. Finally, the desulfurized, dearomatized distillate product is steam stripped in the product stripping column to remove H_2S, dissolved gases and a small amount of naphtha formed.

Like the conventional Topsøe hydrotreating process, the HDS/HDA process uses Topsøe's graded bed loading and high-efficiency patented reactor internals to provide optimum reactor performance and catalyst use leading to the longest possible catalyst cycle lengths. Topsøe's high efficiency internals have a low sensitivity to unlevelness and are designed to ensure the most effective mixing of liquid and vapor streams and maximum utilization of catalyst. These internals are effective at high of liquid loadings, thereby enabling high turndown ratios. Topsøe's graded-bed technology and the use of shape-optimized inert topping and catalysts minimize the build-up of pressure drop, thereby enabling longer catalyst cycle length.

Operating conditions: Typical operating pressures range from 20 to 60 barg (300 to 900 psig), and typical operating temperatures range from 320°C to 400°C (600°F to 750°F) in the first stage reactor, and from 260°C to 330°C (500°F to 625°F) in the second stage reactor. The Topsøe HDS/HDA treatment of a heavy straight-run gas oil feed yielded these product specifications:

	Feed	**Product**
Specific gravity	0.86	0.83
Sulfur, ppmw	3,000	1
Nitrogen, ppmw	400	<1
Total aromatics, wt%	30	<10
Cetane index, D-976	49	57

References: Cooper, Hannerup and Søgaard-Andersen, "Reduction of aromatics in diesel," Oil and Gas, September 1994

Søgaard-Andersen, Cooper and Hannerup, "Topsøe's process for improving diesel quality," NPRA Annual Meeting, 1992.

de la Fuente, E., P. Christensen, and M. Johansen, "Options for meeting EU year 2005 fuel specifications."

Installation: A total of five, two in Europe and three in North America.

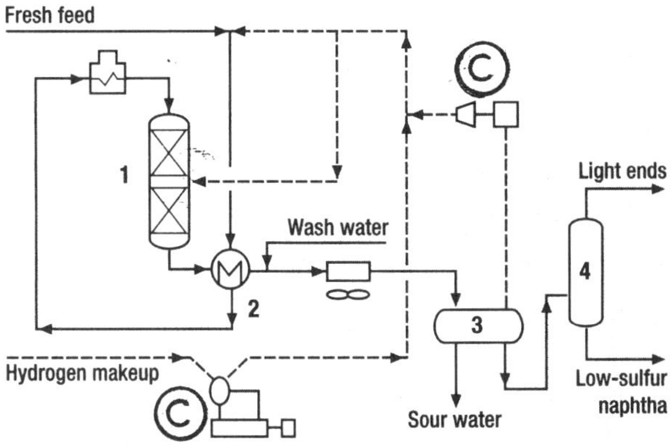

Hydrodesulfurization

Application: The ISAL process enables refiners to meet the world's most stringent specifications for gasoline sulfur while also controlling product octane.

This moderate-pressure, fixed-bed hydroprocessing technology desulfurizes gasoline-range olefinic feedstocks and selectively reconfigures lower octane components to control product octane. This process can be applied as a standalone unit or as part of an overall integrated flow scheme for gasoline desulfurization.

Description: The flow scheme of the ISAL process is very similar to that of a conventional hydrotreating process. The naphtha feed is mixed with hydrogen-rich recycle gas and processed across fixed catalyst beds at moderate temperatures and pressures. Following heat exchange and separation, the reactor effluent is stabilized. The similarity of an ISAL unit to a conventional naphtha hydrotreating unit makes new unit and revamp implementation both simple and straightforward. The technology can be applied to idle reforming and hydrotreating units.

Product quality: The ISAL unit's operation can be adjusted to achieve various combinations of desulfurization, product octane and yield. Typical yield/octane relationships for an integrated flow scheme processing an FBR FCC naphtha containing 400 wppm sulfur and 20 wt% olefins are:

%Desulfurization, wt% C_5^+ product	**93%**		**98%**	
Yield, vol%	99.5	97.4	99.5	97.2
Sulfur, wppm	30	30	10	10
Olefins, wt%	15.3	15.7	14.9	15.3
(R + M)/2 change	–1.9	0	–2.1	0

Economics: The capital and operating costs of an ISAL unit are slightly higher than those of a typical naphtha hydrotreating unit. With this process, refiners benefit from the ability to produce a higher-octane product at incremental additional operating cost primarily related to additional hydrogen consumption.

Installation: Two ISAL applications have been implemented in the U.S. Engineering work has also been completed on three additional ISAL units, with an additional two ISAL units currently in the process design stage.

Licensor: UOP LLC (in cooperation with PDVSA-INTEVEP).

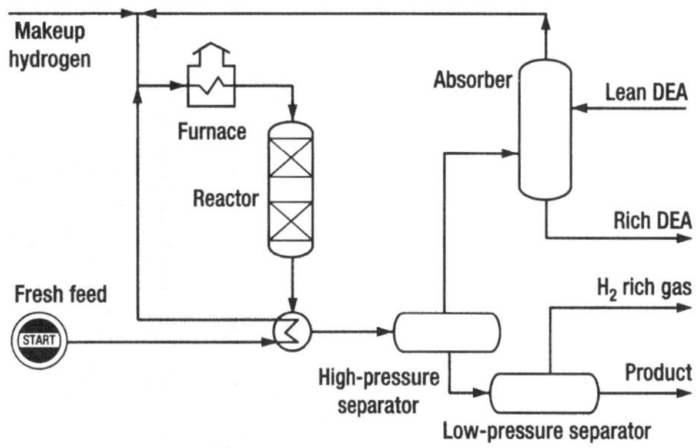

Hydrodesulfurization, ultra-low-sulfur diesel

Application: Topsøe ULSD process is designed to produce ultra-low-sulfur diesel (ULSD)—5 wppm S—from cracked and straight-run distillates. By selecting the proper catalyst and operating conditions, the process can be designed to produce 5 wppm S diesel at low reactor pressures (<500 psig) or at higher reactor pressure when products with improved density, cetane, and polyaromatics are required.

Description: Topsøe ULSD process is a hydrotreating process that combines Topsøe's understanding of deep-desulfurization kinetics, high-activity catalyst, state-of-the-art reactor internal, and engineering expertise in the design of new and revamped ULSD units. The ULSD process can be applied over a very wide range of reactor pressures.

Our highest activity CoMo catalyst is specifically formulated with high-desulfurization activity and stability at low reactor pressure (~ 500 psig) to produce 5 wppm diesel. This catalyst is suitable for revamping existing low-pressure hydrotreaters or in new units when minimizing hydrogen consumption.

The highest activity NiMo catalyst is suitable at higher pressure when secondary objectives such as cetane improvement and density reduction are required. Topsøe offers a wide range of engineering deliverables to meet the needs of the refiners. Our offerings include process scoping study, reactor design package, process design package, or engineering design package.

Installation: Topsøe has licensed 21 ULSD hydrotreaters with 11 revamps. Our reactor internals are installed in more than 60 units.

References: "Cost-Effectively Improve Hydrotreater Designs," *Hydrocarbon Processing,* November 2001 pp. 43–46.

"The importance of good liquid distribution and proper selection of catalyst for ultra deep diesel HDS," JPI Petroleum Refining Conference, October 2000, Tokyo.

Licensor: Haldor Topsøe A/S.

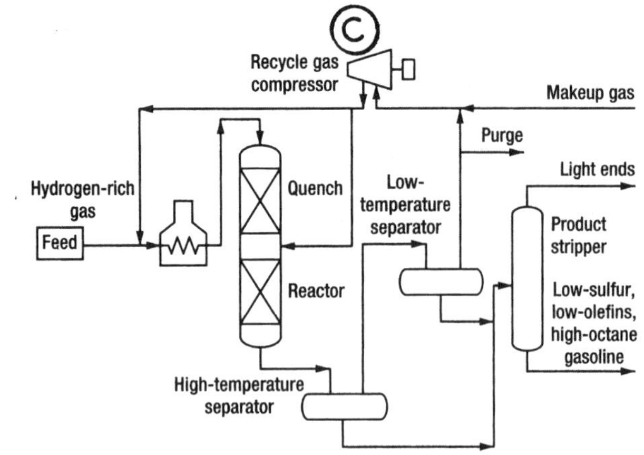

Hydrodesulfurization

Application: Reduce sulfur in gasoline to less than 10 ppm by hydrodesulfurization followed by cracking and isomerization to recover octane.

Description: The basic flow scheme is similar to that of a conventional naphtha hydrotreater. Feed and recycle hydrogen mix is preheated in feed/effluent exchangers and a fired heater then introduced into a fixed-bed reactor. Over the first catalyst bed, the sulfur in the feed is converted to hydrogen sulfide with near complete olefin saturation. In the second bed, over a different catalyst, octane is recovered by cracking and isomerization reactions. The reactor effluent is cooled and the liquid product separated from the recycle gas using high- and-low temperature separators.

The vapor from the separators is combined with makeup gas, compressed and recycled. The liquid from the separators is sent to the product stripper where the light ends are recovered overhead and desulfurized naphtha from the bottoms. The product sulfur level can be as low as 5 ppm. The OCTGAIN process can be retrofitted into existing refinery hydrotreating units. The design and operation permit the desired level of octane recovery and yields.

Yields: Yield depends on feed olefins and desired product octane.

Installations: Commercial experience with two operating units.

Reference: Halbert, T., et al., "Technology Options For Meeting Low Sulfur Mogas Targets," NPRA Annual Meeting, March, 2000.

Licensor: ExxonMobil Research and Engineering Co.

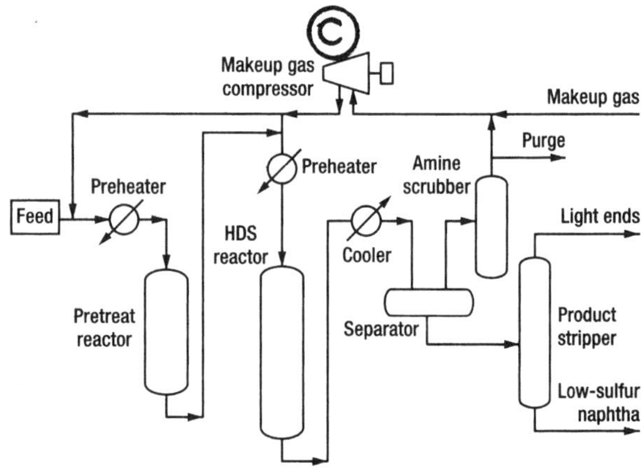

Hydrodesulfurization

Application: Reduce sulfur in FCC gasoline to less than 10 ppm by selective hydrotreating to maximize octane preservation.

Description: The feed is mixed with recycle hydrogen, heated with reactor effluent and passed through the pretreat reactor for diolefin saturation. After further heat exchange with reactor effluent and preheat using a utility, the hydrocarbon/hydrogen mixture enters the HDS reactor containing proprietary RT-225 catalyst. In the reactor, the sulfur is converted to H_2S under conditions which strongly favor hydrodesulfurization while minimizing olefin saturation.

The reactor effluent is cooled and the liquid separated from gas that is amine scrubbed and recycled to the reactor along with makeup gas. The liquid product is stabilized in a product stripper before being sent to storage/blending.

For high-sulfur feeds and/or very low-sulfur product, variations in the design are available to minimize octane loss and hydrogen consumption. The feed may be full range, intermediate or heavy fractions. SCANfining can be retrofitted to existing refinery units such as naphtha or diesel hydrotreaters and reformers.

Yields: Yield of C_5 plus liquid product typically over 100 LV%.

Installations: Twenty-four units under design, construction or operation.

Reference: Sweed, N., et al., "Low sulfur technology," *Hydrocarbon Engineering,* July 2002.

Ellis,E., et al., "Meeting the low sulfur mogas challenge," World Refining Association Third European Fuels Conference, March 2002.

Licensor: ExxonMobil Research and Engineering Co.

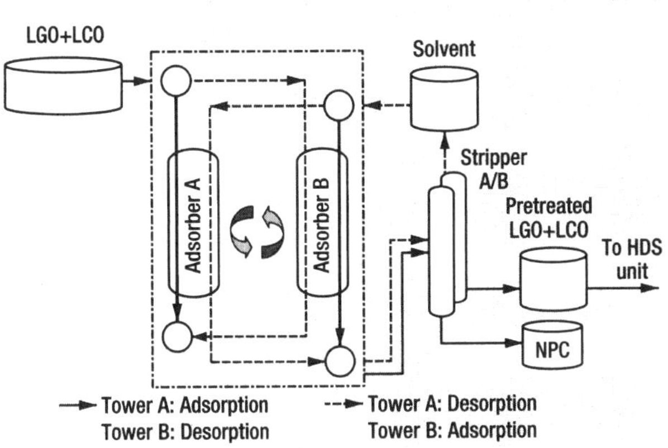

Hydrodesulfurization—pretreatment

Application: The SK HDS pretreatment process allows a refiner to use the existing diesel HDS unit with minor modifications to produce ultra-low-sulfur diesel (ULSD) at less than 10 ppm sulfur. Modifications to older, existing low-pressure HDS units usually require catalyst replacement and installation of a recycle gas scrubber, if not already in place.

Description: The primary goal of this process is to improve performance of the HDS unit by removing most of the nitrogen-bearing natural polar compounds (NPC) from the mid-boiling range distillate streams. This is achieved by selectively adsorbing the NPC in an adsorption bed followed by desorption of NPC by a solvent. The process uses twin adsorbers, a desorption-solvent circulation system, two stripping columns and associated pumps and an overhead system.

Diesel blend feedstock is passed through one adsorbent vessels, which is then stripped to remove a small amount of desorption solvent. The adsorbed NPC is removed from the adsorber bed by desorption solvent and is stripped in the second stripper column.

Pretreated diesel blend from diesel stripper is sent to the downstream HDS unit for sulfur removal. The byproduct, NPC stream, from the NPC stripper may be used as a blend stock in either marine diesel or in high-sulfur fuel oil.

Economics: Total investment cost (ISBL) is $400 to $450/bbl for a 30,000-bpsd unit, U.S.GC, 3Q 2002.

Utilities per barrel of feed:

Fuel gas, fired, FOEB	0.01
Water, cooling, mt	1.5
Electricity, kWh	0.4

An alternate steam stripping design can be provided if required.

Cost of adsorbent: Less than $0.10/bbl

Hydrogen: Approximately 10–20% less hydrogen consumption in hydrotreating unit for feed processed in the SK Pretreatment Unit.

Installation: One 1,000-bpd demonstration unit has been operating successfully since May 2002. A larger commercial-scale unit is currently being studied for SK Corp.

Licensor: The Badger Technology Center of Washington Group International, on behalf of SK Corp.

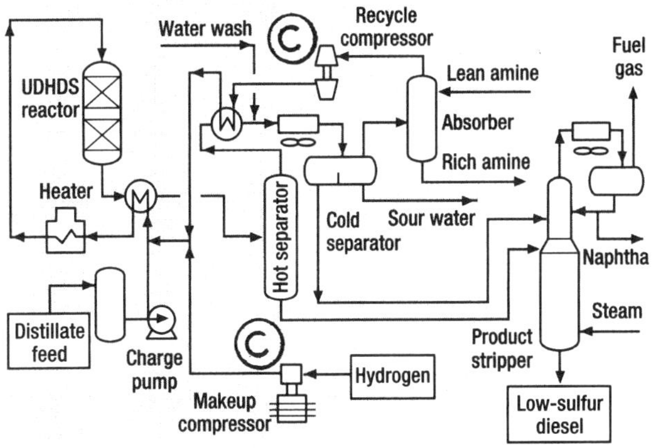

Hydrodesulfurization—UDHDS

Application: A versatile family of premium distillates technologies is used to meet all current and possible future premium diesel upgrading requirements. Ultra-deep hydrodesulfurization (UDHDS) process can produce distillate products with sulfur levels below 10 wppm from a wide range of distillate feedstocks.

Products: High volume yield of ultra-low-sulfur distillate is produced. Cetane and API gravity uplift together with the reduction of polyaromatics to less than 6 wt% or as low as 2 wt% can be economically achieved.

Description: The UDHDS reactor and catalyst technology is offered through Akzo Nobel Catalysts bv. A single-stage, single-reactor process incorporates proprietary high-performance-distribution and quench internals. Feed and combined recycle and makeup gas are preheated and contact the catalyst in a downflow-concurrent-fixed-bed reactor. The reactor effluent is flashed in a high- and a low-pressure separator. An amine-absorber tower is used to remove H_2S from the recycle gas. In the example shown, a steam stripper is used for final product recovery. The UDHDS technology is equally applicable to revamp and grassroots applications.

Economics:

Investment (basis: 25,000 to 35,000 bpsd, 1Q 2000 U.S. Gulf Coast)

New unit, $ per bpsd	1,000 to 1,800

Installation: Over 60 distillate-upgrading units have applied the Akzo Nobel HDS Technology. Eleven of these applications produce or will produce <10ppm sulfur, using UDHDS technology.

Reference: "Technology for premium distillates," ERTC Low Sulfur Fuels Workshop, February/March 2002, London.

Licensor: Akzo Nobel Catalysts bv.

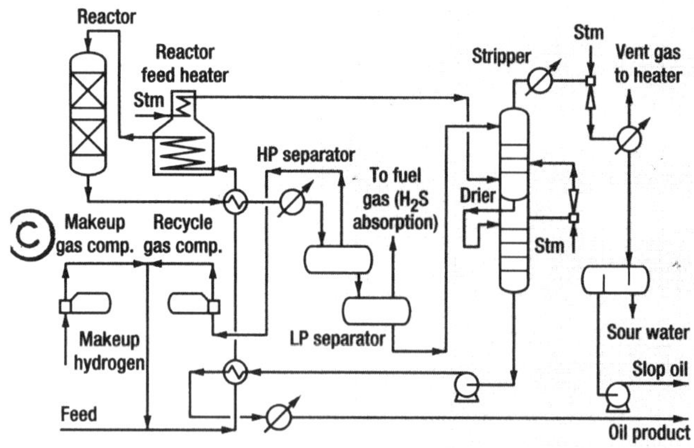

Hydrofinishing/hydrotreating

Application: Process to produce finished lube-base oils and special oils.

Feeds: Dewaxed solvent or hydrogen-refined lube stocks or raw vacuum distillates for lubricating oils ranging from spindle oil to machine oil and bright stock.

Products: Finished lube oils (base grades or intermediate lube oils) and special oils with specified color, thermal and oxidation stability.

Description: Feedstock is fed together with make-up and recycle hydrogen over a fixed-bed catalyst at moderate temperature and pressure. The treated oil is separated from unreacted hydrogen, which is recycled. Very high yields product are obtained.

For lube-oil hydrofinishing, the catalytic hydrogenation process is operated at medium hydrogen pressure, moderate temperature and low hydrogen consumption. The catalyst is easily regenerated with steam and air.

Operating pressures for hydrogen-finishing processes range from 25 to 80 bar. The higher-pressure range enables greater flexibility with regard to base-stock source and product qualities. Oil color and thermal stability depend on treating severity. Hydrogen consumption depends on the feed stock and desired product quality.

Utility requirements, (typical, Middle East Crude), units per m^3 of feed:

Electricity, kWh	15
Steam, MP, kg	25
Steam, LP, kg	45
Fuel oil, kg	3
Water, cooling, m^3	10

Installation: Numerous installations using the Uhde Edeleanu proprietary technology are in operation worldwide. The most recent reference is a complete lube-oil production facility licensed to the state of Turkmenistan, which successfully passed performance testing in 2002.

Licensor: Uhde Edeleanu GmbH.

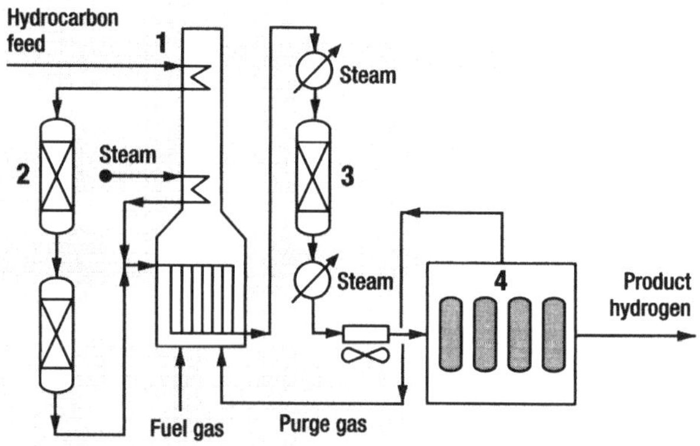

Hydrogen

Application: To produce hydrogen from light hydrocarbons using steam-methane reforming.

Feedstock: Natural gas, refinery gas, LPG and naphtha.

Product: High-purity hydrogen and steam.

Description: Light hydrocarbon feed (1) is heated prior to passing through two fixed-catalyst beds. Organic sulfur compounds present in feed gas (e.g., mercaptans) are converted to hydrogen sulfide (H_2S) and mono-olefins in the gas phase are hydrogenated in the first bed of cobalt molybdenum oxide catalyst (2). The second bed contains zinc oxide to remove H_2S by adsorption. This sulfur-removal stage is necessary to avoid poisoning of the reforming catalysts. Treated feed gas is mixed with steam and heated before passing to the reformer where the hydrocarbons and steam react to form synthesis gas (syngas).

Foster Wheeler supplies proprietary side-fired Terrace Wall reformers, with natural draft mode for increased reliability, compact plot layout with convection section mounted directly above the radiant section and modular fabrication option. Top-fired reformers are options for large capacity plants.

Syngas containing hydrogen, methane, carbon dioxide (CO_2), carbon monoxide (CO) and water leaves the reformer and passes through the waste-heat boiler to the shift reactor (3) where most of the CO is converted to CO_2 and hydrogen by reaction with steam. For heavier feedstocks, pre-reforming is used for conversion of feedstock upstream of the reformer. The syngas is cooled through a series of heat-recovery exchangers before free water is recovered in a knockout drum. The resultant raw hydrogen stream passes to the pressure swing adsorption (PSA) unit for purification (4) to 99.9% hydrogen product quality. Tail gas from the PSA unit provides a substantial proportion of the firing duty for the reformer. The remaining fuel is supplied from feed gas or other sources (e.g. refinery fuel gas).

Saturated and superheated steam is raised by heat exchange with the reformed gas and flue gas in the convection section of the reformer. Steam export quantities can be varied between 1,250 and 5,750 lb/ MMscfd of hydrogen produced using air pre-heat and auxiliary firing options.

Economics: Plant design configurations are optimized to suit the clients' economic requirements, using discounted cash-flow modeling to establish the lowest lifecycle cost of hydrogen production.

Investment: 10–100 MMscfd, 3rd Q 2002, U.S.GC$9–55 million

Utilities, typical per MMscfd of hydrogen produced (natural gas feedstock):

Feed + fuel, lb	960
Water, demineralized, lb	4,420
Steam, export, lb	3,320
Water, cooling, U.S. gal	1,180
Electricity, kWh	12

Reference: Ward, R. D. and N. Sears, "Hydrogen plants for the new millenium," *Hydrocarbon Engineering,* Vol. 7, June 2002.

Installation: Over 100 plants, ranging from 1 MMscfd to 95 MMscfd in a single-train configuration and numerous multi-train configurations.

Licensor: Foster Wheeler.

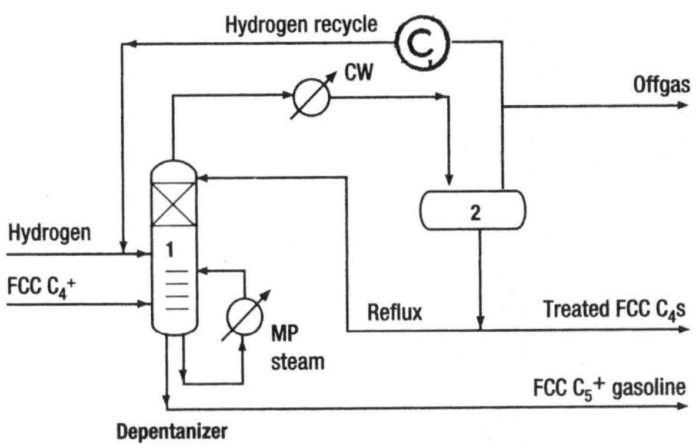

Hydrogenation

Application: CDHydro is used to selectively hydrogenate diolefins in the top section of a hydrocarbon distillation column. Additional applications—including mercaptan removal, hydroisomerization and hydrogenation of olefins and aromatics are also available.

Description: The patented process CDHydro combines fractionation with hydrogenation. Proprietary devices containing catalyst are installed in the fractionation column's top section (1). Hydrogen is introduced beneath the catalyst zone. Fractionation carries light components into the catalyst zone where the reaction with hydrogen occurs. Fractionation also sends heavy materials to the bottom. This prevents foulants and heavy catalyst poisons in the feed from contacting the catalyst. In addition, clean hydrogenated reflux continuously washes the catalyst zone. These factors combine to give a long catalyst life. Additionally, mercaptans can react with diolefins to make heavy, thermally-stable sulfides. The sulfides are fractionated to the bottoms product. This can eliminate the need for a separate mercaptan removal step. The distillate product is ideal feedstock for alkylation or etherification processes.

The heat of reaction evaporates liquid, and the resulting vapor is condensed in the overhead condenser (2) to provide additional reflux. The natural temperature profile in the fractionation column results in a virtually isothermal catalyst bed rather than the temperature increase typical of conventional reactors.

The CDHydro process can operate at much lower pressure than conventional processes. Pressures for CDHydro are typically set by the fractionation requirements. Additionally, the elimination of a separate hydrogenation reactor and hydrogen stripper offer significant capital cost reduction relative to conventional technologies.

Feeding CDHydro with reformate and light-straight run for benzene saturation provides the refiner with increased flexibility to produce RFG. Isomerization of the resulting C_5/C_6 overhead stream provides higher octane and yield due to reduced benzene and C_7^+ content compared to typical isomerization feedstocks.

Economics: Fixed-bed hydrogenation requires a distillation column followed by a fixed-bed hydrogenation unit. CDHydro eliminates the fixed-bed unit by incorporating catalyst in the column. When a new distillation column is used, capital cost of the column is only 5% to 20% more than for a standard column depending on the CDHydro application. Elimination of the fixed-bed reactor and stripper can reduce capital cost by as much as 50%.

Installation: Eighteen CDHydro units are in commercial operation for C_4, C_5, C_6 and benzene hydrogenation applications. Ten units have been in operation for more than five years and total commercial operating time now exceeds 80 years for CDHydro technologies. Seventeen additional units are currently in engineering/construction.

Licensor: CDTECH.

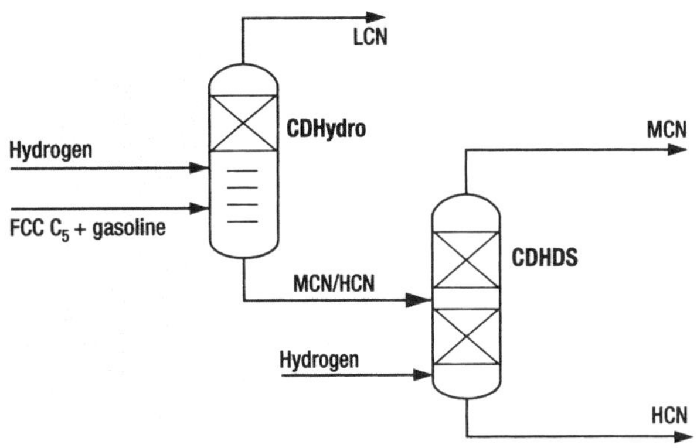

Hydrotreating

Application: CDHydro and CDHDS are used to selectively desulfurize FCC gasoline with minimum octane loss.

Products: Ultra-low-sulfur FCC gasoline with maximum retention of olefins and octane.

Description: The light, mid and heavy cat naphthas (LCN, MCN, HCN) are treated separately, under optimal conditions for each. The full-range FCC gasoline sulfur reduction begins with fractionation of the light naphtha overhead in a CDHydro column. Mercaptan sulfur reacts quantitatively with excess diolefins to product heavier sulfur compounds, and the remaining diolefins are partially saturated to olefins by reaction with hydrogen. Bottoms from the CDHydro column, containing the reacted mercaptans, are fed to the CDHDS column where the MCN and HCN are catalytically desulfurized in two separate zones. HDS conditions are optimized for each fraction to achieve the desired sulfur reduction with minimal olefin saturation. Olefins are concentrated at the top of the column, where conditions are mild, while sulfur is concentrated at the bottom where the conditions result in very high levels of HDS.

No cracking reactions occur at the mild conditions, so that yield losses are easily minimized with vent-gas recovery. The three product streams are stabilized together or separately, as desired, resulting in product streams appropriate for their subsequent use. The two columns are heat integrated to minimize energy requirements. Typical reformer hydrogen is used in both columns without makeup compression. The sulfur reduction achieved will allow the blending of gasoline that meets current and future regulations.

Catalytic distillation essentially eliminates catalyst fouling because the fractionation removes heavy-coke precursors from the catalyst zone before coke can form and foul the catalyst pores. Thus, catalyst life in catalytic distillation is increased significantly beyond typical fixed-bed life. The CDHydro/CDHDS units can operate throughout an FCC turnaround cycle up to five years without requiring a shutdown to regenerate or to replace catalyst. Typical fixed-bed processes will require a mid FCC shutdown to regenerate/replace catalyst, requiring higher capital cost for feed, storage, pumping and additional feed capacity.

Economics: The estimated ISBL capital cost for a 35,000 bpd CDHydro/CDHDS unit with 95% desulfurization is $26 million (2000 U.S. Gulf Coast). Direct operating costs—including utilities, catalyst, hydrogen and octane replacement—are estimated at $0.04/gal of full-range FCC gasoline.

Installation: Five CDHydro units are in operation treating FCC gasoline and 17 more units are currently in engineering/construction. Three CDHDS units are in operation with 17 additional units currently in engineering/construction.

Licensor: CDTECH.

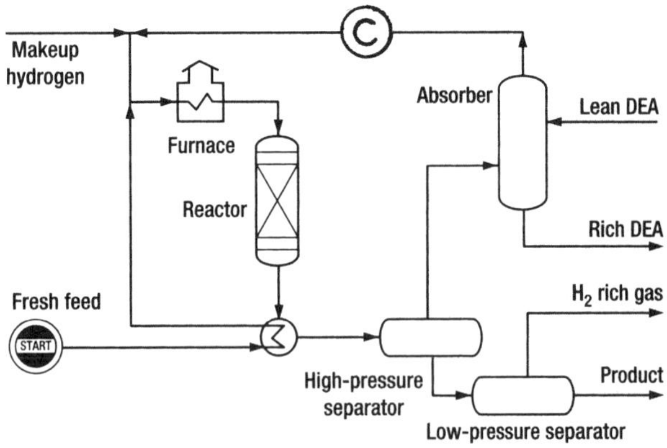

Hydrotreating

Application: Topsøe hydrotreating technology has a wide range of applications, including the purification of naphtha, distillates and residue, as well as the deep desulfurization and color improvement of diesel fuel and pretreatment of FCC and hydrocracker feedstocks.

Products: Ultra-low-sulfur diesel fuel, and clean feedstocks for FCC and hydrocracker units.

Description: Topsøe's hydrotreating process design incorporates our industrially proven high-activity TK catalysts with optimized graded-bed loading and high-performance, patented reactor internals. The combination of these features and custom design of hydrotreating units result in process solutions that meet the refiner's objectives in the most economic way.

In the Topsøe hydrotreater, feed is mixed with hydrogen, heated and partially evaporated in a feed/effluent exchanger before it enters the reactor. In the reactor, Topsøe's high-efficiency internals have a low sensitivity to unlevelness and are designed to ensure the most effective mixing of liquid and vapor streams and the maximum utilization of the catalyst volume. These internals are effective at a high range of liquid loadings, thereby enabling high turndown ratios. Topsøe's graded-bed technology and the use of shape-optimized inert topping and catalysts minimize the build-up of pressure drop, thereby enabling longer catalyst cycle length. The hydrotreating catalysts themselves are of the Topsøe TK series, and have proven their high activities and outstanding performance in numerous operating units throughout the world. The reactor effluent is cooled in the feed-effluent exchanger, and the gas and liquid are separated. The hydrogen gas is sent to an amine wash for removal of hydrogen sulfide and is then recycled to the reactor. Cold hydrogen recycle is used as quench gas between the catalyst beds, if required. The liquid product is steam stripped in a product stripper column to remove hydrogen sulfide, dissolved gases and light ends.

Operating conditions: Typical operating pressures range from 20 to 80 barg (300 to 1,200 psig), and typical operating temperatures range from 320°C to 400°C (600°F to 750°F).

References: Cooper, B. H. and K. G. Knudsen, "Production of ULSD: Catalyst, kinetics and reactor design," World Petroleum Congress, 2002.

Bingham, Muller, Christensen and Moyse, "Performance focused reactor design to maximize benefits of high activity hydrotreating catalysts," European Refining Technology Conference, 1997.

Cooper, "Meeting the challenge for middle distillates: scientific and industrial aspects," Petrotech, 1997.

de la Fuente, E., P. Christensen and M. Johansen, "Options for meeting EU year 2005 fuel specifications."

Installation: More than 35 Topsøe hydrotreating units for the various applications above are in operation or in the design phase.

Licensor: Haldor Topsøe A/S.

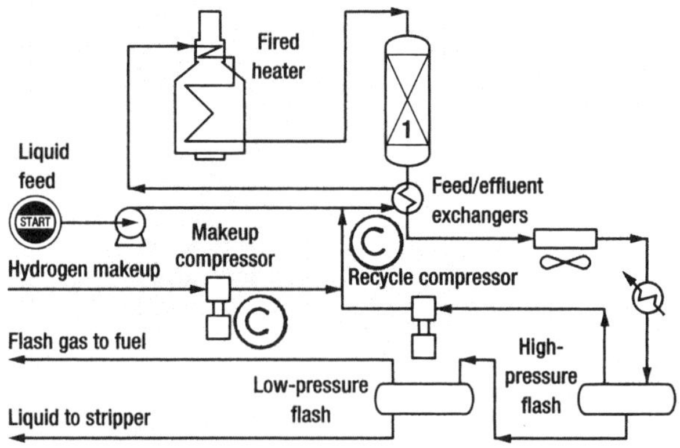

Hydrotreating

Application: Reduction of the sulfur, nitrogen, and metals content of naphthas, kerosines, diesel or gas oil streams.

Products: Low-sulfur products for sale or additional processing.

Description: Single or multibed catalytic treatment of hydrocarbon liquids in the presence of hydrogen converts organic sulfur to hydrogen sulfide and organic nitrogen to ammonia. Naphtha treating normally occurs in the vapor phase, and heavier oils usually operate in mixed-phase. Multiple beds may be placed in a single reactor shell for purposes of redistribution and/or interbed quenching for heat removal. Hydrogen-rich gas is usually recycled to the reactor(s) (1) to maintain adequate hyrogen-to-feed ratio. Depending on the sulfur level in the feed, H_2S may be scrubbed from the recycle gas. Product stripping is done with either a reboiler or with steam. Catalysts are cobalt-molybdenum, nickel-molybdenum, nickel-tungsten or a combination of the three.

Operating conditions: 550°F to 750°F and 400 to 1,500 psig reactor conditions.

Yields: Depend on feed characteristics and product specifications. Recovery of desired product almost always exceeds 98.5 wt% and usually exceeds 99%.

Economics:

Utilities, (per bbl feed)	Naphtha	Diesel
Fuel, 10^3 Btu release	48	59.5
Electricity, kWh	0.65	1.60
Water, cooling (20°F rise), gal	35	42

Licensor: Howe-Baker Engineers, Ltd., a subsidiary of Chicago Bridge & Iron Co.

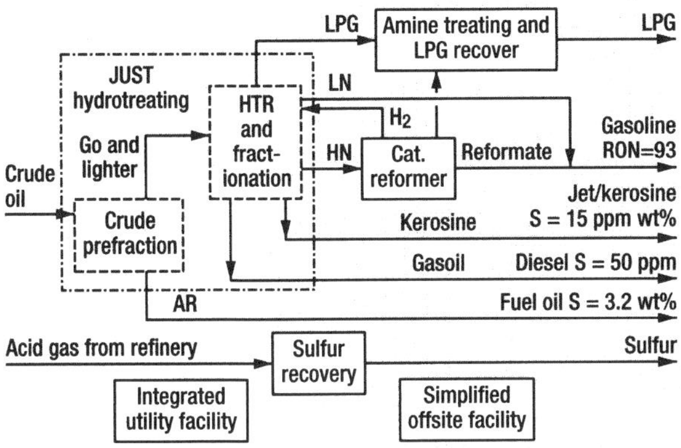

Hydrotreating

Application: The JUST Refinery process is a stand-alone refinery concept that uses new technologies to enable a 30% reduction of CAPEX over conventional refineries with similar functions; thus, improving ROI by 5%. The JUST concept is a low-cost technology option for new refineries. Benefits include:

- Minimizes the initial investment cost through a phased approach
- Improves the competitiveness of small- to medium-sized refineries.

Description: The key technologies for the JUST program include JUST Hydrotreating, which consists of a crude prefractionator, single hydrotreator and product fractionator. The simplified crude prefractionator separates crude into only two fractions: gasoil/lighter fraction and a residue. The gasoil/lighter fraction is topped off by crude prefractionator and is hydrotreated, in its entirety, by one hydrotreator. All products are hydrotreated to meet individual sulfur specification. The single hydrodesulfurization step of the combined streams tremedously reduces the CAPEX and OPEX; it eliminates using multiple (naphtha, kerosine and gasoil) hydrodesulfurization systems.

The heavy-naphtha fraction is processed through a catalytic-reforming unit and produces reformate for gasoline blending. The hydrotreater (HTR) uses hydrogen from the catalytic reformer and is, therefore, hydrogen self-sufficient.

The integrated utility unit and simplified offsite facility also lower total refinery CAPEX and OPEX expenses. A comparison illustrating the JUST Refinery advantages is presented:

	Conventional scheme	JUST Refinery
Process		
No. of process units	9	6
No. of major equip't in HTR	12	4
Utility		
Power and steam system	BTG/Boiler	GTG/HRSG*
Surface condenser	Yes	No
Independent air compressor	Yes	No
Offsite (no.of tanks)	13	28
Initial CAPEX	Base	–30%
Area requirement	Base	–34%
Economics (IRR)	Base	+5%

* HRSG: heat recovery steam generator

Reference: *The Piping Engineering,* Extra Edition, 1997, p. 39.

30th Petroleum-Petrochemical Symposium of Japan Petroleum Institute, C38, 2000.

Licensor: JGC.

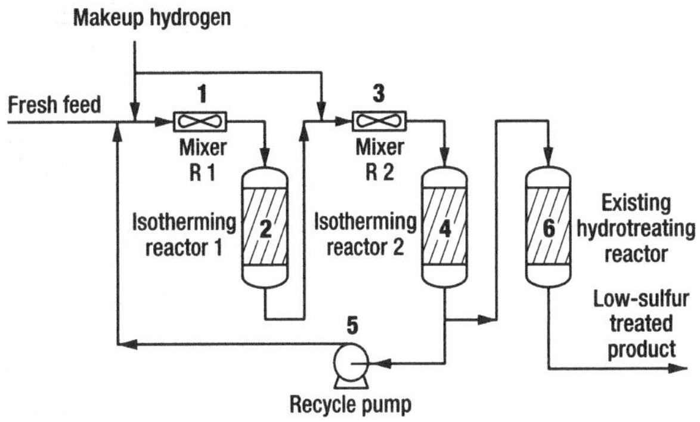

Hydrotreating

Application: The IsoTherming process, when installed in a pre-treat configuration ahead of an existing hydrotreating reactor, provides refiners an economical means to produce ultra-low-sulfur diesel (ULSD). In addition, cat-cracker feeds can be desulfurized to the extent that gasoline post-treating is not required to meet low-sulfur gasoline specifications.

Products: Ultra-low-sulfur diesel, low-sulfur cat feed and low-sulfur gasoline.

Description: This process uses a novel approach to introduce hydrogen into the reactor; it enables much higher space velocities than conventional hydrotreating reactors. The IsoTherming process removes the hydrogen mass transfer limitation and operates in a kinetically limited mode since hydrogen is delivered to the reactor in the liquid phase as soluble hydrogen.

The technology can be installed as a simple pre-treat unit ahead of an existing hydrotreater reactor. Fresh feed, after heat exchange, is combined with hydrogen in Reactor One mixer (1). The feed liquid with soluble hydrogen is fed to IsoTherming Reactor One (2) where partial desulfurization takes place. The stream is combined with additional hydrogen in Reactor Two Mixer (3), and fed to IsoTherming Reactor Two (4) where further desulfurization takes place. Treated oil is recycled (5) back to the inlet of Reactor One. This recycle stream is used to deliver more hydrogen to the reactors and also acts as a heat sink, which results in a nearly isothermal reactor operation. The treated oil from IsoTherming Reactor Two (4) is fed to the existing Hydrotreating Reactor (6) which functions in a polishing mode to produce an ultra-low-sulfur product. The process can also be configured as a grass roots hydrotreater.

Operating conditions: Typical diesel IsoTherming conditions are:

	Diesel feed	IsoTherming pre-treat reactor	Treated product from existing conventional reactor
LCO, vol%	40		
SR, vol%	60		
Sulfur, ppm	1985	140	10
Nitrogen, ppm	388	24	3
Cetane	46.50	47.80	48.60
Relative hydrogen consumption *		100%	18%
LHSV, Hr-1 **		15	3
Relative catalyst volume *		25%	100%
Reactor ΔT		15	10
Reactor pressure, psig		600	600

* Based on existing reactor producing 500 ppm sulfur diesel.
** Based on fresh feedrate without recycle.

Economics: Investment (Basis 15,000–20,000 bpsd, 2Q 2002, U.S. Gulf Coast)

$ per bpsd diesel	500

Installation: First commercial diesel unit onstream October 2002.

Licensor: Linde BOC Process Plants, LLC and Process Dynamics.

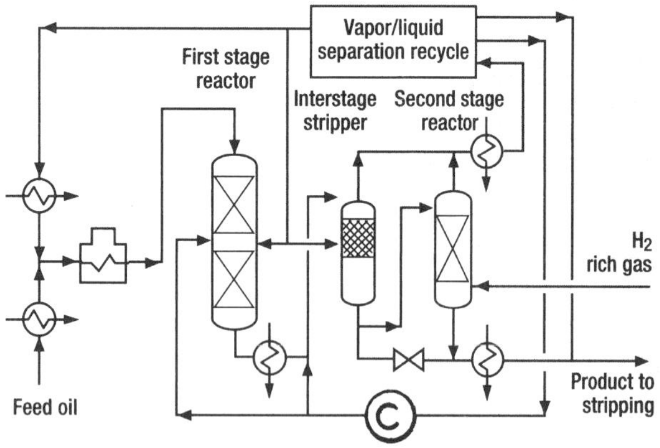

Hydrotreating

Application: Hydroprocessing of middle distillates, including cracked materials (coker/visbreaker gas oils and LCO) using SynTechnology, maximizes distillate yield while producing ultra-low-sulfur diesel with improved cetane and API gain, reduced aromatics, T95 reduction and cold-flow improvement through selective ring opening, saturation and/or isomerization. Various process configurations are available for revamps and new unit design to stage investments to meet changing diesel specifications.

Products: Maximum yield of improved quality distillate while minimizing fuel gas and naphtha. Diesel properties include less than 10-ppm sulfur, with aromatics content (total and/or PNA), cetane, density and T95 dependent on product objectives and feedstock.

Description: SynTechnology includes SynHDS for ultra-deep desulfurization and SynShift/SynSat for cetane improvement, aromatics saturation and density/T95 reduction. SynFlow for cold flow improvement can be added as required. The process combines ABB Lummus Global's cocurrent and/or patented countercurrent reactor technology with special SynCat catalysts from Criterion Catalyst Co. LP. It incorporates design and operations experience from Shell Global Solutions, to maximize reactor performance by using advanced reactor internals.

A single-stage or integrated two-stage reactor system provides various process configuration options and revamp opportunities. In a two-stage reactor system, the feed, makeup and recycle gas are heated and fed to a first-stage cocurrent reactor. Effluent from the first stage is stripped to remove impurities and light ends before being sent to the second-stage countercurrent reactor. When a countercurrent reactor is used, fresh makeup hydrogen can be introduced at the bottom of the catalyst bed to achieve optimum reaction conditions.

Operating conditions: Typical operating conditions range from 500–1,000 psig and 600°F–750°F. Feedstocks range from straight-run gas oils to feed blends containing up to 70% cracked feedstocks that have been commercially processed. For example, the SynShift upgrading of a feed blend containing 72% LCO and LCGO gave these performance figures:

	Feed blend	Product
Gravity, °API	25	33.1
Sulfur, wt% (wppm)	1.52	(2)
Nitrogen, wppm	631	<1
Aromatics, vol%	64.7	34.3
Cetane index	34.2	43.7
Liquid yield on feed, vol%		107.5

Economics: SynTechnology encompasses a family of low-to-moderate pressure processes. Investment cost will be greatly dependent on feed quality and hydroprocessing objectives. For a 30,000 to 35,000-bpsd unit, the typical ISBL investment cost in U.S.$/bpsd (U.S. Gulf Coast 2000) are:

Revamp existing unit	450–950
New unit for deep HDS	1,100–1,200
New unit for cetane improvement and HDA	1,500–1,600

Installation: SynTechnology has been selected for more than 30 units, with half of the projects being revamps. Seven units are in operation.

Licensor: ABB Lummus Global, Inc., on behalf of the SynAlliance, which includes Criterion Catalyst and Technologies Co., and Shell Global Solutions.

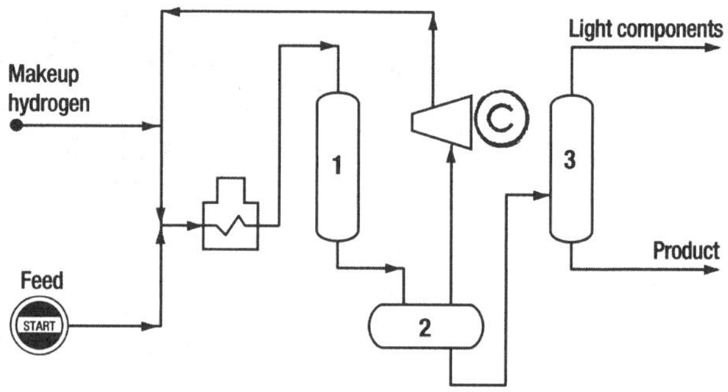

Hydrotreating

Application: Hydrodesulfurization, hydrodenitrogenation and hydrogenation of petroleum and chemical feedstocks using the Unionfining and MQD Unionfining processes.

Products: Ultra-low-sulfur diesel fuel; feed for catalytic reforming, FCC pretreat; upgrading distillates (higher cetane, lower aromatics); desulfurization, denitrogenation and demetallization of vacuum and atmospheric gas oils, coker gas oils and chemical feedstocks.

Description: Feed and hydrogen-rich gas are mixed, heated and contacted with regenerable catalyst (1). Reactor effluent is cooled and separated (2). Hydrogen-rich gas is recycled or used elsewhere. Liquid is stripped (3) to remove light components and remaining hydrogen sulfide, or fractionated for splitting into multiple products.

Operating conditions: Operating conditions depend on feedstock and desired level of impurities removal. Pressures range from 500 to 2,000 psi. Temperatures and space velocities are determined by process objectives.

Yields:

Purpose	**FCC feed**	**Desulf.**	**Desulf.**	**Desulf.**
Feed, source	**VGO + Coker**	**AGO**	**VGO**	**DSL**
Gravity, °API	17.0	25.7	24.3	32.9
Boiling range, °F	400/1,000	310/660	540/1,085	380/700
Sulfur, wt%	1.37	1.40	3	1.1
Nitrogen, ppmw	6,050	400	1,670	102
Bromine number	—	26	—	—
Naphtha, vol%	4.8	4.2	3.9	1.6
Gravity, °API	45.0	50.0	54.0	51
Boiling range, °F	180/400	C_4/325	C_4/356	C_5/300
Sulfur, ppmw	50	<2	<2	<1
Nitrogen, ppmw	30	<1	<2	<0.5
Distillate, vol%	97.2	97.6	98.0	99.0
Gravity, °API	24.0	26.9	27.8	35.2
Boiling range, °F	400+	325/660	300+	300
Sulfur, wt%	0.025	0.001	0.002	0.001
H_2 consump., scf/bbl	700	350	620	300

Economics:

Investment, $ per bpsd	1,200–2,000
Utilities, typical per bbl feed:	
Fuel, 10^3 Btu	40–100
Electricity, kWh	0.5–1.5

Installation: Several hundred units installed.

Reference: UOP, LLC., "Diesel fuel specifications and demand for the 21st Century," 1998 by UOP LLC.

Licensor: UOP LLC.

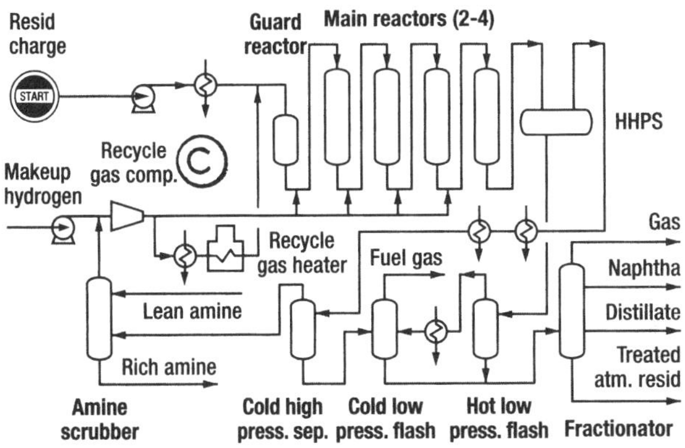

Hydrotreating

Application: RCD Unionfining process reduces the sulfur, nitrogen, Conradson carbon, asphaltene and organometallic contents of heavier residue-derived feedstocks to allow them to be used as either specification fuel oils or as feedstocks for downstream processing units such as hydrocrackers, fluidized catalytic crackers, resid catalytic crackers and cokers.

Feed: Feedstocks range from solvent-derived materials to atmospheric and vacuum residues.

Description: The process uses a fixed-bed catalytic system that operates at moderate temperatures and moderate to high hydrogen partial pressures. Typically, moderate levels of hydrogen are consumed with minimal production of light gaseous and liquid products. However, adjustments can be made to the unit's operating conditions, flowscheme configuration or catalysts to increase conversion to distillate and lighter products.

Fresh feed is combined with makeup hydrogen and recycled gas, and then heated by exchange and fired heaters before entering the unit's reactor section. Simple downflow reactors incorporating a graded bed catalyst system designed to accomplish the desired reactions while minimizing side reactions and pressure drop buildup are used. Reactor effluent flows to a series of separators to recover recycle gas and liquid products. The hydrogen-rich recycle gas is scrubbed to remove H_2S and recycled to the reactors while finished products are recovered in the fractionation section. Fractionation facilities may be designed to simply recover a full-boiling range product or to recover individual fractions of the hydrotreated product.

Economics:

Investment (basis: 15,000 to 20,000 bpsd, 2Q 2002, U.S. Gulf Coast)

$ per bpsd	2,000–3,500

Utilities, typical per barrel of fresh feed (20,000 bpsd basis)

Fuel, MMBtu/hr	46
Electricity, kWh	5,100
Steam, HP, lb/hr	8,900
Steam, LP, lb/hr	1,500

Installation: Twenty-six licensed units with a combined licensed capacity of approximately 900,000 bpsd. Commercial applications have included processing of atmospheric and vacuum residues and solvent-derived feedstocks.

Reference: Thompson, G. J., "UOP RCD Unionfining Process," R. A. Meyers, Ed., *Handbook of Petroleum Refining Processes,* 2nd ed., New York, McGraw-Hill, 1996.

Licensor: UOP LLC.

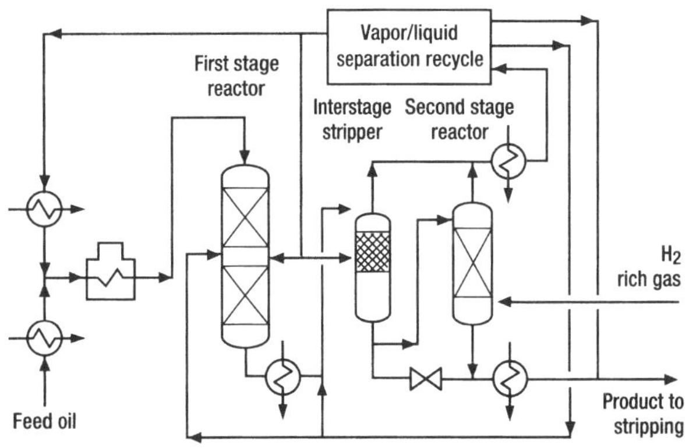

Hydrotreating–aromatic saturation

Application: Hydroprocessing of middle distillates, including cracked materials (coker/visbreaker gas oils and LCO) using SynTechnology, maximizes distillate yield while producing ultra-low-sulfur diesel with improved cetane and API gain, reduced aromatics, T95 reduction and cold-flow improvement through selective ring opening, saturation and/or isomerization. Various process configurations are available for revamps and new unit design to stage investments to meet changing diesel specifications.

Products: Maximum yield of improved quality distillate while minimizing fuel gas and naphtha. Diesel properties include less than 10-ppm sulfur, with aromatics content (total and/or PNA), cetane, density and T95 dependent on product objectives and feedstock.

Description: SynTechnology includes SynHDS for ultra-deep desulfurization and SynShift/SynSat for cetane improvement, aromatics saturation and density/T95 reduction. SynFlow for cold flow improvement can be added as required. The process combines ABB Lummus Global's cocurrent and/or patented countercurrent reactor technology with special SynCat catalysts from Criterion Catalyst Co. LP. It incorporates design and operations experience from Shell Global Solutions, to maximize reactor performance by using advanced reactor internals.

A single-stage or integrated two-stage reactor system provides various process configuration options and revamp opportunities. In a two-stage reactor system, the feed, makeup and recycle gas are heated and fed to a first-stage cocurrent reactor. Effluent from the first stage is stripped to remove impurities and light ends before being sent to the second-stage countercurrent reactor. When a countercurrent reactor is used, fresh makeup hydrogen can be introduced at the bottom of the catalyst bed to achieve optimum reaction conditions.

Operating conditions: Typical operating conditions range from 500–1,000 psig and 600°F–750°F. Feedstocks range from straight-run gas oils to feed blends containing up to 70% cracked feedstocks that have been commercially processed. For example, the SynShift upgrading of a feed blend containing 72% LCO and LCGO gave these performance figures:

	Feed blend	Product
Gravity, °API	25	33.1
Sulfur, wt% (wppm)	1.52	(2)
Nitrogen, wppm	631	<1
Aromatics, vol%	64.7	34.3
Cetane index	34.2	43.7
Liquid yield on feed, vol%		107.5

Economics: SynTechnology encompasses a family of low-to-moderate pressure processes. Investment cost will be greatly dependent on feed quality and hydroprocessing objectives. For a 30,000 to 35,000-bpsd unit, the typical ISBL investment cost in U.S.$/bpsd (U.S. Gulf Coast 2002) are:

Revamp existing unit	450–950
New unit for deep HDS	1,100–1,200
New unit for cetane improvement and HDA	1,500–1,600

Installation: Eleven SynTechnology units are in operation with an additional seven units in design and construction.

Licensor: ABB Lummus Global, Inc., on behalf of the SynAlliance, which includes Criterion Catalyst Co., LP, and Shell Global Solutions.

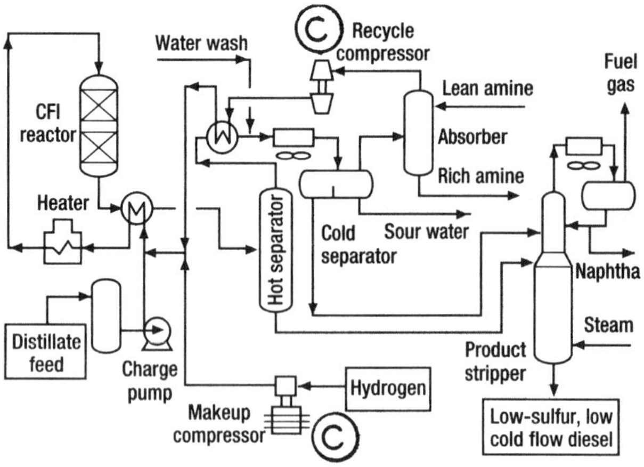

Hydrotreating—catalytic dewaxing

Application: A versatile family of premium distillates technologies is used to meet all current and possible future premium diesel upgrading requirements. The addition of selective normal paraffin hydrocracking (CFI) function to the deep hydrodesulfurization (UDHDS) reactor will improve the diesel product cold flow properties for a wide range of waxy distillate feedstocks.

Products: Ultra-low-sulfur distillate is produced with modest amounts of lighter products. Low-cloud point, or pour point product quality diesel can be achieved with the CFI processes.

Description: This Akzo-Fina CFI technology is offered through the alliance between Akzo Nobel Catalysts and Fina Research S.A. When the distillate product must meet stringent fluidity specifications, Akzo Nobel can offer this selective normal paraffin cracking based CFI dewaxing technology. Dewaxing is generally a higher cost process but delivers higher total product quality. This technology can be closely integrated with UD-HDS and other functions to achieve the full upgrading requirements in low-cost-integrated designs. The CFI process uses a single-stage design even with high levels of heteroatoms in the feed. The Akzo Fina CFI Technology is equally applicable to revamp and grassroots applications.

Economics:

Investment (basis: 15,000 to 25,000 bpsd, 1Q 2000 U.S. Gulf Coast)
Grassroots unit, $ per bpsd 1,000 to 2,000

Installation: Over 15 distillate-upgrading units have applied the Akzo Fina CFI Technology.

Reference: "MAKFining-Premium Distillates Technology: The future of distillate upgrading," NPRA Annual Meeting, March 2000, San Antonio.

Licensor: Akzo Nobel Catalysts bv and Fina Research S.A.

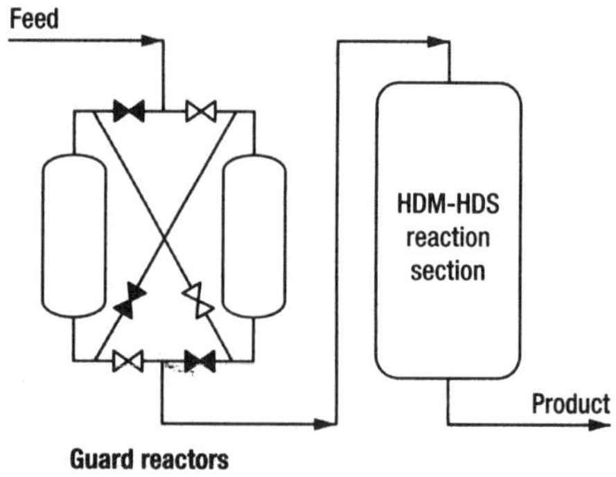

Hydrotreating—resid

Application: Upgrade or convert atmospheric and vacuum residues using the Hyvahl fixed-bed process.

Products: Low-sulfur fuels (0.3% to 1.0% sulfur) and RFCC feeds (removal of metals, sulfur and nitrogen, reduction of carbon residue). Thirty percent to 50% conversion of the 565°C+ fraction into distillates.

Description: Residue feed and hydrogen, heated in a feed/effluent exchangers and furnace, enter a reactor section—typically comprising of a guard-reactor section, main HDM and HDS reactors.

The guard reactors are onstream at the same time in series, and they protect downstream reactors by removing or converting sediment, metals and asphaltenes. For heavy feeds, they are permutable in operation (PRS technology) and allow catalyst reloading during the run. Permutation frequency is adjusted according to feed-metals content and process objectives. Regular catalyst changeout allows a high and constant protection of downstream reactors.

Following the guard reactors, the HDM section carries out the remaining demetallization and conversion functions. With most of the contaminants removed, the residue is sent to the HDS section where the sulfur level is reduced to the design specification.

The PRS technology associated with the high stability of the HDS catalytic system leads to cycle runs exceeding a year even when processing VR-type feeds to produce ultra-low- sulfur fuel oil.

Yields: Typical HDS and HDM rates are above 90%. Net production of 12% to 25% of diesel + naphtha.

Economics:

Investments (basis: 40,000 bpsd, AR to VR feeds, 2002 Gulf coast), U.S.$/ bpsd	3,500–5,500
Utilities, per bbl feed:	
Fuel, equiv. fuel oil, kg	0.3
Power, kWhr	10
Steam production, MP, kg	25
Steam consumption, HP, kg	10
Water, cooling, m^3	1.1

Installation: Two units are in operation (one on atmospheric-residue feed, the other on vacuum residue) and a third unit using VR feed will come onstream at the end of 2002; thus, the total installed capacity will reach 134,000 bpsd.

References: "Option for Resid Conversion," BBTC, Oct. 8–9, 2002, Istanbul.

"Maintaining On-spec products with residue hydroprocessing," 2000 NPRA Annual Meeting, March 26–28, 2000, San Antonio.

Licensor: Axens, Axens NA.

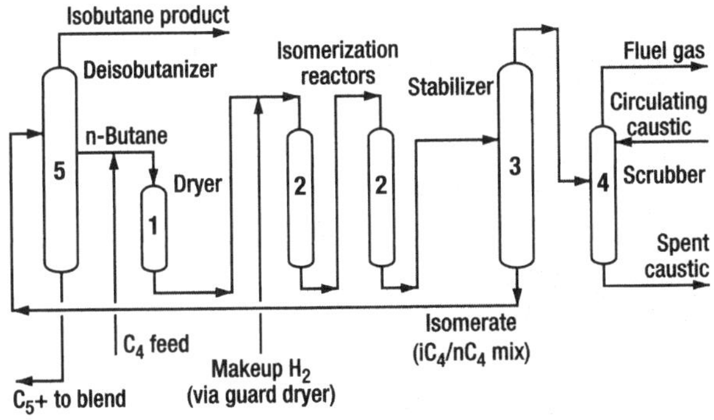

Isomerization

Application: Converting n-butane to isobutene using the Lummus butane isomerization process. Isobutane is typically the feedstock for downstream alkylation units or MTBE complexes.

Products: Isobutane, fuel gas. The Lummus butane isomerization process has high per pass conversion (>60%) and selectivity (>99%), which provides the highest total yield of isobutene.

Description: The Lummus butane process uses Akzo Nobel's AT series catalyst to isomerize n-butane into isobutene. The high-activity chlorided alumina catalyst allows operation at low temperature, which increases both conversion and selectivity while minimizing capital costs. The reaction is vapor phase at mild temperatures with the presence of a small amount of hydrogen. The high stability of the catalyst at low H_2:HC ratios allows operation without a recycle compressor.

The n-butane feed and makeup hydrogen streams are dried over molecular sieves (1), combined, heated to reaction temperatures in feed/effluent exchangers followed by a trim heater and sent to two reactors in series (2). The two reactors are used to allow operational flexibility and lower the temperature in the second reactor for higher conversion. The reactor effluent is sent to a stabilizer column to remove hydrogen and light ends (3). The stabilizer overhead is directed to fuel gas via a caustic scrubber (4). The stabilized bottoms is sent to the deisobutanizer which produces the final isobutene product, recycles n-butane back to the reactors, and removes any C_5^+ material that entered the unit in the feed (5).

Economics:

Investment (basis 10,000 bpsd unit) $/bpsd	1,900

Installation: 12,000 bpd DUGAS Dubai, United Arab Emirates.

Licensor: ABB Lummus Global Inc.

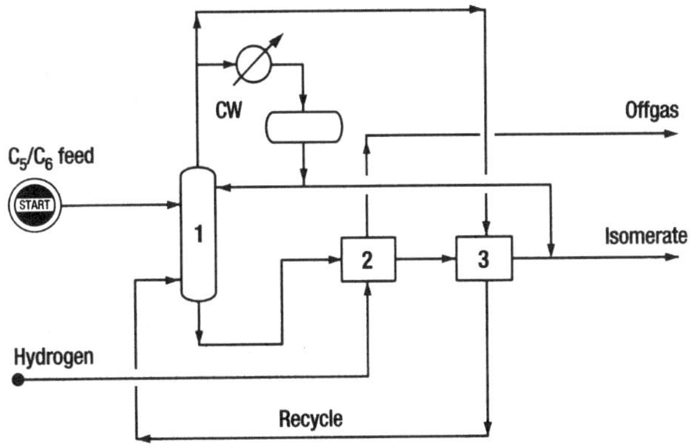

Isomerization

Application: C_5/C_6 paraffin-rich hydrocarbon streams are isomerized to produce high RON and MON product suitable for addition to the gasoline pool.

Description: Several variations of the C_5/C_6 isomerization process are available. With either a zeolite or chlorinated alumina catalyst, the choice can be a once-through reaction for an inexpensive-but-limited octane boost, or, for substantial octane improvement, the Ipsorb Isom scheme shown above to recycle the normal paraffins for their complete conversion. The Hexorb Isom configuration achieves a complete normal paraffin conversion plus substantial conversion of low (75) octane methyl pentanes gives the maximum octane results. The product octanes from five process schemes for treating a light naphtha feed (70 RON) containing a 50/50 mixture of C_5/C_6 paraffins are:

Process configuration	Zeolite catalyst	Chlorinated alumina catalyst
Once-through	80	83
Deisopentanizer and once-through	82	84
Deisohexanizer and recycle	86	88
Normal recycle-Ipsorb	88	90
Normal and deisohex. recycle-Hexorb	92	92

Operating conditions: The Ipsorb Isom process uses a deisopentanizer (1) to separate the isopentane from the reactor feed. A small amount of hydrogen is also added to reactor (2) feed. The isomerization reaction proceeds at moderate temperature producing an equilibrium mixture of normal and isoparaffins. The catalyst has a long service life. The reactor products are separated into isomerate product and normal paraffins in the Ipsorb molecular sieve separation section (3) which features a novel vapor phase PSA technique. This enables the product to consist entirely of branched isomers.

Economics: (basis: Ipsorb "A" Isomerization unit with a 5,000-bpsd 70 RONC feed needing a 20 point octane boost):

Investment*, million U.S.$	13
Utilities:	
Steam, HP, tph	1.0
Steam, MP, tph	8.5
Steam, LP, tph	6.8
Power, kWh/h	310
Cooling water, m^3/h	100

* Mid-2002, Gulf coast, excluding cost of noble metals.

Installation: Of 24 licenses issued for C_5/C_6 isomerization plants, 11 units are operating including one Ipsorb unit.

Licensor: Axens, Axens NA.

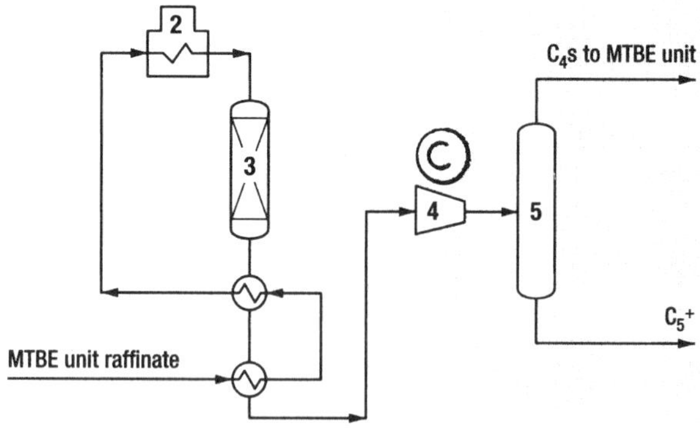

Isomerization

Application: Convert normal olefins to isoolefins.

Description:

C_4 olefin skeletal isomerization (IsomPlus)

A zeolite-based catalyst especially developed for this process provides near equilibrium conversion of normal butenes to isobutylene at high selectivity and long process cycle times. A simple process scheme and moderate process conditions result in low capital and operating costs. Hydrocarbon feed containing n-butenes, such as C_4 raffinate, can be processed without steam or other diluents, nor the addition of catalyst activation agents to promote the reaction. Near-equilibrium conversion levels up to 44% of the contained n-butenes are achieved at greater than 90% selectivity to isobutylene. During the process cycle, coke gradually builds up on the catalyst, reducing the isomerization activity. At the end of the process cycle, the feed is switched to a fresh catalyst bed, and the spent catalyst bed is regenerated by oxidizing the coke with an air/nitrogen mixture. The butene isomerate is suitable for making high purity isobutylene product.

C_5 olefin skeletal isomerization (IsomPlus)

A zeolite-based catalyst especially developed for this process provides near-equilibrium conversion of normal pentenes to isoamylene at high selectivity and long process cycle times. Hydrocarbon feeds containing n-pentenes, such as C_5 raffinate, are processed in the skeletal isomerization reactor without steam or other diluents, nor the addition of catalyst activation agents to promote the reaction. Near-equilibrium conversion levels up to 72% of the contained normal pentenes are observed at greater than 95% selectivity to isoamylenes.

Economics: The Lyondell isomerization process offers the advantages of low capital investment and operating costs coupled with a high yield of isobutylene. Also, the small quantity of heavy byproducts formed can easily be blended into the gasoline pool. Capital costs (equipment, labor and detailed engineering) for three different plant sizes are:

Total installed cost:	**Feedrate,** Mbpd	**ISBL cost,** $MM
	10	8
	15	11
	30	30

Utility costs: per barrel of feed (assuming an electric-motor-driven compressor) are:

Power, kWh	3.2
Fuel gas, MMBtu	0.44
Steam, MP, MMBtu	0.002
Water, cooling, MMBtu	0.051
Nitrogen, scf	57–250

Installation: One plant is in operation. Three licensed units are in various stages of design.

Licensor: CDTECH and Lyondell Chemical Co.

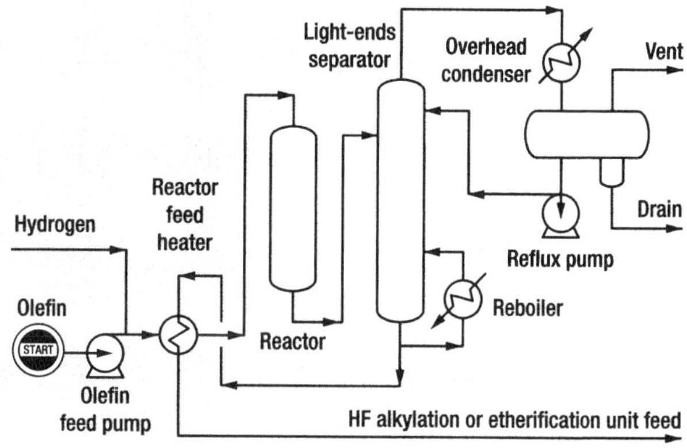

Isomerization

Application: Hydrisom is ConocoPhillips Co.'s selective diolefin hydrogenation process, with specific isomerization of butene-1 to butene-2 and 3-methyl-butene-1 to 2-methyl-butene-1 and 2-methyl-butene-2. The Hydrisom process uses a liquid-phase reaction over a commercially available catalyst in a fixed-bed reactor.

Description: The Hydrisom process is a once-through reaction and, for typical cat cracker streams, requires no recycle or cooling. Hydrogen is added downstream of the olefin feed pump on ratio control and the feed mixture is preheated by exchange with the fractionator bottoms and/or low-pressure steam. The feed then flows downward over a fixed bed of commercial catalyst.

The reaction is liquid-phase, at a pressure just above the bubble point of the hydrocarbon/hydrogen mixture. The rise in reactor temperature is a function of the quantity of butadiene in the feed and the amount of butene saturation that occurs.

The Hydrisom process can also be configured using a proprietary catalyst to upgrade streams containing diolefins up to 50% or more, e.g., steam cracker C_4 steams, producing olefin-rich streams for use as chemical, etherification and/or alkylation feedstocks.

Installation of a Hydrisom unit upstream of an etherification and/or alkylation unit can result in a very quick payout of the investment due to:

- Improved etherification unit operations
- Increased ether production
- Increased alkylate octane number
- Increased alkylate yield
- Reduced chemical and HF acid costs
- Reduced ASO handling
- Reduced alkylation unit utilities
- Upgraded steam cracker or other high diolefin streams (30% to 50%) for further processing.

Installation: Ten units licensed worldwide, including an installation at ConocoPhillips Refinery, Sweeny, Texas.

Licensor: Fuels Technology Division of ConocoPhillips Co.

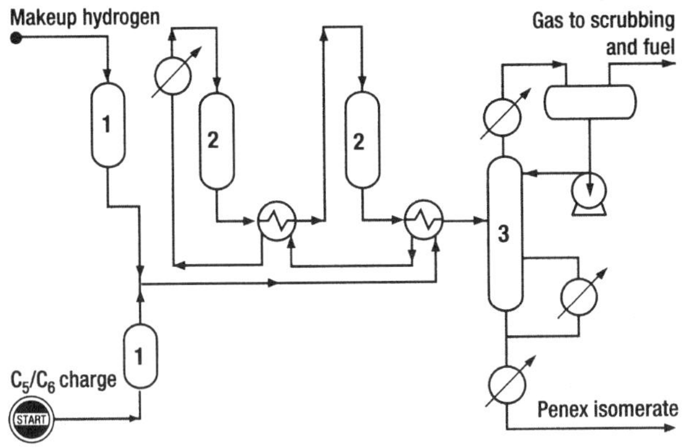

Isomerization

Application: Paraffin isomerization technology for light naphtha offers a wide variety of processing options that allow refiners to tailor performance to their specific needs. Applications include octane enhancement and benzene reduction. The Penex process is specifically designed for continuous catalytic isomerization of pentanes, hexanes and mixtures of the two. The reactions take place in a hydrogen atmosphere, over a fixed catalyst bed, and at operating conditions that promote isomerization and minimize hydrocracking.

Products: A typical C_5/C_6 light naphtha feedstock can be upgraded to 82-84 RONC in hydrocarbon once-through operation. This can be increased to about 87-93 RONC by recycling unconverted normal pentane, normal hexane and/or methylpentanes. Some systems for separating the components for recycle are: vapor phase adsorptive separation (IsoSiv process), liquid phase adsorptive separation (Molex process), fractionation in a deisohexanizer column or a combination of fractionation and selective adsorption. The Par-Isom process is a lower cost isomerization option. It provides a 1–2 lower octane-number product with regenerable catalyst. Dryers are not required; recycle hydrogen is needed. The metal oxide catalyst is an ideal replacement for zeolitic catalyst. This process is a cost-effective revamp option.

Description: Hydrogen recycle is not required for the Penex process, and high conversion is achieved at low temperature with negligible yield loss. A fired heater is not required. The flow diagram represents the Hydrogen-Once-Through (HOT) Penex process. A two reactor in series flow configuration is normally used with the total required catalyst being equally distributed between the two vessels. This allows the catalyst to be fully utilized.

Feed and makeup hydrogen are dried (1) over adsorbent and then mixed. The mixture is heated against reactor effluent and sent to the reactors (2). Reactor effluent passes directly to the stabilizer (3) after heat exchange. Stabilizer bottoms are sent to gasoline blending in a once-through operation or to separation (adsorption or fractionation) in a recycle operation. The light ends are sent to a caustic-scrubber column and then to fuel.

Economics: The typical estimated erected costs for 2Q 2002 ISBL, U.S. Gulf Coast for a 10,000-bpsd unit are:

Flowscheme	EEC, $MM
Penex	10.1
Penex/Molex	25
Penex/DIH	17.1

Installation: UOP is the world's leading licensor in C_5/C_6 isomerization technology. The first Penex unit was placed on stream in 1958. Over 188 UOP C_5/C_6 isomerization units have been commissioned as of 2Q 2002.

Licensor: UOP LLC.

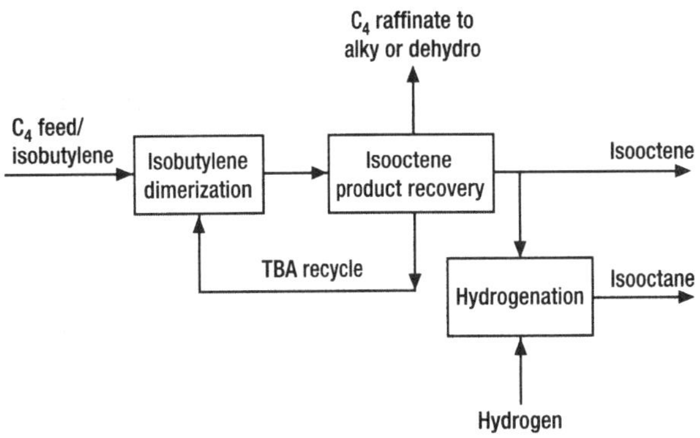

Isooctane

Application: Conversion of isobutylene contained in mixed-C_4 feeds to isooctane (2,2,4 tri-methyl pentane) to produce a high-quality gasoline blendstock. The full range of MTBE plant feeds can be processed—from refinery FCC, olefin-plant raffinate and isobutane dehydrogenation processes. The NExOCTANE process is specifically developed to minimize conversion costs of existing MTBE units and offers a cost-effective alternative to MTBE production.

Products: Isooctene and isooctane can be produced, depending on the refiner's gasoline pool. Typical product properties are:

	Isooctene	Isooctane
RONC	101–103	99–100
MONC	85–87	96–99
Specific gravity	0.701–0.704	0.726–0.729
Vapor pressure, psia	1.8	1.8
ASTM EP, °F	380–390	370–380

Description: In the NExOCTANE process, reuse of existing equipment from the MTBE unit is maximized. The process consists of three sections. First, isobutylene is dimerized to isooctene in the reaction section. The dimerization reaction occurs in the liquid phase over an acidic ion-exchange resin catalyst, and it uses simple liquid-phase-fixed-bed reactors. The isooctene product is recovered in a distillation system, for which generally the existing fractionation equipment can be reused. The recovered isooctene product can be further hydrogenated to produce isooctane. A highly efficient trickle-bed hydrogenation technology is offered with the NExOCTANE process. This compact and cost-effective technology does not require recirculation of hydrogen. In the refinery, the NExOCTANE process fits as a replacement to MTBE production, thus associated refinery operations are mostly unaffected.

Economics:

Investment cost for revamps depend on the existing MTBE plant design, capacity and feedstock composition. Typical utility requirements per bbl product:

Steam, 150-psig, lb	700
Electricity, kWh	2.3
Water, cooling, ft^3	1.2

Installation: Process has been commercially demonstrated.

Licensor: Kellogg Brown & Root, Inc., and Fortum Oil and Gas OY.

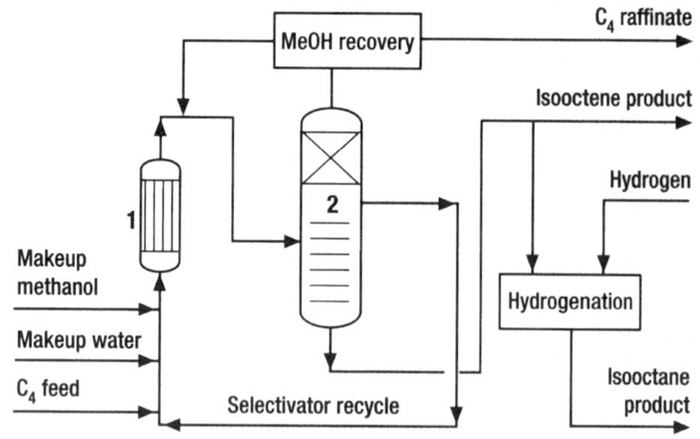

Isooctane/isooctene

Application: CDIsoether is used for manufacture of high-octane, low-vapor pressure, "MTBE-free" isooctene and/or isooctane for gasoline blending. Coproduction of MTBE and isooctene/isooctane in the desired ratio is also possible.

Feed: Hydrocarbon streams containing reactive tertiary olefins such as: FCC C_4s, steamcracker C_4s or isobutane dehydrogenation product.

Products: Isooctene or isooctane stream containing at least 85% of C_8s, with less than 5,000 ppm oligomers higher than C_{12}s.

Description: Depending on conversion and investment requirements, various options are available. CDIsoether can provide isobutylene conversion of up to 99%. The C_4 feed is mixed with a recycle stream containing oxygenates (such as TBA and MTBE), used as "selectivator" and heated before entering the reactor. The reactor (1) is a water-cooled tubular reactor (WCTR) or a boiling-point reactor (BPR).

The heat of reaction is removed by circulating water through the shell of the WCTR, while the heat of reaction remains in the two-phase BPR effluent. There is no product recycle. The reactor effluent flows, along with the selectivator, to the reaction column (2), where isobutene conversion is maximized using catalytic distillation and isooctene product is fractionated as bottoms product.

Unreacted C_4s are taken as column overhead and the selectivator is drawn as a side stream for recycle together with some C_4 hydrocarbons. The isooctene product can be sent to storage or fed to the "hydrogenation unit" to produce saturated hydrocarbon—isooctane.

Economics:

Investment (basis grassroots CDIsoethers unit, charging FCC C_4s)
5,000–7,000 U.S.$ per bpsd of isooctene product

Investment for retrofitting an existing MTBE unit to isooctene production
500–750 U.S.$ per bpsd of isooctene produced

Utilities, per bbl of isooctene:

Steam, (300 psig), lb	200–250
Water, cooling, gal	1,500–2,000
Power, kWh	1.6–2.0

Licensor: Snamprogetti SpA and CDTECH.

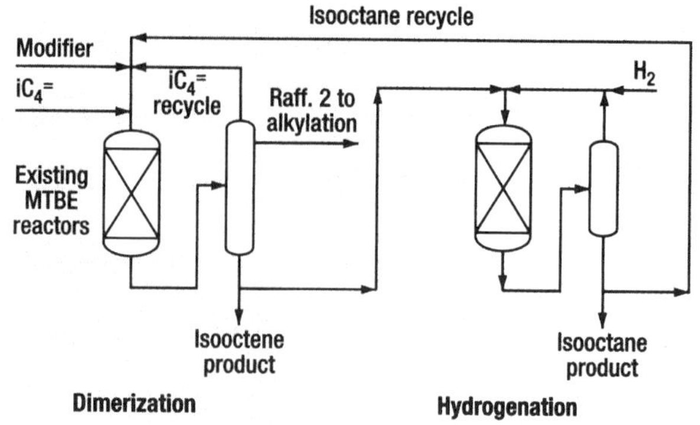

Isooctene/Isooctane/ETBE

Application: To produce isooctene or isooctane from isobutylene both steps—via catalytic dimerization followed by hydrogenation; with intermediate and final fractionation as required to meet final product specifications. Ideally, it is a "drop-in" to an existing MTBE reactor with patented use of modifier to improve selectivity and prolong catalyst life.

The process can be easily modified to make ETBE from ethanol and isobutylene as well.

Description: The process produces an isooctene intermediate or final product starting with either a mixed C_4 feed or on-purpose isobutylene production. It is based on a highly selective conversion of isobutylene to isooctene followed by hydrogenation, which will convert over 99.5% of the isooctene to isooctane. The product has high-gasoline blending quality with superior octane rating and low Rvp. The design has the added advantage of being inter-convertible between isooctene/isooctane and MTBE production.

Economics: The "drop-in" design capability offers an efficient and cost-effective approach to conversion of existing MTBE units. In retro-fit applications, this feature allows for maximum utilization of existing equipment and hardware, thus reducing the capital costs of conversion to an alternate process/production technology. For the production of isooctane, the process uses low-risk conventional hydrogenation with slight design enhancements for conversion of isooctene.

The unit can be designed to be inter-convertible between MTBE, isooctene/isooctane and/or ETBE operations. Thus, economics, as well as changes in regulations, can dictate changes in the mode of operation over time.

Commercial plants: Preliminary engineering and licensing is under evaluation at several MTBE producers worldwide.

Licensor: Lyondell Chemical and Aker Kvaerner.

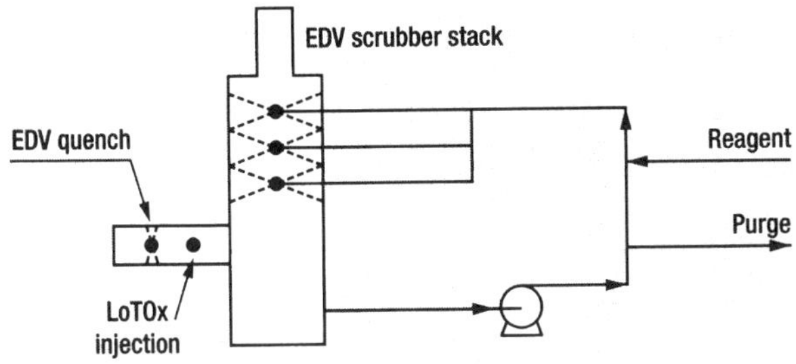

Low-temperature NO_x reduction

Application: The LoTOx low-temperature oxidation process removes NO_x from flue gases in conjunction with BELCO's EDV wet scrubbing system. Ozone is a very selective oxidizing agent; it converts relatively insoluble NO and NO_2 to higher, more soluble nitrogen oxides. These oxides are easily captured in a wet scrubber that is controlling sulfur compounds and/or particulates simultaneously.

Description: In the LoTOx process, ozone is added to oxidize insoluble NO and NO_2 to highly oxidized, highly soluble species of NO_x that can be effectively removed by a variety of wet or semi-dry scrubbers. Ozone, a highly effective oxidizing agent, is produced onsite and ondemand by passing oxygen through an ozone generator—an electric corona device with no moving parts. The rapid reaction rate of ozone with NO_x results in high selectivity for NO_x over other components within the gas stream.

Thus, the NO_x in the gas phase is converted to soluble ionic compounds in the aqueous phase; the reaction is driven to completion, thus removing NO_x with no secondary gaseous pollutants. The ozone is consumed by the process or destroyed within the system scrubber. All system components are proven, well-understood technologies with a history of safe and reliable performance.

Operating conditions: Ozone injection typically occurs in the flue-gas stream upstream of the scrubber, near atmospheric pressure and at temperatures up to roughly 160°C. For higher-temperature streams, the ozone is injected after a quench section of the scrubber, at adiabatic saturation, typically 60°C to 75°C. High-particulate and sulfur loading (SO_x or TRS) do not cause problems.

Economics: The costs for NO_x control using this technology are especially low when used as a part of a multi-pollutant control scenario. Sulfurous and particulate-laden streams can be treated attractively as no pretreatment is required by the LoTOx system. Typical costs range from \$1,500 to \$4,000/t of NO_x controlled.

Installation: The technology has been developed and commercialized over the past seven years, winning the prestigious 2001 Kirkpatrick Chemical Engineering Technology Award. As of early 2002, four full-scale commercial installations are operating successfully. Pilot-scale demonstrations have been completed on coal- and petroleum-coke fired boilers, as well as, many other combustion and process sources. An FCCU pilot demonstration is contracted to occur in the fall 2002.

Licensor: Belco Technologies Corp., as a sub-licensor for The BOC Group, Inc.

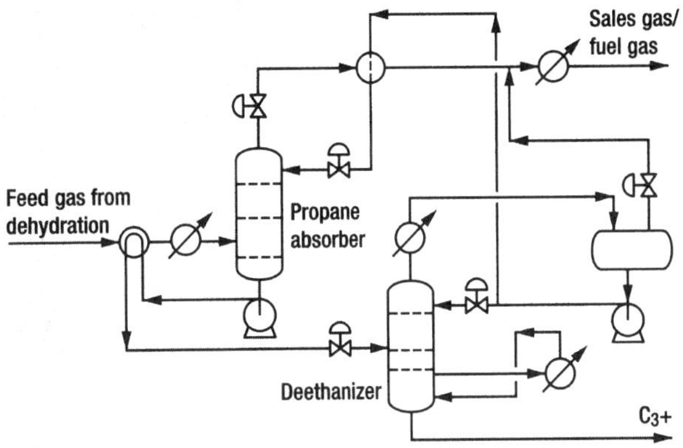

LPG recovery

Application: Recovery of propane and heavier components from various refinery offgas streams and from low-pressure associated natural gas. Propane recovery levels approaching 100% are typical.

Description: Low-pressure hydrocarbon gas is compressed and dried before being chilled by cross-exchange and propane refrigerant. The chilled feed stream is then contacted with a recycled liquid ethane stream in the propane absorber. The absorber bottoms is pumped to the deethanizer, which operates at higher pressure than the absorber. The tower overhead is condensed with propane refrigerant to form a reflux stream composed primarily of ethane. A slip stream of the reflux is withdrawn and recycled back to the propane absorber. The deethanizer bottoms stream contains the valuable propane and heavier components which may be further processed as required by conventional fractionation.

Economics: Compared to other popular LPG recovery processes, PRO-MAX typically requires 10-25% less refrigeration horsepower.

Installation: First unit under construction for Pertamina.

Reference: U.S. Patent 6,405,561 issued June 18, 2002.

Licensor: Black & Veatch Pritchard, Inc.

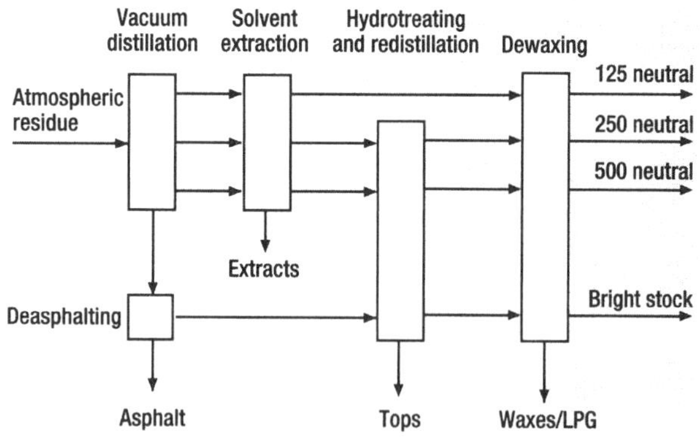

Lube hydroprocessing

Application: The Hybrid base oil manufacturing process is an optimized combination of solvent extraction and one-stage hydroprocessing. It is particularly suited to the revamping/debottlenecking of existing solvent extraction lube oil plants; capacity increases as great as 60% can be achieved. Solvent extraction also liensed by Shell.

Feed: Derived from a wider range of crudes than can be used with solvent extraction. Yields and capacity are less sensitive to feedstock than when the solvent extraction process is applied.

Description: Two separate upgrading units are used; solvent extraction and one-stage hydroprocessing. Individual waxy distillate streams are either mildly solvent-extracted and then hydroprocessed or are solvent extracted at normal severity. Deasphalted oil is either mildly solvent extracted and then hydroprocessed or is only hydroprocessed. The choice of solvent extraction and hydroprocessing depends upon the feedstock and the objectives of debottlenecking (min. capital expenditure or max. capacity increase).

When the Shell process is used to debottleneck a lube oil plant, it is necessary to construct two new units: a hydrotreating/redistillation unit and an additional dewaxing unit (the existing dewaxing unit usually has insufficient spare capacity). There is normally no need to construct additional vacuum distillation/deasphalting/solvent extraction units. However, some modifications will be required to the existing vacuum distillation and solvent extraction units.

Yields: Depend upon the grade of base oil and the crude origin of the feedstock. Shell Hybrid gives a significantly higher yield of base oil crude and the yield is much less sensitive to feedstock origin than with solvent extraction process.

Base oils obtained via the Shell Hybrid process are lighter in color and have lower Conradson carbon residue contents than their solvent extracted counterparts and can more advantageously be used in a number of special applications. Moreover, the low-sulfur, low-pour-point gas oil byproducts from the hydrotreating unit can have enhanced value in special markets, while the quantity of low-value byproducts (e.g., extracts) is substantially reduced.

Economics: The following table compares the economics of debottlenecking a 300 ktpy solvent extraction complex to 500 ktpy with the economics of a new 200 ktpy solvent extraction complex.

	Solvent extraction 200 ktpy grass-roots	Hybrid debottlenecking (from 300 to 500 ktpy)
Capital charge	36% of total	24–36% of solvex total
Fixed costs	20% of total	7–9% of solvex total
Variable costs	8% of total	8% of solvex total
Hydrocarbon cost	36% of total	11% of solvex total
Total	100% of total	50–64% of solvex total

Installation: The process has been commercially applied in Shell's Geelong refinery since 1980. Pertamina is applying the Shell Hybrid technology to debottleneck its Cilacap refinery—the successful start occurred in the second half of 1998.

Licensor: Shell Global Solutions International B.V.

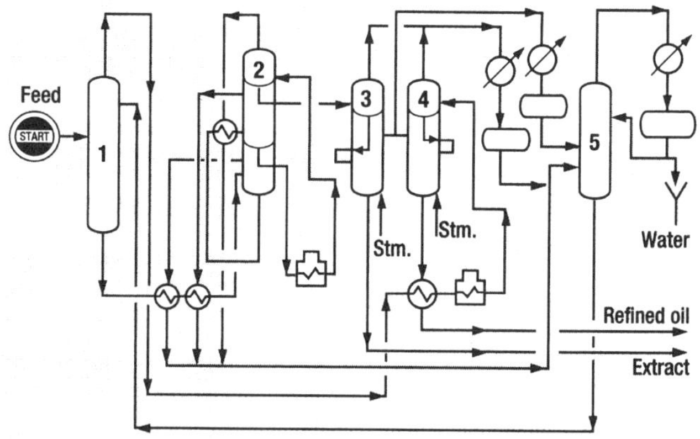

Lube treating

Application: Bechtel's MP Refining process is a solvent extraction process that uses N-methyl-2-pyrrolidone (NMP) as the solvent to selectively remove undesirable components of low lubrication oil quality, which are naturally present in crude oil distillate and residual stocks. The unit produces paraffinic or naphthenic raffinates that are suitable for further processing into lube-base stocks. This process selectively removes aromatics and compounds containing heteroatoms (e.g., oxygen, nitrogen and sulfur).

Products: A raffinate that may be dewaxed to produce a high-quality, lube-base oil, characterized by high-viscosity index, good thermal and oxidation stability, light color and excellent additive response. The byproduct extracts, having a high aromatic content, can be used, in some cases, for carbon black feedstocks, rubber extender oils and other nonlube applications where this feature is desirable.

Description: The distillate or residual feedstock and solvent are contacted in the extraction tower (1) at controlled temperatures and flowrates required for optimum countercurrent, liquid-liquid extraction of the feedstock. The extract stream, containing the bulk of the solvent, exits the bottom of the extraction tower. It is routed to a recovery section to remove solvent contained in this stream. The solvent is separated from the extract oil by multiple-effect evaporation (2) at various pressures, followed by vacuum flashing and steam stripping (3) under vacuum. The raffinate stream exits the overhead of the extraction tower and is routed to a recovery section for removal of the NMP solvent contained in this stream by flashing and steam stripping (4) under vacuum.

Overhead vapors from the steam strippers are condensed and combined with solvent condensate from the recovery sections and are distilled at low pressure to remove water from the solvent (5). Solvent is recovered in a single tower because NMP does not form an azeotrope with water, as does furfural. The water is drained to the oily-water sewer. The solvent is cooled and recycled to the extraction section.

Economics:

Investment (basis: 10,000-bpsd feedrate capacity, 2002 U.S. Gulf Coast), \$/bpsd	2,500
Utilities, typical per bbl feed:	
Fuel, 10^3 Btu (absorbed)	130
Electricity, kWh	0.8
Steam, lb	8
Water, cooling (25°F rise), gal	550

Installation: This process is being used in 15 licensed units to produce high-quality lubricating oils. Of this number, eight are units converted from phenol or furfural, with another two units being planned for conversion from furfural. Presently, two new units that will refine used oil have been designed.

Licensor: Bechtel Corp.

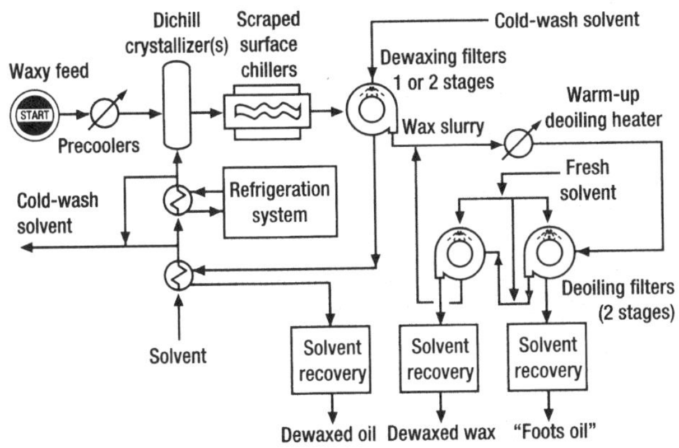

Lube treating

Application: Lube raffinates from extraction are dewaxed to provide basestocks having low pour points (as low as –35°C). Basestocks range from light stocks (60N) to higher viscosity grades (600N and bright stock). Byproduct waxes can also be upgraded for use in food applications.

Feeds: DILCHILL dewaxing can be used for a wide range of stocks that boil above 550°F, from 60N up through bright stock. In addition to raffinates from extraction, DILCHILL dewaxing can be applied to hydrocracked stocks and to other stocks from raffinate hydroconversion processes.

Processes: Lube basestocks having low pour points. Although slack waxes containing 2–10 wt.% residual oil are the typical byproducts, lower-oil-content waxes can be produced by using additional dewaxing and/or "warm-up deoiling" stages.

Description: DILCHILL is a novel dewaxing technology in which wax crystals are formed by cooling waxy oil stocks, which have been diluted with ketone solvents, in a proprietary crystallizer tower that has a number of mixing stages. This nucleation environment provides crystals that filter more quickly and retain less oil. This technology has the following advantages over conventional incremental dilution dewaxing in scraped-surface exchangers: less filter area is required, less washing of the filter cake to achieve the same oil-in-wax content is required, refrigeration duty is lower, and only scraped surface chillers are required which means that unit maintenance costs are lower. No wax recrystallization is required for deoiling.

Warm waxy feed is cooled in a prechiller before it enters the DILCHILL crystallizer tower. Chilled solvent is then added in the crystallizer tower under highly agitated conditions. Most of the crystallization occurs in the crystallizer tower. The slurry of wax/oil/ketone is further cooled in scraped-surface chillers and the slurry is then filtered in rotary vacuum filters. Flashing and stripping of products recover solvent. Additional filtration stages can be added to recover additional oil or produce low-oil content saleable waxes.

Economics: Depend on the slate of stocks to be dewaxed, the pour point targets and the required oil-in-wax content.

Utilities: Depend on the slate of stocks to be dewaxed, the pour point targets and the required oil-in-wax content.

Installation: The first application of DILCHILL dewaxing was the conversion of an ExxonMobil affiliate unit on the U.S. Gulf Coast in 1972. Since that time, 10 other applications have been constructed. These applications include both grassroots units and conversions of incremental dilution plants. Six applications use "warming-up deoiling."

Licensor: ExxonMobil Research & Engineering Co.

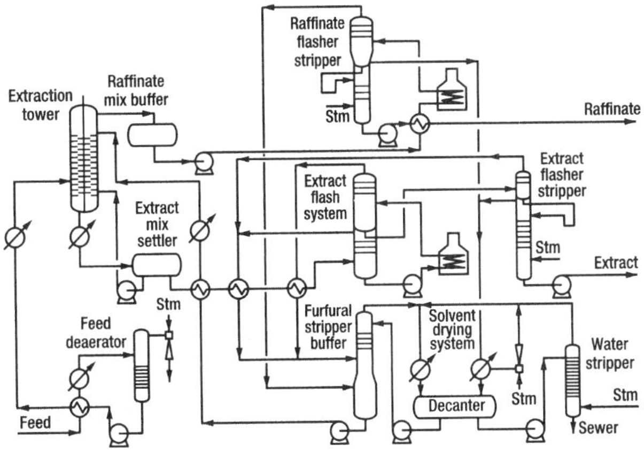

Lube treating

Application: Process to produce lube oil raffinates with high viscosity index from vacuum distillates and deasphalted oil.

Feeds: Vacuum distillate lube cuts and deasphalted oils.

Products: Lube oil raffinates of high viscosity indices. The raffinates contain substantially all of the desirable lubricating oil components present in the feedstock. The extract contains a concentrate of aromatics that may be utilized as rubber oil or cracker feed.

Description: This liquid-liquid extraction process uses furfural as the selective solvent to remove aromatics and other impurities present in the distillates and deasphalted oils. Furfural has a high solvent power for those components that are unstable to oxygen as well as for other undesirable materials including color bodies, resins, carbon-forming constituents and sulfur compounds. In the extraction tower, the feed oil is introduced below the top at a predetermined temperature. The raffinate phase leaves at the top of the tower and the extract, which contains the bulk of the furfural, is withdrawn from the bottom. The extract phase is cooled and a so-called "pseudo raffinate" may be sent back to the extraction tower. Multi-stage solvent recovery systems for raffinate and extract solutions secure energy efficient operation.

Utility requirements, (typical, Middle East Crude), units per m^3 of feed:

Electricity, kWh	10
Steam, MP, kg	10
Steam, LP, kg	35
Fuel oil, kg	20
Water, cooling, m^3	20

Installation: Numerous installations using the Uhde Edeleanu proprietary technology are in operation worldwide. The most recent is a complete lube-oil production facility licensed to the state of Turkmenistan, which successfully passed performance testing in 2002.

Licensor: Uhde Edeleanu GmbH.

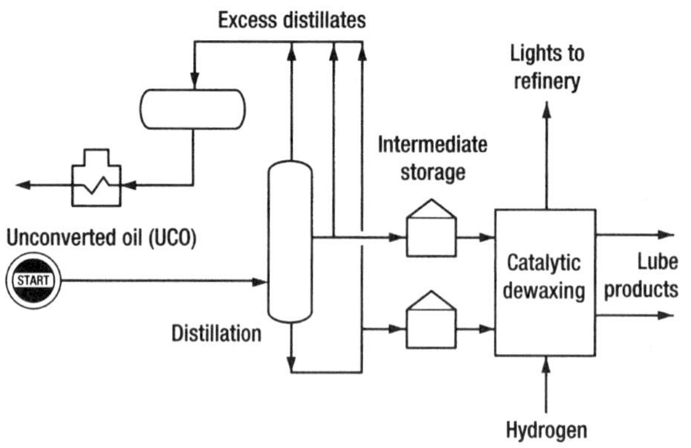

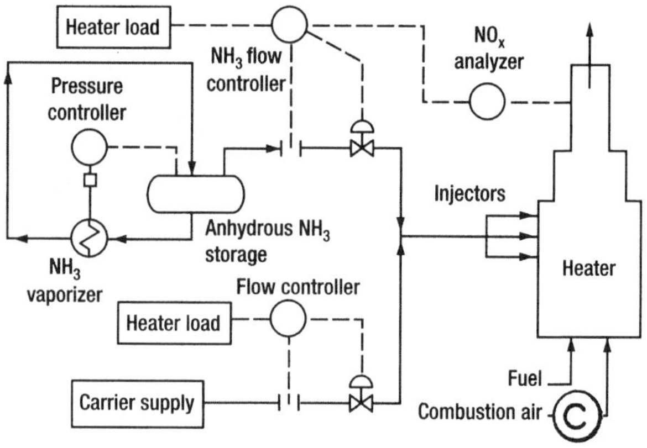

Lube treating

Application: Unconverted oil from a fuels hydrocracker is used to produce higher quality lube base stocks at lower investment and operating costs than either solvent refining or lube oil hydrocracking utilizing the SK UCO Lube Process.

Description: The base oils manufactured by the SK UCO Lube Process have many desirable properties as lube base stocks over those produced by conventional solvent-refining or lube hydrocracking processes.

Unconverted oil from the existing fractionator in a fuels hydrocracker is processed and separated into grades having the desired viscosity, which are then cooled and sent to intermediate storage. The various grades of base oil are then catalytically dewaxed and isomerized in blocked operation. Excess distillates are sent back to the hydrocracker.

Since the withdrawn UCO can usually be replaced with an equal amount of additional fresh vacuum distillate feed, the hydrocracker fuels production is maintained. The hydrocracking and catalytic dewaxing steps are not included in the SK UCO Lube Process, but are readily available from others.

Properties:

Test item	Test method	Solvent refining	Lube hydro-cracking	SK UCO Lube Process
Viscosity @100°C, cSt	ASTM D 445	5.2	5.1	6.0
Viscosity index	ASTM D 2270	97	99	130
Pour point, °C	ASTM D 97	–12	–12	–12
CCS vis @–20°C, cP	ASTM D 2602	2,100	2,000	1,440
Flash point, °C	ASTM D 92	218	220	234
NOACK volatility, wt%	DIN 51581	17.0	16.6	7.8
Aromatics, wt%	ASTM D 2549	27.7	3.5	1.0
Sulfur content, wt%	ANTEC	0.58	0.03	0.00

Economics: Investment (Basis: 5,000 bpd of lube base oils excluding fuels hydrocracker, 1998 U.S. Gulf Coast) $80 million.

Installation: 5,000 bpd of VHVI lube base oils at SK Corporation's Ulsan, Korea refinery.

Reference: Andre, J. P., S. H. Kwon and S. K. Hahn, "Yukong's new lube base oil plant," *Hydrocarbon Engineering*, November 1997.

"An economical route to high quality lubricants," NPRA 1996 Annual Meeting, March 1996.

Licensor: The Badger Technology Center of Washington Group International, under exclusive arrangement with SK Corp.

NO_x abatement

Application: Flue gases are treated with ammonia via ExxonMobil's proprietary selective noncatalytic NO_x reduction technology— Thermal $DeNO_x$. NO_x plus ammonia (NH_3) are converted to elemental nitrogen and water if temperature and residence time are appropriate. The technology has been widely applied since it was first commercialized in 1974.

Products: If conditions are appropriate, the flue gas is treated to achieve NO_x reductions of 40% to 70%+ with minimal NH_3 slip or leakage.

Description: The technology involves the gas-phase reaction of NO with NH_3 (either aqueous or anhydrous) to produce elemental nitrogen if conditions are favorable. Ammonia is injected into the flue gas using steam or air as a carrier gas into a zone where the temperature is 1,600°F to 2,000°F. This range can be extended down to 1,300°F with a small amount of hydrogen added to the injected gas. For most applications, wall injectors are used for simplicity of operation.

Yield: Cleaned flue gas with 40% to 70%+ NO_x reduction and less than 10-ppm NH_3 slip.

Economics: Considerably less costly than catalytic systems but relatively variable depending on scale and site specifics. Third-party studies have estimated the all-in cost at about 600 U.S.$/ton of NO_x removed.

Installation: Over 135 applications on all types of fired heaters, boilers and incinerators with a wide variety of fuels (gas, oil, coal, coke, wood and waste).

Reference: McIntyre, A. D., "Applications of the THERMAL $DeNO_x$ process to utility and independent power production boilers," ASME Joint International Power Generation Conference, Phoenix, 1994.

McIntyre, A. D., "The THERMAL $DeNO_x$ process: Liquid fuels applications," International Flame Research Foundation's 11th Topic Oriented Technical Meeting, Biarritz, France, 1995.

McIntyre, A. D., "Applications of the THERMAL $DeNO_x$ process to FBC boilers," CIBO 13th Annual Fluidized Bed Conference, Lake Charles, Louisiana, 1997.

Licensor: ExxonMobil Research & Engineering Co.

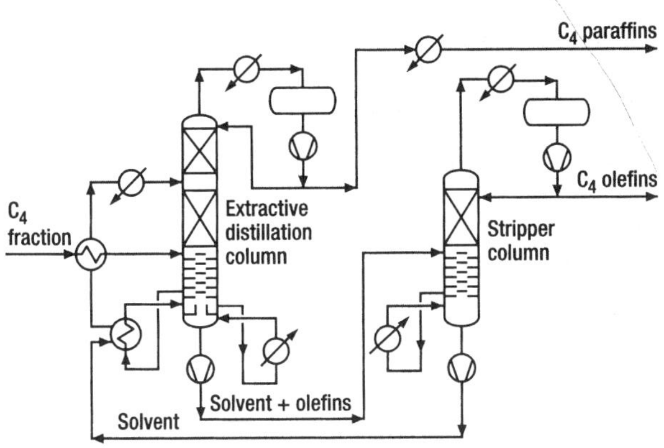

Olefins

Application: Separation of pure C_4 olefins from olefinic/paraffinic C_4 mixtures via extractive distillation using a selective solvent. BUTENEX is the Uhde technology to separate light olefins from various C_4 feedstocks, which include ethylene cracker and FCC sources.

Description: In the extractive distillation (ED) process, a single-compound solvent, N-Formylmorpholine (NFM), or NFM in a mixture with further morpholine derivatives, alters the vapor pressure of the components being separated. The vapor pressure of the olefins is lowered more than that of the less soluble paraffins. Paraffinic vapors leave the top of the ED column, and solvent with olefins leave the bottom of the ED column.

The bottom product of the ED column is fed to the stripper to separate pure olefins (mixtures) from the solvent. After intensive heat exchange, the lean solvent is recycled to the ED column. The solvent, which can be either NFM, or a mixture including NFM, perfectly satisfies the solvent properties needed for this process, including high selectivity, thermal stability and a suitable boiling point.

Economics:

Consumption per ton of FCC C_4 fraction feedstock:

Steam, t/t	0.5–0.8
Water, cooling (ΔT = 10°C), m³/t	15.0
Electric power, kWh/t	25.0
Product purity:	
n-Butene content	99.+ wt.–% min.
Solvent content	1 wt.–ppm max.

Installation: Two commercial plants for the recovery of n-butenes have been installed since 1998.

Reference: Preusser, G., "Separation of n-Butanes and Butene-2 by extractive distillation," Achema, June 1986, Frankfurt.

Licensor: Uhde GmbH.

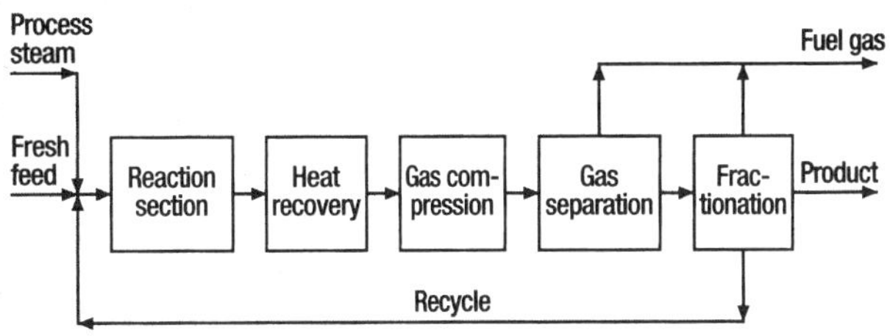

Olefins

Application: Dehydrogenation of C_4 or C_3 paraffins to pure olefins using steam-active reforming over a noble metal catalyst. STAR, the steam active reforming process, is the Uhde technology to dehydrogenate light paraffins into olefins.

Description: Fresh paraffin feed is combined with internally generated steam and passed after preheating to the reactor—a fixed-bed, tubular top-fired reformer type. Dehydrogenation reactions occurs at 4 to 6 bar at 500°C to 580°C. In a subsequent fixed-bed reactor, oxygen (or air) is admixed to enhance olefins yield by partial combustion of the hydrogen generated in the upstream reactor. The reaction section operates in sequential mode (7 hours on-stream, 1 hour regeneration). Product flow is balanced by a parallel reactor arrangement for continuous production.

After heat recovery, the gas is compressed, and the pure olefin product is separated from non-converted paraffins and light ends. Apart from fuel gas, which is used within the unit, high-purity olefin is the only product.

Economics:

Consumption per ton of propylene product based on standard grade propane feedstock:

Feedstock, t/t	1.20
Fuel, Gcal/t	1.18
Water, cooling (ΔT = 10 °C), m³/t	200
Electric power, kWh/t	170
Product purity:	
Propylene	99.70 wt.–% min.

Installation: Two commercial plants for the dehydrogenation of butane have been commissioned since 1992.

Reference: Thiagarajan, N., Ranke, U. and Ennenbach, F., "Propane /butane dehydrogenation by steam active reforming," Achema, May 2000, Frankfurt.

Licensor: Uhde GmbH.

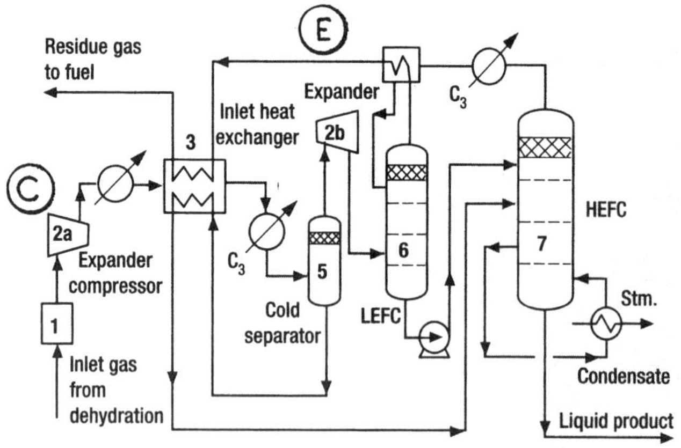

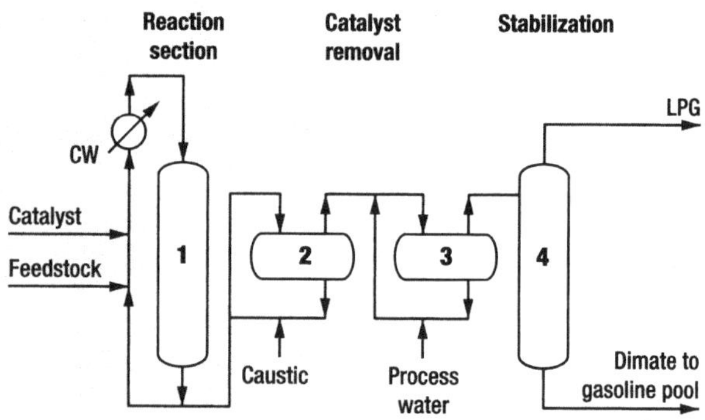

Olefins recovery

Application: Linde BOC Process Plants Cryo-Plus process recovers ethylene or propylene and heavier components from refinery offgas streams. Typical applications are on cat crackers, cokers or reformers downstream of the existing gas-recovery systems. Incremental valuable hydrocarbons that are currently being lost to the refinery fuel system can now be economically recovered.

Description: Refinery offgases from cat crackers, cokers or other sources are first dehydrated by molecular sieve (1). The expander/compressor (2a) compresses the gas stream, which is then cooled by heat exchange with internal process streams (3). Depending on the richness of the feed gas, supplemental refrigeration (4) may be used to further cool the gas stream prior to primary vapor/liquid separation (5). Light gases are fed to a turboexpander (2b) where the pressure is reduced resulting in a low discharge temperature. The expander discharge is fed to the bottom of the LEFC (6). The HEFC (7) overhead is cooled and fed to the top of the LEFC. The recovered ethylene or propylene and heavier liquid stream exit the bottom of the HEFC (7). Process advantages include:

- Low capital cost
- High propylene or high ethylene recovery (up to 99%)
- Low energy usage
- Small footprint—can be modularized
- Simple to operate
- Wide range of turndown capability.

Economics: Typically, the payback time for plant investment is one to two years.

Installation: Sixteen plants operating in U.S. refineries, with two under construction. The first plant was installed in 1984.

References: Buck, L., "Separating hydrocarbon gases," U.S. Patent No. 4,617,039, Oct. 14, 1986.

Key, R., and Z. Malik, "Technology advances improve liquid recovery from refinery offgases," NPRA Annual Meeting, San Antonio, March 26–28, 2000.

Licensor: Linde BOC Process Plants LLC, a member of Linde Engineering Division.

Oligomerization of C_3C_4 cuts

Application: To dimerize light olefins such as ethylene, propylene and butylenes using the Dimersol process. The main applications are:

- Dimerization of propylene, producing a high-octane, low-boiling point gasoline called Dimate
- Dimerization of n-butylene producing C_8 olefins for plasticizer synthesis.

The C_3 feeds are generally the propylene cuts from catalytic cracking units. The C_4 cut source is mainly the raffinate from butadiene and isobutylene extraction.

Description: Dimerization is achieved in the liquid phase at ambient temperature by means of a soluble catalytic complex. One or several reactors (1) in series are used. After elimination of catalyst (2, 3), the products are separated in an appropriate distillation section (4).

Product quality: For gasoline production, typical properties of the Dimate are:

Specific gravity, @15°C	0.70
End point, °C	205
70% vaporized, °C	80
Rvp, bar	0.5
RONC	96
MONC	81
RON blending value, avg.	103

Economics: For a plant charging 100,000 tpy of C_3 cut (% propylene) and producing 71,000 tpy of Dimate gasoline:

Investment for a 2002 ISBL Gulf Coast erected cost, excluding engineering fees, U.S.	$7 million
Utilities per ton of feed	
Electric power, kWh	10.8
Steam, HP, t	0.14
Water, cooling, t	28.5
Catalyst + chemicals, U.S.D	9.3

Installation: Twenty-seven units have been built or are under construction.

Reference: "Olefin oligomerization with homogeneous catalysis," 1999 Dewitt Petrochemical Conference, Houston.

Licensor: Axens, Axens NA.

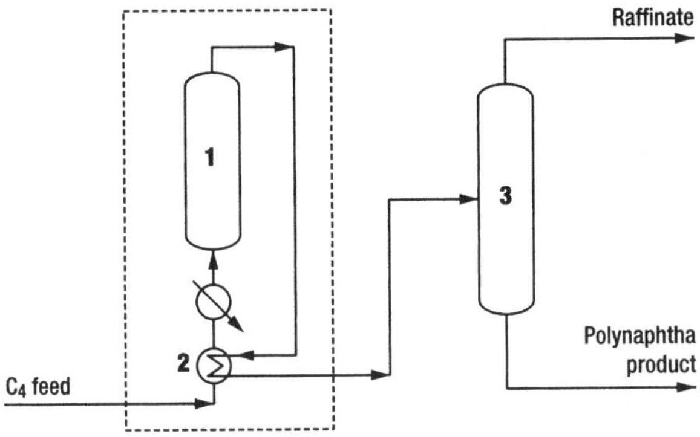

Oligomerization—polynaphtha

Application: To produce C_6+ isoolefin fractions that can be used as high-octane blending stocks for the gasoline pool and high-smoke-point blending stocks for kerosine and jet fuel. The Polynaphtha and Selectopol processes achieve high conversions of light olefinic fractions into higher value gasoline and kerosine from propylene and mixed-butene fractions such as C_3 and C_4 cuts from cracking processes.

Description: Propylene or mixed butenes (or both) are oligomerized catalytically in a series of fixed-bed reactors (1). Conversion and selectivity are controlled by reactor temperature adjustment while the heat of reaction is removed by intercooling (2). The reactor section effluent is fractionated (3) producing raffinate, gasoline and kerosine.

The Selectopol process is a variant of the polynaphtha process where the operating conditions are adjusted to convert selectively the isobutene portion of an olefinic C_4 fraction to high-octane, low-Rvp gasoline blending stock. It provides a low cost means of debottlenecking existing alkylation units by converting all of the isobutene and a small percentage of the n-butenes, without additional isobutane.

Polynaphtha and Selectopol processes have the following features: low investment, regenerable solid catalyst, no catalyst disposal problems, long catalyst life, mild operating conditions, versatile product range, good quality motor fuels and kerosine following a simple hydrogenation step and the possibility of retrofitting old phosphoric acid units.

The polygasoline RON and MON obtained from FCC C_4 cuts are significantly higher than those of FCC gasoline and, in addition, are sulfur-free. Hydrogenation improves the MON, whereas the RON remains high and close to that of C_4 alkylate.

Kerosine product characteristics such as oxidation stability, freezing point and smoke point are excellent after hydrogenation of the polynaphtha product. The kerosine is also sulfur-free and low in aromatics.

The Polynaphtha process has operating conditions very close to those of phosphoric acid poly units. Therefore, an existing unit's major equipment items can be retained with only minor changes to piping and instrumentation. Some pretreatment may be needed if sulfur, nitrogen, or water contents in the feed warrant; however, the equipment cost is low.

Economics: Typical ISBL Gulf Coast investments for 5,000-bpd of FCC C_4 cut for polynaphtha (of maximum flexibility) and Selectopol (for maximum gasoline) units are U.S.\$8.5 and \$3.0 million, respectively. Respective utility costs are U.S.\$4.4 and 1.8 per ton of feed while catalyst costs are U.S.\$0.2 per ton of feed for both processes.

Installations: Five Selectopol and polynaphtha units have been licensed (four in operation), with a cumulative operating experience exceeding 40 years.

Licensor: Axens, Axens NA.

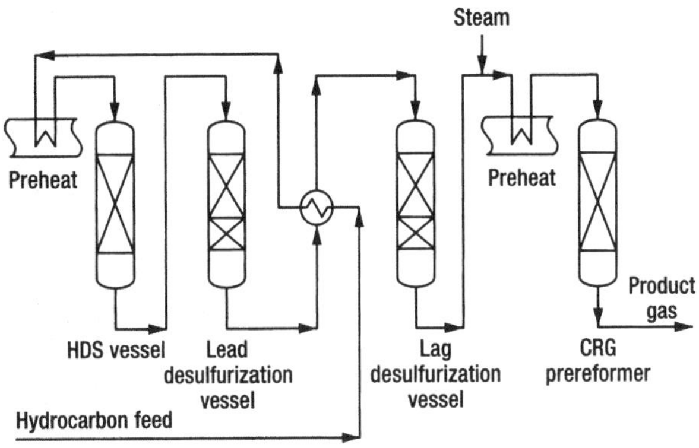

Prereforming with feed ultrapurification

Application: Ultra-desulfurization and adiabatic-steam reforming of hydrocarbon feed from refinery offgas or natural gas through LPG to naphtha feeds as a prereforming step in the route to hydrogen production.

Description: Sulfur components contained in the hydrocarbon feed are converted to H_2S in the HDS vessel and then fed to two desulfurization vessels in series. Each vessel contains two catalyst types—the first for bulk sulfur removal and the second for ultrapurification down to sulfur levels of less than 1 ppb.

The two-desulfurization vessels are arranged in series in such a way that either may be located in the lead position allowing online change out of the catalysts. The novel interchanger between the two vessels allows for the lead and lag vessels to work under different optimized conditions for the duties that require two catalyst types. This arrangement may be retrofitted to existing units.

Desulfurized feed is then fed to a fixed bed of nickel-based catalyst that converts the hydrocarbon feed, in the presence of steam, to a product stream containing only methane together with H_2, CO, CO_2 and unreacted steam which is suitable for further processing in a conventional fired reformer. The CRG prereformer enables capital cost savings in primary reforming due to reductions in the radiant box heat load. It also allows high-activity gas-reforming catalyst to be used. The ability to increase preheat temperatures and transfer radiant duty to the convection section of the primary reformer can minimize involuntary steam production.

Operating conditions: The desulfurization section typically operates between 170°C and 420°C and the CRG prereformer will operate over a wide range of temperatures from 250°C to 650°C and at pressures up to 75 bara.

Installation: CRG process technology covers 40 years of experience with over 150 plants built and operated. Ongoing development of the catalyst has lead to almost 50 such units since 1990.

Catalyst: The CRG catalyst is manufactured under license by Synetix.

Licensor: The process and CRG catalyst are licensed by Davy Process Technology.

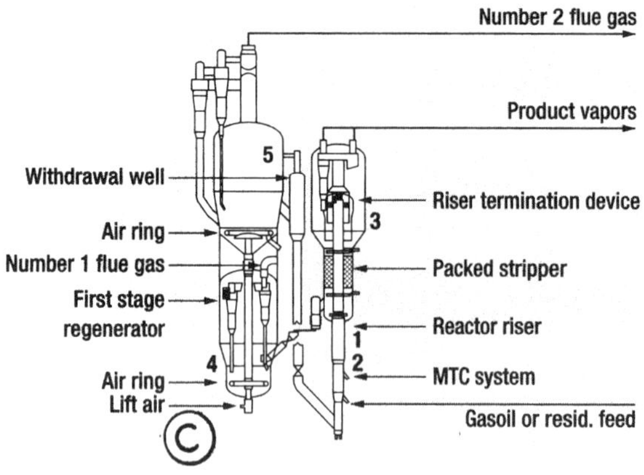

Resid catalytic cracking

Application: Selective conversion of gas oil and heavy residual feedstocks.

Products: High-octane gasoline, distillate and C_3–C_4 olefins.

Description: For residue cracking the process is known as R2R (reactor–2 regenerators). Catalytic and selective cracking occurs in a short-contact-time riser (1) where oil feed is effectively dispersed and vaporized through a proprietary feed-injection system. Operation is carried out at a temperature consistent with targeted yields. The riser temperature profile can be optimized with the proprietary mixed temperature control (MTC) system (2).

Reaction products exit the riser-reactor through a high-efficiency, close-coupled, proprietary riser termination device RS^2 (riser separator stripper) (3). Spent catalyst is pre-stripped followed by an advanced high-efficiency packed stripper prior to regeneration. The reaction product vapor may be quenched using Amoco's proprietary technology to give the lowest dry gas and maximum gasoline yield. Final recovery of catalyst particles occurs in cyclones before the product vapor is transferred to the fractionation section.

Catalyst regeneration is carried out in two independent stages (4, 5) equipped with proprietary air and catalyst distribution systems resulting in fully regenerated catalyst with minimum hydrothermal deactivation, plus superior metals tolerance relative to single-stage systems. These benefits are derived by operating the first-stage regenerator in a partial-burn mode, the second-stage regenerator in a full-combustion mode and both regenerators in parallel with respect to air and flue gas flows. The resulting system is capable of processing feeds up to about 6 wt% ConC without additional catalyst cooling means, with less air, lower catalyst deactivation and smaller regenerators than a single-stage regenerator design. Heat removal for heavier feedstocks (above 6 CCR) may be accomplished by using a reliable dense-phase catalyst cooler, which has been commercially proven in over 24 units and is licensed exclusively by Stone & Webster/Axens.

The converter vessels use a cold-wall design that results in minimum capital investment and maximum mechanical reliability and safety. Reliable operation is ensured through the use of advanced fluidization technology combined with a proprietary reaction system. Unit design is tailored to refiner's needs and can include wide turndown flexibility. Available options include power recovery, wasteheat recovery, flue-gas treatment and slurry filtration.

Existing gas oil units can be easily retrofitted to this technology. Revamps incorporating proprietary feed injection and riser termination devices and vapor quench result in substantial improvements in capacity, yields and feedstock flexibility within the mechanical limits of the existing unit.

Installation: Stone & Webster and Axens have licensed 26 full-technology R2R units and performed more than 100 revamp projects.

Reference: Letzsch, W. S., "Commercial performance of the latest FCC technology advances," NPRA Annual Meeting, March 2000.

Licensor: Stone & Webster Inc., a Shaw Group Co., and Axens, IFP Group Technologies.

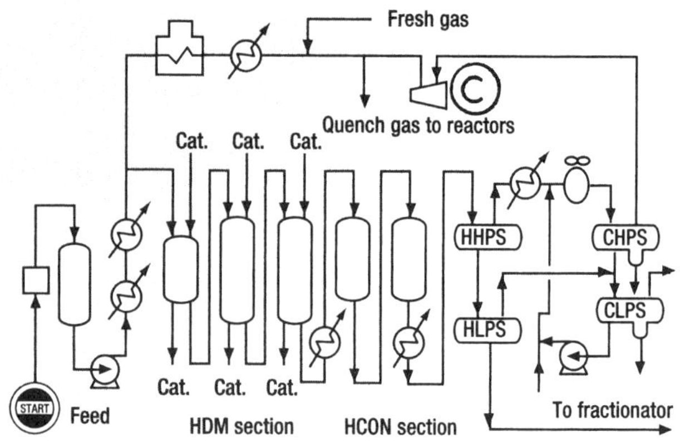

Residue hydroprocessing

Application: Produces maximum distillates and low-sulfur fuel oil, or low-sulfur LR-CCU feedstock, with very tight sulfur, vanadium and CCR specifications, using moving bed "bunker" and fixed-bed technologies. Bunker units are available as a retrofit option to existing fixed-bed residue HDS units.

Description: At limited feed metal contents, the process typically uses all fixed-bed reactors. With increasing feed metal content, one or more moving-bed "bunker" reactors are added up-front of the fixed-bed reactors to ensure a fixed-bed catalyst life of at least one year. A steady state is developed by continuous catalyst addition and withdrawal: the catalyst aging is fully compensated by catalyst replacement, at typically 0.5 to 2 vol% of inventory per day.

An all bunker option, which eliminates the need for catalyst change-out, is also available. A hydrocracking reactor, which converts the synthetic vacuum gasoil into distillates, can be efficiently integrated into the unit. A wide range of residue feeds, like atmospheric or vacuum residues and deasphalted oils, can be processed using Shell residue hydroprocessing technologies.

Operating conditions:

Reactor pressures:	100–200 bar
	1,450–3,000 psi
Reactor temperatures:	370–420°C
	700–790°F

Yields: Typical yields for an SR HYCON unit on Kuwait feed:

Feedstock Yields:		SR (95% 520C+) [%wof]	with integrated HCU [%wof]
Gases	C_1–C4	3	5
Naphtha	C_5–165°C	4	18
Kero + gasoil	165–370°C	20	43
VGO	370–580°C	41	4
Residue	580°C+	29	29
H2 cons.		2	3

Economics: Investment costs for the various options depend strongly on feed properties and process objectives of the residue hydroprocessing unit. Investment costs for a typical new single string 5,000 tpsd SR-Hycon unit will range from 200–300 MM US $, the higher figure includes an integrated hydrocracker.

Installation: There is one unit with both bunker reactors and fixed-bed reactors, operating on short residue (vacuum residue) at 4,300 tpd or 27 kbpsd capacity, and two all-fixed bed units of 7,700 and 7,000 tpd (48 and 44 kbpsd resp.), the latter one in one single string. Commercial experiences range from low-sulfur atmospheric residues to high-metal, high-sulfur vacuum residues with over 300 ppmw metals.

Reference: Scheffer, B., et al, "The Shell Residue Hydroconversion Process: Development and achievements," The European Refining Technology Conference, London, November 1997.

Licensor: Shell Global Solutions International B.V.

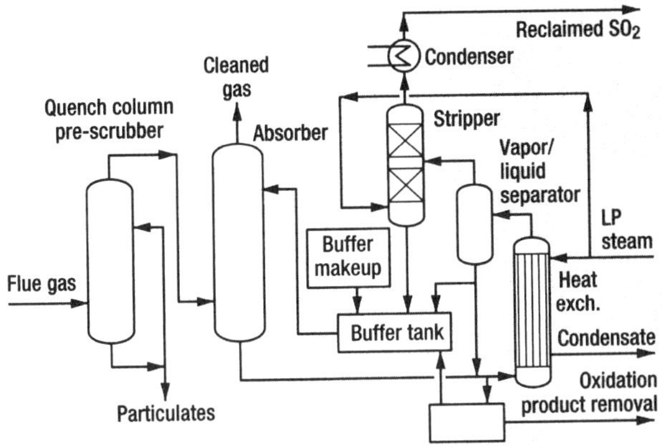

SO_2 removal

Application: Regenerative scrubbing system to recover SO_2 from flue gas containing high SO_2 levels such as gas from FCC regenerator or incinerated SRU tail gas and other high SO_2 applications. The LABSORB process is a low-pressure drop system and is able to operate under varying conditions and not sensitive to variations in the upstream processes.

Products: The product from the LABSORB process is a concentrated SO_2 stream consisting of approximately 90% SO_2 and 10% moisture. This stream can be sent to the front of the SRU to be mixed with H_2S and form sulfur, or it can be concentrated for other marketable uses.

Description: Hot dirty flue gas is cooled in a flue-gas cooler or waste-heat recovery boiler prior to entering the systems. Steam produced can be used in the LABSORB plant. The gas is then quenched to adiabatic saturation (typically 50°C–75°C) in a quencher/pre-scrubber; it proceeds to the absorption tower where the SO_2 is removed from the gas. The tower incorporates multiple internal and re-circulation stages to ensure sufficient absorption.

A safe, chemically stable and regenerable buffer solution is contacted with the SO_2-rich gas for absorption. The rich solution is then piped to a LABSORB buffer regeneration section where the solution is regenerated for re-use in the scrubber. Regeneration is achieved using low-pressure steam and conventional equipment such as strippers, condensers and heat exchangers.

Economics: This process is very attractive at higher SO_2 concentrations or when liquid or solid effluents are not allowed. The system's buffer loss is very low, contributing to a very low operating cost. Additionally, when utilizing LABSORB as an SRU tail-gas treater, many components normally associated with the SCOT process are not required thus saving considerable capital.

Installations: One SRU tail-gas system and two FCCU scrubbing systems.

Reference: Confuorto, Weaver and Pedersen, "LABSORB regenerative scrubbing operating history, design and economics," Sulfur 2000, San Francisco, October 2000.

Confuorto, Eagleson and Pedersen, "LABSORB, A regenerable wet scrubbing process for controlling SO_2 emissions," Petrotech-2001, New Delhi, January 2001.

Licensor: Belco Technologies Corp. (LABSORB was developed by Prof. Olav Erga of NTNU in Trondheim, Norway.)

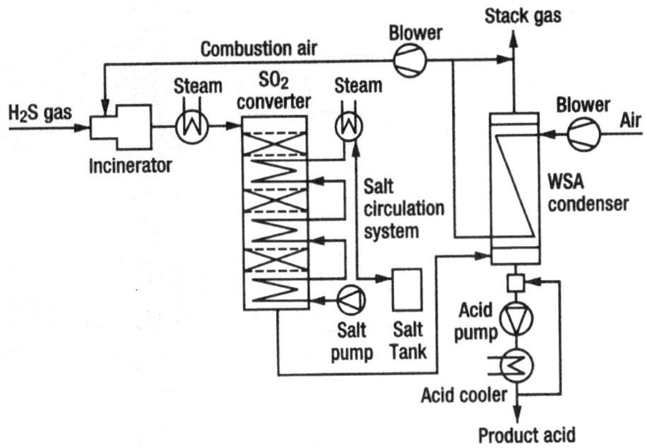

Sour gas treatment

Application: The WSA process (Wet gas Sulfuric Acid) process treats all types of sulfur-containing gases such as amine and Rectisol regenerator offgas, SWS gas and Claus plant tail gas in refineries, gas treatment plants, petrochemicals and coke chemicals plants. This process can also be applied for SO_x removal and regeneration of spent sulfuric acid.

Sulfur, in any form, is efficiently recovered as concentrated commercial-quality sulfuric acid.

Description: Feed gas is combusted and cooled to approximately 420°C in a waste-heat boiler. The gas then enters the SO_2 converter containing one or several beds of SO_2 oxidation catalyst to convert SO_2 to SO_3. The gas is cooled in the gas cooler whereby SO_3 hydrates to H_2SO_4 (gas), which is finally condensed as concentrated sulfuric acid (typically 98% w/w).

The WSA condenser is cooled by ambient air; the heated air may be used as combustion air in the burner for increased thermal efficiency. The heat of reaction released in the reactor and gas cooler is transferred by a heat displacement system to a boiler where it is recovered as steam. The process operates without removing water from the gas. Therefore, the number of equipment items is minimized, and generation of waste condensate is eliminated. Cleaned process gas leaving the WSA condenser is sent to stack without further treatment.

The WSA process is characterized by:

- More than 99% recovery of sulfur as commercial-grade sulfuric acid.
- No generation of waste solids or wastewater.
- No consumption of absorbents or auxiliary chemicals.
- Efficient heat recovery thus, ensuring economical operation.
- Simple and fully automated operation allows adaptation to variations in feed gas flow and composition.

Installation: More than 40 units worldwide

Licensor: Haldor Topsøe A/S.

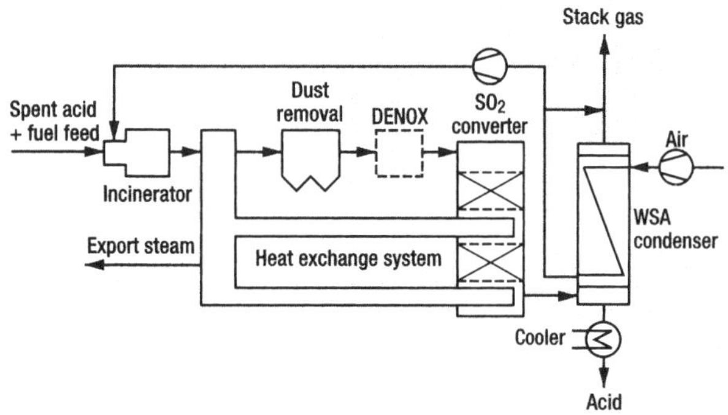

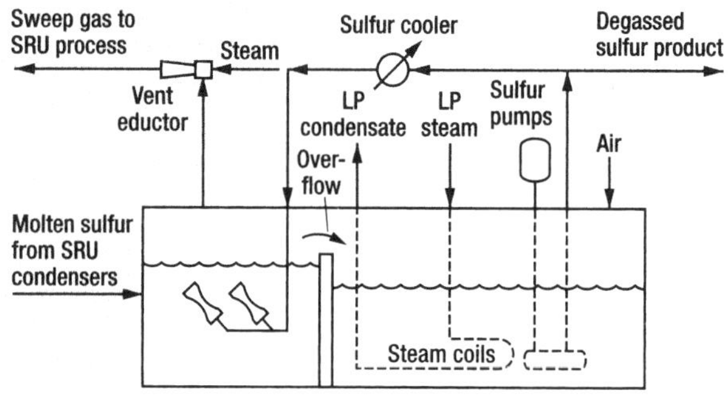

Spent acid recovery

Application: The WSA process (Wet gas Sulfuric Acid) treats spent sulfuric acid from alkylation as well as other types of waste sulfuric acid in the petrochemical and coke chemicals industry. Amine regenerator offgas and /or refinery gas may be used as auxiliary fuel. The regenerated acid will contain min. 98% H_2SO_4 and can be recycled directly to the alkylation process.

The WSA process is also applied for conversion of H_2S and removal of SO_x.

Description: Spent acid is decomposed to SO_2 and water in a burner using amine regenerator offgas or refinery gas as fuel. The SO_2 containing flue gas is cooled in a waste-heat boiler; solid matter originating from the acid feed is separated in an electrostatic precipitator. By adding preheated air, the process gas temperature and oxygen content are adjusted before the catalytic converter when converting SO_2 to SO_3. The gas is cooled in the gas cooler whereby SO_3 is hydrated to H_2SO_4 (gas), which is finally condensed as 98% sulfuric acid.

The WSA condenser is cooled by ambient air; the heated air may be used as combustion air in the burner for increased thermal efficiency. The heat of reaction released in the reactor and the gas cooler is transferred by a heat displacement system to a boiler where it is recovered as steam.

The process operates without removing water from the gas. Therefore, the number of equipment items is minimized and generation of condensate is eliminated. This is especially important in spent-acid regeneration where SO_3 formed by the acid decomposition will otherwise be lost with the condensate as wastewater.

The WSA process is characterized by:

- More than 99% recovery of sulfuric acid
- No generation of waste solids or wastewater
- No consumption of absorbents or auxiliary chemicals
- Efficient heat recovery ensuring economical operation
- Simple and fully automated operation enables variations in feed flow and composition.

Installation: More than 40 WSA units worldwide, including six for spent-acid recovery.

Licensor: Haldor Topsøe A/S.

Sulfur degassing

Application: Hydrogen sulfide (H_2S) removal from sulfur.

Description: Sulfur, as produced by the Claus process, typically contains from about 200 to 500 ppmw H_2S. The H_2S may be contained in the molten sulfur as H_2S or as hydrogen polysulfides (H_2S_x). The dissolved H_2S separates from the molten sulfur readily, but the H_2S_x does not.

The sulfur degassing process accelerates the decomposition of hydrogen polysulfides to H_2S and elemental sulfur (S). The dissolved H_2S gas is released in a controlled manner. Sulfur temperature, residence time, and the degree of agitation all influence the degassing process. Chemical catalysts, including oxygen (air) that accelerate the rate of H_2S_x decomposition are known to improve the degassing characteristics.

In fact, the majority of successful commercial degassing processes use compressed air, in some fashion, as the degassing medium. Research performed by Alberta Sulphur Research Ltd. has demonstrated that air is a superior degassing agent when compared to nitrogen, steam or other inert gases. Oxygen present in air promotes a level of direct oxidation of H_2S to elemental S, which reduces the gaseous H_2S partial pressure and increases the driving force for H_2S_x decomposition to the more easily removed gaseous phase H_2S.

The MAG degassing system concept was developed to use the benefits of degassing in the presence of air without relying on a costly compressed air source. With the MAG system, motive pressure from a recirculated degassed sulfur stream is converted to energy in a mixing assembly within the undegassed sulfur. The energy of the recirculated sulfur creates a high air-to-sulfur interfacial area by generating intense turbulence within the jet plume turning over the contents many times, thus exposing the molten sulfur to the sweep air. Intimate mixing is achieved along with turbulence to promote degassing. This sulfur degassing system can readily meet a 10 ppmw total H_2S (H_2S + H_2S_x) specification.

Tests show degassing rate constants nearly identical to traditional air sparging for well-mixed, air-swept degassing systems. Thus, comparable degassing to air sparging can be achieved without using a compressed air source. The assemblies are designed to be self-draining of molten sulfur and to be easily slipped in and out for maintenance through the pit nozzles provided. The mixing assemblies require no moving parts or ancillary equipment other than the typical sulfur-product-transfer pump that maximizes unit reliability and simplifies operations.

The process is straightforward; it is inherently safer than systems using spray nozzles and/or impingement plates because no free fall of sulfur is allowed.

Economics: Typically does not require changes to existing sulfur processing infrastructure.

Installation: Several units are in design.

Reference: U.S. Patent 5935548 issued Aug. 10, 1999.

Licensor: Black & Veatch Pritchard, Inc.

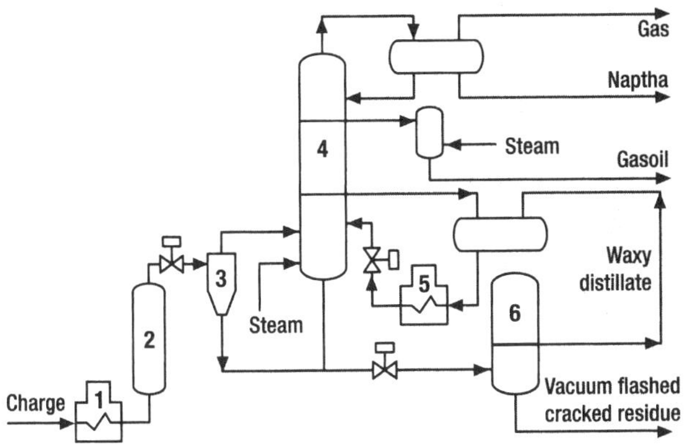

Thermal gasoil process

Application: The Shell Thermal Gasoil process is a combined residue and waxy distillate conversion unit. The process is an attractive low-cost conversion option for hydroskimming refineries in gasoil-driven markets or for complex refineries with constrained waxy distillate conversion capacity. The typical feedstock is atmospheric residue, which eliminates the need for an upstream vacuum flasher. This process features Shell Soaker Visbreaking technology for residue conversion and an integrated recycle heater system for the conversion of waxy distillate.

Description: The preheated atmospheric (or vacuum) residue is charged to the visbreaker heater (1) and from there to the soaker (2). The conversion takes place in both the heater and soaker and is controlled by the operating temperature and pressure. The soaker effluent is routed to a cyclone (3). The cyclone overheads are charged to an atmospheric fractionator (4) to produce the desired products including a light waxy distillate. The cyclone and fractionator bottoms are routed to a vacuum flasher (6), where waxy distillate is recovered. The combined waxy distillates are fully converted in the distillate heater (5) at elevated pressure.

Yields: Depend on feed type and product specifications.

Feed atmospheric residue	**Middle East**
Viscosity, cSt @ 100°C	31
Products, % wt.	
Gas	6.4
Gasoline ECP 165°C	12.9
Gasoil ECP 350°C	38.6
Residue ECP 520°C+	42.1
Viscosity 165°C plus, cSt @100°C	7.7

Economics: The investment amounts to 1,400–1,600 U.S.$/bbl installed excluding treating facilities and depending on capacity and configuration (basis: 1998)

Utilities, typical consumption per bbl of feed @ 180°C:

Fuel, 10^3 cal	34
Electricity, kWh	0.8
Net steam production, kg	29
Water, cooling, m^3	0.17

Installation: Thirteen Shell thermal gasoil units have been built or are under construction. Post startup services and technical services on existing units are available from Shell.

Reference: "Thermal Conversion Technology in Modern Power Integrated Refinery Schemes," 1999 NPRA Annual Meeting.

Licensor: Shell Global Solutions International B.V., and ABB Lummus Global B.V.

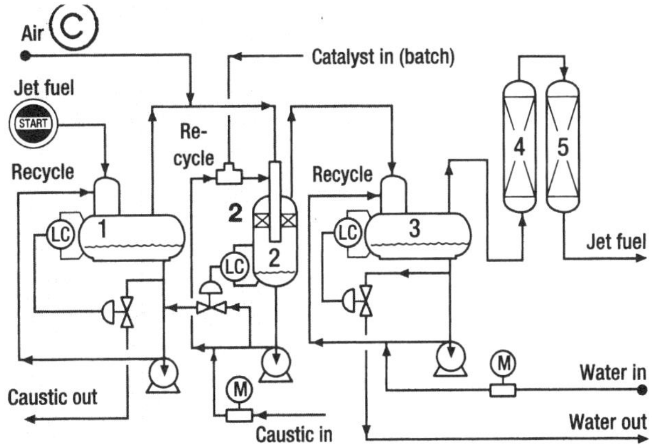

Treating

Application: Treating gas, LPG, butane, propane, gasoline, condensate, kerosine, diesel and light crude oil with caustic, amine, water and acid using the following technologies that use the FIBER-FILM contactor.

Description: A proprietary FIBER-FILM contactor is used in treating processes to achieve co-current contacting between the hydrocarbon feed and aqueous solution. The FIBER-FILM contactor is comprised of a bundle of long, continuous, small diameter fibers contained in a cylinder. The large, interfacial area created by the contactor greatly enhances mass transfer without dispersion of one phase into the other as is necessary for typical conventional mixing systems. Process advantages include:

- Low capital costs
- Flexibility to operate over a wide range of hydrocarbon flowrates
- Small sized equipment and low space requirement
- Low pressure drop
- Can be retrofitted into existing systems or skid mounted for easy system installation
- Low guaranteed aqueous carryover.

AQUAFINING uses water to remove amines and caustic contaminants from hydrocarbon streams.

THIOLEX removes H_2S, COS and mercaptans from gas, LPG, butane and gasoline with caustic.

AMINEX removes H_2S, COS and CO_2 from gas, LPG, propane and butane with amine.

THIOLEX coupled with REGEN, a caustic regeneration process, is used for mercaptan extraction with minimal caustic consumption. One or more stages of caustic extraction are used to remove H_2S, COS and mercaptans from gas, LPG, propane, butane and gasoline. The catalyst-containing caustic solution then is sent to a tower for regeneration with air. The disulfides formed are either gravity separated and/or solvent extracted. The regenerated solution is then reused in the extraction unit.

MERICAT uses a catalyst-containing caustic solution and air to oxidize mercaptans to disulfides in gasoline.

MERICAT II sweetens kerosine/jet fuel by combining MERICAT with a catalyst impregnated carbon bed.

NAPFINING uses caustic to reduce the acidity of kerosine/jet fuel and heavier middle distillates.

CHLOREX uses dilute caustic to remove HCl and NH_4Cl from reformer gas and liquid products.

ESTEREX uses sulfuric acid to remove neutral and acidic esters from alkylation reactor effluent streams.

MERICON oxidizes and/or neutralizes spent caustics containing sulfides, mercaptans, napthenic acids and phenols.

EXOMER removes recombinant mercaptans from selectively hydrotreated gasoline with a proprietary treating solution to reduce its total sulfur content.

Installation: Over 540 installations treating 6.0 million bpsd and 21 million scfd in 39 countries.

Licensor: Merichem Chemicals & Refinery Services LLC.

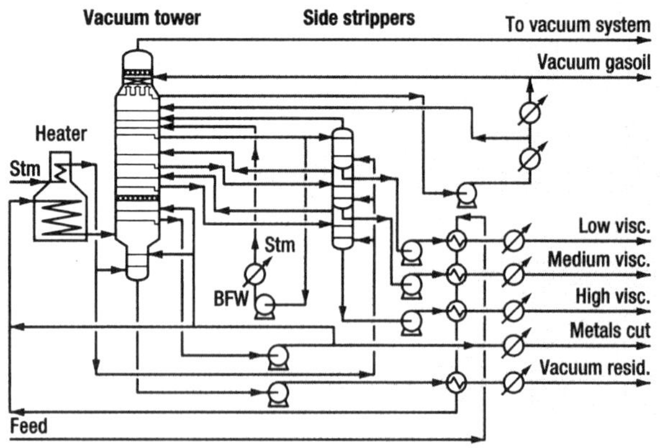

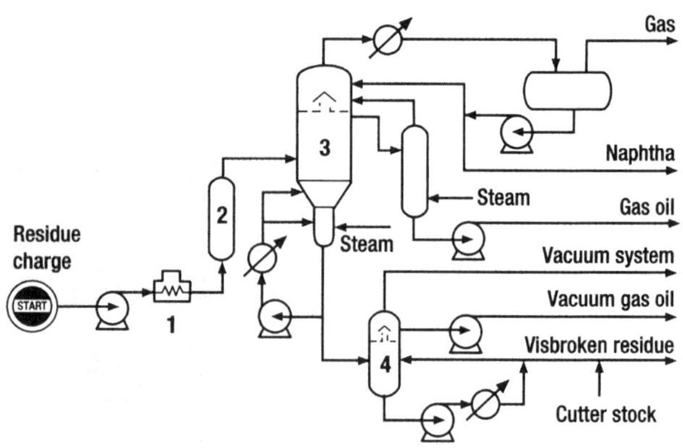

Vacuum distillation

Application: Process to produce vacuum distillates that are suitable for lubricating oil production by downstream units.

Feeds: Atmospheric bottoms from crude oils (atmospheric residue) or hydrocracker bottoms.

Products: Vacuum distillates of precisely defined viscosities and flash points as well as vacuum residue with specified softening point, penetration and flash point.

Description: Feed is preheated in a heat-exchanger train and fed to the fired heater. The heater-coil temperature is controlled to produce the required quality of vacuum distillates and residue. Uhde Edeleanu-designed units ensure that vaporization occurs in the furnace coils to minimize superheating the residue. Circulating reflux streams enable maximum heat recovery and reduced column diameter.

Wash trays minimze the metals content in the heaviest vacuum distillate to avoid difficulties in downstream lubricating oil production plants. Heavy distillate from the wash trays is recycled to the heater inlet or withdrawn as metals cut.

When processing naphthenic residues, a neutralization section may be added to the fractionator.

Utility requirements, (typical, Middle East Crude), units per m^3 of feed:

Electricity, kWh	7
Steam, MP, kg	30
Steam production, LP, kg	35
Fuel oil, kg	15
Water, cooling, m^3	10

Installation. Numerous installations using the Uhde Edeleanu proprietary technology are in operation worldwide. The most recent reference is a complete lube-oil production facility licensed to the state of Turkmenistan, which successfully passed performance testing in 2002.

Licensor: Uhde Edeleanu GmbH.

Visbreaking

Application: The Shell Soaker Visbreaking process is most suitable to reduce the viscosity of vacuum (and atmospheric) residues in (semi) complex refineries. The products are primarily distillates and stable fuel oil. The total fuel oil production is reduced by decreasing the quantity of cutter stock required. Optionally, a Shell vacuum flasher may be installed to recover additional gas oil and waxy distillates as cat cracker or hydrocracker feed from the cracked residue. The Shell Soaker Visbreaking technology has also proven to be a very cost-effective revamp option for existing units.

Description: The preheated vacuum residue is charged to the visbreaker heater (1) and from there to the soaker (2). The conversion takes place in both the heater and the soaker. The operating temperature and pressure are controlled such as to reach the desired conversion level and/or unit capacity. The cracked feed is then charged to an atmospheric fractionator (3) to produce the desired products like gas, LPG, naphtha, kerosine, gas oil, waxy distillates and cracked residue. If a vacuum flasher is installed, additional gas oil and waxy distillates are recovered from the cracked residue.

Yields: Vary with feed type and product specifications.

Feed, vacuum residue	**Middle East**
Viscosity, cSt @100°C	770
Products, wt%	
Gas	2.3
Gasoline, 165°C EP	4.7
Gas oil, 350°C EP	14.0
Waxy distillate, 520°C EP	20.0
Residue, 520°C+	59.0
Viscosity, 165°C plus, cSt @100°C	97

Economics: The investment amounts to 1,000 to 1,400 U.S.$/bbl installed excluding treating facilities and depending on capacity and the presence of a vacuum flasher (basis: 1998).

Utilities, typical consumption per bbl feed @180°C:

Fuel, 10^3 kcal	16
Electricity, kWh	0.5
Net steam production, kg	18
Water, cooling, m3	0.1

Installation: Eighty-six Shell Soaker Visbreaking units have been built or are under construction. Post startup services and technical services for existing units are available from Shell.

Reference: *Visbreaking Technology,* Erdöl und Kohle, January 1986.

Licensor: Shell Global Solutions International B.V. and ABB Lummus Global B.V.

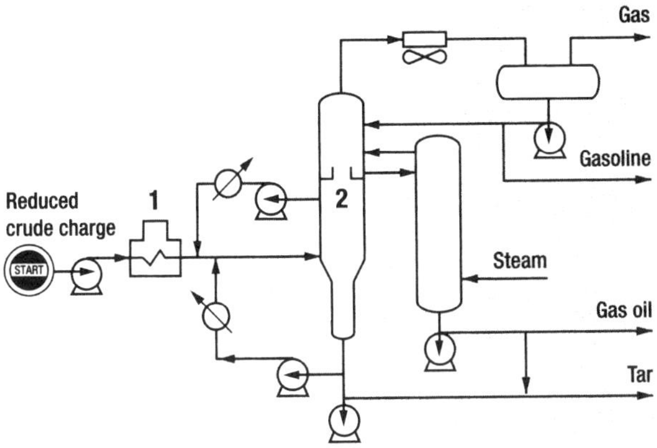

Visbreaking

Application: Manufacture incremental gas and distillate products and simultaneously reduce fuel oil viscosity and pour point. Also, reduce the amount of cutter stock required to dilute the resid to meet the fuel oil specifications. Foster Wheeler/UOP offer "coil" type visbreaking process.

Products: Gas, naphtha, gas oil, visbroken resid (tar).

Description: In a "coil" type operation, charge is fed to the visbreaker heater (1) where it is heated to a high temperature, causing partial vaporization and mild cracking. The heater outlet stream is quenched with gas oil or fractionator bottoms to stop the cracking reaction. The vapor-liquid mixture enters the fractionator (2) to be separated into gas, naphtha, gas oil and visbroken resid (tar).

Operating conditions: Typical ranges are:

Heater outlet temperature, °F	850–910
Quenched temperature, °F	710–800

An increase in heater outlet temperature will result in an increase in overall severity.

Yields:

Feed, source	**Light Arabian**	**Light Arabian**
Type	Atm. resid	Vac. resid
Gravity, °API	15.9	7.1
Sulfur, wt%	3.0	4.0
Concarbon, wt%	8.5	20.3
Viscosity, CKS @ 130°F	150	30,000
CKS @ 210°F	25	900
Products, wt%		
Gas	3.1	2.4
Naphtha (C_5–330°F)	7.9	6.0
Gas oil (330–600°F)	14.5	15.5(1)
Visbroken resid (600°F+)	74.5	76.1(2)

(1) 330–662°F cut for Light Arabian vacuum residue.
(2) 662°F+ cut for Light Arabian vacuum residue.

Economics:

Investment (basis: 40,000–10,000 bpsd, 4Q 1999, U.S. Gulf), $ per bpsd 785–1,650

Utilities, typical per bbl feed:

Fuel, MMBtu	0.1195
Power, kW/bpsd	0.0358
Steam, MP, lb	6.4
Water, cooling, gal	71.0

Installation: Over 50 units worldwide.

Reference: *Handbook of Petroleum Refining Processes,* 2nd Ed., McGraw-Hill, 1997, pp. 12.83–12.97.

Licensor: Foster Wheeler/UOP LLC.

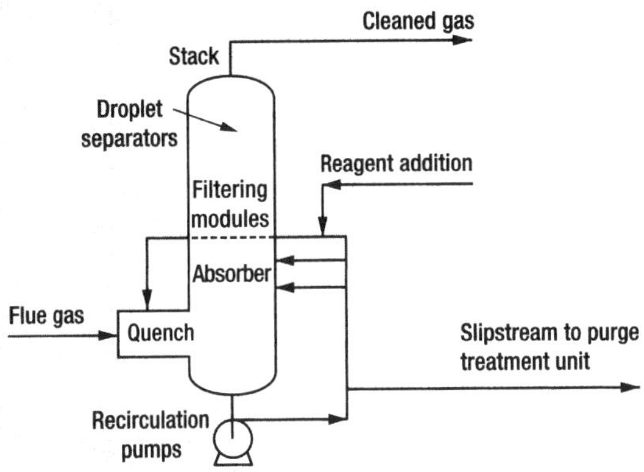

Wet scrubbing system

Application: EDV Technology is a low-pressure drop scrubbing system, to scrub particulate matter (including PM2.5), SO_2 and SO_3 from flue gases. It is especially well suited where the application requires high reliability, flexibility and the ability to operate for 3–6 years continuously without maintenance shutdowns. The EDV technology is highly suited for FCCU regenerator flue-gas applications.

Products: The effluents from the process will vary based on the reagent selected for use with the scrubber. In the case where a sodium-based reagent is used, the product will be a solution of sodium salts. Similarly, a magnesium-based reagent will result in magnesium salts. A lime/limestone-based system will produce a gypsum waste. The EDV technology can also be used with the LABSORB buffer thus making the system regenerative. The product, in that case, would be a usable condensed SO_2 stream.

Description: The flue gas enters the spray tower through the quench section where it is immediately quenched to saturation temperature. It proceeds to the absorber section for particulate and SO_2 reduction. The spray tower is an open tower with multiple levels of BELCO-G-Nozzles. These nonplugging and abrasion-resistant nozzles remove particulates by impacting on the water/reagent curtains. At the same time, these curtains also reduce SO_2 and SO_3 emissions. The BELCO-G-Nozzles are designed not to produce mist; thus a conventional mist eliminator is not required—these units can be prone to plugage.

Upon leaving the absorber section, the saturated gases are directed to the EDV filtering modules to remove the fine particulates and additional SO_3. The filtering module is designed to cause condensation of the saturated gas onto the fine particles and onto the acid mist, thus allowing it to be collected by the BELCO-F-nozzle located at the top.

To ensure droplet-free stack, the flue gas enters a droplet separator. This is an open design that contains fixed-spin vanes that induce a cyclonic flow of the gas. As the gases spiral down the droplet separator, the centrifugal forces drive any free droplets to the wall, separating them from the gas stream.

Economics: The EDV wet scrubbing system has been extremely successful in the incineration and refining industries due to the very high scrubbing capabilities, very reliable operation and reasonable price.

Installation: More than 200 applications worldwide on various processes including 25 FCCU applications, 3 CDU applications and 1 fluidized coker application to date.

Reference: Confuorto and Weaver, "Flue gas scrubbing of FCCU regenerator flue gas—performance, reliability, and flexibility—a case history," *Hydrocarbon Engineering,* 1999.

Eagleson and Dharia, "Controlling FCCU emissions," 11th Refining Technology Meeting, HPCL, Hyderabad, 2000.

Licensor: Belco Technologies Corp.

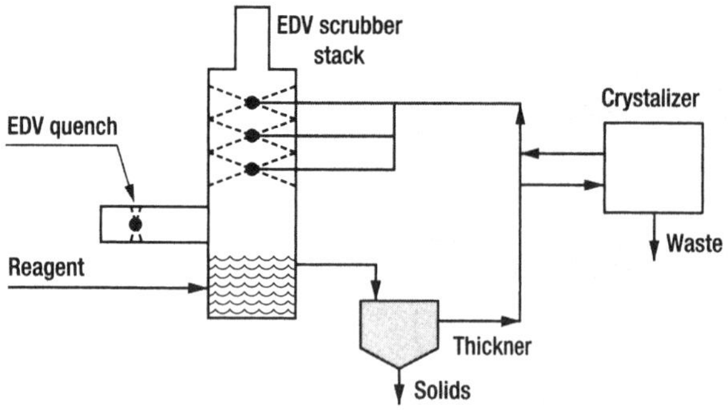

Wet-chemistry NO_x reduction

Application: When AquaNO_x is applied to BELCO's EDV wet scrubbing system, one can achieve simultaneous high-efficiency SO_x and NO_x removal from FCCU regenerator gas or flue gases.

Description: A NO_x absorbing additive to a caustic scrubbing DeSOx unit enables simultaneous removal of both pollutants. The SO_2 of the feed gas, plus additional reducing reagent at low SO_2 levels, converts NO to N_2 and gives a sodium-salt-waste byproduct. This byproduct is removed from the recirculating solvent as a solid by low-temperature crystallization. The process operating conditions are chosen to regenerate the additive for continued high-efficiency scrubbing. SO_2 removal is nearly complete, while the NO removal can be >90% if sufficient mass transfer capability is provided in the absorber. The process can be easily retrofitted to an existing caustic scrubber by installing the liquid side equipment to maintain solvent quality. The equipment footprint is small.

Operating conditions: Most applications would use atmospheric pressure scrubbing, at the adiabatic saturation temperature of the feed gas in the range up to 160°F.

Economics: The costs additional to caustic DeSO_x costs for a unit treating 440,000 scfm of 115 ppm NO and 230 ppm SO_2 are about $4 million capital + royalty fee and $800,000 operating cost/yr.

Development status: The process has been pilot tested at small scale on FCCU regenerator offgas. One or more large-scale pilot plant tests are planned for late 2002.

Licensor: Belco Technologies Corp., and Cansolv Technologies Inc.

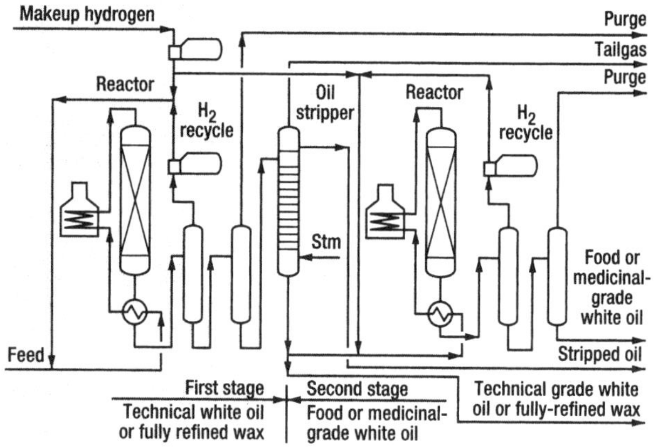

White oil and wax hydrotreating

Application. Process to produce white oils and waxes.

Feeds: Non-refined as well as solvent- or hydrogen-refined naphthenic or paraffinic vacuum distillates or deoiled waxes.

Products: Technical- and medical-grade white oils and waxes for plasticizer, textile, cosmetic, pharmaceutical and food industries. Products are in accordance with the U.S. Food and Drug Administration (FDA) regulations and the German Pharmacopoeia (DAB 8 and DAB 9) specifications.

Description: This catalytic hydrotreating process uses two reactors. Hydrogen and feed are heated upstream of the first reaction zone (containing a special presulfided NiMo/ alumina catalyst) and are separated downstream of the reactors into the main product and byproducts (hydrogen sulfide and light hydrocarbons). A stripping column permits adjusting product specifications for technical-grade white oil or feed to the second hydrogenation stage.

When hydrotreating waxes, however, medical quality is obtained in the one-stage process. In the second reactor, the feed is passed over a highly active hydrogenation catalyst to achieve a very low level of aromatics, especially of polynuclear compounds. This scheme permits each stage to operate independently and to produce technical- or medical-grade white oils separately. Yields after the first stage range from 85% to 99% depending on feedstock. Yields from the second hydrogenation step are nearly 100%. When treating waxes, the yield is approximately 98%.

Utility requirements, (typical, Middle East Crude), units per m^3 of feed:

	1st stage for techn. white oil	2nd stage for med. white oil	Food-grade wax
Electricity, kWh	197	130	70
Steam, LP, kg	665	495	140
Water, cooling, m^3	48	20	7
Hydrogen, kg	10.0	2.6	1.6

Installation: Four installations use the Uhde Edeleanu proprietary technology, one of which has the largest capacity worldwide.

Licensor: Uhde Edeleanu GmbH and through sublicense agreement with BASF.

Segment 3

Petrochemical Processes

PROCESS INDEX

LICENSOR INDEX

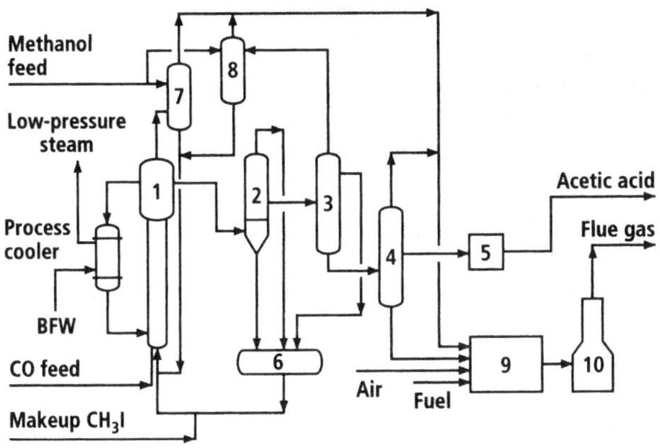

Acetic acid

Application: To produce acetic acid using the process, ACETICA. Methanol and carbon monoxide (CO) are reacted with the carbonylation reaction using a heterogeneous Rd catalyst.

Description: Fresh methanol is split into two streams and is contacted with reactor offgas in the high-pressure absorber (7) and light gases in the low-pressure absorber (8). The methanol, exiting the absorbers, are recombined and mixed with the recycle liquid from the recycle-surge drum (6). This stream is charged to a unique bubble-column reactor (1).

Carbon monoxide is compressed and sparged into the reactor riser. The reactor has no mechanical moving parts, and is free from leakage/maintenance problems. The ACETICA Catalyst is an immobilized Rd-complex catalyst on solid support; it offers higher activity and operates under less water condition in the system due to heterogeneous system. Result: the system is much less corrosivity.

Reactor effluent liquid is withdrawn and flash-vaporized in Flash Column (2) to remove the net acetic acid product and reject the CO dissolved in the liquid. The crude acetic acid is sent to the dehydration tower (3) to remove the remaining water and any residual organic materials. Dried acetic acid is routed to the finishing column (4), where heavy byproducts are removed in the bottom draw off. The finished acetic-acid product is treated to remove color bodies.

Vapor streams from the flash vessel and dehydration column overhead are combined and contacted with methanol in the low-pressure absorber. Unconverted CO, methane, acetaldehyde, other light byproducts exiting in the vapor outlets of the high- and low-pressure absorbers and heavy byproducts from the finishing column are sent to the incinerator (9) and flue-gas scrubber (10).

Economics: The Gulf Coast ISBL cost is estimated as US$ 94.1 million for a 200,000-mtpy:

Feed and utility consumption:

Methanol, mt/mt	0.539
CO,mt/mt	0.517
Power (@CO Supply 0 K/G), kWh/mt	129
Water, cooling, m^3/mt	137
Steam @100psig, mt/mt	1.7

Commercial plant: Contract for license and preparation of basic design package has been signed for a Chinese client on September 2002.

Reference: "The Chiyoda/UOP ACETICA process for the production of acetic acid," 8th Annual Saudi-Japanese Symposium on Catalysts in Petroleum Refining and Petrochemicals, KFUPM-RI, Dhahran, Saudi Arabia, Nov. 29–30, 1998.

Elsevier, "Recent advances in process and catalysts for the production of acetic acid," *Applied Catalysis: Industrial Catalytic Processes,* Vol. 221,1–2, 2001, p. 253.

Licensor: Chiyoda Corp.

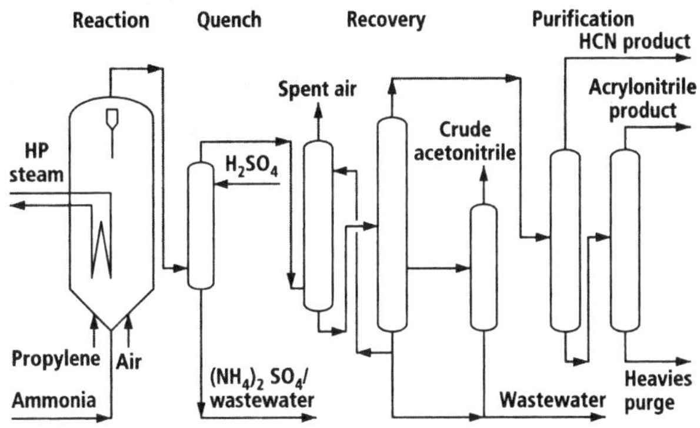

Acrylonitrile

Application: A process to produce high-purity acrylonitrile and high-purity hydrogen cyanide from propylene, ammonia and air. Recovery of byproduct acetonitrile is optional.

Description: Propylene, ammonia, and air are fed to a fluidized bed reactor to produce acrylonitrile (ACRN) using DuPont's proprietary catalyst system. Other useful products from the reaction are hydrogen cyanide (HCN) and acetonitrile (ACE). The reaction is highly exothermic and heat is recovered from the reactor by producing high-pressure steam. The reactor effluent is quenched and neutralized with a sulfuric solution to remove the excess ammonia.

The product gas from the quench is absorbed with water to recover the ACRN, HCN, and ACE. The aqueous solution of ACRN, HCN, and ACE is then fractionated and purified into high-quality products. The products' recovery and purification is a highly efficient and low-energy consumption process. This ACRN technology minimizes the amount of aqueous effluent, a major consideration for all acrylonitrile producers.

This ACRN technology is based on a high-activity, high-throughput catalyst. The propylene conversion is 99% with a selectivity of 85% to useful products of ACRN, HCN, and ACE. The DuPont catalyst is a mechanically superior catalyst, resulting in a low catalyst loss. DuPont has developed a Catalyst Bed Management Program (CBMP) to maintain the properties of the catalyst bed inside the reactor at optimal performance throughout the operation. The catalyst properties, the CBMP and proprietary reactor internals provide an optimal performance of the ACRN reactor, resulting in high yields.

With over 30 years of operating experience, DuPont has developed know-how to increase the onstream factor of the plant. This know-how includes the effective use of inhibitors to reduce the formation of cyanide and nitrile polymers and effective application of an antifoulant system to increase onstream time for equipment.

Commercial plants: DuPont Chemical Solution Enterprise, Beaumont, Texas (200,000 mtpy).

Licensor: Kellogg Brown & Root, Inc.

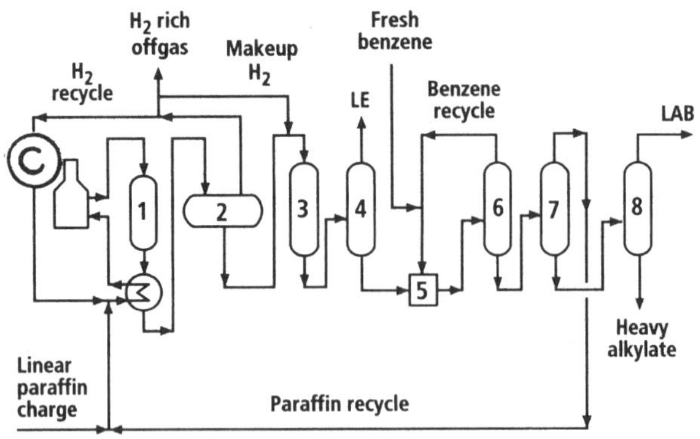

Alkylbenzene, linear

Application: To produce linear alkylbenzene (LAB) from C_{10} to C_{14} linear paraffins by alkylating benzene with olefins made by the Pacol dehydrogenation and the DeFine selective hydrogenation processes. The alkylation reaction is carried out over a solid, heterogeneous catalyst in the Detal process unit.

Description: The Pacol reactor (1) dehydrogenates the feed into corresponding linear olefin. Reactor effluent is separated into gas and liquid phases in a separator (2). Diolefins in the Pacol separator liquid are selectively converted back to mono-olefins in the DeFine reactor (3). Light ends are removed from the reactor effluent in a stripper (4). The olefin-paraffin mixture is then alkylated with benzene in the fixed-bed Detal reactor (5). Product from the reactor flows to the fractionation section (6) for separation and recycle of unreacted benzene to the reactor, and unreacted paraffins are separated (7) and recycled to the Pacol section. A rerun column (8) separates the LAB product from the heavy alkylate bottoms stream.

An exisiting LAB producer can increase production by using UOP's new Pacol catalyst or by adding a PEP unit to remove aromatics and increase the alkylation reaction efficiency.

The process is nonpolluting. No process waste streams are produced. The catalysts used are noncorrosive and require no special handling.

Yields: Based on 100 weight parts of LAB, 81 parts of linear paraffins and 34 parts of benzene are charged to process. The LAB product has a typical Bromine Index of less than 10 and is 99% sulfonable.

Economics: US Gulf Coast battery limits for the production of 80,000 tpy of LAB:

Investment, $/tpy	650
Typical utilities consumption, per metric ton of LAB:	
Catalysts and chemicals, $	17
Power, kWh	287
Water, cooling, m^3	12
Fuel fired, 10^6 kcal	4.2

Commercial plants: Twenty-five LAB complexes based on the Pacol process are in operation. Three of these complexes use the Detal process.

Reference: Banerji, A., et al., *Growth and Developments in LAB Technologies: 30 Years of Innovation and More to Come,* 1993 World Surfactant Congress, Montreaux, Switzerland, September 1993.

Licensor: UOP LLC.

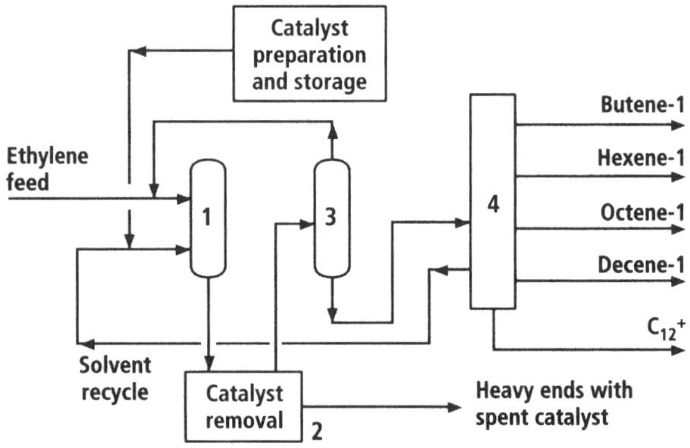

Alpha olefins, linear

Application: To produce high-purity alpha olefins (C_4–C_{10}) suitable as copolymers for LLDPE production and as precursors for plasticizer alcohols and polyalphaolefins using the AlphaSelect process.

Description: Polymer-grade ethylene is oligomerized in the liquid-phase reactor (1) with a catalyst/solvent system designed for high activity and selectivity. Liquid effluent and spent catalyst are then separated (2); the liquid is distilled (3) for recycling unreacted ethylene to the reactor, then fractionated (4) into high-purity alpha-olefins. Spent catalyst is treated to remove volatile hydrocarbons and recovered. The table below illustrates the superior purities attainable (wt%) with the AlphaSelect process:

n-Butene-1	>99
n-Hexene-1	>98
n-Octene-1	>96
n-Decene-1	>92

The process is simple; it operates at mild operating temperatures and pressures and only carbon steel equipment is required. The catalyst is nontoxic and easily handled.

Yields: Yields are adjustable to meet market requirements and very little high boiling polymer is produced as illustrated:

Alpha-olefin product distribution, wt%

n-Butene-1	33–43
n-Hexene-1	30–32
n-Octene-1	17–21
n-Decene-1	9–14

Economics: Typical case for a 2002 ISBL investment at a Gulf Coast location producing 65,000 tpy of C_4–C_{10} alpha-olefins is:

Investment, million US$	36
Raw material	
Ethylene, tons per ton of product	1.15
Byproducts, ton/ton of main products	
C_{12}^+ olefins	0.1
Fuel gas	0.03
Heavy ends	0.02
Utilities cost, US$/ton product	51
Catalyst + chemicals, US$/ton product	32

Commercial plants: The AlphaSelect process is strongly backed by extensive Axens industrial experience in homogeneous catalysis, in particular, the Alphabutol process for producing butene-1 for which there are 20 units producing 318,000 tpy.

Licensor: Axens, Axens NA.

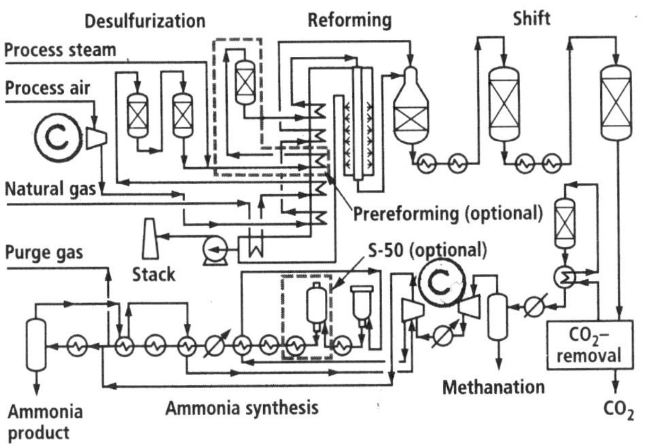

Ammonia

Application: To produce ammonia from a variety of hydrocarbon feedstocks ranging from natural gas to heavy naphtha using Topsøe's low-energy ammonia technology.

Description: Natural gas or another hydrocarbon feedstock is compressed (if required), desulfurized, mixed with steam and then converted into synthesis gas. The reforming section comprises a prereformer (optional, but gives particular benefits when the feedstock is higher hydrocarbons or naphtha), a fired tubular reformer, and a secondary reformer, where process air is added. The amount of air is adjusted to obtain an H_2/N_2 ratio of 3.0 as required by the ammonia synthesis reaction. The tubular steam reformer is Topsøe's proprietary side-wall-fired design. After the reforming section, the synthesis gas undergoes high- and low-temperature shift conversion, carbon dioxide removal and methanation.

Synthesis gas is compressed to the synthesis pressure, typically ranging from 140 to 220 kg/cm²g and converted into ammonia in a synthesis loop using radial flow synthesis converters, either the two-bed S-200, the three-bed S-300, or the S-250 concept using an S-200 converter followed by a boiler or steam superheater, and a one-bed S-50 converter. Ammonia product is condensed and separated by refrigeration. This process layout is flexible, and each ammonia plant will be optimized for the local conditions by adjustment of various process parameters. Topsøe supplies all catalysts used in the catalytic process steps for ammonia production.

Features, such as the inclusion of a prereformer, installation of a ring-type burner with nozzles for the secondary reformer and upgrading to an S-300 ammonia converter, are all features that can be applied for existing ammonia plants. These features will ease maintenance and improve plant efficiency.

Commercial plants: More than 60 plants use the Topsøe process concept. Since 1990, 50% of the new ammonia production capacity has been based on the Topsøe technology. Capacities of the plants constructed within the last decade range from 650 mtpd up to 2,050 mtpd being the world's largest ammonia plant. Design of new plants with even higher capacities are available.

Licensor: Haldor Topsøe A/S.

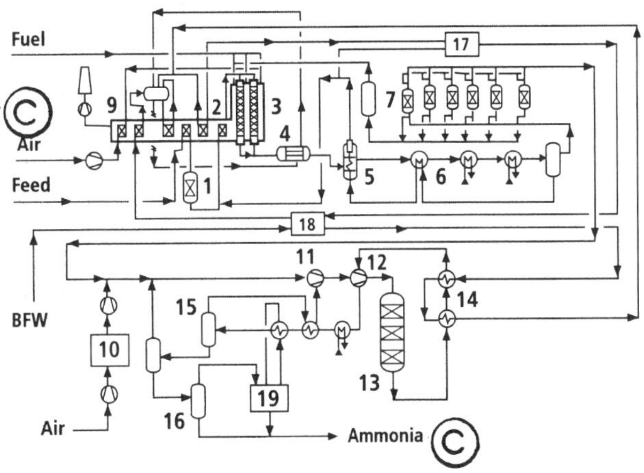

Ammonia

Application: The Linde ammonia concept (LAC) produces ammonia from light hydrocarbons. The process is a simplified route to ammonia, consisting of a modern hydrogen plant, standard nitrogen unit and a high-efficiency ammonia synthesis loop.

Description: Hydrocarbon feed is preheated and desulfurized (1). Process steam, generated from process condensate in the isothermal shift reactor (5) is added to give a steam ratio of about 2.7; reformer feed is further preheated (2). Reformer (3) operates with an exit temperature of 850°C.

Reformed gas is cooled to the shift inlet temperature of 250°C by generating steam (4). The CO shift reaction takes place in a single stage in the tube-cooled isothermal shift reactor (5), where process steam is generated from condensate. No process condensate effluent from the LAC plant is generated, thus eliminating a condensate treatment system.

After further heat recovery, final cooling and condensate separation (6), the gas is sent to the pressure swing adsorption (PSA) unit (7). Loaded adsorbers are regenerated isothermally using a controlled sequence of depressurization and purging steps.

Nitrogen is produced by the low-temperature air separation in a cold box (10). Air is filtered, compressed and purified before being supplied to the cold box. Pure nitrogen product is further compressed and mixed with the hydrogen to give a pure ammonia synthesis gas. The synthesis gas is compressed to ammonia-synthesis pressure by the syngas compressor (11), which also recycles unconverted gas through the ammonia loop. Pure syngas eliminates the loop purge and associated purge gas treatment system.

The ammonia loop is based on the Ammonia Casale axial-radial three-bed converter with internal heat exchangers (13), giving a high conversion. Heat from the ammonia synthesis reaction is used to generate HP steam (14), preheat feed gas (12) and the gas is then cooled and refrigerated to separate ammonia product (15). Unconverted gas is recycled to the syngas compressor (11) and ammonia product chilled to –33°C (16) for storage. Utility units in the LAC plant are the power-generation system (17), which provides power for the plant from HP superheated steam, BFW purification unit (18) and the refrigeration unit (19).

Economics: Simplification over conventional processes gives important savings such as: investment, catalyst-replacement costs, maintenance costs, etc. Total feed requirement (process feed plus fuel) is approximately 7 Gcal/metric ton (mt) ammonia (25.2 MMBtu/short ton) depending on plant design and location.

Commercial plants: The first complete LAC plant, for 1,350-mtd ammonia, has been built for GSFC in India. Two other LAC plants, for 230- and 600-mtd ammonia, are under construction in Australia. There are extensive reference lists for Linde hydrogen and nitrogen plants and Ammonia Casale synthesis systems.

References: "A Combination of Proven Technologies," *Nitrogen*, March–April 1994.

Licensor: Linde AG.

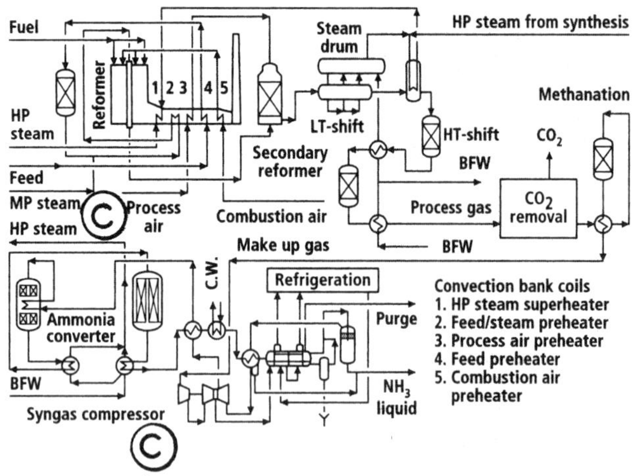

Ammonia

Application: To produce ammonia from natural gas, LNG, LPG or naphtha. Other hydrocarbons—coal, oil, resides or methanol pure gas—are possible feedstocks with an adapted front-end. The process uses conventional steam reforming synthesis gas generation (front-end) and a medium-pressure ammonia synthesis loop. It is optimized with respect to low energy consumption and maximum reliability. The largest single-train plant built by Uhde has a nameplate capacity of 1,800 metric tons per day (mtpd) and an energy consumption of 6.65 Gcal per metric ton (mt) of ammonia. Since a revamp, it is operating beyond 2,000 mtpd.

Description: The feedstock (natural gas as an example) is desulfurized, mixed with steam and converted into synthesis gas over nickel catalyst at approximately 40 bar and 800°C to 850°C in the primary reformer. The Uhde steam reformer is a top-fired reformer with tubes made of centrifugal high alloy steel and a proprietary "cold outlet manifold" system, which enhances reliability.

In the secondary reformer, process air is admitted to the syngas via a special nozzle system that provides a perfect mixture of air and gas. Subsequent high-pressure steam generation and superheating guarantee maximum process heat usage to achieve an optimized energy efficient process.

CO is converted to CO_2 in the HT and LT shift over standard catalysts. CO_2 is removed in a scrubbing unit, which is normally either the BASF-aMDEA or the UOP-Benfield process. Remaining carbonoxides are reconverted to methane in the catalytic methanation to trace ppm levels.

The ammonia synthesis loop uses two ammonia converters with three catalytic beds. Waste heat is used for steam generation downstream the second and third bed. Waste-heat steam generators with integrated boiler feedwater preheater are supplied with a special cooled tube sheet to minimize skin temperatures and material stresses. The converters themselves have radial catalyst beds with standard small grain iron catalyst. The radial flow concept minimizes pressure drop in the synthesis loop and allows maximum ammonia conversion rates.

Liquid ammonia is separated by condensation from the synthesis loop and is either subcooled and routed to storage, or conveyed at moderate temperature to subsequent consumers.

Ammonia flash and purge gases are treated in a scrubbing system and a hydrogen recovery unit (not shown), and the remains are used as fuel.

Commercial plants: Fourteen ammonia plants were commissioned between 1990 and 2000, with capacities ranging from 500 mtpd to 1,800 mtpd.

Licensor: Uhde GmbH.

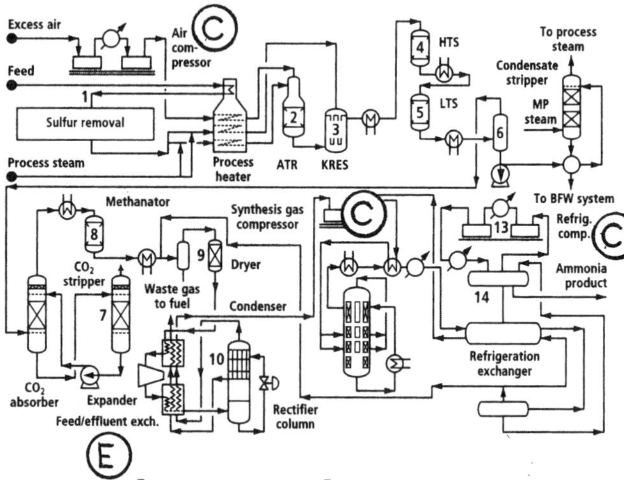

Ammonia, KAAP*plus*

Application: To produce ammonia from hydrocarbon feedstocks using a high-pressure heat exchange-based steam reforming process integrated with a low-pressure advanced ammonia synthesis process.

Description: The key steps in the KAAP*plus* process are reforming using the KBR reforming exchanger system (KRES), cryogenic purification of the synthesis gas and low-pressure ammonia synthesis using KAAP catalyst.

Following sulfur removal (1), the feed is mixed with steam, heated, and split into two streams. One stream flows to the autothermal reformer (ATR) (2) and the other to the tube side of the reforming exchanger (3), which operates in parallel with the ATR. Both convert the hydrocarbon feed into raw synthesis gas using conventional nickel catalyst. In the ATR, feed is partially combusted with excess air to supply the heat needed to reform the remaining hydrocarbon feed. The hot autothermal reformer effluent is fed to the shell side of the KRES reforming exchanger, where it combines with the reformed gas exiting the catalyst-packed tubes. The combined stream flows across the shell side of the reforming exchanger where it supplies heat to the reforming reaction inside the tubes.

Shell-side effluent from the reforming exchanger is cooled in a waste heat boiler, where high-pressure steam is generated, and then flows to the CO shift converters containing two catalyst types: one (4) is a high-temperature catalyst and the other (5) is a low-temperature catalyst. Shift reactor effluent is cooled, condensed water separated (6) and then routed to the gas purification section. CO_2 is removed from synthesis gas using a wet CO_2 scrubbing system such as hot potassium carbonate or MDEA (methyl diethanolamine) (7).

After CO_2 removal, final purification includes methanation (8), gas drying (9), and cryogenic purification (10). The resulting pure synthesis gas is compressed in a single-case compressor and mixed with a recycle stream (11). The gas mixture is fed to the KAAP ammonia converter (12), which uses a ruthenium-based, high-activity ammonia synthesis catalyst. It provides high conversion at the relatively low pressure of 90 bar with a small catalyst volume. Effluent vapors are cooled by ammonia refrigeration (13) and unreacted gases are recycled. Anhydrous liquid ammonia is condensed and separated (14) from the effluent.

Energy consumption of KBR's KAAP*plus* process is less than 25 MM Btu (LHV)/short-ton. Elimination of the primary reformer combined with low-pressure synthesis provides a capital cost savings of about 10% over conventional processes.

Commercial plants: Over 200 large-scale, single-train ammonia plants of KBR design are onstream or have been contracted worldwide. The KAAP*plus* advanced ammonia technology provides a low-cost, low-energy design for ammonia plants, minimizes environmental impact, reduces maintenance and operating requirements and provides enhanced reliability. Two plants use KRES technology and 17 plants use Purifier technology. Three 1,850-mtpd grassroots KAAP plants in Trinidad are in full operation, and a fourth is under construction.

Licensor: Kellogg Brown & Root, Inc.

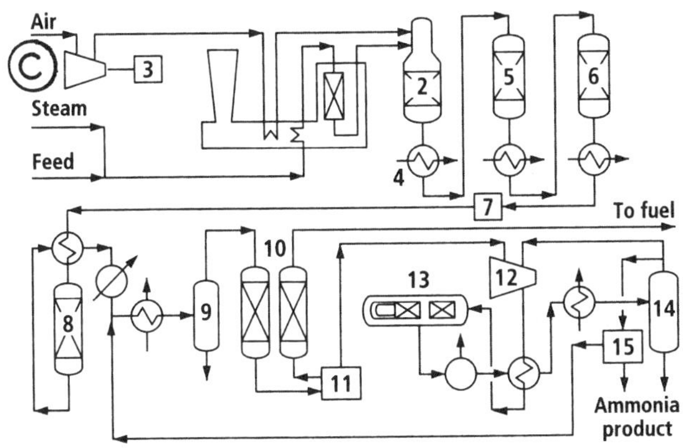

Ammonia, KBR purifier

Application: To produce ammonia from hydrocarbon feedstocks and air.

Description: The key features of the KBR Purifier Process are mild primary reforming, secondary reforming with excess air, cryogenic purification of syngas, and synthesis of ammonia over magnetite catalyst in a horizontal converter.

Desulfurized feed is reacted with steam in the primary reformer (1) with exit temperature at about 700°C. Primary reformer effluent is reacted with excess air in the secondary reformer (2) with exit at about 900°C. The air compressor is normally a gas-driven turbine (3). Turbine exhaust is fed to the primary reformer and used as preheated combustion air. An alternative to the above described conventional reforming is to use KBR's reforming exchanger system (KRES), as described in KBR's KAAP*plus*.

Secondary reformer exit gas is cooled by generating high-pressure steam (4). The shift reaction is carried out in two catalytic steps—high-temperature (5) and low-temperature shift (6). Carbon dioxide removal (7) uses licensed processes. Following CO_2 removal, residual carbon oxides are converted to methane in the methanator (8). Methanator effluent is cooled, and water is separated (9) before the raw gas is dried (10).

Dried synthesis gas flows to the cryogenic purifier (11), where it is cooled by feed/effluent heat exchange and fed to a rectifier. The syngas is purified in the rectifier column, producing a column overhead that is essentially a 75:25 ratio of hydrogen and nitrogen. The column bottoms is a waste gas that contains unconverted methane from the reforming section, excess nitrogen and argon. Both overhead and bottoms are reheated in the feed/effluent exchanger. The waste gas stream is used to regenerate the dryers and then is burned as fuel in the primary reformer. A small, low-speed expander provides the net refrigeration.

The purified syngas is compressed in the syngas compressor (12), mixed with the loop-cycle stream and fed to the converter (13). Converter effluent is cooled and then chilled by ammonia refrigeration. Ammonia product is separated (14) from unreacted syngas. Unreacted syngas is recycled back to the syngas compressor. A small purge is scrubbed with water (15) and recycled to the dryers.

Commercial plants: Over 200 single-train plants of KBR design have been contracted worldwide. Seventeen of these plants use the KBR Purifier process.

Licensor: Kellogg Brown & Root, Inc.

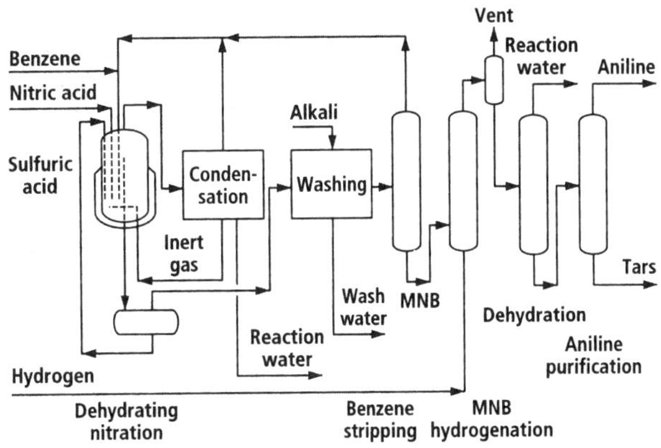

Aniline

Application: A process for the production of high-quality aniline from benzene and nitric acid.

Description: Aniline is produced by the nitration of benzene with nitric acid to mononitrobenzene (MNB) which is subsequently hydrogenated to aniline. In the DuPont/KBR process, benzene is nitrated with mixed acid (nitric and sulfuric) at high efficiency to produce mononitrobenzene (MNB) in the unique dehydrating nitration (DHN) system. The DHN system uses an inert gas to remove the water of nitration from the reaction mixture, thus eliminating the energy-intensive and high-cost sulfuric acid concentration system.

As the inert gas passes through the system, it becomes humidified, removing the water of reaction from the reaction mixture. Most of the energy required for the gas humidification comes from the heat of nitration. The wet gas is condensed and the inert gas is recycled to the nitrator. The condensed organic phase is recycled to the nitrator while the aqueous phase is sent to effluent treatment. The reaction mixture is phase separated and the sulfuric acid is returned to the nitrator.

The crude MNB is washed to remove residual acid and the impurities formed during the nitration reaction. The product is then distilled and residual benzene is recovered and recycled. Purified MNB is fed, together with hydrogen, into a liquid phase plug-flow hydrogenation reactor that contains a DuPont proprietary catalyst. The supported noble metal catalyst has a high selectivity and the MNB conversion per pass is 100%.

The reaction conditions are optimized to achieve essentially quantitative yields and the reactor effluent is MNB-free. The reactor product is sent to a dehydration column to remove the water of reaction followed by a purification column to produce high-quality aniline product.

Commercial plants: DuPont produces aniline using this technology for the merchant market with a total production capacity of 160,000 tpy at a plant located in Beaumont, Texas. In addition, DuPont's aniline technology has been licensed to two companies in the last eight years with a total aniline capacity of 150,000 tpy.

Licensor: Kellogg Brown & Root, Inc.

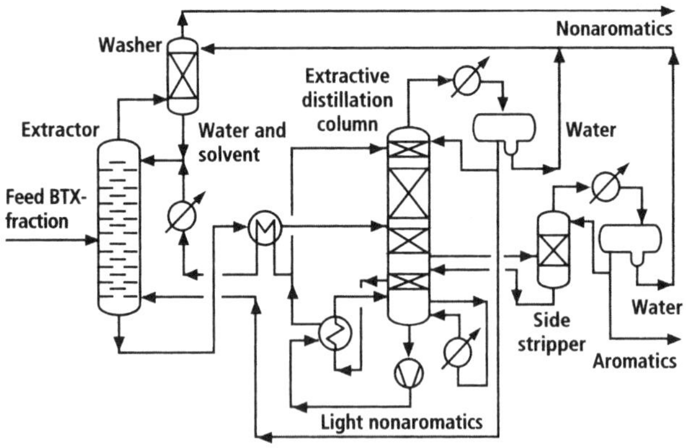

Aromatics extraction

Application: Simultaneous recovery of benzene, toluene and xylenes (BTX) from reformate or pyrolysis gasoline (pygas) using liquid-liquid extraction.

Description: At the top of extractor operating at 30°C to 50°C and 1 to 3 bar, the solvent, N-Formylmorpholin with 4% to 6% water, is fed as a continuous phase. The feedstock—reformate or pygas—enters several stages above the base of the column. Due to density differences, the feedstock bubbles upwards, countercurrent to the solvent. Aromatics pass into the solvent, while the nonaromatics move to the top, remaining in the light phase. Low-boiling nonaromatics from the top of the extractive distillation (ED) column enter the base of the extractor as countersolvent.

Aromatics and solvent from the bottom of the extractor enter the ED, which is operated at reduced pressure due to the boiling-temperature threshold. Additional solvent is fed above the aromatics feed containing small amounts of nonaromatics that move to the top of the column. In the bottom section, as well as in the side rectifier, aromatics and practically water-free solvent are separated.

The water is produced as a second subphase in the reflux drum after azeotropic distillation in the top section of the ED. This water is then fed to the solvent-recovery stage of the extraction process.

Economics:

Consumption per ton of feedstock	
Steam (20 bar), t/t	0.46
Water, cooling (T=10°C), m^3/t	12
Electric power, kWh/t	18
Production yield	
Benzene, %	~100
Toluene, %	99.7
EB, Xylenes,%	94.0
Purity	
Benzene, wt%	99.999
Toluene, wt%	>99.99
EB, Xylenes, wt%	>99.99

Commercial plants: One Morphylex plant was erected.

Reference: Emmrich, G., F. Ennenbach and U. Ranke, "Krupp Uhde Processes for Aromatics Recovery," European Petrochemical Technology Conference, June 21–22, 1999, London.

Licensor: Uhde GmbH.

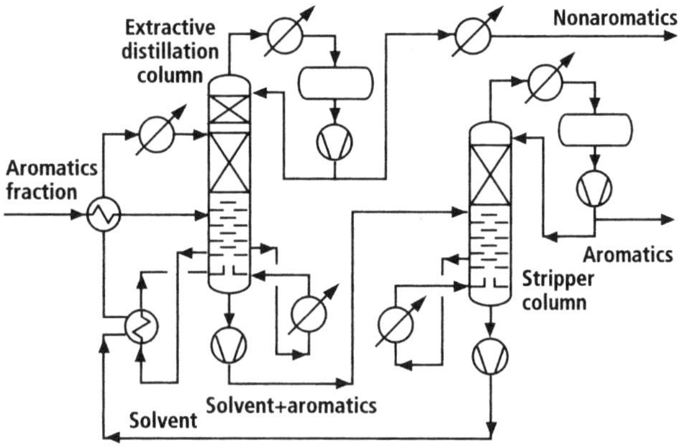

Aromatics extractive distillation

Application: Recovery of high-purity aromatics from reformate, pyrolysis gasoline or coke-oven light oil using extractive distillation.

Description: In the extractive distillation (ED) process, a single-compound solvent, N-Formylmorpholin (NFM) alters the vapor pressure of the components being separated. The vapor pressure of the aromatics is lowered more than that of the less-soluble nonaromatics.

Nonaromatics vapors leave the top of the ED column with some solvent, which is recovered in a small column that can either be mounted on the main column or installed separately.

Bottom product of the ED column is fed to the stripper to separate pure aromatics from the solvent. After intensive heat exchange, the lean solvent is recycled to the ED column. NFM perfectly satisfies the necessary solvent properties needed for this process including high selectivity, thermal stability and a suitable boiling point.

Economics:

Pygas feedstock:		
	Benzene	**Benzene/toluene**
Production yield		
Benzene	99.95%	99.95%
Toluene	–	99.98%
Quality		
Benzene	30 wt ppm NA*	80 wt ppm NA*
Toluene	–	600 wt ppm NA*
Consumption		
Steam	475 kg/t ED feed	680 kg/t ED feed**

Reformate feedstock with low aromatics content (20wt%):	
	Benzene
Quality	
Benzene	10 wt ppm NA*
Consumption	
Steam	320 kg/t ED feed

*Maximum content of nonaromatics
**Including benzene/toluene splitter

Commercial plants: 45 Morphylane plants (total capacity of more than 6 MMtpy).

Reference: Emmrich, G., F. Ennenbach and U. Ranke, "Krupp Uhde Processes for Aromatics Recovery," European Petrochemical Technology Conference, June 21–22, 1999, London.

Licensor: Uhde GmbH.

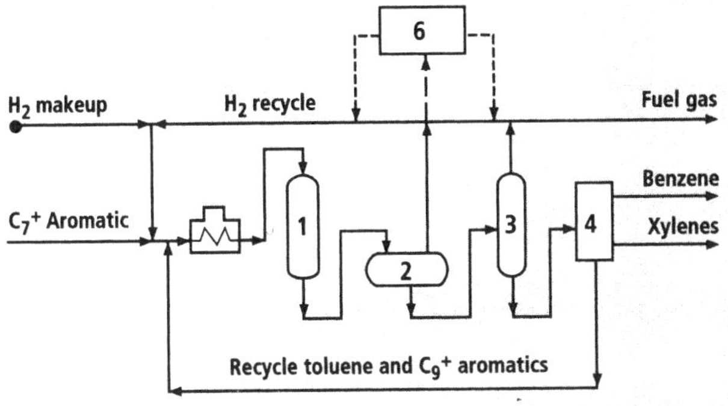

Benzene

Application: To produce high-purity benzene and heavier aromatics from toluene and heavier aromatics using the Detol process.

Description: Feed and hydrogen are heated and passed over the catalyst (1). Benzene and unconverted toluene and/or xylene and heavier aromatics are condensed (2) and stabilized (3).

To meet acid wash color specifications, stabilizer bottoms are passed through a fixed-bed clay treater, then distilled (4) to produce the desired specification benzene. The cryogenic purification of recycle hydrogen to reduce the make-up hydrogen requirement is optional (6).

Unconverted toluene and/or xylenes and heavier aromatics are recycled.

Yields: Aromatic yield is 99.0 mol% of fresh toluene or heavier aromatic charge. Typical yields for production of benzene and xylenes are:

Type production	Benzene	Xylene
Feed, wt%		
Nonaromatics	3.2	2.3
Benzene	—	11.3
Toluene	47.3	0.7
C_8 aromatics	49.5	0.3
C_9^+ aromatics	—	85.4
Products, wt% of feed		
Benzene*	75.7	36.9
C_8 aromatics**	—	37.7

* 5.45°C minimum freeze point
** 1,000 ppm nonaromatics maximum

Economics: Basis of ISBL US Gulf Coast:

Estimated investment, $/bpsd	6,700
Typical utility requirements, per bbl feed:	
Electricity, kWh	5.8
Fuel, MMBtu	0.31 *
Water, cooling, gal	450
Steam, lb	14.4

* No credit taken for vent gas streams

Commercial plants: Twelve plants with capacities ranging from about 12 million to 100 million gal/y have been licensed.

Licensor: ABB Lummus Global.

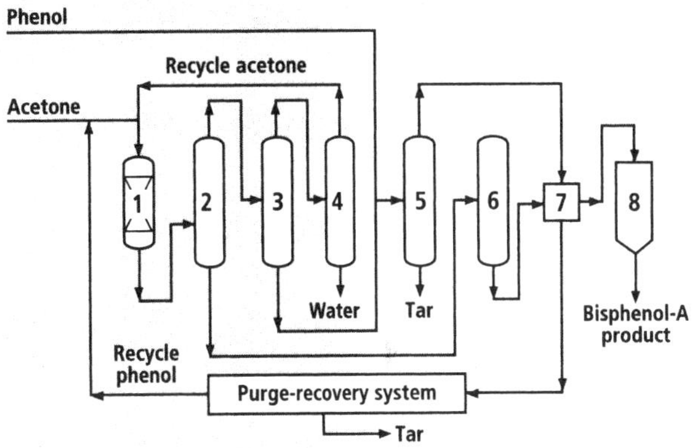

Bisphenol-A

Application: The Mitsubishi BPA process uses ion-exchange resin catalysts to react phenol and acetone to produce high-quality Bisphenol-A (BPA).

Description: Phenol and acetone are reacted in a BPA synthesis reactor (1), which is packed with the proprietary cation-exchange resin-base catalyst (named "4PET"). This catalyst has a higher acetone conversion, higher BPA selectivity and longer life than conventional catalysts. Unreacted acetone, water and some phenol are separated from the reactor effluent by distillations (2, 3, 4). Acetone is recycled to the BPA reactor (1); water is purged as waste. Phenol is mixed with fresh phenol and purified by distillation (5). The stream, which contains BPA, phenol and other impurities are fed to the crystallizer (6), where phenol-BPA adduct crystal grows. Separated adduct crystals are washed with purified phenol in a solid-liquid separator (7).

Crystals are then melted and sent to the prilling tower (8), where spherical prills are produced as the final products. Solidified BPA is conveyed to storage and shipping facilities.

The mother liquor is recycled back to the BPA reactor from solid-liquid separator (7). Part of the mother liquor is fed to the purge-recovery system where impurities are partially decomposed and recombined to form BPA. Undesired impurities are condensed and discharged as tarry materials. The optimal purge ratio from the mother liquor controls product quality, while minimizing raw material unit consumption.

Product quality: Typical values for BPA prills:

BPA purity, wt%	99.9–99.98
Freezing point, °C	156.7–156.8
Melt color @175°C, [APHA]	5–10
Free phenol, wtppm	5–20
2,4-BPA isomer, wtppm	50–150

Commercial plants: In 2001, Mitsubishi Chemical Corp. (MCC) acquired from Chiyoda Corp., all intellectual property for BPA technology, including licensing rights. MCC is improving this process and has developed a new catalyst—4PET—which enhances the performance of this process.

MCC's first BPA plant at Kashima, Japan, started up in 1998 with 70,000-tpy (70-Mtpy) capacity. In 2000, the unit was revamped to 100 Mtpy by incorporating the improved process. In 2002, MCC's second BPA plant at Kurosaki, Japan, started up with 100-Mtpy capacity and uses 4PET catalyst. In addition to these plants, two additional plants are operating, and one plant is now under construction:

Company	Capacity, Mtpy	Startup
Shin Nippon Bisphenol (Japan)	95	1991
Taiwan Prosperity Chemical Corp. (Taiwan)	25	1995
Mitsubishi Chemical Corp. (Japan)	100	1998
Mitsubishi Chemical Corp. (Japan)	100	2002
Blue Star Chemical Cleaning Corp. (China)	25	(2003)

Licensor: Mitsubishi Chemical Corp.

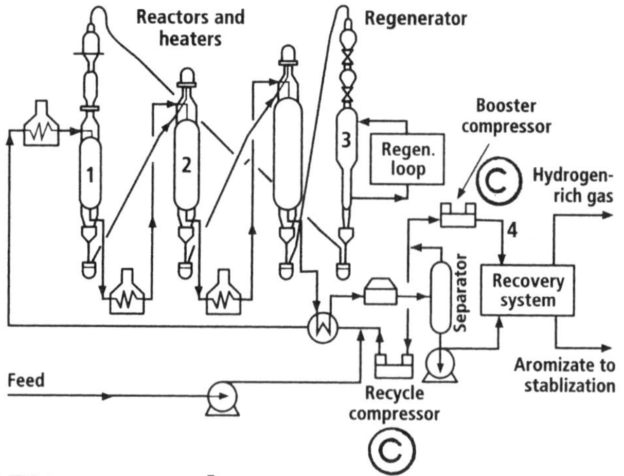

BTX aromatics

Application: To produce high yields of benzene, toluene, xylenes and hydrogen from naphthas via the CCR Aromizing process coupled with RegenC continuous catalyst regeneration technology. Benzene and toluene cuts are fed directly to an aromatics extraction unit. The xylenes fraction, obtained by fractionation and subsequent treatment by the Arofining process for diolefins and olefins removal, is ideal for paraxylene and orthoxylene production.

Description: This process features moving bed reactors and a continuous catalyst regeneration system coupled with a hard, smooth-flowing catalyst. Feed enters the reactor (1), passes radially through the moving catalyst bed, exits at the reactor bottom and proceeds in the same manner through the 2–3 remaining reactors (2). The robust (latest generation AR 501 & 505) catalyst moves downward through each reactor. Leaving the reactor, the catalyst is gas-lifted to the next reactor's feed hopper where it is distributed for entry. After the last reactor, an inert gas lift system isolates and transports the catalyst to the recently-introduced RegenC regeneration section (3). Coke is removed; catalyst is returned to its original state and sent to the first reactor; the cycle begins again. A recovery system (4) separates hydrogen for use in downstream units and the Aromizate is sent to a stabilization section. The unit is fully automated and operating controls are integrated into a DCS, requiring only a minimum of supervisory and maintenance effort.

Yields: (%)

	Feed		Products
TBP cut, °C	80–150	Hydrogen	4.1
Paraffins	57	C_5+	87
Naphthenes	37	Benzene	8.5
Aromatics	6	Toluene	26.3
		Xylenes	26.1
		Total aromatics	74.3

Economics: The ISBL investment for a typical 25,000-bpsd CCR Aromizing unit with a RegenC regenerator, 2002 Gulf Coast location.

Investment including initial catalyst inventory,* US$ million	52
Typical utility requirements:	
Fuel, 10^6 kcal/h	76
Steam, HP t/h (net export)	(17)
Electricity, kWh/h	5,900
Catalyst operating cost, $/ton feed	0.5

* Exclusive of noble metals

Commercial plants: Sixty CCR reforming units have been licensed including seven plants in operation and four under design.

Licensor: Axens, Axens NA.

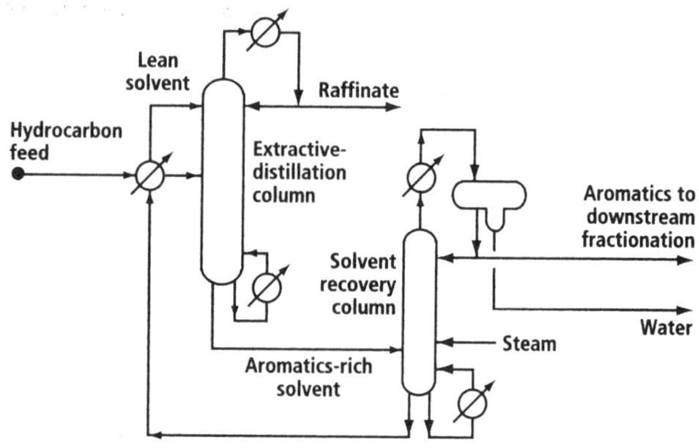

BTX aromatics

Application: An aromatics process based on extractive distillation, GT-BTX efficiently recovers benzene, toluene and xylenes from refinery or petrochemical aromatics streams, such as catalytic reformate or pyrolysis gasoline.

Description: Hydrocarbon feed is preheated with hot circulating solvent and fed at a midpoint into the extractive distillation column (EDC). Lean solvent is fed at an upper point to selectively extract the aromatics into the column bottoms in a vapor/liquid distillation operation. Nonaromatic hydrocarbons exit the column top and pass through a condenser. A portion of the overhead stream is returned to the column top as reflux to wash out any entrained solvent. The balance of the overhead stream is the raffinate product, requiring no further treatment.

Rich solvent from the bottom of the EDC is routed to the solvent-recovery column (SRC), where the aromatics are stripped overhead. Stripping steam from a closed-loop water circuit facilitates hydrocarbon stripping. The SRC operates under vacuum to reduce the boiling point at the column base.

Lean solvent from the bottom of the SRC is passed through heat exchange before returning to the EDC. A small portion of the lean circulating solvent is processed in a solvent-regeneration step to remove heavy decomposition products, which are purged daily.

The process advantages over conventional liquid-liquid extraction processes include lower capital and operating costs and simplicity of operation. Advantages over other extractive processes include: superior solvent system, fewer equipment pieces, small equipment and expanded feedstock range. Design flexibility allows use for grassroots aromatics recovery units or debottlenecking conventional systems in many revamp configurations.

Economics:

	New unit	Expansion of conventional BTX recovery unit
Feed, bpd	3,500 Lt. reformate	4,000 incremental
Capital cost, $MM	6.5	3.5
Simple pretax payout, yr	2.2	1.2
ROI, %	44	85

Commercial plants: Three grassroots applications.

Reference: "Benzene reduction in motor gasoline—obligation or opportunity," *Hydrocarbon Processing* Process Optimization Conference, April 1997. "Improve BTX processing economics," *Hydrocarbon Processing*, March 1998.

Licensor: GTC Technology Inc.

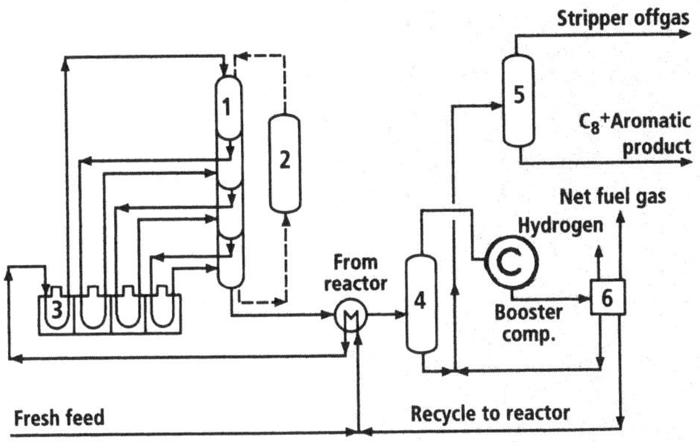

BTX aromatics

Application: To produce petrochemical-grade benzene, toluene and xylenes (BTX) via the aromatization of propane and butanes using the BP-UOP Cyclar process.

Description: The process consists of a reactor section, continuous catalyst regeneration (CCR) section and product-recovery section. Stacked radial-flow reactors (1) facilitate catalyst transfer to and from the CCR catalyst regeneration section (2). A charge heater and interheaters (3) achieve optimum conversion and selectivity for the endothermic reaction. Reactor effluent is separated into liquid and vapor products (4). The liquid product is sent to a stripper column (5) to remove light saturates from the C_6^- aromatic product. Vapor from the separator is compressed and sent to a gas recovery unit (6). The compressed vapor is then separated into a 95% pure hydrogen coproduct, a fuel-gas stream containing light byproducts and a recycled stream of unconverted LPG.

Yields: Total aromatics yields as a wt% of fresh feed range from 61% for propane to 66% for mixed butanes feed. Hydrogen yield is approximately 7-wt% fresh feed. Typical product distribution is 27% benzene, 43% toluene, 22% C_8 aromatics and 8% C_9^+ aromatics.

Economics: US Gulf Coast inside battery limits basis, assuming gas turbine driver is used for product compressor.

Investment, US$ per metric ton (mt) of feed 175–208

Typical utility requirements, unit per mt of feed

Electricity, kWh	0.013
Steam, MP, mt (credit)	(0.7)
Steam, LP, mt	0.13
Water, cooling, mt	19
Fuel, MMkcal	2
Boiler feedwater, mt	0.55

Commercial plants: In 1990, the first Cyclar unit was commissioned at the BP refinery at Grangemouth, Scotland. This unit was designed to process 1,000 bpd of C_3 or C_4 feedstock at either high- or low-pressure over a wide range of operating conditions. A second unit capable of processing C_3 and C_4 feedstock was commissioned in 2000, and operates at design capacities.

Reference: Doolan, P. C., and P. R. Pujado, "Make aromatics from LPG," *Hydrocarbon Processing,* September 1989, pp. 72–76. Gosling, C. D., et al., "Process LPG to BTX products," *Hydrocarbon Processing,* December 1991.

Licensor: UOP LLC.

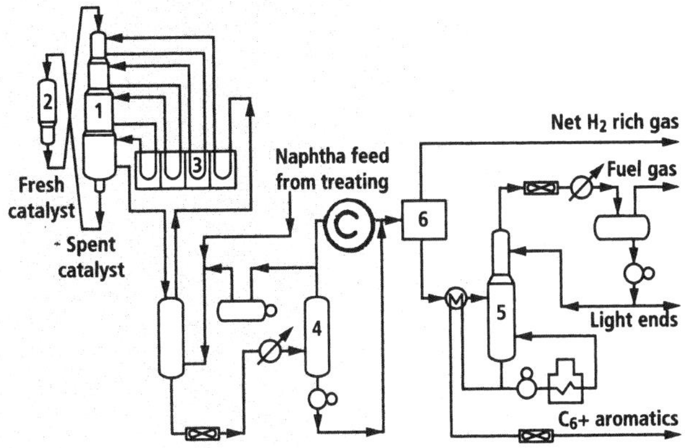

BTX aromatics

Application: To produce reformate, which is concentrated in benzene, toluene and xylenes (BTX) from naphtha and condensate feedstocks via a high-severity reforming operation with a hydrogen byproduct. The CCR Platforming Process is licensed by UOP.

Description: The process consists of a reactor section, continuous catalyst regeneration section (CCR) and product recovery section. Stacked radial flow reactors (1) facilitate catalyst transfer to and from the CCR catalyst regeneration section (2). A charge heater and interheaters (3) are used to achieve optimum conversion and selectivity for the endothermic reaction.

Reactor effluent is separated into liquid and vapor products (4). Liquid product is sent to a stabilizer (5) to remove light ends. Vapor from the separator is compressed and sent to a gas-recovery section (6) to separate 90%-pure hydrogen byproduct. A fuel gas byproduct of LPG can also be produced. UOP's latest R-270 series catalyst maximizes aromatics yields.

Yields: Typical yields from lean Middle East naphtha:

H_2, wt%	4.3
Benzene, wt%	1.7
Toluene, wt%	29.9
Xylenes, wt%	30.4
A_9^+, wt%	13.1

Economics: Capital investment per mtpy of feed:

US$	50–75

Utilities per metric ton feedrate

Electricity, kWh	12
Steam, HP, mt	0.16
Water, cooling m^3	20
Fuel, MMkcal	0.13

Commercial plants: There are 173 units in operation and 30 additional units in design and construction. Total operating capacity represents over 3.9 million-bpd.

Licensor: UOP LLC.

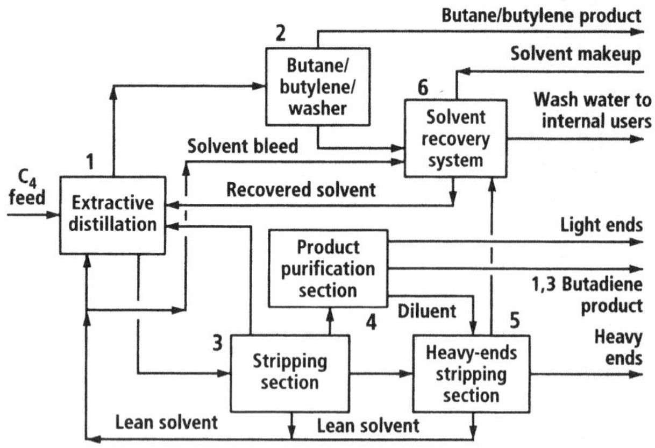

Butadiene extraction

Application: To produce a polymer-grade butadiene product from mixed-C_4 streams by extractive distillation using acetonitrile (ACN) as the solvent.

Description: This butadiene extraction process was originally developed by Shell Chemicals. It is offered under license agreement by Kellogg Brown & Root, who has updated and optimized the process to reduce capital and operating costs.

The process scheme consists of contacting mixed-C_4 feed with lean solvent in the extractive distillation column (1). The raffinate, butenes and butanes, which are not absorbed, flow overhead to the wash column (2) for solvent recovery. The butadiene-rich solvent flows to the stripper system (3) where the butadiene is separated from the solvent. Raw butadiene is purified to meet specifications in the purification section (4). Heavy ends (C_4 acetylenes) are also separated in the stripper system (3) as a side product and further processed in the heavy-ends stripping section (5). The solvent recovery step (6) maintains solvent quality and recovers solvent from various product streams.

Use of acetonitrile is advantageous to other solvent systems for a number of reasons. ACN's lower boiling point results in lower operating temperatures resulting in low fouling rates and long run-lengths. Only low-pressure steam is required for reboilers. The low molecular weight and low molar volume of ACN, combined with its high selectivity to butadiene, results in low solvent circulation rates and smaller equipment sizes. The low viscosity of ACN increases tower efficiencies and reduces column size and cost. ACN is also very stable, noncorrosive and biodegradable. The basic process is noncorrosive and requires only carbon steel materials of construction.

Yields: This process can exceed 98% recovery of the butadiene contained in the feed as product. This product will meet all butadiene derivative requirements with typical specifications shown below.

Component	Value	Units
1,3-Butadiene	99.5	% wt. Minimum
Total acetylenes	20	ppmwt maximum
Methyl acetylene	10	ppmwt maximum
Vinyl acetylenes	10	ppmwt maximum
Propadiene	10	ppmwt maximum
1,2-Butadiene	10	ppmwt maximum
C_5 hydrocarbons	200	ppmwt maximum

Commercial plants: Over 35 butadiene units have been constructed using the Shell ACN technology. Unit capacities range from 20 Mtpy to over 225 Mtpy.

Licensor: Kellogg Brown & Root, Inc.

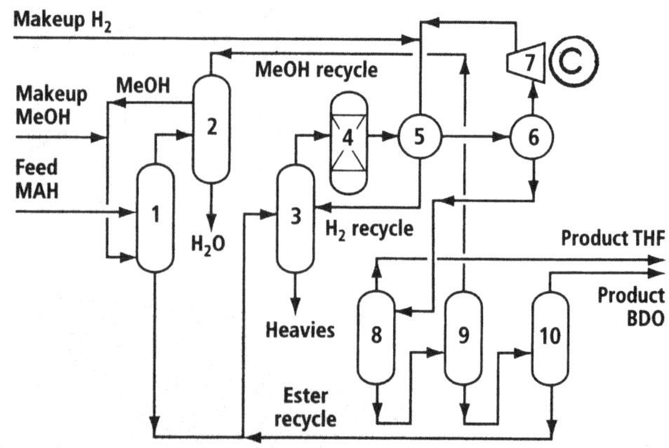

Butanediol, 1,4-

Application: To produce 1,4 butanediol (BDO) from butane via maleic anhydride and hydrogen using ester hydrogenation.

Description: Maleic anhydride is first esterified with methanol in a reaction column (1) to form the intermediate dimethyl maleate. The methanol and water overhead stream is separated in the methanol column (2) and water discharged.

The ester is then fed directly to the low-pressure, vapor-phase hydrogenation system where it vaporized into an excess of hydrogen in the vaporizer (3) and fed to a fixed-bed reactor (4), containing a copper catalyst. The reaction product is cooled (5) and condensed (6) with the hydrogen being recycled by the centrifugal circulator (7).

The condensed product flows to the lights column (8) where it is distilled to produce a small co-product tetrahydrofuran (THF) stream. The heavies column (9) removes methanol, which is recycled to the methanol column (2). The product column (10) produces high-quality butanediol (BDO). Unreacted ester and gamma butyralactone (GBL) are recycled to the vaporizer (3) to maximize process efficiency.

The process can be adapted to produce higher quantities of co-product THF and to extract the GBL as a co-product if required.

Economics: per ton of BDO equivalent

Maleic anhydride	1.125
Hydrogen	0.116
Methanol	0.050
Electric power, kWh	164
Steam, tons	3.6
Water, cooling, m^3	326

Commercial plants: Since 1989, six plants have been licensed with a total capacity of 300,000 tpy.

Licensor: Davy Process Technology, UK.

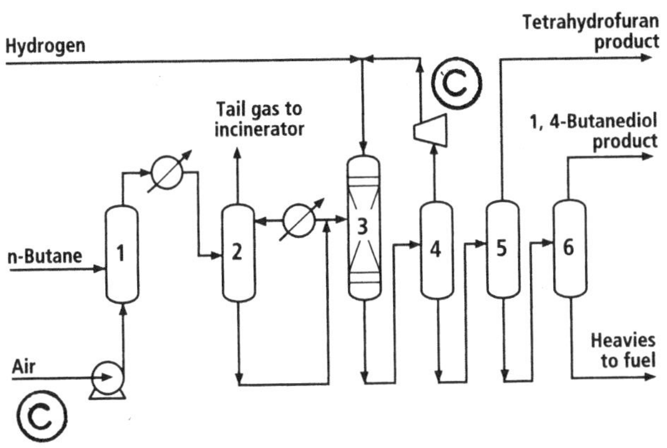

Butanediol, 1,4-

Application: To produce 1,4-butanediol (BDO), or mixture of BDO with tetrahydrofuran (THF) and/or gamma-butyrolactone (GBL) from normal butane using a fluid-bed oxidation and fixed-bed hydrogenation reactor combination.

Description: BP Chemicals has combined its 40 years of experience in fluid-bed oxidation technology with Lurgi Oel-Gas-Chemie GmbH's 30 years of hydrogenation expertise to jointly develop a direct, dual-reactor process, called the GEMINOX BDO.

Air and n-butane are introduced into a fluid-bed, catalytic reactor (1). The fluid-bed reactor provides a uniform temperature profile for optimum catalyst performance. Reaction gases are cooled and filtered to remove small entrained catalyst particles and then routed to the recovery section. Reactor effluent is contacted in an aqueous scrubber (2), where essentially 100% of the reactor-made maleic anhydride is recovered as maleic acid. The process has the capability of co-producing maleic anhydride (MAH) with the addition of the appropriate purification equipment. Scrubber overhead gases are sent to an incinerator for safe disposal.

The resulting maleic acid from the scrubber is then sent directly to the fixed-bed, catalytic hydrogenation reactor (3). Reactor yields exceed 94% BDO. By adjustments to the hydrogenation reactor and recovery-purification sections, mixtures of BDO with THF and/or GBL can be directly produced at comparable, overall yields and economics.

The hydrogenation reactor effluent is then sent through a series of distillation steps (4, 5 and 6) to produce final market quality product(s). Two unique process features are:

- No continuous liquid waste stream to treat—the water separated in the product purification section is recycled back to the aqueous MAH scrubber (2).
- No pretreatment nor post-treatment of the two catalysts is necessary; no other chemicals are added.

Economics: The GEMINOX BDO technology uses fewer processing steps, as found in competing BDO technologies, leading to significant capital cost savings and lower operating costs. Overall, 25–40% cost of production savings are possible compared to competing technologies. The unique product flexibility afforded by this process also allows the user to quickly meet changing customer and market needs.

Commercial plants: BP's first world-scale 60,000-tpd GEMINOX BDO plant in Lima, Ohio, has been successfully operating since July 2000.

Licensor: BP Chemicals and Lurgi Oel Gas Chemie GmbH.

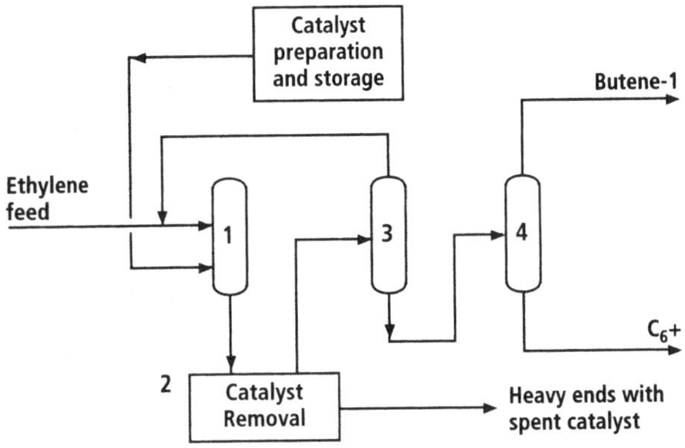

Butene-1

Application: To produce high-purity butene-1 that is suitable for copolymers in LLDPE production via the Alphabutol ethylene dimerization process developed by IFP/Axens in cooperation with SABIC.

Description: Polymer-grade ethylene is oligomerized in the liquid-phase reactor (1) with a catalyst system that has high activity and selectivity. Liquid effluent and spent catalyst are then separated (2); the liquid is distilled (3) for recycling of unreacted ethylene to the reactor and fractionated (4) into high-purity butene-1. Spent catalyst is treated to remove volatile hydrocarbons and recovered.

The Alphabutol process features are: simple processing, high turndown, ease of operation, low operating pressure and temperature, liquid-phase operation and carbon steel equipment. The technology has advantages over other production or supply sources: uniformly high-quality product, low impurities, reliable feedstock source, low capital costs, high turndown and ease of production.

Yields: LLDPE copolymer grade butene-1 is produced with a purity exceeding 99.5 wt%. Typical product specification is:

Other C_4s (butenes + butanes)	<0.3 wt%
Ethane	<0.15 wt%
Ethylene	<0.05 wt%
C_6 olefins	<100 ppmw
Ethers (as DME)	<2ppmw
Sulfur, chlorine	<1ppmw
Dienes, acetylenes	<5ppmw each
CO, CO_2, O_2, H_2O, MeOH	<5ppmw each

Economics: Case for a 2002 ISBL investment at a Gulf Coast location for producing 20,000-tpy of butene-1 is:

Investment, million US$	8
Raw material	
Ethylene, tons per ton of butene-1	1.1
Byproducts, C_6^+ tons per ton of butene-1	0.08
Typical operating cost, US$ per ton of butene-1	38

Commercial plants: There are 20 licensed units producing 318,000 tpy. Sixteen units are in operation; the others are under design or construction.

Licensor: Axens, Axens NA.

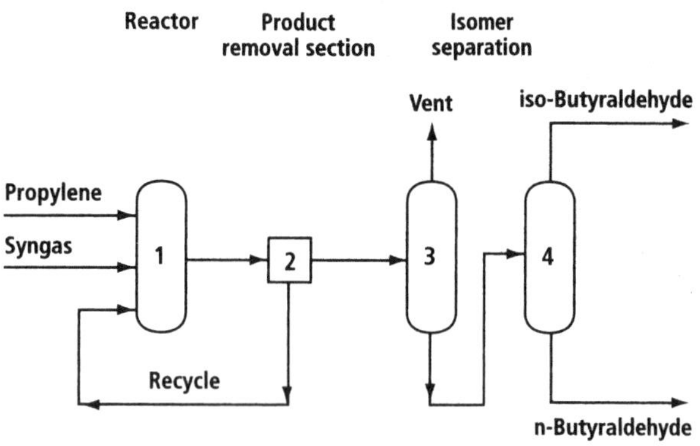

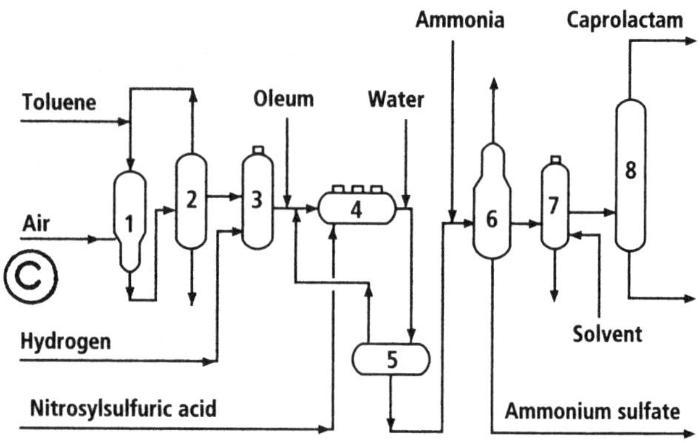

Butyraldehyde, n and i

Application: To produce normal and iso-butyraldehyde from propylene and synthesis gas ($CO + H_2$) using the LP Oxo process, which is a low-pressure, rhodium-catalyzed oxo process.

Description: The process reacts propylene with a 1:1 syngas at low pressure (<20 kg/cm^{2}g) in the presence of a rhodium catalyst complexed with a ligand (1). The oxonation reaction produces normal and isobutyraldehyde in a n/i ratio, which is typically 10:1, but ratios of up to 30:1 and down to 1:1 are possible with alternate ligands. The butyraldehyde product is removed from the catalyst solution (2) and purified by distillation (3). N-butyraldehyde is separated from the iso (4).

The process is characterized by its simple flow sheet and low-operating pressure. This results in low capital and maintenance expenses and product cost, and high plant availability. Mild reaction conditions minimize byproduct formation. Low byproduct formation also contributes to higher process efficiencies and product qualities.

Technology for hydrogenation to normal or iso-butanols or aldolization and hydrogenation to 2-ethylhexanol exists and has been widely licensed. A version of the LP Oxo process has been licensed to produce a mixture of C_{10} alcohols (predominantly 2 propylheptanol) from an n-butene feedstock.

Economics: Typical performance data (per ton of mixed butyraldehyde):

Feedstocks	
Propylene, kg (contained in chemical grade)	600
Synthesis Gas ($CO + H_2$), Nm^3	639

Commercial plants: The LP Oxo process has been licensed for 19 plants worldwide and is now used to produce more than 60% of the world's butyraldehyde capacity. Plants range in size from 30,000 to 350,000 tpy. The rhodium-based catalyst has a long life, and spent catalysts can be reactivated onsite. The technology is also practiced by Union Carbide Corp., at its Texas City and Taft plants.

Licensees: Nineteen worldwide since 1978.

Licensor: Davy Process Technology, UK/Dow Chemical Co., US.

Caprolactam

Application: A process to manufacture caprolactam from nitration-grade toluene. Fiber-grade flaked or molten caprolactam and white ammonium sulfate crystals are produced.

Description: Toluene and air are fed to the reactor (1) in which the oxidation to benzoic acid is carried out at 160°C and 10 atm. The reaction product is a 30% solution of benzoic acid in toluene plus a small quantity of byproducts. Fractionation (2) separates unconverted toluene for recycle, pure benzoic and a bottom fraction of heavy byproducts.

Benzoic acid is hydrogenated under pressure in presence of palladium catalyst in a series of continuous stirred tank reactors (3) at 170°C and 16 atm.

Conversion is complete in a single pass. Cyclohexane-carboxylic acid is blended with oleum and fed to a multistage reactor (4) where it is converted to caprolactam by reaction with nitrosylsulfuric acid.

This acid is produced in a conventional ammonia oxidation plant, where the nitrogen oxides are absorbed in oleum.

Reactor effluent is diluted with water (5), and unconverted cyclohexane carboxylic acid is recycled to the process, while the lactam solution flows to the crystallization plant (6) where it is neutralized with ammonia. Ammonium sulfate crystallizes at bottom and the top organic layer of caprolactam is recovered and purified through a two-solvent (toluene and water) extraction (7) and a continuous fractionation (8).

Economics: Based on a 90,000-tpy battery limits plant, W. Europe, 1990:

Investment, $/tpy	2,800
Typical raw material and utility requirements, per kg of caprolactam:	
Toluene, kg	1.08
Ammonia, kg	1.14
Sulfur, kg	0.9
Hydrogen, Nm^3	0.8
Ammonium sulfate, kg, (credit)	(3.5)
Electricity, kWh	0.8
Steam, kg	11
Fuel gas, Nm^3	0.25
Water, cooling, m^3	0.6
Water, chilled, m^3	0.13

Commercial plants: 20,000-tpy SNIA BPD factory in Torviscosa, Italy; 80,000-tpy ANIC factory in Manfredonia, Italy; expansion to 88,000-tpy ELECTROCHIMPROM factory in Chirchik, URSS; 50,000-tpy plant, Shijiazhuang, China.

Reference: Taverna, M., and M. Chiti, *Hydrocarbon Processing*, November 1970, p. 137.

Licensor: SNIA BPD S.p.A., exclusive contractor: SNAICO Engineering S.p.A.

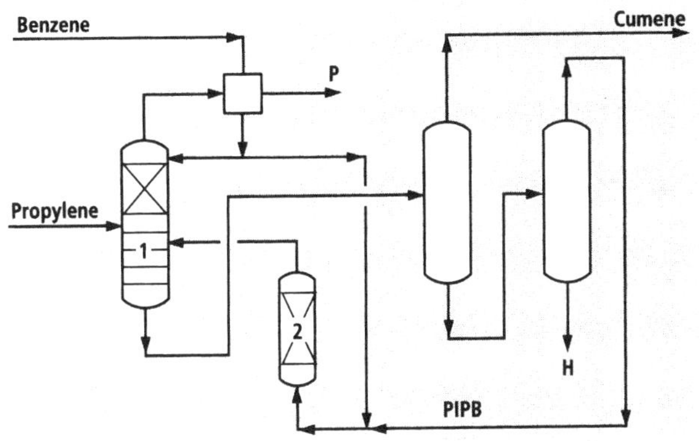

Cumene

Application: Advanced technology to produce high-purity cumene from propylene and benzene using patented catalytic distillation (CD) technology. The CDCumene process uses a specially formulated zeolite alkylation catalyst packaged in a proprietary CD structure and another specially formulated zeolite transalkylation catalyst in loose form.

Description: The CD column (1) combines reaction and fractionation in a single-unit operation. Alkylation takes place isothermally and at low temprature. CD also promotes the continuous removal of reaction products from reaction zones. These factors limit byproduct impurities and enhance product purity and yield. Low operating temperatures and pressures also decrease capital investment, improve operational safety and minimize fugitive emissions.

In the mixed-phase CD reaction system, propylene concentration in the liquid phase is kept extremely low (<0.1 wt%) due to the higher volatility of propylene to benzene. This minimizes propylene oligomerization, the primary cause of catalyst deactivation and results in catalyst run lengths of 3 to 6 years. The vapor-liquid equilibrium effect provides propylene dilution unachievable in fixed-bed systems, even with expensive reactor pumparound and/or benzene recycle arrangements.

Overhead vapor from the CD column (1) is condensed and returned as reflux after removing propane and lights (P). The CD column bottom section strips benzene from cumene and heavies. The distillation train separates cumene product and recovers polyisopropylbenzenes (PIPB) and some heavy aromatics (H) from the net bottoms. PIPB reacts with benzene in the transalkylator (2) for maximum cumene yield. Operating conditions are mild and noncorrosive; standard carbon steel can be used for all equipment.

Yields: 100,000 metric tons (mt) of cumene are produced from 65,000 mt of benzene and 35,300 mt of propylene giving a product yield of over 99.7%. Cumene product is at least 99.95% pure and has a Bromine Index of less than 2, without clay treatment.

Economics: Based on a 300,000-mtpy cumene plant located in the US Gulf Coast, the ISBL investment is about US$15 million.

Typical operating requirements, per metric ton of cumene:	
Propylene	0.353
Benzene	0.650
Yield	99.7%
Utilities:	
Electricity, kWh	8
Heat (Import), 10^6 kcal	0.5
Steam (export), mt	1.0
Water, cooling, m^3	12

Commercial plants: Formosa Chemicals & Fibre Corporation, Taiwan—270,000 mtpy

Licensor: CDTECH, a partnership between ABB Lummus Global and Chemical Research & Licensing.

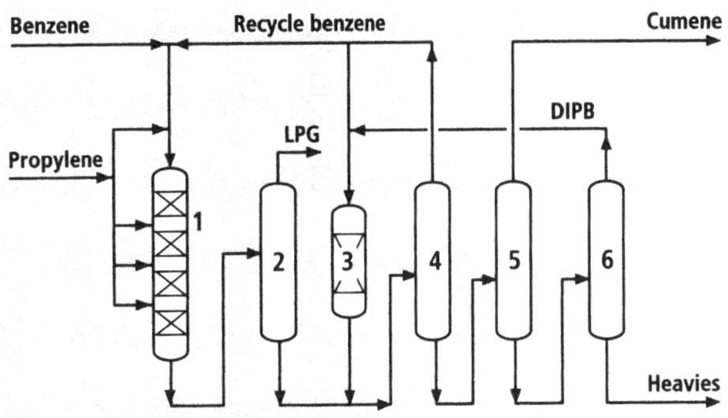

Cumene

Application: To produce high-quality cumene (isopropylbenzene) by alkylating benzene with propylene (typically refinery or chemical grade) using liquid-phase Q-Max process based on zeolitic catalyst technology.

Description: Benzene is alkylated to cumene over a zeolite catalyst in a fixed-bed, liquid-phase reactor. Fresh benzene is combined with recycle benzene and fed to the alkylation reactor (1). The benzene feed flows in series through the beds, while fresh propylene feed is distributed equally between the beds. This reaction is highly exothermic, and heat is removed by recycling a portion of reactor effluent to the reactor inlet and injecting cooled reactor effluent between the beds.

In the fractionation section, propane that accompanies the propylene feedstock is recovered as LPG product from the overhead of the depropanizer column (2), unreacted benzene is recovered from the overhead of the benzene column (4) and cumene product is taken as overhead from the cumene column (5). Di-isopropylbenzene (DIPB) is recovered in the overhead of the DIPB column (6) and recycled to the transalkylation reactor (3) where it is transalkylated with benzene over a second zeolite catalyst to produce additional cumene. A small quantity of heavy byproduct is recovered from the bottom of the DIPB column (6) and is typically blended to fuel oil. The cumene product has a high purity (99.96–99.97 wt%), and cumene yields of 99.7 wt% and higher are achieved.

The zeolite catalyst is noncorrosive and operates at mild conditions; thus, carbon-steel construction is possible. Catalyst cycle lengths are two years and longer. The catalyst is fully regenerable for an ultimate catalyst life of six years and longer. Existing plants that use SPA or $AlCl_3$ catalyst can be revamped to gain the advantages of Q-Max cumene technology while increasing plant capacity.

Economics: basis: ISBL US Gulf Coast

Investment, US$/tpy	40–90
Raw materials & utilities, per metric ton of cumene	
Propylene, tons	0.35
Benzene, tons	0.66
Electricity, kW	12
Steam, tons (import)	0.7
Water, cooling, m^3	3

The Q-Max design is typically tailored to provide optimal utility advantage for the plant site, such as minimizing heat input for standalone operation or recovering heat as steam for usage in a nearby phenol plant.

Commercial plants: Seven Q-Max units are in operation with a total cumene capacity of 2.3 million tpy, and two additional units are either in design or under construction.

Licensor: UOP LLC.

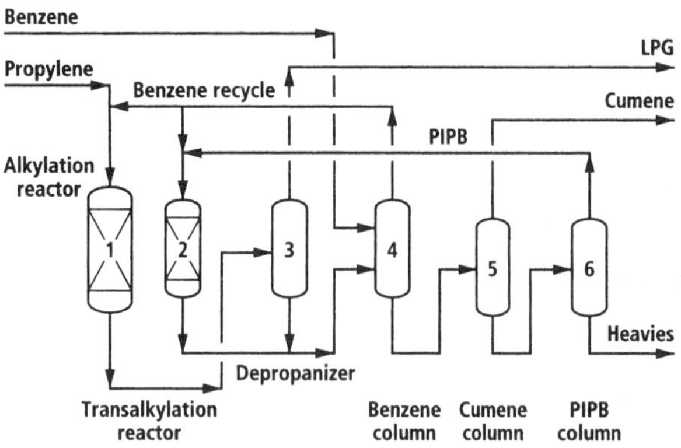

Cumene

Application: To produce cumene from benzene and any grade of propylene—including lower-quality refinery propylene-propane mixtures—using the Mobil/Badger process and a new generation of zeolite catalysts from ExxonMobil.

Description: The process includes: a fixed-bed alkylation reactor, a fixed-bed transalkylation reactor and a distillation section. Liquid propylene and benzene are premixed and fed to the alkylation reactor (1) where propylene is completely reacted. Separately, recycled polyisopropylbenzene (PIPB) is premixed with benzene and fed to the transalkylation reactor (2) where PIPB reacts to form additional cumene. The transalkylation and alkylation effluents are fed to the distillation section. The distillation section consists of as many as four columns in series. The depropanizer (3) recovers propane overhead as LPG. The benzene column (4) recovers excess benzene for recycle to the reactors. The cumene column (5) recovers cumene product overhead. The PIPB column (6) recovers PIPB overhead for recycle to the transalkylation reactor.

Process features: The process allows a substantial increase in capacity for existing SPA, $AlCl_3$, or other zeolite cumene plants while improving product purity, feedstock consumption, and utility consumption. The new catalyst is environmentally inert, does not produce byproduct oligomers or coke and can operate at extremely low benzene to propylene ratios with proven commercial cycle lengths of over five years. Expected catalyst life is well over five years.

Yield and product purity: This process is essentially stoichiometric and product purity above 99.97% weight has been regularly achieved in commercial operation.

Economics: Estimated ISBL investment for a 300,000-mtpy unit on the US Gulf Coast (2002 construction basis), is US$15 million.

Utility requirements, per ton of cumene product:

Heat, MMkcal (import)	0.32
Steam, ton (export)	(0.60)

The utilities can be optimized for specific site conditions/economics and integrated with an associated phenol plant.

Commercial plants: The first commercial application of this process came onstream in 1996. At present, there are 11 plants operating with a combined capacity exceeding 4.7 million mtpy. In addition, two grassroots plants and an SpA plant revamp are in the design phase. Fifty percent of the worldwide cumene production is from plants using the Mobil/Badger process.

Licensor: The Badger Technology Center of Washington Group International, Inc.

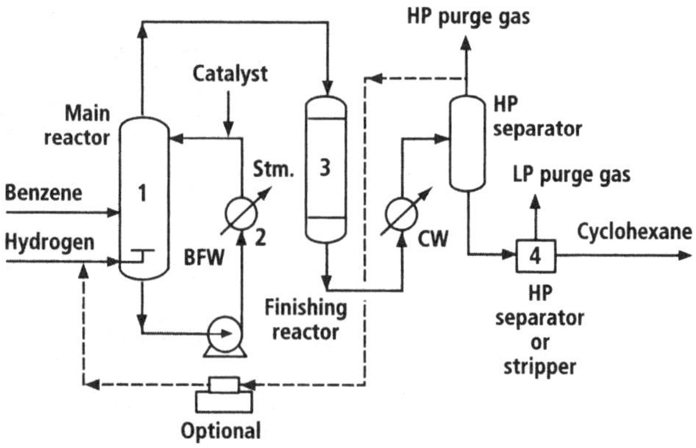

Cyclohexane

Application: Produce high-purity cyclohexane by liquid-phase catalytic hydrogenation of benzene.

Description: The main reactor (1) converts essentially all the feed isothermally in the liquid phase at a thermodynamically-favorable low temperature using a continuously-injected soluble catalyst. The catalyst's high activity allows use of low hydrogen partial pressure, which results in fewer side reactions, e.g., isomerization or hydrocracking. The heat of reaction vaporizes cyclohexane product and, using pumparound circulation through an exchanger, also generates steam (2). With the heat of reaction being immediately removed by vaporization, accurate temperature control is assured. A vapor-phase fixed-bed finishing reactor (3) completes the catalytic hydrogenation of any residual benzene. This step reduces residual benzene in the cyclohexane product to very low levels. Depending on the purity of the hydrogen make-up gas, the stabilization section includes either an LP separator (4) or a small stabilizer to remove the light ends.

A prime advantage of the liquid-phase process is its substantially lower cost compared to vapor phase processes: investment is particularly low because a single, inexpensive main reactor chamber is used compared to multiple-bed or tubular reactors used in vapor phase processes. Quench gas and unreacted benzene recycles are not necessary and better heat recovery generates both the cyclohexane vapor for the finishing step and a greater amount of steam. These advantages result in lower investment and operating costs. Operational flexibility and reliability are excellent; changes in feedstock quality and flows are easily handled. Should the catalyst be deactivated by feed quality upsets, fresh catalyst can be injected without shutting down.

Yield: 1.075 kg of cyclohexane is produced from 1 kg of benzene.

Economics: Basis: 200,000-tpy cyclohexane complex, ISBL 2002 Gulf Coast location with PSA hydrogen is US$ 7.8 million. Catalyst cost is US$ 1.2/metric ton of product.

Commercial plants: Thirty-one cyclohexane units have been licensed.

Licensor: Axens, Axens NA.

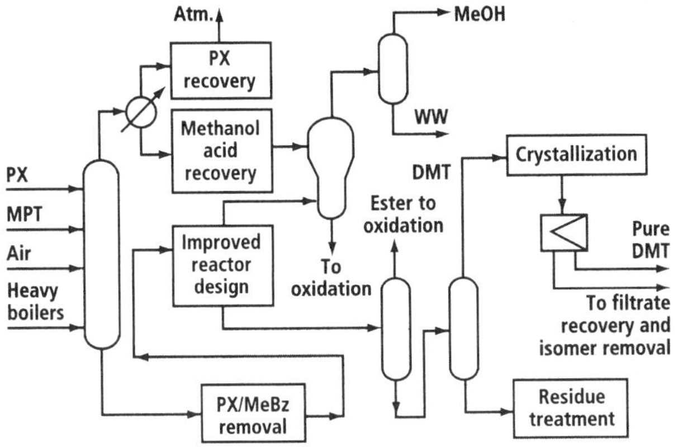

Dimethyl terephthalate

Application: To increase capacity and reduce energy usage of existing or grassroots dimethyl terephthalate (DMT) production facilities using variations of GT-DMT proprietary technology.

Description: The common production method of DMT from paraxylene and methanol is through successive oxidations in four major steps: oxidation, esterification, distillation and crystallization. A mixture of p-xylene and methyl p-toluate (MPT) is oxidized with air using a heavy-metal catalyst. All organics are recovered from the offgas and recycled to the system. The acid mixture from the oxidation is esterified with methanol and produces a mixture of esters. The crude ester mixture is distilled to remove all heavy boilers and residue produced; lighter esters are recycled to the oxidation section. Raw DMT is then sent to the crystallization section to remove DMT isomers and aromatic aldehydes.

The technology improvements enhance the traditional processing in each section. The adaptations include: changes in process configurations and operating conditions, alterating the separation schemes, revising the recovery arrangement, increasing the value of the byproducts and reducing the overall plant recycles.

GTC Technology offers complete implementation of the technology and overall plant reviews for selective improvements to reduce operating and overall production costs. Some separate improvements available are:

1. Oxidation optimization reduces byproduct formation, thus lowering p-xylene consumption
2. Recoveries of byproducts for sale such as methyl benzoate (MeBz) and acetic and formic acid
3. Improved esterifier reactor design enables higher throughputs and improves methanol usage
4. Enhanced isomer removal minimizes DMT losses
5. Improved crystallization schemes for reduced energy, lowers methanol handling and losses, improves purity and operating flexibility
6. Integration of steam usage in the plant for considerable savings on operating costs
7. Operating reviews to reduce operating downtime and extend online factors
8. Advanced control models for improved operability.

Economics: Based on process modifications, an existing DMT plant can increase production with an investment of $200 to $600/ton/yr of additional capacity. A new plant will have an investment reduction of about 20% equipment cost. Raw material consumption per ton of product (with the complete modification) is 605 ton of paraxylene and 360 ton of methanol.

Commercial plants: GT-DMT technology is used by seven DMT producers.

Licensor: GTC Technology Inc.

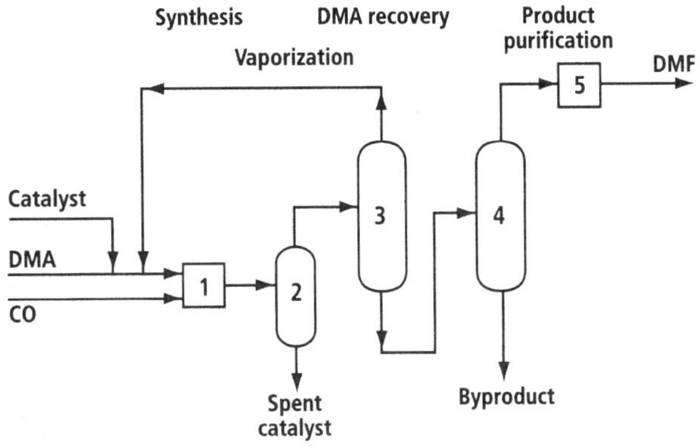

Dimethylformamide

Application: To produce dimethylformamide (DMF) from dimethylamine (DMA) and carbon monoxide (CO).

Description: Anhydrous DMA and CO are continuously fed to a specialized reactor (1), operating at moderate conditions and containing a catalyst dissolved in solvent. The reactor products are sent to a separation system where crude product is vaporized (2) to separate the spent catalyst. Excess DMA and catalyst solvent are stripped (3) from the crude product and recycled back to the reaction system. Vacuum distillation (4) followed by further purification (5) produces a high-quality solvent and fiber-grade DMF product. A saleable byproduct stream is also produced.

Yields: Greater than 95% on raw materials. CO yield is a function of its quality.

Economics: Typical performance data per ton of product:

Dimethylamine, t	0.63
Carbon monoxide, t	0.41
Steam, t	1.3
Water, cooling, m^3	100
Electricity, kWh	10

Commercial plants: Thirteen plants in eight countries use this process with a production capacity exceeding 100,000 mtpy.

Licensor: Davy Process Technology, UK.

EDC via oxygen lean oxychlorination

Application: The modern Vinnolit oxychlorination process produces EDC by an exothermic reaction from feedstocks including ethylene, anhydrous hydrogen chloride and oxygen. Anhydrous hydrogen chloride can be used from the VCM process as well as from other processes like isocyanates (MDI, TDI), chlorinated methanes, chlorinated ethanes, epichlorohydrin, etc.

Description: Oxygen can be used from air separation plant, as well as from the cost-effective pressure swing adsorption (PSA) process. The Vinnolit oxychlorination process is also able to handle ethylene and/or anhydrous hydrogen chloride containing vent streams from direct chlorination, acetaldehyde, monochloro acetic acid and other processes.

The exothermic reaction is catalyzed by a copper chloride catalyst in a single-step, fluidized-bed reactor at temperatures of 200°C–220°C. The heat of reaction is recovered by producing 10-barg steam or heating other heat-transfer fluids.

Catalyst fines are removed with a newly developed hot-gas catalyst filter or alternatively by wastewater treatment that meets even the strictest regulations for copper, dioxins and furanes. The environmentally friendly process uses recycle gas, which is fed back to the reactor after condensing EDC and water. After EDC removal, the process water meets even the most stringent environmental requirements.

After removal of carbon dioxide (CO_2) and chloral/chloroethanol, the crude EDC is purified in the EDC distillation unit; it can be used as furnace feed or sales EDC. The important process features and advantages are:

- **Reliability:** A stable temperature control combined with an excellent heat transfer and a uniform temperature profile (no hot spots) in the fluidized bed easily achieves an onstream time >99% per year. A specially designed raw-material sparger system allows operation spans of two years without maintenance. Larger heat-transfer area allows a higher steam temperature and pressure in the cooling coils, which improves the safety margin to the critical surface temperature where hydrochloric acid dewpoint corrosion may occur.
- **Safety:** The oxygen is mixed with anhydrous hydrochloride outside the reactor and is fed independently of the ethylene into the fluidized bed. The oxygen concentration in the recycle stream is approx. 0.5 vol%, which are well outside the explosion range.
- **Flexibilty:** A turndown ratio as low as 20% capacity utilization can be achieved, as well as, quick load changes.
- **Low manufacturing costs:** The unlimited catalyst service is combined with the low losses via the highly efficient cyclone system (less than 20g catalyst per ton of produced EDC). High raw-material yields (98.5% ethylene, 99% anhydrous hydrochloride and 94% oxygen) and the possibility to use low-cost oxygen from PSA units ensure a highly competitive process with low production costs.
- **Environmentally friendly:** Yields of 99% are the key for a low byproduct formation. A highly efficient, hot-gas filtration system separates the small quantities of catalyst fines. Besides the EDC removal via steam stripping, no additional wastewater treatment is required. The charter for European Council for Vinyl Manufacturers (ECVM) is easily met (EDC < 5 g/t of EDC purification capacity, copper < 1 g/t of Oxychlorination capacity, dioxin-like components < 1 μg TEQ/t of Oxychlorination capacity.

Process data and crude EDC purity:

Operating temperature, °C:	200–225
Operating pressure, barg:	2.5–4
Steam pressure, barg:	10

Commercial plants: The process is used in 15 reactors at 12 sites, alone as HCl-consuming plant or as part of the balanced VCM process. In some cases, it has replaced oxychlorination technologies by modifying existing reactors that do not use the Vinnolit process. Two new oxychlorination trains are under design.

Licensor: Vin Tec; **preferred contractor:** Uhde GmbH.

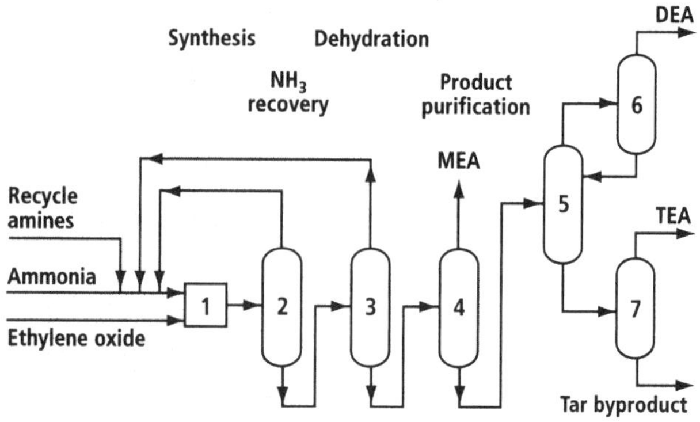

Ethanolamines

Application: To produce mono-(MEA), di-(DEA) and tri-ethanolamines (TEA) from ethylene oxide and ammonia.

Description: Ammonia solution, recycled amines and ethylene oxide are fed continuously to a reaction system (1) that operates under mild conditions and simultaneously produces MEA, DEA and TEA. Product ratios can be varied to maximize MEA, DEA or TEA production. The correct selection of the NH_3/EO ratio and recycling of amines produces the desired product mix. The reactor products are sent to a separation system where ammonia (2) and water are separated and recycled to the reaction system. Vacuum distillation (4,5,6,7) is used to produce pure MEA, DEA and TEA. A saleable heavies tar byproduct is also produced. Technical grade TEA (85 wt%) can also be produced if required.

Yields: Greater than 98% on raw materials.

Economics: Typical performance data per ton amines MEA/DEA/TEA product ratio of ⅓ : ⅓ : ⅓

Ethylene oxide, t	0.82
Ammonia, t	0.19
Steam, t	5
Water, cooling , m^3	300
Electricity, kWh	30

Commercial plants: One 20,000-mtpy original capacity facility.

Licensor: Davy Process Technology, UK.

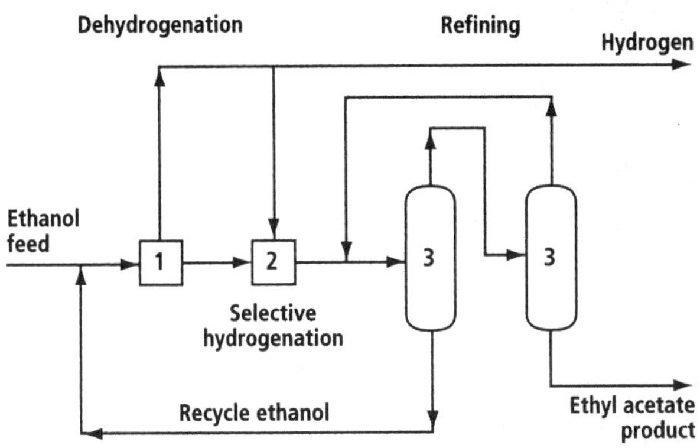

Ethyl acetate

Application: To produce ethyl acetate from ethanol without acetic acid or other co-feeds.

Description: Ethanol is heated and passed through a catalytic dehydrogenation reactor (1) where part of the ethanol is dehydrogenated to form ethyl acetate and hydrogen. The product is cooled in an integrated heat-exchanger system; hydrogen is separated from the crude product. The hydrogen is mainly exported. Crude product is passed through a second catalytic reactor (2) to allow "polishing" and remove minor byproducts such as carbonyls.

The polished product is passed to a distillation train (3) where a novel distillation arrangement allows the ethanol/ethyl acetate water azeotrope to be broken. Products from this distillation scheme are unreacted ethanol, which is recycled, and ethyl acetate product.

The process is characterized by low-operating temperatures and pressures, which allow all equipment to be constructed from either carbon steel or low-grade stainless steels. It allows ethyl acetate to be made without requiring acetic acid as a feed material. The process is appropriate for both synthetic ethanol and fermentation ethanol as the feed. The synthetic ethanol can be impure ethanol without significantly affecting the conversion or selectivity. The product ethyl acetate is greater than 99.95%.

Economics: Typical performance data per ton of ethyl acetate produced:

Feedstock	1.12 tons of ethanol
Product	45 kg of hydrogen

Commercial plants: The technology has been developed during the mid to late 1990s. The first commercial plant is a 50,000-tpy plant in South Africa, using synthetic ethanol.

Licensees: One since 1998.

Licensor: Davy Process Technology, UK.

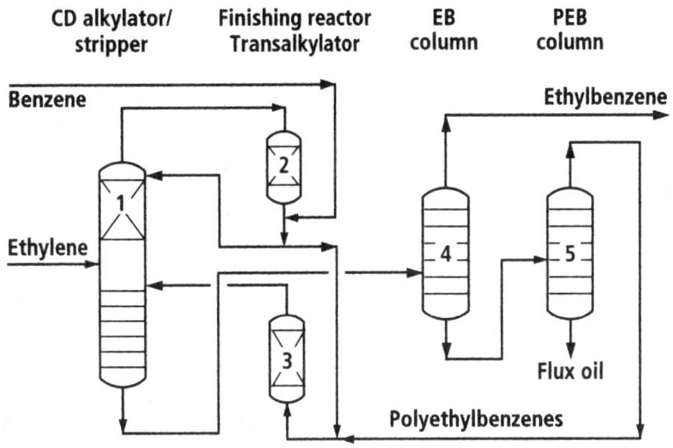

Ethylbenzene

Application: Advanced technology to produce high-purity ethylbenzene (EB) alkylating benzene with ethylene using patented catalytic distillation (CD) technology. The CDTECH EB process uses a specially formulated zeolite alkylation catalyst packaged in a proprietary CD structure. The process is able to handle a wide range in ethylene feed composition—-from 10% to 100% ethylene.

Description: The CD alkylator stripper (1) operates as a distillation column. Alkylation and distillation occur in the alkylator in the presence of a zeolite catalyst packaged in patented structured packing. Unreacted ethylene and benzene vapor from the alkylator top are condensed and fed to the finishing reactor (2) where the remaining ethylene reacts over zeolite catlayst pellets. The alkylator stripper bottoms is fractionated (4, 5) into EB product, polyethylbenzenes and flux oil. The polyethylbenzenes are transalkylated with benzene over zeolite catalyst pellets in the transalkylator (3) to produce additional EB. The ethylene can be polymer grade or, with only minor differences in the process scheme, dilute ethylene containing as little as 10 mol% ethylene as in FCC offgas. Reactors are designed for 3 to 6 years of uninterrupted runlength. The process does not produce any hazardous effluent. Low operating temperatures allow using carbon steel for all equipment.

Yields and product quality: Both the alkylation and transalkylation reactions are highly selective—producing few byproducts. The EB product has high purity (99.9 wt% minimum) and is suitable for styrene-unit feed. Xylene make is less than 10 ppm. The process has an overall yield of 99.7%.

Economics: The EB process features consistent product yields, high-product purity, low-energy consumption, low investment cost and easy, reliable operation.

Investment (500,000 tpy, ISBL Gulf Coast), US$: 17 million
Raw materials and utilities, based on one metric ton of EB:

Ethylene, kg	265
Benzene, kg	738
Electricity, kWh	20
Water, cooling m^3	3
Steam, mtons (export)	1.3
Hot oil, 10^6 kcal	0.6

Commercial plants: Two commercial plants are in operation in Argentina and Canada with capacities from 140,000 to 477,000 mtpy. They process ethylene feedstocks with purities ranging from 75% ethylene to polymer-grade ethylene. Two additional units are either in design or under construction—the largest is an 850,000-mtpy unit.

Licensor: CDTECH, a partnership between ABB Lummus Global and Chemical Research & Licensing.

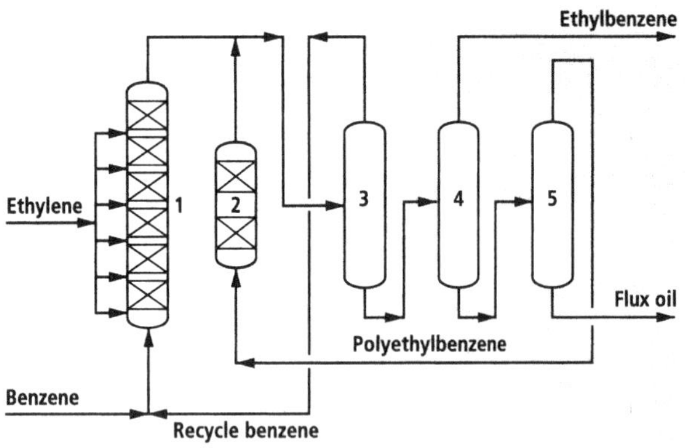

Ethylbenzene

Application: State-of-the-art technology to produce high-purity ethylbenzene (EB) by liquid-phase alkylation of benzene with ethylene. The Lummus/UOP EBOne process uses specially formulated, proprietary zeolite catalyst from UOP. The process can handle a wide range of ethylene feed compositions ranging from chemical (70%) to polymer grade (100%).

Description: Benzene and ethylene are combined over a proprietary zeolite catalyst in a fixed-bed, liquid-phase reactor. Fresh benzene is combined with recycle benzene and fed to the alkylation reactor (1). The combined benzene feed flows in series through the beds, while fresh ethylene feed is distributed equally between the beds. The reaction is highly exothermic, and heat is removed between the reaction stages by generating steam. Unreacted benzene is recovered from the overhead of the benzene column (3), and EB product is taken as overhead from the EB column (4). A small amount of polyethylbenzene (PEB) is recovered in the overhead of the PEB column (5) and recycled back to the transalkylation reactor (2) where it is combined with benzene over a second proprietary zeolite catalyst to produce additional EB product. A small amount of flux oil is recovered from the bottom of the PEB column (5) and is usually burned as fuel. The catalysts are non-corrosive and operate at mild conditions, allowing for all carbon-steel construction. The reactors can be designed for 2–6 year catalyst cycle length, and the catalyst is fully regenerable. The process does not produce any hazardous effluent

Yields and product quality: Both the alkylation and transalkylation reactions are highly selective, producing few byproducts. The EB product has a high purity (99.9 wt% minimum) and is suitable for styrene-unit feed. Xylene make is less than 10 ppm. The process has an overall yield of 99.7%

Economics: The EBOne process features consistently high product yields over the entire catalyst life cycle, high-product purity, low-energy consumption, low investment cost, and simple, reliable operation.

Investment, ISBL Gulf Coast, US$/mtpy	30–45
Raw material and utilities, per metric ton of EB	
Ethylene, mtons	0.265
Benzene, mtons	0.738
Utilities, US$	1

Additional utility savings can be realized via heat integration with downstream Lummus/UOP Classic SM or SMART SM styrene unit.

Commercial plants: Fourteen EBOne units are in operation throughout the world, with a total EB capacity of 3.7 million mtpy. Unit capacities range from 65,000 to 530,000 mtpy. Ethylene feedstock purity ranges from 80 to 100%. Seven additional units are either in design or under construction—the largest unit is 725,000 mtpy.

Licensor: ABB Lummus Global and UOP LLC.

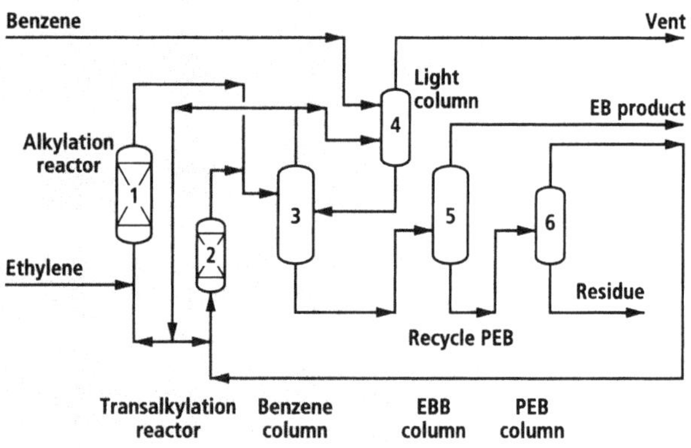

Ethylbenzene

Application: To produce ethylbenzene (EB) from benzene and a polymer-grade ethylene or an ethylene/ethane feedstock using the Mobil/Badger EBMax process and proprietary ExxonMobil alkylation and transalkylation catalysts. The technology can be applied in the design of grassroots units, upgrading of existing vapor-phase technology plants, or conversion of aluminum chloride technology EB plants to zeolite technology.

Description: Ethylene reacts with benzene in either a totally liquid-filled or mixed-phase alkylation reactor (1) containing multiple fixed-beds of ExxonMobil's proprietary catalyst, forming EB and very small quantities of polyethylbenzenes (PEB). In the transalkylation reactor (2), PEB is converted to EB by reaction with benzene over ExxonMobil's transalkylation catalyst. PEB and benzene recovered from the crude EB enter the transalkylation reactor. Effluents from the alkylation and transalkylation reactors are fed to the benzene column (3), where unreacted benzene is recovered from crude EB. The fresh benzene feedstock and a small vent stream from the benzene column are fed to the lights column (4) to reject light impurities. The lights column bottoms is returned to the benzene column. The bottoms from the benzene column is fed to the EB column (5) to recover EB product. The bottoms from the EB column is fed to the PEB column (6) where recyclable alkylbenzenes are recovered as a distillate and diphenyl compounds are rejected in a bottoms stream that can be used as fuel.

Catalysts: Cycle lengths in excess of four years are expected for the alkylation and transalkylation catalysts. Process equipment is fabricated entirely from carbon steel. Capital investment is reduced as a consequence of the high activity and extraordinary selectivity of the alkylation catalyst and the ability of both the alkylation and transalkylation catalysts to operate with very low quantities of excess benzene.

Product quality: The EB product contains less than 100 ppm of C_8 plus C_9 impurities. Product purities of 99.95% to 99.99% are expected.

Economics:

Raw materials and steam, tons per ton of EB product:	
Ethylene	0.265
Benzene	0.739
Steam, high-pressure used	0.98
Steam, medium- and low-pressured generated	1.39

Utilities can be optimized for specific-site conditions.

Commercial plants: Since the commercialization of the Mobil/Badger EB technology in 1980, 44 licenses have been granted. The total licensed capacity for the Mobil/Badger EB technology exceeds 17 million mtpy. The capacity for the EBMax technology exceeds 7.2 million mtpy.

Licensor: The Badger Technology Center of Washington Group International, Inc.

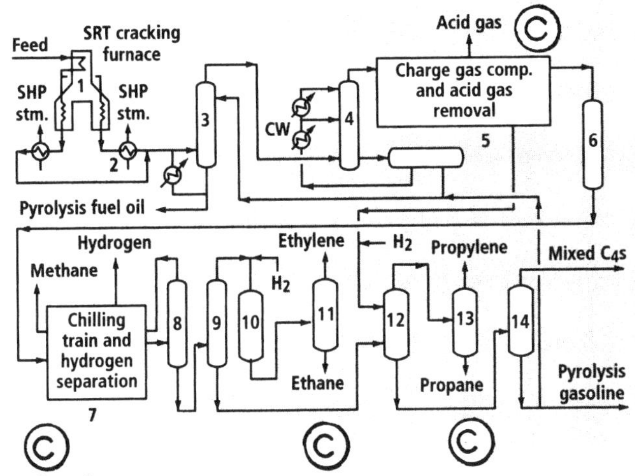

Ethylene

Application: To produce polymer-grade ethylene (99.95 vol%). Major byproducts are propylene (chemical or polymer-grade), a butadiene-rich C_4 stream, C_6 to C_8 aromatics-rich pyrolysis gasoline and high-purity hydrogen.

Description: Hydrocarbon feedstock is preheated and cracked in the presence of steam in tubular SRT (short residence time) pyrolysis furnaces (1). This approach features extremely high olefin yields, long runlength and mechanical integrity. The products exit the furnace at 1,500°F to 1,600°F and are rapidly quenched in the transfer line exchangers (2) that generate super high-pressure (SHP) steam. The latest generation furnace design is the SRT VI.

Furnace effluent, after quench, flows to the gasoline fractionator (3) where the heavy oil fraction is removed from the gasoline and lighter fraction (liquids cracking only). Further cooling of furnace effluents is accomplished by a direct water quench in the quench tower (4). Raw gas from the quench tower is compressed in a multistage centrifugal compressor (5) to greater than 500 psig. The compressed gas is then dried (6) and chilled. Hydrogen is recovered in the chilling train (7), which feeds the demethanizer (8). The demethanizer operates at about 100 psia, providing increased energy efficiency. The bottoms from the demethanizer go to the deethanizer (9).

Acetylene in the deethanizer overhead is hydrogenated (10) or recovered. The ethylene-ethane stream is fractionated (11) and polymer-grade ethylene is recovered. Ethane leaving the bottom of the ethylene fractionator is recycled and cracked to extinction.

The deethanizer bottoms and condensate stripper bottoms from the charge compression system are depropanized (12). Methylacetylene and propadiene are hydrogenated in the depropanizer using CDHydro catalytic distillation hydrogenation technology. The depropanizer bottoms is separated into mixed C_4 and light gasoline streams (14). Polymer-grade propylene is recovered in a propylene fractionator (13).

A revised flow scheme eliminates ~25% of the equipment from this conventional flowsheet. It uses CDHydro hydrogenation for the selective hydrogenation of C_2 through C_4 acetylenes and dienes in a single tower; reduces the cracked-gas discharge pressure to 250 psig; uses a single refrigeration system to replace the three separate systems; and applies metathesis to produce up to 1/3 of the propylene product catalytically rather than by thermal cracking, thereby lowering energy consumption by ~15%.

Energy consumption: Energy consumptions are 3,300 kcal/kg of ethylene produced for ethane cracking and 5,000 kcal/kg of ethylene for naphtha feedstocks. Energy consumption can be as low as 4,000 kcal/kg of ethylene for naphtha feedstocks with gas turbine integration. As noted above, the new flow scheme reduces energy consumption by 14%.

Commercial plants: Approximately 40% of the world's ethylene plants use Lummus' ethylene technology. Many existing units have been significantly expanded (above 150% of nameplate) using Lummus' MCET (maximum capacity expansion technology) approach.

Licensor: ABB Lummus Global.

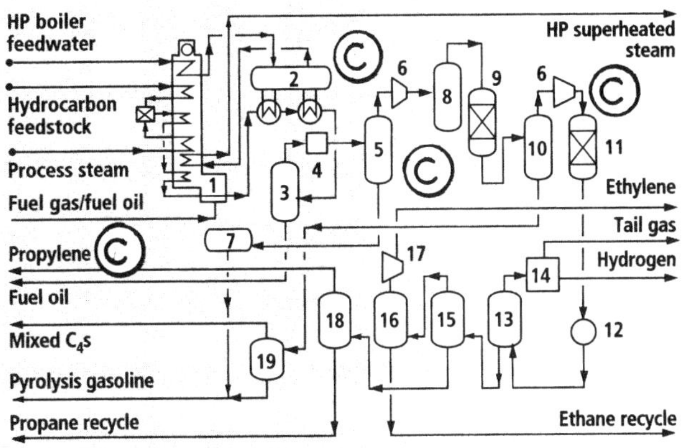

Ethylene

Application: High performance steam-cracking and recovery to produce polymer-grade ethylene and propylene, butadiene-rich mixed C_4s, aromatic-rich pyrolysis gasoline, hydrogen and fuel streams. Cracking feedstocks range from ethane through vacuum gas oils.

Description: Kellogg Brown & Root's proprietary Selective Cracking Optimum REcovery (SCORE) olefins technology represents the integration of the technologies of the former M.W. Kellogg and Brown & Root companies combined with olefins technology developed by ExxonMobil Chemical Co., through a long-term, worldwide licensing agreement. ExxonMobil brings creating technology as well as the benefits of extensive operating experience to further improve operability, reliability, and reduce production costs.

The SCORE pyrolysis furnace portfolio features the straight tube SC-1 design, which has a low reaction time (in the range of 0.1 seconds), and low operating pressures. The design and operating conditions produce higher olefin yields. The portfolio includes a range of designs to satisfy any requirements.

The pyrolysis furnace (1) effluent is processed for heat and product recovery in an efficient, reliable low cost recovery section. The recovery section design can be optimized for specific applications and/or selected based on operating company preferences. Flow-schemes based on deethanizer-first, depropanizer-first and demethanizer-first configurations are available. The depropanizer-first flow-scheme, primarily applicable to liquid crackers, is shown above. The similar, but simpler, deethanizer-first scheme is appropriate for ethane through ethane/propane gas crackers. These two schemes use front-end acetylene converter systems, which minimize green-oil production, and allow using low-pressure recovery towers. KBR also has extensive experience with the demethanizer-first flow-scheme, which can be offered to clients preferring that technology.

Cracked gases are cooled and fractionated to remove fuel oil and water (2–5) then compressed (6), processed for acid-gas removal (8) and dried (9). The C_3 and lighter material is separated as an overhead product in the depropanizer (10) and acetylene is hydrogenated in the acetylene converter (11). The acetylene converter effluent is processed in the demethanizer system (12–14) to separate the fuel gas and hydrogen products. The demethanizer bottoms is sent to the deethanizer (15) from which the overhead flows to the C_2-splitter (16), which produces the polymer-grade ethylene product and the ethane stream, which is typically recycled to the furnaces as a feedstock. The deethanizer bottoms flows to the C_3-splitter (18) where the polymer-grade propylene is recovered as the overhead product. The C_3-splitter bottoms product, propane, is typically recycled to the furnaces as a feedstock. The depropanizer bottoms product, C_4s and heavier, flow to the debutanizer (19) for recovery of the mixed-C_4 product and aromatic-rich pyrolysis gasoline.

Yields: Ethylene yields to 84% for ethane, 38% for naphtha and 32% for gas oils may be achieved depending upon feedstock characteristics.

Commercial plants: KBR has been involved in over 140 ethylene projects worldwide with single-train ethylene capacities up to 1.3 million tpy including 21 new grassroots ethylene plants since 1990.

Licensor: Kellogg Brown & Root, Inc.

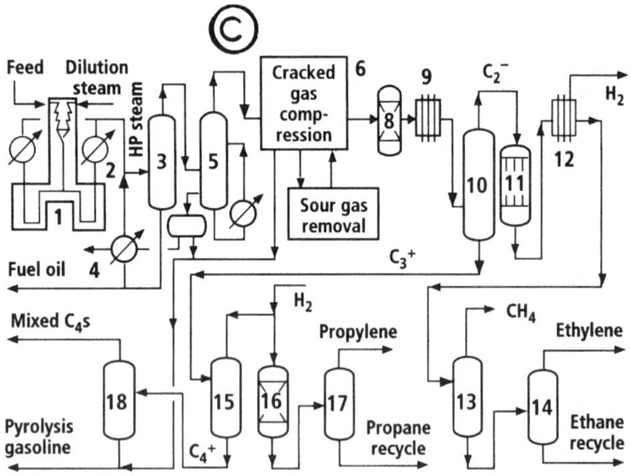

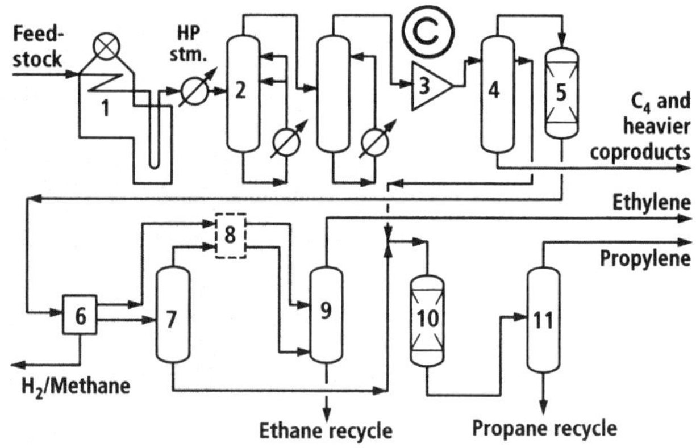

Ethylene

Application: To produce polymer-grade ethylene and propylene by thermal cracking of hydrocarbon fractions—from ethane through naphtha up to hydrocracker residue. Byproducts are a butadiene-rich C_4 stream, a C_6–C_8 gasoline stream rich in aromatics and fuel oil.

Description: Fresh feedstock and recycle streams are preheated and cracked in the presence of dilution steam in highly selective PyroCrack furnaces (1). PyroCrack furnaces are optimized with respect to residence time, temperature and pressure profiles for the actual feedstock and the required feedstock flexibility, thus achieving the highest olefin yields. Furnace effluent is cooled in transfer line exchangers (2), generating HP steam, and by direct quenching with oil for liquid feedstocks.

The cracked gas stream is cooled and purified in the primary fractionator (3) and quench water tower (5). Waste heat is recovered by a circulating oil cycle, generating dilution steam (4) and by a water cycle (5) to provide heat to reboilers and process heaters. The cracked gas from the quench tower is compressed (6) in a 4- or 5-stage compressor and dried in gas and liquid adsorbers (8). CO_2 and H_2S are removed in a caustic-wash system located before the final compressor stage.

The compressed cracked gas is further cooled (9) and fed to the recovery section: front-end deethanizer (10), isothermal front-end C_2 hydrogenation (11), cold train (12), demethanizer (13) and the heat-pumped low-pressure ethylene fractionatior (14), which is integrated with the ethylene refrigeration cycle. This well-proven Linde process is highly optimized, resulting in high flexibility, easy operation, low energy consumption, low investment costs and long intervals between major turnarounds (typically five years).

The C_3 from the deethanizer bottoms (10) is depropanized (15), hydrogenated (16) to remove methyl acetylene and propadiene (16) and fractionated to recover polymer grade propylene. C_4 components are separated from heavier components in the debutanizer (18) to recover a C_4 product and a C_5 stream. The C_5, together with the hydrocarbon condensates from the hot section, forms an aromatic-rich gasoline product.

Economics: Ethylene yields vary between 25%, 35%, 45% and 83% for gas oils, naphtha, LPG and ethane respectively. The related specific energy consumption range is 6,000/5,400/4,600 and 3,800 kcal/kg ethylene. Typical installation costs for a world-scale ISBL gas (naphtha) cracker on a Gulf Coast basis are 500 (750) US$/ton installed ethylene capacity.

Commercial plants: Over 15 million tons of ethylene are produced in more than 40 plants worldwide. Many plants have been expanded in capacity up to 50% and more.

Recent awards for world-scale ethylene plants include Borouge in Abu Dhabi, Optimal in Malaysia, Amir Kabir and Marun in Iran and TVK II in Hungary. The Marun plant is one of the world's largest crackers with a capacity of 1.1 million mtpy ethylene and 200,000 mtpy propylene.

Licensor: Linde AG.

Ethylene

Application: To produce polymer-grade ethylene and propylene by thermally cracking paraffinic feedstocks (ethane through hydrocracked residue). Two main process technologies are used:

1. USC (ultra selective cracking)—Pyrolysis and quench systems
2. ARS (advanced recovery system)—Cold fractionation.

Designs have been incorporated to meet environmental restrictions.

Description: Feeds are sent to USC cracking furnaces (1). Contaminants removal may be installed upstream. A portion of the cracking heat may be supplied by gas turbine exhaust. Pyrolysis occurs within the temperature-time requirements specific to the feedstock and product requirements. Rapid quenching preserves high-olefin yield and the waste heat generates high-pressure steam. Lower temperature waste heat is recovered and pyrolysis fuel oil and gasoline distillate fractionated (2). Cracked gas (C_4 and lighter) is then compressed (3), scrubbed with caustic to remove acid gases and dried prior to fractionation. C_2 and lighter components are separated from the C_4 and heavier components in the low fouling front-end dual pressure depropanizer (4). Overhead vapor is hydrogenated to remove acetylene (5) and is routed to the ARS (6).

ARS minimizes refrigeration energy by using distributed distillation and simultaneous heat and mass transfer in the dephlegmator (exclusive arrangement with Air Products) or HRS system. Two C_2 streams of varying composition are produced. Hydrogen and methane are separated overhead.

The heavier C_2 stream is deethanized (7) and C_2 overhead passes to the ethylene-ethane fractionator (9). The lighter C_2 stream is routed directly to the ethylene-ethane fractionator (9). Polymer-grade ethylene product is sent overhead from the ethylene-ethane fractionator. Acetylene recovery may optionally be installed upstream of the ethylene-ethane fractionator (8).

C_3s are combined and hydrogenated to remove methyl acetylene and propadiene (10). Polymer or chemical-grade propylene is then produced overhead from the C_3 superfractionator (11).

C_4 and heavier coproducts are further separated in a sequence of distillation steps. Ethane and propane are typically recycle cracked. Refrigeration is supplied by cascade ethylene/propylene systems.

Specific advantages of ARS technology are: 1. reduced chilling train refrigeration requirements due to chilling/prefractionation in the dephlegmator or HRS system, 2. reduced methane content in feed to demethanizer, 3. partial deethanizer bypassing, 4. dual feed ethylene fractionator (lower reflux ratio), and 5. reduced refrigeration demand (approx. 75%).

Economics: Ethylene yields range from 57% (ethane, high conversion) to 28% (heavy hydrogenated gas oils). Corresponding specific energy consumptions range from 3,000 kcal/kg to 6,000 kcal/kg.

Commercial plants: Over 120 ethylene units have been built by Stone & Webster. Expansion techniques based on ARS technology have increased original capacities by over 100%.

Licensor: Stone & Webster Inc., a Shaw Group Co.

Ethylene

Application: To produce polymer-grade ethylene and propylene, a butadiene-rich C_4 cut, an aromatic C_6–C_8 rich-raw pyrolysis gasoline and high-purity hydrogen by steam pyrolysis of hydrocarbons ranging from ethane to vacuum gas oils.

Description: For either gaseous (ethane/propane) or liquid (C_4/naphtha/gasoil) feeds, this technology is based on Technip's proprietary pyrolysis furnaces and progressive separation. It produces olefins at low energy consumption and particularly low environmental impact.

Hydrocarbon feedstocks are preheated (also to recover heat) and then cracked with steam in a tubular pyrolysis furnace with outlet temperatures ranging from 1,500°F to 1,600°F. The furnace technology can be either an SMK type (for gas cracking) or GK style (for liquid cracking). The GK design can be oriented to a high olefins yield with very flexible propylene/ethylene ratios (GK5 or the latest generation GK6), or to a high BTX production (GK3 type). This approach allows long run length, excellent mechanical integrity and attractive economics.

The hydrocarbon mixture at the furnace outlet is quenched rapidly in the transfer line exchangers (TLE or SLE), generating high-pressure steam. In liquid crackers, cracked gas flows to a primary fractionator, after direct quench with oil, where fuel oil is separated from gasoline and lighter components, and then to a quench-water tower for water recovery (to be used as dilution steam) and heavy gasoline production (end-point control).

A multistage compressor, driven by a steam turbine, compresses the cooled gas. LP and HP condensate are stripped in two separate strippers where medium gasoline is produced and part of the C_3^+ cut is recovered respectively. A caustic scrubber removes acid gases.

Compressed gas at 450 psig is dried and then chilled. A double demethanizing stripping system operating at medium pressure and reboiled by cracked gas minimizes the refrigeration required (heat integration), as well as, the investment for separating methane (top) and C_2^+ cut (bottoms). A dual-column concept (absorber concept) is applied between the secondary demethanizer overheads and the chilled cracked that minimizes the ethylene losses with a low energy requirement. The high-purity hydrogen is produced in a cold box.

The bottoms from the two demethanizers (of different quality) are sent to the deethanizer. The progressive separation allows the deethanizer reflux ratio to be reduced. The deethanizer overhead is selectively hydrogenated for acetylene conversion before the ethylene splitter where ethylene is separated from ethane. The residual ethane is recycled for further cracking.

The HP stripper and deethanizer bottoms (of different quality) are fed to a two-column dual pressure depropanizing system for C_3-cut separation from the C_4 cut and heavies, thus providing low fouling at minimum energy consumption.

The methyl-acetylene and propadiene in the C_3 cut are hydrogenated to propylene in a liquid-phase reactor. Polymer-grade propylene is separated from propane in a C_3-splitter. The residual propane is either recycled for further cracking, or exported. C_4s and light gasoline are separated in a debutanizer. Gas expansion (heat recovery) and external cascade using ethylene and propylene systems supply refrigeration.

Economics: Ultimate range of ethylene yields vary from 83% (ethane) to around 25% (vacuum gas oils), 35% for the intermediate full-range naphtha. These correspond to the respective total olefins yields (ethylene and propylene) from 84% (ethane) to 38% (vacuum gas oils), and 49% for an intermediate full-range naphtha. The specific energy consumption range is 3,000 kcal/kg ethylene (ethane) to 5,500 kcal/kg ethylene (gas oil), and 4,700 kcal/kg ethylene for an intermediate full-range naphtha.

Commercial plants: In mid 2001, Technip was awarded a 1.4 million-tpy ethane/naphtha-feed cracker. Technip has built over 450 cracking furnaces with 15 units operating worldwide. Many expansions, over the nominal capacity based on progressive separation techniques, are under way, with up to an 80% increase in capacity.

Licensor: Technip-Coflexip.

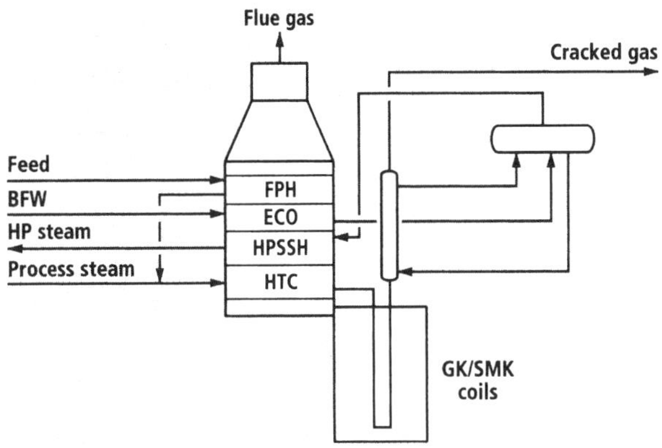

Ethylene

Application: Thermal cracking of a wide range of feedstocks into light olefins and aromatics using proprietary cracking coils.

Feedstocks: Ethane through to heavy feeds up to 600°C EP.

Products: Cracked gas rich in ethylene, propylene, butadiene and BTX.

Description: Thermal cracking occurs in presence of steam at high temperatures in cracking coils located centrally in the firebox. Coil outlet temperatures vary up to 880°C depending on feed quality and cracking severity. The proprietary cracking coils are the GK5, GK6 and SMK coils. They feature high selectivity to ethylene and propylene, together with low coking rates (long run lengths).

Cracked gases from the furnace pass through a transferline exchanger (TLE) system, where heat is recovered to generate high-pressure steam. The primary TLEs are linear or special S and T type exchangers. The selected exchanger type ensures low to very low fouling rates and, thus extends run lengths. Heat from the flue gases is recovered in the convection section to preheat feed and process steam, and to superheat generated HP Steam. The technology may be applied to retrofit furnaces. Furnace performance is optimized using proprietary SPYRO programs. NO_x abatement technology is incorporated.

Performance data:

Ethane conversion, %		65–75
Naphtha cracking severity (as P/E)		0.40–0.70
Overall thermal efficiency		92–95
Coil residence time, sec	—GK5/GK6 coils	0.15–0.25
	—SMK coil	0.35–0.40

Once-through ethylene yields depend on feed characteristics and severity, and range from 58% for ethane to 36% for liquid feeds.

Commercial plants: Over 450 installations since the mid-1960s.

Licensor: Technip-Coflexip.

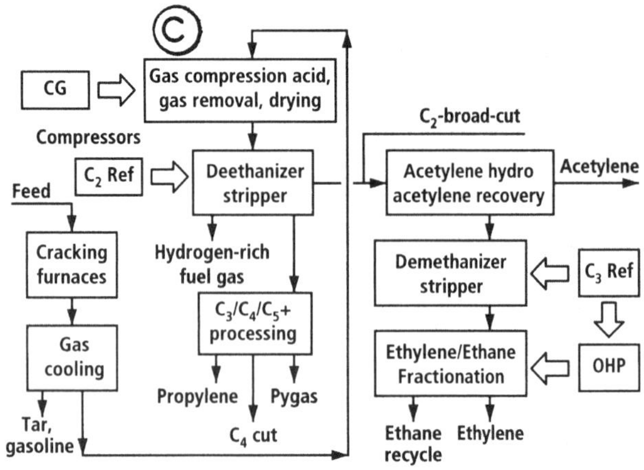

Ethylene

Application: To produce polymer-grade ethylene and propylene, a butadiene-rich C_4 cut, an aromatic C_6–C_8 rich-raw pyrolysis gasoline and high-purity hydrogen by using the T-PAR process for gas separation and product purification from raw cracked gas.

Description: Effluents from cracking furnaces are cooled and processed for tar and heavy-gasoline removal.

A multistage compressor, driven by a steam turbine, compresses the cooled gas. LP and HP condensates are stripped in two separate strippers where medium gasoline is produced and part of the C_3+ cut is recovered respectively. A caustic scrubber removes acid gases.

Compressed gas at 450 psig is dried and then chilled. A multi-stream heat exchanger chills the tail gas to –265°F. Liquid condensates are separated at various temperatures, such as –30°F, -65°F, -100°F and –140°F, and are reheated against incoming cracked gas. The partially vaporized streams are sent to a deethanizer stripper operating at about 320 psig. The bottoms C_3+ stream is sent to propylene and heavys recovery.

The overhead is reheated and enters an adiabatic acetylene hydrogenation reactor, which transforms the acetylene selectively to ethylene and ethane. As an alternate, a solvent-recovery process can be applied without reheating the gas.

Reactor effluent is chilled and light-ends are separated from the C_2-hydrocarbons. The demethanizer overhead is processed for ethylene recovery while the bottoms is sent to ethylene/ethane separation. An open heat-pump splitter is applied, thus sending ethylene product to the gas pipeline from the discharge of the ethylene-refrigerant compressor.

Dilute ethylene for chemical applications, such as styrene production, can be withdrawn downstream of the hydrogenation reactor. The ethylene content is typically 60 vol%. Catalyst suppliers have tested the hydrogenation step, and commercially available front-end catalysts are suitable for this application.

Economics: The advantages of this process are low equipment costs (viz. the deethanizer system and ethylene/ethane separation) and reliability of the acetylene hydrogenation due to low excess hydrogen at the reactor inlet. The refrigeration compressor benefits from low specific power and suction volume, while the cracked-gas compressor processes above-ambient-temperature gas.

Commercial plants: Technip is commercializing the T-PAR process on a case-by-case basis. Patent is pending with worldwide applications.

Licensor: Technip-Coflexip.

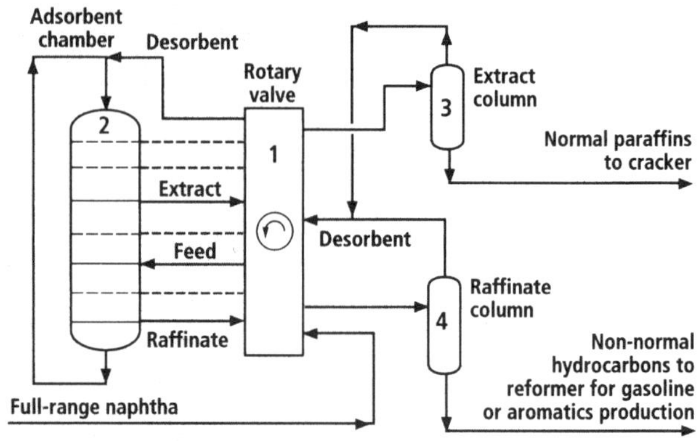

Ethylene

Application: The MaxEne process increases the ethylene yield from naphtha crackers by raising the concentration of normal paraffins (n-paraffins) in the naphtha-cracker feed. The MaxEne process is the newest application of UOP's Sorbex technology. The process uses adsorptive separation to separate C_5–C_{11} naphtha into a rich n-paraffins stream and a stream depleted of n-paraffins.

Description: The separation takes place in an adsorption chamber (2) that is divided into a number of beds. Each bed contains proprietary shape-selective adsorbent. Also, each bed in the chamber is connected to a rotary valve (1). The rotary valve is used along with the shape-selective adsorbent to simulate a moving bed counter-current adsorptive separation. Four streams are distributed by the rotary valve to and from the adsorbent chamber. The streams are as follows:

- **Feed:** The naphtha feed contains a mixture of hydrocarbons.
- **Extract:** This stream contains n-paraffin and a liquid desorbent. Naphtha, rich in n-paraffin, is recovered by fractionation (3) and is sent to the naphtha cracker.
- **Raffinate:** This stream contains non-normal paraffin and a liquid desorbent. Naphtha, depleted in n-paraffin, is recovered by fractionation (4) and is sent to a refinery or an aromatics complex.
- **Desorbent:** This stream contains a liquid desorbent that is recycled from the fractionation section to the chamber.

The rotary valve is used to periodically switch the position of the liquid feed and withdrawal points in the adsorbent chamber. The process operates in a continuous mode at low temperatures in a liquid phase.

Economics: For increase in naphtha-cracker ethylene production from 247,000 mtpy to 330,000 mtpy, based on US Gulf Coast:

Investment cost, $ million	80
Increased margin, $ million/yr	28.6
Simple ROI, %	36

Licensor: UOP LLC.

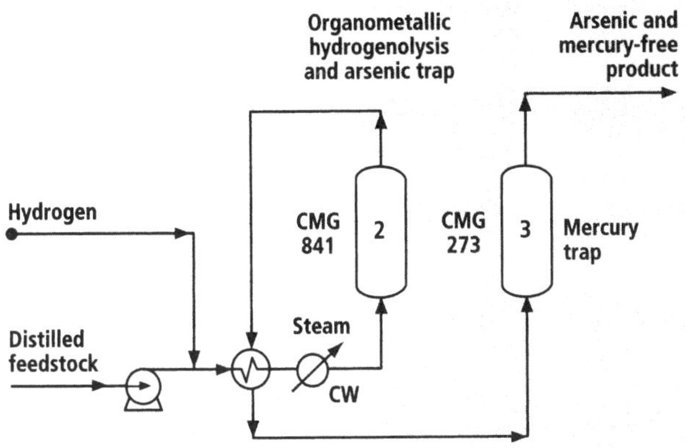

Ethylene feed pretreatment—mercury, arsenic and lead removal

Application: Upgrade natural gas condensate and other contaminated streams to higher-value ethylene plant feedstocks. Mercury, arsenic and lead contamination in potential ethylene plant feedstocks precludes their use, despite attractive yield patterns. The contaminants poison catalysts, cause corrosion in equipment and have undesirable environmental implications. For example, mercury compounds poison hydrotreating catalysts and, if present in the steam-cracker feed, are distributed in the C_2–C_5^+ cuts. A condensate containing mercury may have negative added-value as a gas field product.

Description: Three RAM processes are available to remove arsenic (RAM I); arsenic, mercury and lead (RAM II); and arsenic, mercury and sulfur from liquid hydrocarbons (RAM III). Described above is the RAM II process. Feed is heated by exchange with reactor effluent and steam (1). It is then hydrolyzed in the first catalytic reactor (2) in which organometallic mercury compounds are converted to elemental mercury, and organic arsenic compounds are converted to arsenic-metal complexes and trapped in the bed. Lead, if any, is also trapped on the bed. The second reactor (3) contains a specific mercury-trapping mass. There is no release of the contaminants to the environment, and spent catalyst and trapping material can be disposed of in an environmentally acceptable manner.

Typical RAM II Performance

Contaminant	Feedstock	Product
Mercury, ppb	2000	< 1*
Arsenic, ppb	100	< 1*

* 3 ppb is the threshold limit of the analytical procedure commonly used. With provisions for eliminating solid matter, water and free oxygen and using a more sensitive method, levels of less than one ppb can be achieved.

Economics: The ISBL 2002 investment at a Gulf Coast location for two condensates each containing 50-ppb average mercury content (max. 500 ppb), 10 ppb arsenic and 120 ppb lead excluding basic engineering, detailed engineering, offsites, contractor fees:

	Clear, oxygen-free condensate	Aerated condensate with particulate matter
Investment, US$/bpd	129	175
Utilities, US$/bpd	0.08	0.23
Catalyst cost, US$/bpd	0.03	0.03

Commercial plants: Fourteen RAM units have been licensed worldwide.

References: Didillon, B., L. Savary, J. Cosyns, Q. Debuisschert, and P. Travers, "Mercury and Arsenic Removal from Ethylene Plant Feedstocks," Second European Petrochemicals Technology Conference, Prague, 2000.

Licensor: Axens, Axens NA.

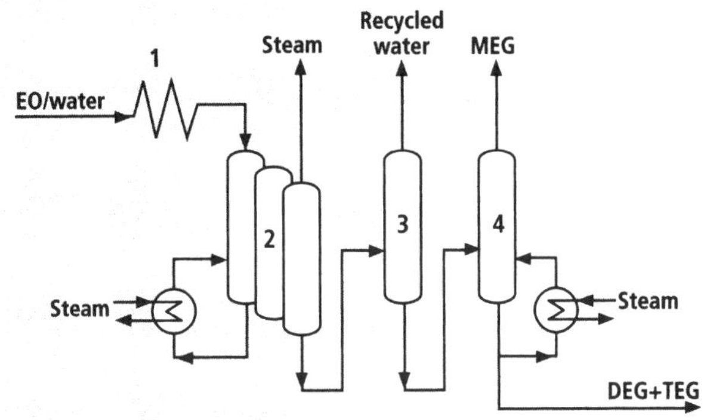

Ethylene glycol

Application: To produce ethylene glycols (MEG, DEG, TEG) from ethylene oxide (EO) using Dow's Meteor process.

Description: In the Meteor Process, an EO/water mixture is preheated and fed directly to an adiabatic reactor (1), which can operate with or without a catalyst. An excess of water is provided to achieve high selectivities to monoethylene glycol (MEG). Diethylene (DEG) and triethylene (TEG) glycols are produced as coproducts. In a catalyzed mode, higher selectivities to MEG can be obtained, thereby reducing DEG production to one-half that produced in the uncatalyzed mode. The reactor is specially designed to fully react all of the EO and to minimize back-mixing, which promotes enhanced selectivity to MEG.

Excess water from the reactor effluent is efficiently removed in a multi-effect evaporation system (2). The last-effect evaporator overhead produces low-pressure steam, which is a good low-level energy source for other chemical units or other parts of the EO/MEG process. The concentrated water/glycols stream from the evaporation system is fed to the water column (3) where the remaining water and light ends are stripped from the crude glycols. The water-free crude glycol stream is fed to the MEG refining column (3) where polyester-grade MEG, suitable for polyester fiber and PET production, is recovered. DEG and TEG exiting the base of the MEG refining column can be recovered as high-purity products by subsequent fractionation.

Economics: The conversion of EO to glycols is essentially complete. The reaction not only generates the desired MEG, but also produces DEG and TEG that can be recovered as coproducts. The production of more DEG and TEG may be desirable if the manufacturer has a specific use for these products or if market conditions provide a good price for DEG and TEG relative to MEG. A catalyzed process will produce less heavy glycols. The ability to operate in catalyzed or uncatalyzed mode provides flexibility to the manufacturer to meet changing market demands.

Commercial plants: Since 1954, 18 UCC-designed glycol plants have been started up or are under construction.

Licensor: Union Carbide Corp., a subsidiary of The Dow Chemical Co.

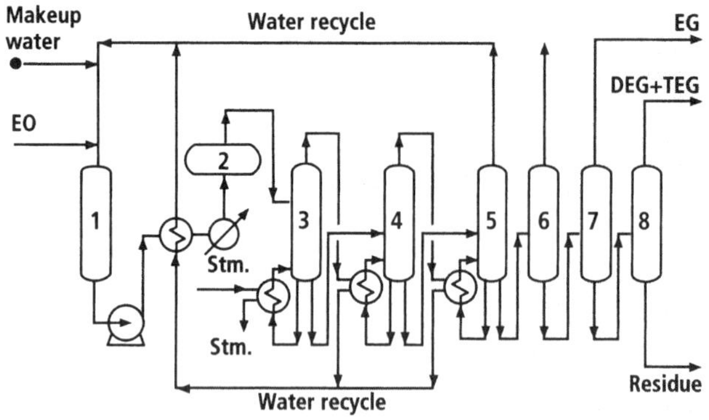

Ethylene glycols

Application: To produce ethylene glycols (MEG, DEG, TEG) from ethylene oxide (EO).

Description: The flowsheet shown is only one of several possible schemes. The raw materials to a free-standing glycol plant are refined ethylene oxide and pure water. These are mixed with recycle waters and pumped from a feed tank (1) to the hydration reactor after being preheated with hot recycle water and steam. When the glycol unit is part of a combined oxide/glycol plant, it is economically desirable to feed it bleed streams from the ethylene oxide unit. Since this stream can affect product quality, additional proprietary treatment is necessary. The SD process is also able to simplify the combined oxide/glycol plants by feeding crude oxide solution (aqueous) from the ethylene oxide reaction unit directly to the glycol reaction and omitting the usual oxide refining step.

In the glycol reactor (2), sufficient residence time is provided to react (noncatalytically) all of the ethylene oxide. Operating pressure of the reaction is controlled at a level that limits or avoids vaporization of ethylene oxide from the aqueous solution.

The water-glycol mixture from the reactor is fed to the first stage (3) of a multiple stage evaporator, which is reboiled using high pressure steam. The remaining stages (4, 5) of the evaporator operate at successively lower pressures, with the final stage (5) normally under vacuum. Evaporated water is recovered as condensate and recycled back to the glycol reaction feed-mixing tank (1)—or to the oxide reaction section in combined oxide/glycol plants.

Concentrated crude glycol solution from the final evaporation stage is stripped of remaining water and light ends in the light-ends column (6). The water-free glycol mixture is then fractionated in a series of vacuum distillation towers (7, 8) to produce purified monoethylene glycol (MEG) of fiber-grade quality and byproduct diethylene glycol (DEG) and triethylene glycol (TEG).

Product quality: The glycol quality produced in these plants has set a new standard in the industry.

Yields: Conversion is essentially complete and quantitative.

Commercial plants: Over 60 ethylene glycol projects have been completed or are in the design stage.

Reference: *Encyclopedia of Chemical Processing and Design.*

Licensor: Scientific Design Co., Inc.

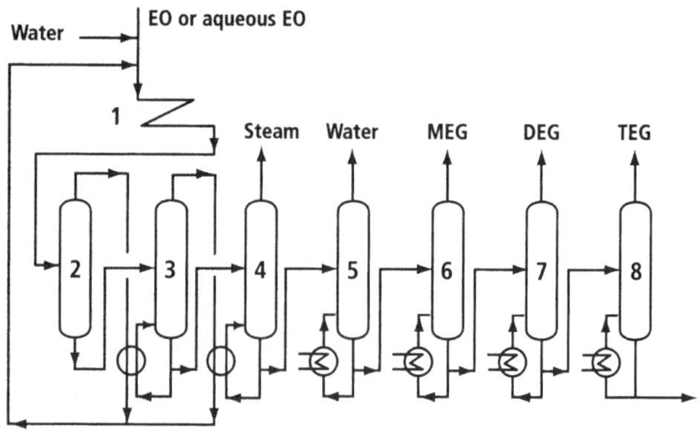

Ethylene glycols

Application: To produce ethylene glycols (MEG, DEG and TEG) from ethylene oxide (EO).

Description: Purified EO or a water/EO mixture is combined with recycle water and heated to reaction conditions. In the tubular reactor (1) essentially all EO is thermally converted into mono-ethylene glycol (MEG) with di-ethylene glycol (DEG) and tri-ethylene glycol (TEG) as co-products in minor amounts. Excess water, required to achieve a high selectivity to MEG, is evaporated in a multi-stage evaporator (2, 3, 4). The last evaporator produces low-pressure steam that is used as a heating medium at various locations in the plant. The resulting crude glycols mixture is subsequently purified and fractionated in a series of vacuum columns (5, 6, 7, 8).

The selectivity to MEG can be influenced by adjusting the glycol reactor feed composition.

Most MEG plants are integrated with EO plants. In such an integrated EO/MEG facility, the steam system can be optimized to fully exploit the benefits of high-selectivity catalyst applied in the EO plant. However, standalone MEG plants have been designed and built.

The quality of glycols manufactured by this process ranks amongst the highest in the world. It consistently meets the most stringent specifications of polyester fiber and PET producers.

Commercial plants: Since 1958, more than 60 Shell-designed MEG plants have been commissioned or are under construction.

Licensor: Shell International Chemicals B.V.

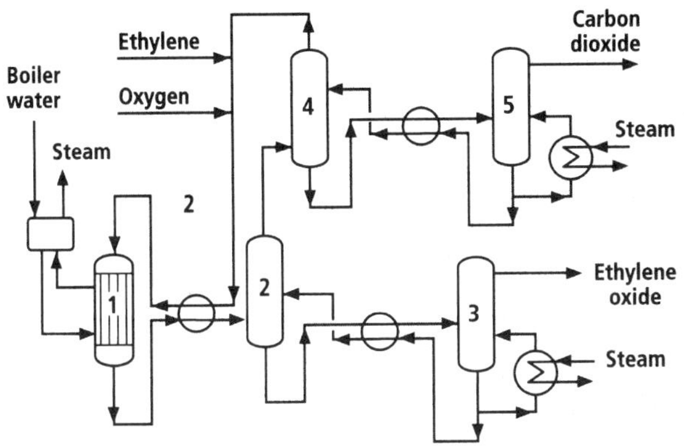

Ethylene oxide

Application: To produce ethylene oxide (EO) from the direct oxidation of ethylene using the Dow Meteor process.

Description: The Meteor Process, a technology first commercialized in 1994, is a simpler, safer process for the production of EO, having lower capital investment requirements and lower operating costs. In the Meteor Process, ethylene and oxygen are mixed with methane-ballast recycle gas and passed through a single-train, multitubular catalytic reactor (1) to selectively produce EO. Use of a single reactor is one example of how the Meteor process is a simpler, safer technology with lower facility investment costs.

The special high-productivity Meteor EO catalyst provides very high efficiencies while operating at high loadings. Heat generated by the reaction is removed and recovered by the direct boiling of water to generate steam on the shell side of the reactor. Heat is recovered from the reactor outlet gas before it enters the EO absorber (2) where EO is scrubbed from the gas by water. The EO-containing water from the EO absorber is concentrated by stripping (3). The cycle gas exiting the absorber is fed to the CO_2 removal section (4,5) where CO_2, which is co-produced in the EO reactor, is removed via activated, hot potassium carbonate treatment. The CO_2 lean cycle gas is recycled by compression back to the EO reactor.

Most EO plants are integrated with glycol production facilities. When producing glycols, the EO stream (3) is suitable for feeding directly to a Meteor glycol process. When EO is the desired final product, the EO stream (3) can be fed to a single purification column to produce high-purity EO. This process is extremely flexible and can provide the full range of product mix between glycols and purified EO.

Economics: The process requires a lower capital investment and has lower fixed costs due to process simplicity and the need for fewer equipment items. Lower operating costs are also achieved through the high-productivity Meteor EO catalyst, which has very high efficiencies at very high loadings.

Commercial plants: Union Carbide was the first to commercialize the direct oxidation process for EO in the 1930s. Since 1954, 18 Union Carbide-designed plants have been started up or are under construction. Three million tons of EO equivalents per year (approximately 20% of total world capacity) are produced in Union Carbide-designed plants.

Licensor: Union Carbide Corp., a subsidiary of The Dow Chemical Co.

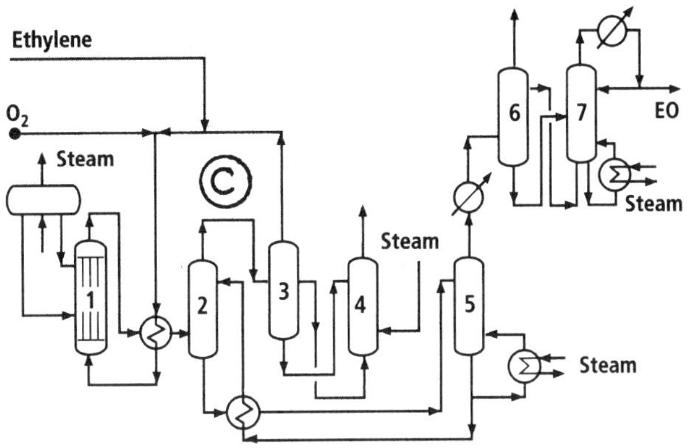

Ethylene oxide

Application: To produce ethylene oxide (EO) from ethylene using oxygen as the oxidizing agent.

Description: The flowsheet for an oxygen-based unit is one of several possible process schemes. Compressed oxygen, ethylene and recycle gas are mixed and fed to a multitubular catalytic reactor (1). The temperature of oxidation is controlled by boiling water in the shell side of the reactor.

From the reactor, the effluent gases containing ethylene oxide are first cooled by recuperative exchange with recycle gases. The gases then pass to a scrubber (2) where the ethylene oxide is absorbed as a dilute aqueous solution. Most of the unabsorbed gases are compressed and returned to the reactor via the recuperative exchanger, thus completing a closed circuit. A portion of the recycle gas is diverted through a CO_2 removal system (3, 4) before being returned to the reaction system.

Ethylene oxide is steam-stripped (5) from the EO-rich absorber bottoms and recovered as refined or purified product in a fractionation train (6, 7).

Yields: The weight yield of purified EO (kg per kg ethylene feed) is in excess of 120%. In addition, a significant amount of technical-grade glycol may be recovered by processing waste streams.

Commercial plants: Over 100 ethylene oxide projects have been completed or are being designed. This represents a total design capacity of about 5 MMtpy of ethylene oxide.

Licensor: Scientific Design Co., Inc.

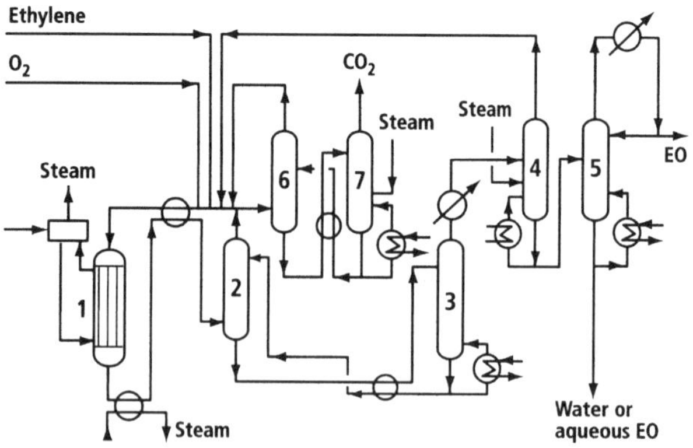

Ethylene oxide

Application: To produce ethylene oxide (EO) from ethylene and oxygen in a direct oxidation process.

Description: In the direct oxidation process, ethylene and oxygen are mixed with recycle gas and passed through a multi-tubular catalytic reactor (1) to selectively produce EO. A special silver-containing high-selectivity catalyst is used that has been improved significantly over the years. Methane is used as ballast gas. Heat generated by the reaction is recovered by boiling water at elevated pressure on the reactor's shell-side; the resulting high-pressure steam is used for heating purposes at various locations within the process.

EO contained in the reactor product-gas is absorbed in water (2) and further concentrated in a stripper (3). Small amounts of co-absorbed ethylene and methane are recovered from the crude EO (4) and recycled back to the EO reactor. The crude EO can be further concentrated into high-purity EO (5) or routed to the glycols plant (as EO/water feed).

EO reactor product-gas, after EO recovery, is mixed with fresh feed and returned to the EO reactor. Part of the recycle gas is passed through an activated carbonate solution (6, 7) to recover CO_2, a byproduct of the EO reaction that has various commercial applications.

Most EO plants are integrated with fiber-grade mono-ethylene glycol (MEG) production facilities. In such an integrated EO/MEG facility, the steam system can be optimized to fully exploit the benefits of high-selectivity catalyst.

When only high-purity EO is required as a product, a small amount of technical-grade MEG inevitably is co-produced.

Yields: Modern plants are typically designed for and operate at a molar EO catalyst selectivity approaching 90% with fresh catalyst and 86–87% as an average over 3 years catalyst life, resulting in an average EO production of about 1.4 tons per ton of ethylene. However, the technology is flexible and the plant can be designed tailor-made to customer requirements or different operating time between catalyst changes.

Commercial plants: Since 1958, more than 60 Shell-designed plants have been commissioned or are under construction. Approximately 40% of the global capacity of EO equivalents is produced in Shell-designed plants.

Licensor: Shell International Chemicals B.V.

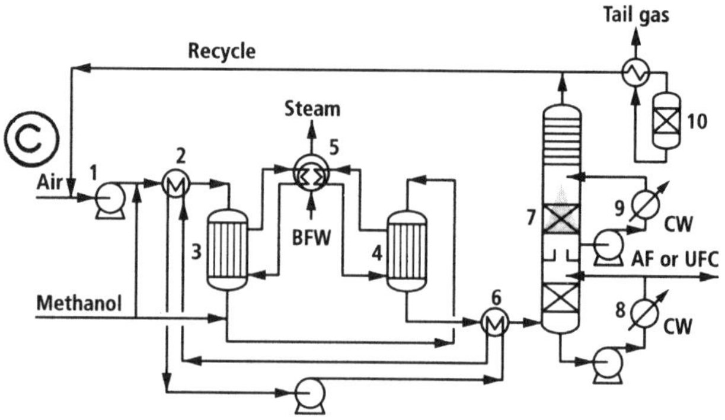

Formaldehyde

Application: To produce aqueous formaldehyde (AF) or urea formaldehyde precondensate (UFC) from methanol using the *Haldor Topsøe Formaldehyde SR process* comprising two reactors in series.

Description: Air and recycle gas are compressed by the blower (1) and then mixed with liquid methanol that is injected through spray nozzles. The mixture is preheated to about 200°C by heat exchange with hot circulating oil in the heat exchanger (2) after which the gas is successively passed to the two series reactors (3 and 4).

Additional methanol is injected into the gas between the two reactors. The reactors contain many tubes filled with FK-2 catalyst, where methanol and oxygen react to make formaldehyde. Reaction heat is removed by a bath of boiling heat-transfer oil. Hot oil vapor is condensed in the waste-heat boiler (5), thus generating steam at up to 40 bar pressure. Before entering the absorber (7), the reacted gas is cooled in the after cooler (6) and reheats the circulating oil from the process-gas heater (2).

In the absorber, the formaldehyde is absorbed in water or urea solution. Heat is removed by one or two cooling circuits (8, 9). From the lower circuit (8), product in the form of either AF or UFC is withdrawn. Scrubbed gas from the absorber is split in two streams—recycle gas and tail gas. The tail gas is vented after any organic impurities are catalytically incinerated in the reactor (10). Thus, the tail-gas purity conforms to the environmental standards for any country.

With regard for the catalyst, the percentage of methanol that can be added to a formaldehyde reactor is limited to about 9-vol%. Using two reactors in series higher production yields are achievable with the same gas flow than what would be possible in a plant with only one reactor (or a plant with two reactors in parallel).

Advantages of series reactors vs. single or parallel reactors are:
- Lower capital cost due to reduced size of equipment and piping
- Longer catalyst life: 30–36 months in reactor I, 18 months in reactor II
- Lower electricity consumption and higher steam production
- Higher conversion of methanol, therefore less methanol in product
- The *Haldor Topsøe Formaldehyde SR process* is well-suited to expand existing formaldehyde plants—up to 100% capacity increase may be achieved.

Utility requirements: Per 1,000 kg of 37-wt% formaldehyde:

Product	55 wt% AF	85 wt% UFC
Methanol, kg	420–425	425–430
70% urea solution, kg	–	220
Process water, kg	250	72
Water, cooling, m^3	42	38
Electricity, kWh	49	52

Commercial plants: Two commercial SR units built, both are operating successfully. One additional unit is under construction.

Licensor: Haldor Topsøe A/S.

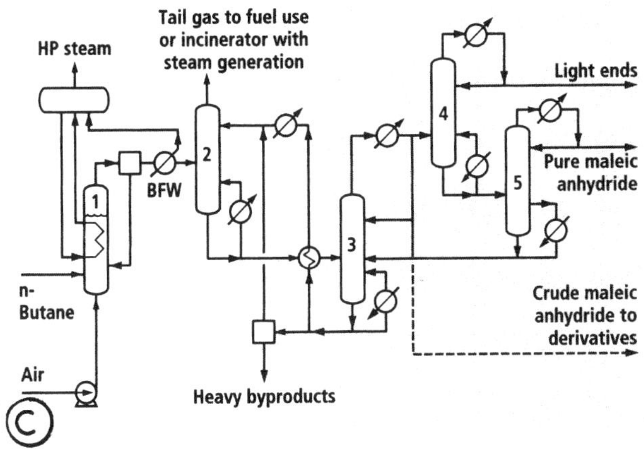

Maleic anhydride

Application: To produce maleic anhydride from n-butane using a fluid-bed reactor system and an organic solvent for continuous anhydrous product recovery.

Description: N-butane and air are fed to a fluid-bed catalytic reactor (1) to produce maleic anhydride. The fluid-bed reactor eliminates hot spots and permits operation at close to the stoichiometric reaction mixture. This results in a greatly reduced air rate relative to fixed-bed processes and translates into savings in investment and compressor power, and large increases in steam generation. The fluid-bed system permits online catalyst addition/removal to adjust catalyst activity and reduces downtime for catalyst change out.

The recovery area uses a patented organic solvent to remove the maleic anhydride from the reactor effluent gas. A conventional absorption (2)/stripping (3) scheme operates on a continuous basis. Crude maleic anhydride is distilled to separate light (4) and heavy (5) impurities. A slipstream of recycle solvent is treated to eliminate any heavy byproducts that may be formed. The continuous nonaqueous product recovery system results in superior product quality and large savings in steam consumption. It also reduces investment, product degradation loss (and byproduct formation) and wastewater.

Economics: The ALMA process produces high-quality product with attractive economics. The fluid-bed process is especially suited for large single-train plants.

Commercial plants: Nine commercial plants have been licensed with a total capacity of 200,000 mtpy. The largest commercial installation is Lonza's 50,000-mtpy plant in Ravenna, Italy. Second generation process optimizations and catalyst have elevated the plant performances since 1998.

Licensor: ABB Lummus Global/Lonza S.p.A.

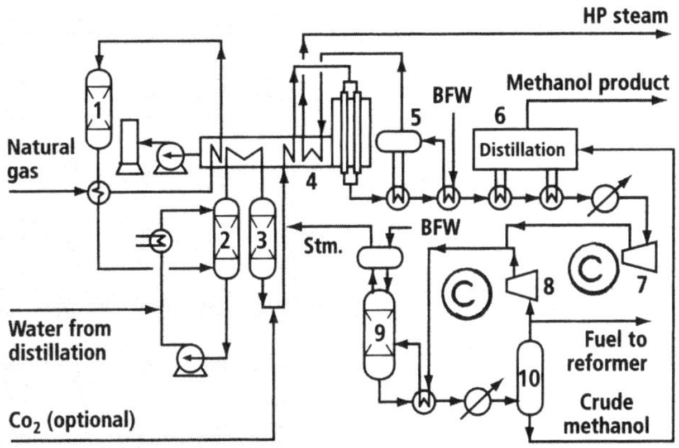

Methanol

Application: Improved low-pressure methanol process to produce methanol from natural or associated gas using two-stage steam reforming followed by compression, synthesis and distillation. Capacities, ranging from 5,000 to 7,000 mtpd, are practical in a single stream. Carbon dioxide (CO_2) can be used as a supplementary feedstock to adjust the stoichiometric ratio of the synthesis gas.

Description: Gas feedstock is compressed (if required), desulfurized (1) and sent to the optional saturator (2) where some process steam is generated. The saturator is used where maximum water recovery is important. Further process steam is added, and the mixture is preheated and sent to the pre-reformer (3), using the Catalytic-Rich-Gas process. Steam raised in the methanol converter is added, along with available CO_2, and the partially reformed mixture is preheated and sent to the reformer (4). High-grade heat in the reformed gas is recovered as high-pressure steam (5), boiler feedwater preheat, and for reboil heat in distillation system (6). The high-pressure steam is used to drive the main compressors in the plant.

After final cooling, the synthesis gas is compressed (7) and sent to the synthesis loop. The loop can operate at pressures between 70 to 100 bar. The converter design does impact the loop pressure, with radial-flow designs enabling low loop pressure even at the largest plant size. Low loop pressure reduces the total energy requirements for the process.

The synthesis loop comprises a circulator (8), the converter operates around 200°C to 270°C, depending on the converter type. Reaction heat from the loop is recovered as steam, and is used directly as process steam for the reformer.

A purge is taken from the synthesis loop to remove inerts (nitrogen, methane), as well as surplus hydrogen associated with non-stoichiometric operation. The purge is used as fuel for the reformer. Crude methanol from the separator contains water, as well as traces of ethanol and other compounds. These impurities are removed in a two-column distillation system (6). The first column removes the light ends such as ethers, esters, acetone and dissolved noncondensable gases. The second column removes water, higher alcohols and similar organic heavy ends.

Economics: Recent trends have been to build methanol plants in regions offering low-cost gas (such as Chile, Trinidad and the Arabian Gulf). In these regions, total economics favor low investment rather than low-energy consumption. Recent plants have an energy efficiency of 7.2–7.8 Gcal/ton. A guideline figure to construct a 5,000-mtpd plant is US$370–400 million.

Commercial plants: Thirteen plants with capacities ranging from 2,000 to 3,000 mtpd, as well as 50 smaller plants have been built using the Synetix LPM methanol technology. Two 5,000-mtpd plants are under construction.

Licensor: Davy Process Technology/Synetix, UK.

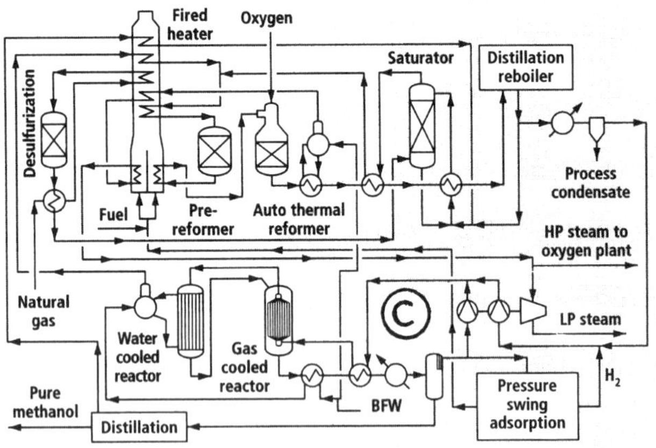

Methanol

Application: To produce methanol in a single-train plant from natural gas or oil-associated gas with capacities up to 10,000 mtpd. It is also well suited to increase capacities of existing steam-reforming-based methanol plants.

Description: Natural gas is preheated and desulfurized. After desulfurization, the gas is saturated with a mixture of preheated process water from the distillation section and process condensate in the saturator. The gas is further preheated and mixed with steam as required for the pre-reforming process. In the pre-reformer, the gas is converted to H_2, CO_2 and CH_4. Final preheating of the gas is achieved in the fired heater. In the autothermal reformer, the gas is reformed with steam and O_2. The product gas contains H_2, CO, CO_2 and a small amount of unconverted CH_4 and inerts together with undercomposed steam. The reformed gas leaving the autothermal reformer represents a considerable amount of heat, which is recovered as HP steam for preheating energy and energy for providing heat for the reboilers in the distillation section.

The reformed gas is mixed with hydrogen from the pressure swing adsorption (PSA) unit to adjust the synthesis gas composition. Synthesis gas is pressurized to 5–10 MPa by a single-casing synthesis gas compressor and is mixed with recycle gas from the synthesis loop. This gas mixture is preheated in the trim heater in the gas-cooled methanol reactor. In the Lurgi water-cooled methanol reactor, the catalyst is fixed in vertical tubes surrounded by boiling water. The reaction occurs under almost isothermal condition, which ensures a high conversion and eliminates the danger of catalyst damage from excessive temperature. Exact reaction temperature control is done by pressure control of the steam drum generating HP steam.

The "preconverted" gas is routed to the shell side of the gas-cooled methanol reactor, which is filled with catalyst. The final conversion to methanol is achieved at reduced temperatures along the optimum reaction route. The reactor outlet gas is cooled to about 40°C to separate methanol and water from the gases by preheating BFW and recycle gas. Condensed raw methanol is separated from the unreacted gas and routed to the distillation unit. The major portion of the gas is recycled back to the synthesis reactors to achieve a high overall conversion. The excellent performance of the Lurgi combined converter (LCC) methanol synthesis reduces the recycle ratio to about 2. A small portion of the recycle gas is withdrawn as purge gas to lessen inerts accumulation in the loop.

In the energy-saving three-column distillation section, low-boiling and high-boiling byproducts are removed. Pure methanol is routed to the tank farm, and the process water is preheated in the fired heater and used as makeup water for the saturator.

Economics: Energy consumption for a stand-alone plant, including utilities and oxygen plant, is about 30 GJ/metric ton of methanol. Total installed cost for a 5,000-mtpd plant including utilities and oxygen plant is about US$350 million, depending on location.

Commercial plants: Thirty-five methanol plants have been built using Lurgi's low-pressure methanol technology. Two 5,000-mtpd plants using Lurgi's mega methanol technology are under construction.

Licensor: Lurgi Oel Gas Chemie GmbH.

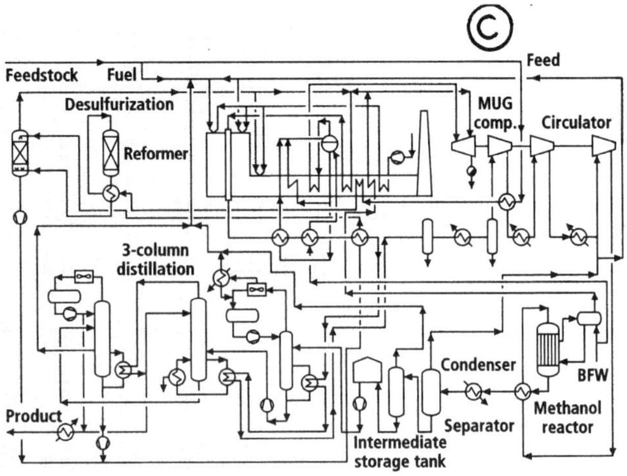

Methanol

Application: Production of high-purity methanol from hydrocarbon feedstocks such as natural gas, process offgases and LPG up to heavy naphtha. The process uses conventional steam-reforming synthesis gas generation and a low-pressure methanol synthesis loop technology. It is optimized with respect to low energy consumption and maximum reliability. The largest single-train plant built by Uhde has a nameplate capacity of 1,250 mtpd.

Description: The methanol plant consists of several processing steps: feed purification, steam reforming, syngas compression, methanol synthesis and crude methanol distillation. The feed is desulfurized and mixed with process steam before entering the steam reformer. This steam reformer is a top-fired box type furnace with a cold outlet header system developed by Uhde. The reforming reaction occurs over a nickel catalyst. Outlet-reformed gas is a mixture of H_2, CO, CO_2 and residual methane. It is cooled from approximately 880°C to ambient temperature. Most of the heat from the synthesis gas is recovered by steam generation, BFW preheating, heating of crude methanol distillation and demineralized water preheating.

Also, heat from the flue gas is recovered by feed/feed-steam preheating, steam generation and superheating as well as combustion air preheating. After final cooling, the synthesis gas is compressed to the synthesis pressure, which ranges from 30–100 bara (depending on plant capacity) before entering the synthesis loop.

The synthesis loop consists of a recycle compressor, feed/effluent exchanger, methanol reactor, final cooler and crude methanol separator. Uhde's methanol reactor is an isothermal tubular reactor with a copper catalyst contained in vertical tubes and boiling water on the shell side. The heat of methanol reaction is removed by partial evaporation of the boiler feedwater, thus generating 1–1.4 tons of MP steam per ton of methanol. Advantages of this reactor type are low byproduct formation due to almost isothermal reaction conditions, high level heat of reaction recovery, and easy temperature control by regulating steam pressure. To avoid inert buildup in the loop, a purge is withdrawn from the recycle gas and is used as fuel for the reformer.

Crude methanol that is condensed downstream of the methanol reactor is separated from unreacted gas in the separator and routed via an expansion drum to the crude methanol distillation. Water and small amount of byproducts formed in the synthesis and contained in the crude methanol are removed by an energy-saving three-column distillation system.

Economics: Typical consumption figures (feed + fuel) range from 7 to 8 Gcal per metric ton of methanol and will depend on the individual plant concept.

Commercial plants: Eleven plants have been built and revamped worldwide using Uhde's methanol technology.

Licensor: Uhde GmbH.

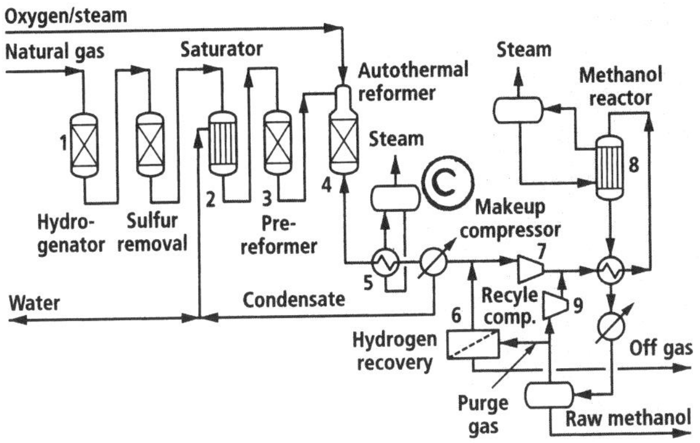

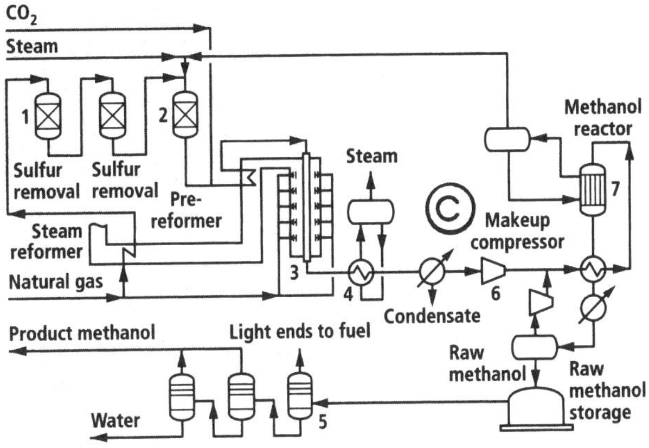

Methanol-autothermal reforming (ATR)

Application: To produce methanol from natural or associated gas feedstocks using autothermal reforming (ATR) followed by boiling water reactor synthesis. This technology is well suited for very large-scale plants as well as for the production of methanol to olefins or fuel-grade methanol. Topsøe also offers technology for smaller methanol facilities and technology to modify ammonia capacity into methanol production.

Description: The gas feedstock is compressed (if required), desulfurized (1) and sent to a saturator (2) where the natural gas is saturated with process condensate and excess water from the distillation section. Recycling of process condensate and excess water minimizes the water requirement. Low-grade medium-pressure steam is used in the saturator, thus saving high-pressure steam. The mixture of natural gas and steam is preheated, prereformed (3) and sent to the autothermal reformer (4). Autothermal reforming features a stand-alone oxygen-fired reformer and, thus, the cost-intensive primary tubular reformer may be omitted completely. The autothermal reformer can operate at any pressure. The operating pressure is normally selected between 30 and 40 kg/cm^{2}g.

Synthesis gas generated in the autothermal reformer is cooled by high-pressure steam generation (5), preheat of boiler feed water, reboiling in the distillation section and preheat of demineralized water.

The synthesis gas obtainable with this technology is typically deficient in hydrogen. Therefore, the synthesis gas composition must be adjusted by recycling recovered hydrogen (6) from the synthesis loop. After final cooling by air or cooling water, the recycle hydrogen is added to the synthesis gas, which is compressed in a single-step compressor (7) and sent to the synthesis loop (8).

The synthesis loop comprises of a straight-tubed boiling water reactor, which is more efficient than adiabatic reactors. Reaction heat is removed from the reactor by generation of medium-pressure steam. This steam is used for heating in the saturator (2). Preheating the reactor feed cools effluent from the synthesis reactor. Further cooling is by air or water cooling. Raw methanol is separated and sent directly to the distillation section featuring a very efficient three-column layout. Recycle gas is sent to the recirculator compressor (9) after a purge to remove inert compound build-up. The purge is sent to a hydrogen recovery unit where hydrogen is separated and recycled to the synthesis gas compressor.

Topsøe supplies a complete range of catalysts for methanol production. The total energy consumption for this process scheme is about 7.1 Gcal/ton methanol. Total energy consumption for production of fuel grade methanol is approximately 6.8 Gcal/ton methanol.

Economics: For large-scale plants, the total investment including an oxygen plant is approximately 10% lower than for a conventional plant based on tubular steam reforming.

Licensor: Haldor Topsøe A/S.

Methanol—steam-methane reforming

Application: To produce methanol from natural or associated gas feedstocks using advanced tubular reforming followed by boiling water reactor synthesis. This technology is an option for capacities up to approximately 3,000 mtpd methanol for cases where carbon dioxide (CO_2) is available. Topsøe also offers technology for larger scale methanol facilities up to 10,000 mtpd per production train and technology to modify ammonia capacity into methanol production.

Description: The gas feedstock is compressed (if required), desulfurized (1) and process steam is added. Process steam used is a combination of steam from the process condensate stripper and superheated medium pressure steam from the header. The mixture of natural gas and steam is preheated, prereformed (2) and sent to the tubular reformer (3). The prereformer uses waste heat from the flue-gas section of the tubular reformer for the reforming reaction, thus reducing the total load on the tubular reformer. Due to high outlet temperature, exit gas from the tubular reformer has a low concentration of methane, which is an inert in the synthesis. The synthesis gas obtainable with this technology typically contains surplus hydrogen, which will be used as fuel in the reformer furnace. If CO_2 is available, the synthesis gas composition can be adjusted hereby minimizing the hydrogen surplus. Carbon dioxide can preferably be added downstream the prereformer.

The flue gas generated in the tubular reformer is used for preheat of reformer and prereformer feed, natural gas preheat, steam superheat and preheat of combustion air. The synthesis gas generated in the tubular reformer is cooled by high-pressure steam generation (4), preheat of boiler feed water and reboiling in the distillation section (5).

After final cooling by air or cooling water, the synthesis gas is compressed (6) and sent to the synthesis loop (7). The synthesis loop comprises of a straight-tubed boiling water reactor, which is more efficient than adiabatic reactors. Reaction heat is removed from the reactor by generating MP steam. This steam is used for stripping of process condensate and thereafter as process steam. Preheating the reactor feed cools effluent from the synthesis reactor. Further cooling is obtained by preheating demineralized water and by air or water cooling. Raw methanol is separated and sent directly to the distillation section (5) featuring a very efficient three-column layout. Recycle gas is sent to the recirculator compressor (8) after a small purge to remove inert compound build-up.

Topsøe supplies a complete range of catalysts for methanol production. The total energy consumption for this process scheme is about 7.2 Gcal/ton methanol without CO_2 addition. With CO_2 addition, the total energy consumption can be reduced to 7.0 Gcal/ton methanol.

Economics: Tubular reforming technology is attractive at capacities >2,500–3,000 mtpd methanol, where the economy of scale of alternative technologies such as 2-step or autothermal reforming cannot be fully utilized.

Commercial plants: The most recent plant is a 3,030-mtpd methanol facility with CO_2 import and is scheduled to come onstream in 2003.

Licensor: Haldor Topsøe A/S.

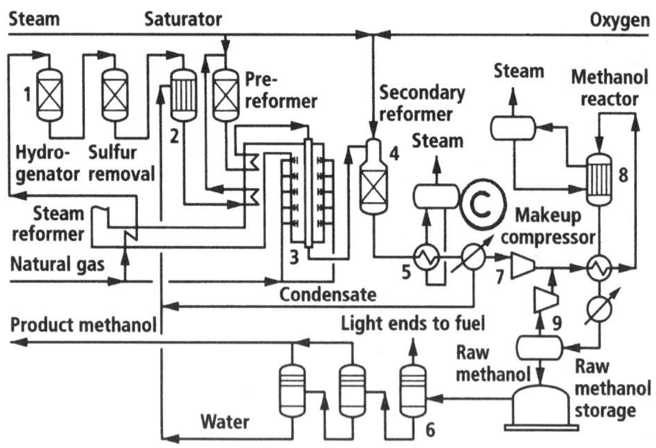

Methanol—two-step reforming

Application: To produce methanol from natural or associated gas feedstocks using two-step reforming followed by low-pressure synthesis. This technology is well suited for world-scale plants. Topsøe also offers technology for smaller as well as very large methanol facilities up to 10,000 tpd, and technology to modify ammonia capacity into methanol production.

Description: The gas feedstock is compressed (if required), desulfurized (1) and sent to a saturator (2) where process steam is generated. All process condensate is reused in the saturator resulting in a lower water requirement. The mixture of natural gas and steam is preheated and sent to the primary reformer (3). Exit gas from the primary reformer goes directly to an oxygen-blown secondary reformer (4). The oxygen amount and the balance between primary and secondary reformer are adjusted so that an almost stoichiometric synthesis gas with a low inert content is obtained. The primary reformer is relatively small and the reforming section operates at about 35 kg/cm^{2}g.

The flue gas' heat content preheats reformer feed. Likewise, the heat content of the process gas is used to produce superheated high-pressure steam (5), boiler feedwater preheating, preheating process condensate going to the saturator and reboiling in the distillation section (6).

After final cooling by air or cooling water, the synthesis gas is compressed in a one-stage compressor (7) and sent to the synthesis loop (8), comprised of three adiabatic reactors with heat exchangers between the reactors. Reaction heat from the loop is used to heat saturator water. Steam provides additional heat for the saturator system. Effluent from the last reactor is cooled by preheating feed to the first reactor, by preheating demineralized water and by air or water cooling. Raw methanol is separated and sent directly to the distillation (6), featuring a very efficient three-column layout. Recycle gas is sent to the recirculator compressor (9) after a small purge to remove inert compound buildup.

Topsøe supplies a complete range of catalysts that can be used in the methanol plant. Total energy consumption for this process scheme is about 7.0 Gcal/ton including energy for oxygen production.

Economics: Total investments, including an oxygen plant, are approximately 10% lower for large plants than for a conventional plant based on straight steam reforming.

Commercial plants: The most recent large-scale plant is a 3,030-tpd facility in Iran and is scheduled to be onstream in 2003.

Licensor: Haldor Topsøe A/S.

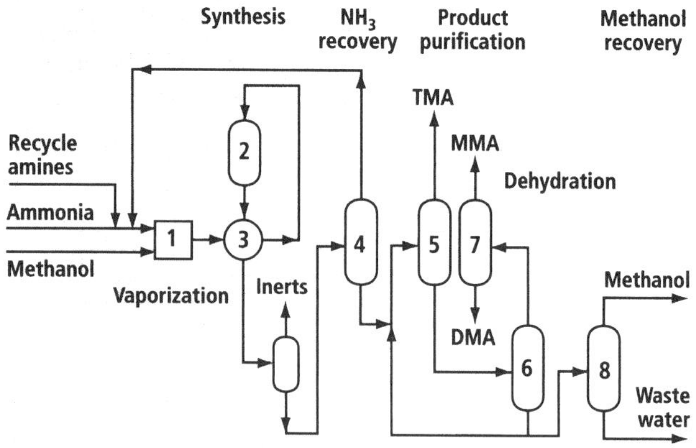

Methylamines

Application: To produce mono- (MMA), di- (DMA) and trimethylamines (TMA) from methanol and ammonia.

Description: Anhydrous liquid ammonia, recycled amines and methanol are continuously vaporized (1), superheated (3) and fed to a catalyst-packed converter (2). The converter utilizing a high-activity, low-byproduct amination catalyst simultaneously produces MMA, DMA and TMA. Product ratios can be varied to maximize MMA, DMA, or TMA production. The correct selection of the N/C ratio and recycling of amines produces the desired product mix. Most of the exothermic reaction heat is recovered in feed preheating (3). The reactor products are sent to a separation system where firstly ammonia (4) is separated and recycled to the reaction system. Water from the dehydration column (6) is used in extractive distillation (5) to break the TMA azeotropes and produce pure anhydrous TMA. The product column (7) separates the water-free amines into pure anhydrous MMA and DMA. Methanol recovery (8) improves efficiency and extends catalyst life by allowing greater methanol slip exit from the converter. Addition of a methanol-recovery column to existing plants can help to increase production rates.

Anhydrous MMA, DMA and TMA, can be used directly in downstream processes such as MDEA, DMF, DMAC, choline chloride and/or diluted to any commercial specification.

Yields: Greater than 98% on raw materials.

Economics: Typical performance data per ton of product amines having MMA/DMA/TMA product ratio of $1/3 : 1/3 : 1/3$

Methanol, t	1.38
Ammonia, t	0.40
Steam, t	8.8
Water, cooling, m^3	500
Electricity, kWh	20

Commercial plants: Twenty-six companies in 18 countries use this process with a production capacity exceeding 300,000 mtpy.

Licensor: Davy Process Technology, UK.

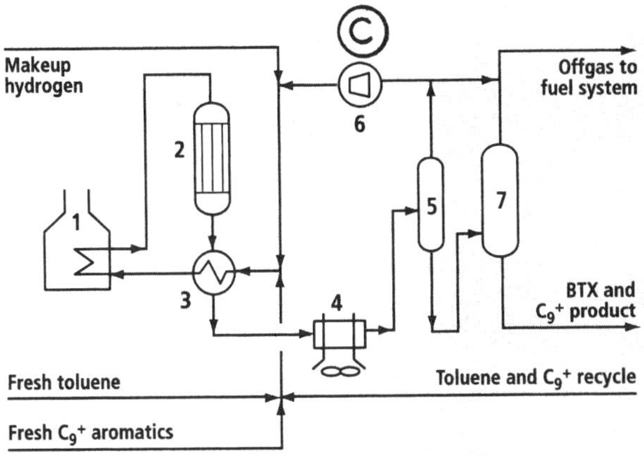

Mixed xylenes

Application: To convert C_9^+ heavy aromatics, alone or in conjunction with toluene or benzene co-feed, primarily to mixed xylenes using ExxonMobil Chemical's TransPlus process.

Description: Fresh feed, ranging from 100% C_9^+ aromatics to mixtures of C_9^+ aromatics with either toluene or benzene, are converted primarily to xylenes in the TransPlus process. Co-boiling C_{11} aromatics components, up to 435°F NBP, can be included in the C_9^+ feed. In this process, liquid feed along with hydrogen-rich recycle gas, are sent to the reactor (2) after being heated to reaction temperature through feed/effluent heat exchangers (3) and the charge heater (1).

Primary reactions occurring are the dealkylation of alkylaromatics, transalkylation and disproportionation, producing benzene/toluene and C_8 aromatics containing over 95% xylenes. The thermodynamic equilibrium of the resulting product aromatics is mainly dependent on the ratio of methyl groups to aromatic rings in the reactor feed. Hydrogen-rich gas from the high-pressure separator (5) is recycled back to the reactor with make-up hydrogen (6). Unconverted toluene and C_9^+ aromatics are recycled to extinction.

The ability of TransPlus to process feeds rich in C_9^+ aromatics enhances the product slate toward xylenes. Owing to its unique catalyst, long cycle lengths are possible.

Economics: Favorable operating conditions, relative to other alternative technologies, will result in lower capital and operating costs for grassroots units and higher throughput potential in retrofit applications.

Commercial plants: The first commercial unit was started up in Taiwan in 1997. Performance of this unit has been excellent.

Licensor: ExxonMobil Chemical, (retrofit applications); Axens, Axens NA (grassroots applications).

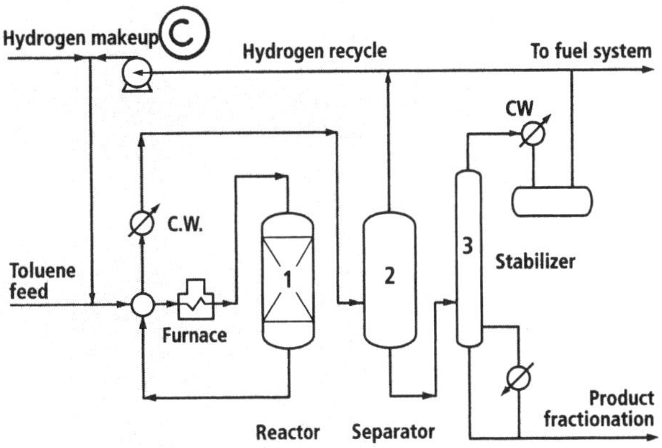

Mixed xylenes

Application: To selectively convert toluene to mixed xylene and high-purity benzene using ExxonMobil Chemical's Toluene DisProportionation 3rd Generation (MTDP-3) process.

Description: Dry toluene feed and up to 25 wt% C_9 aromatics along with hydrogen-rich recycle gas are pumped through feed effluent heat exchangers and the charge heater into the MTDP-3 reactor (1). Toluene disproportionation occurs in the vapor phase to produce the mixed xylene and benzene product. Hydrogen-rich gas from the high-pressure separator (2) is recycled back to the reactor together with makeup hydrogen. Unconverted toluene is recycled to extinction.

Reactor yields, wt%:

	Feed	Product
C_5 and lighter		1.3
Benzene		19.8
Toluene	100.0	52.0
Ethylbenzene		0.6
p-Xylene		6.3
m-Xylene		12.8
o-Xylene		5.4
C_9^+ aromatics		1.8
	100.0	100.0
Toluene conversion, wt%		48

Operating conditions: MTDP-3 operates at high space velocity and low H_2/hydrocarbon mole ratio. These conditions could potentially result in increased throughput without reactor and/or compressor replacement in retrofit applications. The third generation catalyst offers long operating cycles and is regenerable.

Economics: Estimated onsite battery limit investment for 1997 open shop construction at US Gulf Coast location is $1,860 per bpsd capacity.

Typical utility requirements, per bbl feed converted:

Electricity, kWh	3.0
Fuel, 10^3 kcal/hr	87.8
Water, cooling (10°C rise), cm/hr	0.3
Catalyst fill, lb/lb feed converted	1.5 3 10^{25}
Maintenance, per year as % of investment	2.0

Commercial plants: Four MTDP-3 licensees since 1995.

Reference: *Oil & Gas Journal,* Oct. 12, 1992, pp. 60–67.

Licensor: ExxonMobil Chemical (retrofit applications); Axens, Axens NA (grassroots applications).

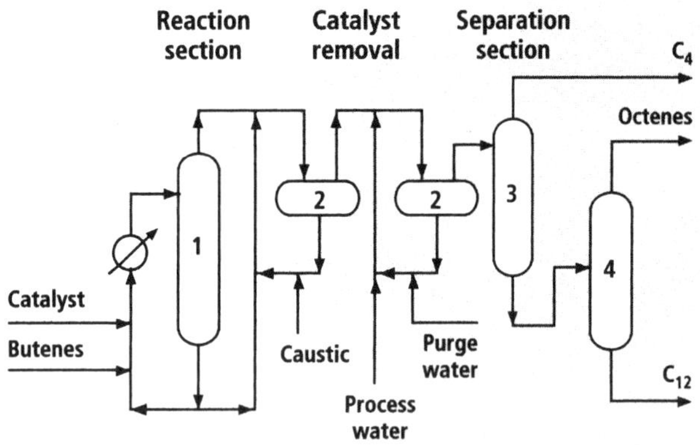

Octenes

Application: The Dimersol-X process transforms butenes to octenes, which are ultimately used in the manufacture of plasticizers via isononanol (isononyl alcohol) and diisononyl phthalate units.

Description: Butenes enter the Dimersol-X process, which comprises three sections. In the reactor section, dimerization takes place in multiple liquid-phase reactors (1) using homogeneous catalysis and an efficient recycle mixing system. The catalyst is generated in situ by the reaction of components injected in the recycle loop. The catalyst in the reactor effluent is deactivated in the neutralization section and separated (2). The stabilization section (3) separates unreacted olefin monomer and saturates from product dimers while the second column (4) separates the octenes. A third column can be added to separate dodecenes.

Yields: Nearly 80% conversion of n-butenes can be attained and selectivities toward octenes are about 85%. The typical C_8 product is a mixture having a minimum of 98.5% octene isomers with the following distribution:

n-Octenes	7%
Methyl-heptenes	58%
Dimethyl-hexenes	35%

Dimersol-X octenes exhibit a low degree of branching resulting in higher downstream oxonation reaction yields and rates, and better plasticizer quality.

Economics: Basis: ISBL 2002 for a Gulf Coast location using 50,000 tpy of a raffinate-2 C_4 cut containing 75% n-butenes.

Investment, US$ million	6
Typical operating cost, US$	60 per metric ton of octenes

Commercial plants: Thirty-five Dimersol units treating various olefinic C_3 and C_4 cuts have been licensed. Typical octenes production capacities range from 20,000 tpy up to 90,000 tpy.

Reference: Convers, A., D. Commereuc, and B. Torck, "Homogeneous Catalysis," IFP Conference.

Licensor: Axens, Axens NA.

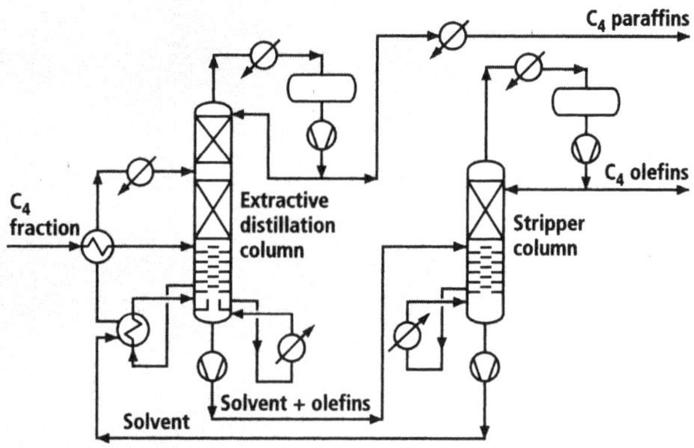

Olefins

Application: Separation of pure C_4 olefins from olefinic/paraffinic C_4 mixtures via extractive distillation using a selective solvent. BUTENEX is the Uhde technology to separate light olefins from various C_4 feedstocks, which include ethylene cracker and FCC sources.

Description: In the extractive distillation (ED) process, a single-compound solvent, N-Formylmorpholine (NFM), or NFM in a mixture with further morpholine derivatives, alters the vapor pressure of the components being separated. The vapor pressure of the olefins is lowered more than that of the less soluble paraffins. Paraffinic vapors leave the top of the ED column, and solvent with olefins leave the bottom of the ED column.

The bottom product of the ED column is fed to the stripper to separate pure olefins (mixtures) from the solvent. After intensive heat exchange, the lean solvent is recycled to the ED column. The solvent, which can be either NFM, or a mixture including NFM, perfectly satisfies the solvent properties needed for this process, including high selectivity, thermal stability and a suitable boiling point.

Economics:

Consumption per ton of FCC C_4 fraction feedstock:

Steam, t/t	0.5–0.8
Water, cooling (ΔT = 10°C), m^3/t	15.0
Electric power, kWh/t	25.0

Product purity:

n-Butene content	99.+ wt.-% min.
Solvent content	1 wt.-ppm max.

Installation: Two commercial plants for the recovery of n-butenes have been installed since 1998.

Reference: Preusser, G., "Separation of n-Butanes and Butene-2 by extractive distillation," Achema, June 1986, Frankfurt.

Licensor: Uhde GmbH.

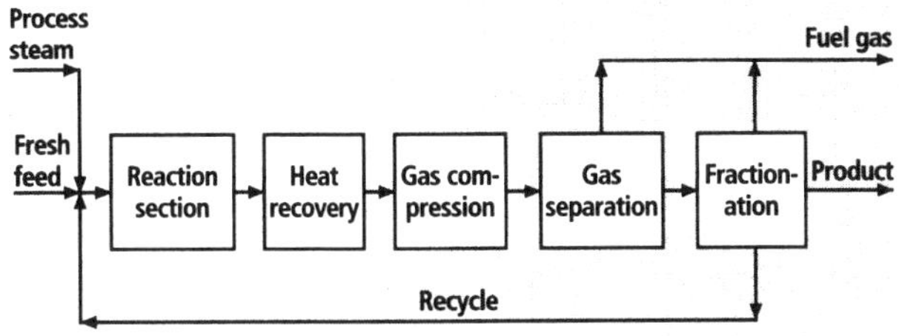

Olefins

Application: Dehydrogenation of C_4 or C_3 paraffins to pure olefins using steam-active reforming over a noble metal catalyst. STAR, the steam active reforming process, is the Uhde technology to dehydrogenate light paraffins into olefins.

Description: Fresh paraffin feed is combined with internally generated steam and passed after preheating to the reactor—a fixed-bed, tubular top-fired reformer type. Dehydrogenation reactions occurs at 4 to 6 bar at 500°C to 580°C. In a subsequent fixed-bed reactor, oxygen (or air) is admixed to enhance olefins yield by partial combustion of the hydrogen generated in the upstream reactor. The reaction section operates in sequential mode (7 hours on-stream, 1 hour regeneration). Product flow is balanced by a parallel reactor arrangement for continuous production.

After heat recovery, the gas is compressed, and the pure olefin product is separated from non-converted paraffins and light ends. Apart from fuel gas, which is used within the unit, high-purity olefin is the only product.

Economics:

Consumption per ton of propylene product based on standard grade propane feedstock:

Feedstock, t/t	1.20
Fuel, Gcal/t	1.18
Water, cooling (ΔT = 10 °C), m^3/t	200
Electric power, kWh/t	170

Product purity:

Propylene	99.70 wt.–% min.

Installation: Two commercial plants for the dehydrogenation of butane have been commissioned since 1992.

Reference: Thiagarajan, N., Ranke, U. and Ennenbach, F., "Propane/butane dehydrogenation by steam active reforming," Achema, May 2000, Frankfurt.

Licensor: Uhde GmbH.

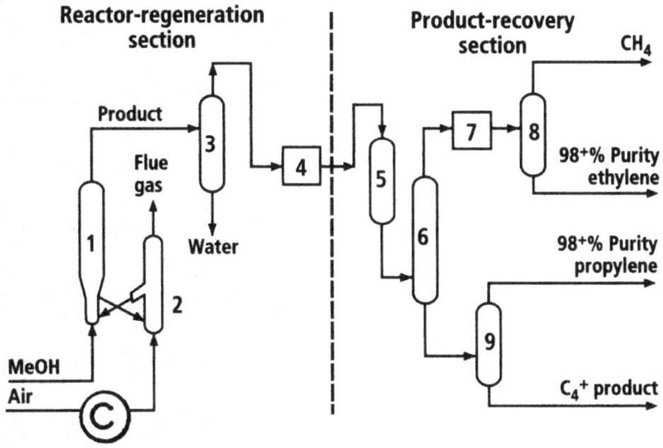

Olefins

Application: To produce ethylene, propylene and butenes from natural gas or equivalent, via raw methanol, using the UOP/Hydro MTO (methanol to olefins) process.

Description: This new process consists of a reactor section, a continuous catalyst regeneration section and product recovery section. One or more fluidized-bed reactors (1) are used with continuous catalyst transfer to and from the continuous catalyst regenerator (2). The robust regenerable MTO-100 catalyst is based on a nonzeolitic molecular sieve. Raw (nondewatered) methanol is fed to the low-pressure reactor (1), which offers very high (99%$^+$) conversion of the methanol with very high selectivity to ethylene and propylene. The recovery section design depends on product use, but will contain a product water recovery and recycle system (3), a CO_2 removal system (4), a dryer (5), a deethanizer (6), an acetylene saturation unit (7), a demethanizer (8), and a depropanizer (9). The process can produce 99$^+$% (polymer-grade) ethylene and propylene by adding simple product splitters to the recovery section.

Yields: The process gives very high total olefins yields. A typical product yield structure is shown based on 5,204 mt/d raw methanol feedrate to an MTO plant:

	Metric tons/day
Ethylene	882
Propylene	882
Total light olefins	1,762
Butenes	272
C_5^+	100
Fuel gas	88
Other (water, coke, CO_x)	2,980

The process is flexible. Ethylene to propylene product weight ratio can be modified between the range of 0.75 to 1.5 by altering reactor operating severity. The total yield of olefins varies slightly throughout this range.

Economics: The MTO process competes favorably with conventional liquid crackers due to lower capital investment. It is also an ideal vehicle to debottleneck existing ethylene plants and, unlike conventional steam crackers, the MTO process is a continuous reactor system with no fired heaters.

Commercial plants: NA.

Licensor: UOP LLC/Hydro.

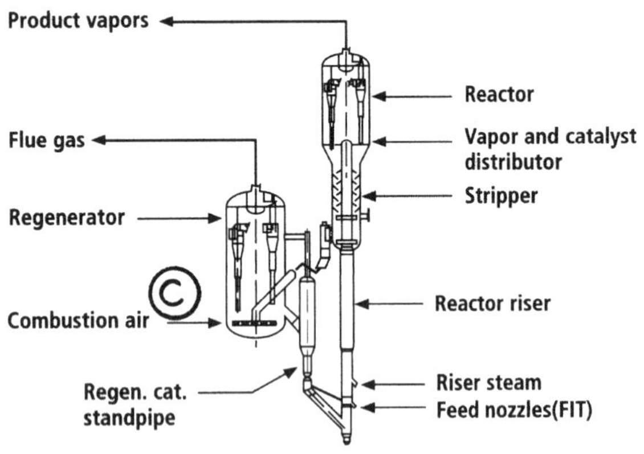

Olefins—catalytic

Application: To selectively convert vacuum gas oils and the resulting blends of each into C_2-C_5 olefins, aromatic-rich, high-octane gasoline and distillate using deep catalytic cracking (DCC) methods.

Description: DCC is a fluidized process to selectively crack a wide variety of feedstocks into light olefins. Propylene yields over 24 wt% are achievable with paraffinic feeds. A traditional reactor/regenerator unit design uses a catalyst with physical properties similar to traditional FCC catalyst. The DCC unit may be operated in two operational modes: maximum propylene (Type I) or maximum iso-olefins (Type II). Each operational mode utilizes unique catalyst as well as reaction conditions. DCC maximum propylene uses both riser and bed cracking at severe reactor conditions, while Type II utilizes only riser cracking like a modern FCC unit at milder conditions.

The overall flow scheme of DCC is very similar to a conventional FCC. However, innovations in catalyst development, process variable selection and severity enables the DCC to produce significantly more olefins than FCC in a maximum olefins mode of operation.

Products, wt% FF	DCC Type I	DCC Type II	FCC
Ethylene	6.1	2.3	0.9
Propylene	20.5	14.3	6.8
Butylene	14.3	14.6	11.0
in which $IC_4^=$	5.4	6.1	3.3
Amylene	—	9.8	8.5
in which $IC_5^=$	—	6.5	4.3

This technology is suitable for revamps as well as grassroot applications.

Commercial plants: Currently seven units are in operation, six in China and one in Thailand.

Reference: Chapin, Letzsch and Swaty, "Petrochemical options from deep catalytic cracking and the FCCU," 1998 NPRA Annual Meeting.

Licensor: Stone & Webster Inc., A Shaw Group Co./Research Institute of Petroleum Processing, Sinopec.

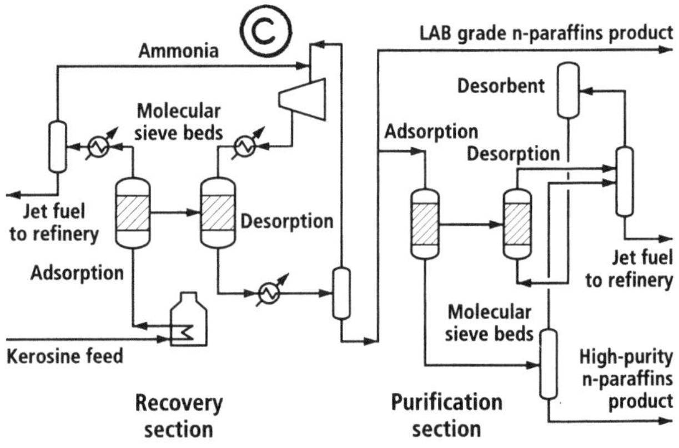

Paraffin, normal

Application: Efficient low-cost recovery and purification processes for the production of Lab-grade and/or high-purity n-paraffin products from kerosine.

Description: The ExxonMobil Chemical (EMC) process offers commercially proven technologies for efficient recovery and purification of high-purity n-paraffin from kerosine feedstock. Kerosine feedstocks are introduced to the recovery section where the n-paraffins are efficiently recovered from the kerosine stream in a vapor-phase fixed-bed molecular sieve adsorption process. In the process, the n-paraffins are selectively adsorbed on molecular sieve and subsequently desorbed with a highly effective desorbent.

The non n-paraffin hydrocarbons are rejected and returned to the refinery. The process provides a unique environment allowing the solid adsorbent to be very tolerant of sulfur compounds, which are typically present in kerosine feedstock. The adsorbent is therefore able to last long cycle lengths with a total life up to 20 years, as commercially demonstrated by ExxonMobil. In most cases, due to the high sulfur tolerance, the kerosine feedstock will not require hydrotreating pretreatment, which significantly reduces capital investment and operating cost. The recovery section produces LAB-grade n-paraffin product.

High-purity specialty grade n-paraffin products are produced in the ExxonMobil Purification process. The LAB-grade product from the recovery process is further processed in a purification section, where residual aromatics and other impurities are further reduced. Purification is accomplished in a liquid-phase, fixed-bed adsorption system. The impurities are selectively adsorbed on a molecular sieve, and subsequently removed with a hydrocarbon desorbent. The high-purity n-paraffins product is the highest quality available in the market. ExxonMobil commercially produces and markets n-paraffin product with aromatics content below 100 wtppm. The ExxonMobil n-paraffin technologies offer the industry's lowest capital and operating cost solutions and highest purity products for n-paraffin producers.

Product quality: Typical properties of high-purity n-paraffin product:

Purity, wt%	99
Aromatics, wt ppm	100
Bromine Index, mg/100g	<20
Sulfur, wt ppm	<1

Yield: Typically, over 99% of the n-paraffin contained in the kerosine stream is recovered.

Commercial plants: ExxonMobil Chemical has 40 years of experience in the production of n-paraffins and is the second largest producer in the world. ExxonMobil's n-paraffin plant at Baytown, Texas, produces high-purity product in a single train at a nameplate capacity of 250 million tpy.

Licensor: Kellogg Brown & Root, Inc.

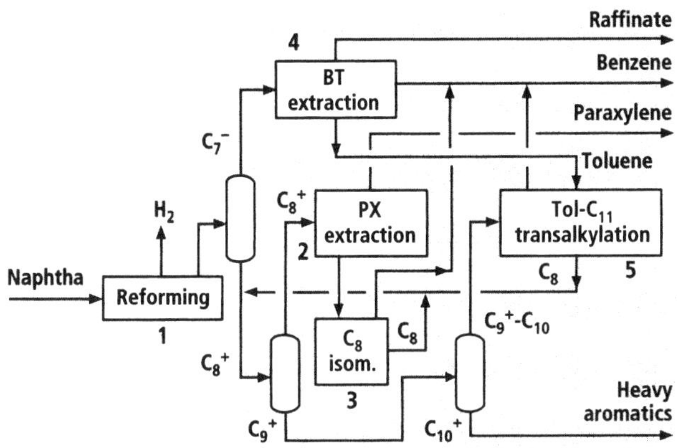

Paraxylene

Application: Suite of advanced aromatics technologies combined in the most effective manner to meet customers' investment and production objectives for paraxylene and benzene and are licensed under the name ParamaX.

Description: Aromatics are produced from naphtha in the Aromizing section (1), and separated by conventional distillation. The xylene fraction is sent to the Eluxyl unit (2), which produces 99.9% paraxylene via simulated countercurrent adsorption. The PX-depleted raffinate is isomerized back to equilibrium in the isomerization section (3) with either EB dealkylation-type (XyMax) processes or EB isomerization-type (Oparis) catalysts. High-purity benzene and toluene are separated from non-aromatic compounds with extractive distillation (Morphylane**) processes (4). Toluene and C_9 to C_{11} aromatics are converted to more valued benzene and mixed xylenes in the TransPlus* process (5), leading to incremental paraxylene production.

Eluxyl technology has the industrially proven ability to meet ultimate single train PX purity and capacities as high as 750,000 mtpy. Proprietary hybrid Eluxyl configurations integrate an intermediate purity adsorption section with a single-stage crystallization, ideal for retrofits. Axens is the licensor of all the technologies involved in the ParamaX suite.

* Mobil and ** Uhde technologies licensed by Axens for grassroots applications

Production: Typical paraxylene single train complex from naphtha to paraxylene, featuring Aromizing, Eluxyl, XyMax and TransPlus units.

	Thousand tpy
Feed—60–175 Arab light naphtha	1,378
Paraxylene	600
Benzene	168
Net producer of hydrogen	

Economics: The ISBL 2002 Gulf Coast location erected cost, including first load of catalysts and chemicals, with 30% allowance for offsites.

Investment, million US$	420
Annual utilities, catalyst and chemical operating cost (million US$/yr)	41

Commercial plants: Eight Eluxyl units have been licensed, accounting for 3 million tpy of paraxylene and three units that are in operation. Isomerization units: 19 EB isomerizing units and 21 ExxonMobil EB dealkylating units have been put into operation. One TransPlus unit is currently in operation.

Reference: Dupraz, C., et al., "Maximizing paraxylene production with ParamaX"; Hotier, G., and Methivier, A., "Paraxylene Production with the Eluxyl Process." AIChE 2002 Spring Meeting, New Orleans, March 2002.

Licensor: Axens, Axens NA.

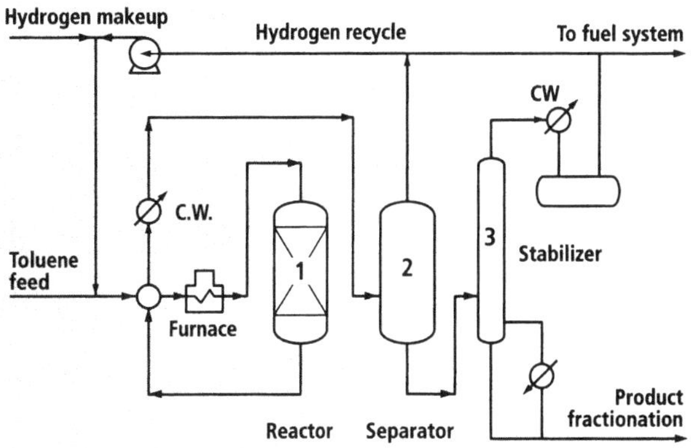

Paraxylene

Application: To selectively convert toluene to high-purity (90%+) paraxylene-rich (PX) xylenes and benzene using ExxonMobil Chemical's technologies—PxMax and ASTDP.

Description: Dry toluene feed and hydrogen-rich recycle gas are pumped through feed/effluent exchangers and charge heater and into the reactor (1). Selective toluene disproportionation (STDP) occurs in the vapor phase to produce the paraxylene-rich xylene and benzene co-product. Byproduct yields are small. Reactor effluent is cooled by heat exchange and liquid products are separated from the recycle gas. Hydrogen-rich gas from the separator (2) is recycled back to the reactor together with makeup hydrogen. Liquid product is stripped of remaining light gas in the stabilizer (3) and sent to product fractionation. Unconverted toluene is recycled to extinction.

The PxMax technology uses catalyst, which is ex-situ selectivated by pretreatment during catalyst manufacture. The ASTDP technology uses catalyst, which is in-situ coke selectivated. Both technologies provide significantly higher selectivity and longer operating cycles than other STDP technologies. Operating costs associated with downstream recovery are also reduced by the high paraxylene purity from PxMax and ASTDP.

Operating conditions: PxMax operates at lower start-of-cycle temperatures and lower hydrogen to hydrocarbon recycle ratios than other STDP technologies, resulting in longer cycles and lower utilities. By eliminating the in-situ selectivation step, the PxMax version of this technology results in simplified operation and lower capital costs. Both catalysts offer long operating cycles and are regenerable.

Commercial plants: There are seven MSTDP units (predecessor technology to PxMax), and ASTDP and four units using PxMax technology. The first two PxMax units started up in 1996 and 1997 at Chalmette Refining's Louisiana Refinery and Mobil Chemical's Beaumont plant, respectively.

Licensor: ExxonMobil Chemical (retrofit applications); Axens, Axens NA (grassroots applications).

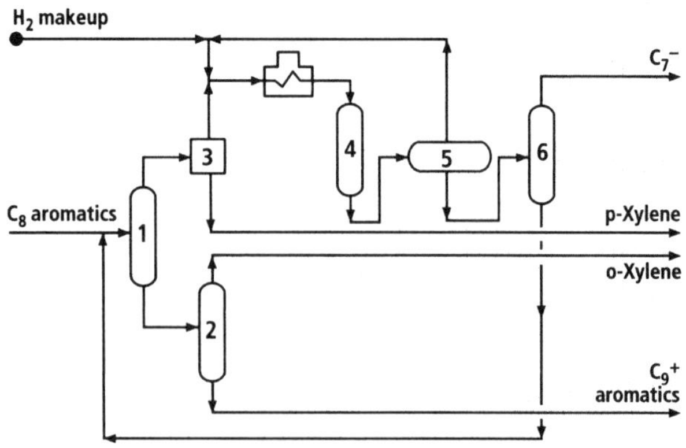

Paraxylene

Application: To produce a desired xylene isomer (or isomers) from a mixture of C_8 aromatics using the UOP Isomar and Parex processes.

Description: Fresh feed containing an equilibrium mixture of C_8 aromatic isomers is fed to a xylene splitter (1). Bottoms from the splitter are then separated (2) into an overhead product of o-xylene and a byproduct of C_9^+ aromatics. Overhead from the splitter is sent to a UOP Parex process unit (3) to recover ultra-high-purity p-xylene. If desired, high-purity m-xylene may also be recovered using the MX Sorbex process. Remaining components are recycled to the UOP Isomar process unit reactor (4) where they are catalytically converted back toward an equilibrium mixture of C_8 aromatic isomers. Hydrogen-rich recycle gas is separated (5) from the reactor effluent before fractionation (6) to remove light-cracked byproducts overhead. The remaining C_8 aromatics are then combined with the fresh feed and sent to the xylene splitter (1).

The feedstock consists of a mixture of C_8 aromatics typically derived from catalytically reformed naphtha, hydrotreated pyrolysis gasoline or an LPG aromatization unit. The feed may contain up to 40% ethylbenzene, which is converted either to xylenes or benzene by the Isomar reactor at a high-conversion rate per pass. Feedstocks may be pure solvent extracts or fractional heartcuts containing up to 25% nonaromatics. Hydrogen may be supplied from a catalytic reforming unit or any suitable source. Chemical hydrogen consumption is minimal.

o-Xylene product purity of up to 99% is possible, depending on the composition of the feed and fractionation efficiency. The Parex unit is capable of producing 99.9% pure p-xylene with per pass recovery greater than 97%.

Operating conditions: Moderate temperature and pressure requirements permit using carbon and low-alloy steel and conventional process equipment.

Yields: Typical mass balance for the Parex-Isomar complex:

Composition	Fresh feed, wt. units	Product, wt. units
Ethylbenzene	25.5	—
p-Xylene	14.0	71.1
m-Xylene	41.0	—
o-Xylene	19.5	19.6

Economics: Estimated inside battery limits (ISBL) erected and utility costs are given for a Parex-Isomar complex which includes the xylene splitter column and the o-xylene column, US Gulf Coast fourth quarter 2002.

Investment, US$ per mt of feed	94–108
Utilities, US$ per mt of p-xylene product	30

Commercial plants: Since 1971, UOP has licensed 73 Parex units and 56 Isomar units.

Licensor: UOP LLC.

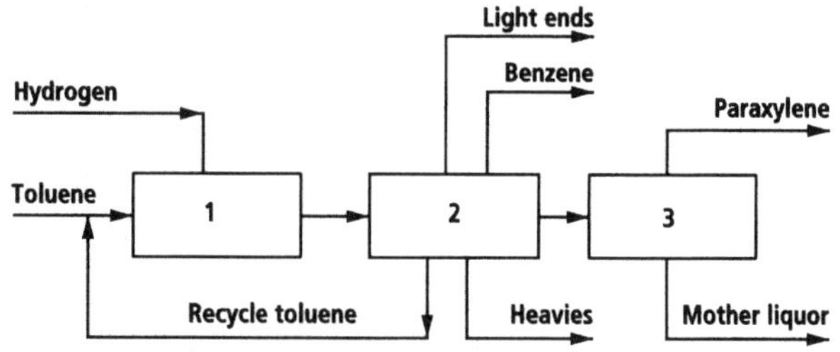

Paraxylene

Application: The PX-Plus XP Process converts toluene to paraxylene and benzene. The paraxylene is purified to 99.9+ wt% via single-stage crystallization and a wash column. The benzene purity is 545-grade by fractionation.

Description: The PX-Plus XP Process is composed of three processing steps:

- (1) Selective toluene disproportionation, via the PX-Plus Process
- (2) Fractionation for recovery of recycle toluene and benzene product
- (3) The Badger/Niro paraxylene crystallization process, where single-stage crystallization and crystal wash columns are used.

In the PX-Plus technology, fresh toluene is combined with recycle gas, heated and fed to a fixed-bed reactor. The para-selective catalyst produces xylene product with 90% paraxylene in the xylenes. Reactor effluent flows to a separator, where the recycle gas is recovered, and the liquid product is sent to a stripper.

In the fractionation section, stripper bottoms are fed to a benzene column, where the benzene product is recovered and the unconverted toluene is fractionated for recycle. The toluene column bottoms are sent to a rerun column where the paraxylene concentrated fraction is taken overhead.

In the Badger/Niro crystallization unit, the xylenes are fed to a single-stage crystallization section that uses continuous suspension crystallization. In this section, the paraxylene is purified with a single refrigerant compressor system, and the mother liquor rejected. The purified paraxylene is fed to a Niro wash column section where ultra-high-purity paraxylene is produced by countercurrent crystal washing.

Components of this flexible technology are especially suited for capacity expansion of existing paraxylene production facilities.

Yields:

Toluene conversion per pass	30%
Paraxylene yield, wt%	40
Benzene yield, wt%	45
Light ends, wt%	<6
Paraxylene recovery	93.5%
Paraxylene purity, wt%	99.9

Economics: Capital investment per mty of paraxylene product

EEC, US$	200
Utilities per mt of paraxylene product	
Electricity, kWh	87
Steam, HP, mt	0.7
Steam, LP, mt	0.07
Water, cooling, m^3	15
Fuel, MMkcal	1.2

Commercial plants: Licenses include: one PX-Plus unit licensed and two Badger/Niro Crystallization Process units. Also, process packages produced for three other Badger/Niro Crystallization projects.

Licensor: UOP LLC, The Badger Technology Center of Washington Group International, and Niro Process Technology B.V.

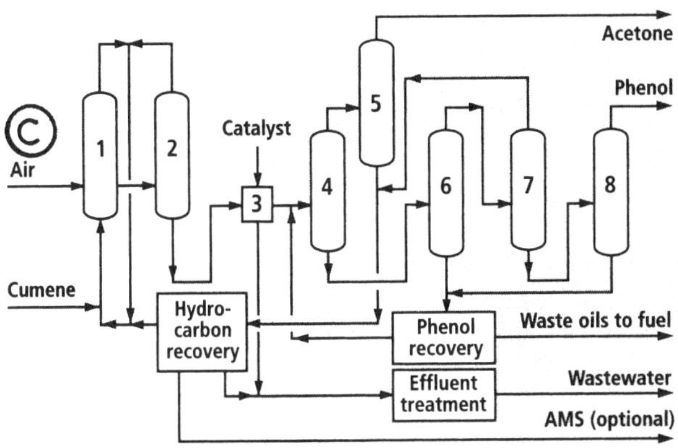

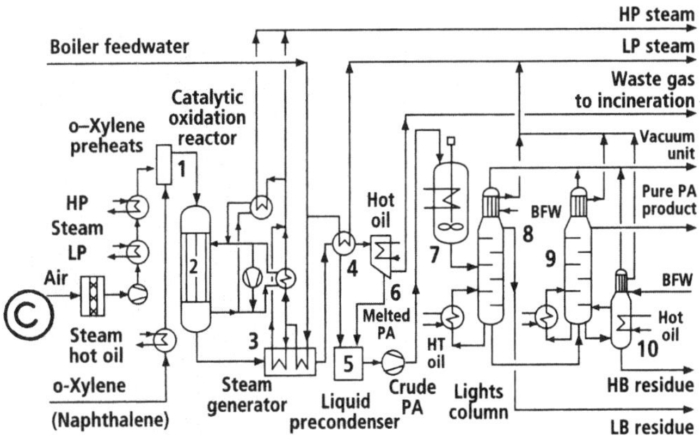

Phenol

Application: A high-yield process to produce high-purity phenol and acetone from cumene with optional byproduct recovery of alpha methylstyrene (AMS) and acetophenone (AP).

Description: Cumene is oxidized (1) with air at high efficiency (+95%) to produce cumene hydroperoxide (CHP), which is concentrated (2) and cleaved (3) under high-yield conditions (+99%) to phenol and acetone in the presence of an acid catalyst. The catalyst is removed and the cleavage mixture is fractionated to produce high-purity products (4–8), suitable for all applications. AMS is hydrogenated to cumene and recycled to oxidation or optionally recovered as a pure byproduct. Phenol and acetone are purified. A small aqueous effluent is pretreated to allow efficient biotreatment of plant wastewater. With AMS hydrogenation, 1.31 tons of cumene will produce 1 ton of phenol and 0.615 tons of acetone. This high-yield process produces very high-quality phenol and acetone products with very little heavy and light-end byproducts. With over 40 years of continuous technological development, the Kellogg Brown & Root (KBR) phenol process features low cumene and energy consumptions, coupled with unsurpassed safety and environmental systems.

Commercial plants: Thirty plants worldwide have been built or are now under construction with a total phenol capacity over 2.8 MMtpy. Four grassroots phenol plants, using KBR's phenol technology were engineered and/or started up in 1999/2002: An additional three plants are in design. More than 50% of the world's phenol is produced via the KBR process.

Reference: *Hydrocarbon Engineering,* December/January 1999.

Licensor: Kellogg Brown & Root, Inc.

Phthalic anhydride

Application: To produce phthalic anhydride (PA) from o-xylene, naphthalene or mixtures of both feedstocks using a fixed-bed vapor phase process, originally known as the von Heyden Process.

Description: Air is heated and loaded with evaporated (1) o-xylene and/or naphthalene. The hydrocarbon/air mixture enters a multitubular reactor (2) containing catalyst. An agitated salt melt removes the heat of reaction and maintains constant temperature conditions. Reaction heat generates high-pressure steam.

Modern plants operate with o-xylene feedstock loadings of 90–100 g/Nm^3 air. The loadings of 100 g/Nm^3 air in an adiabactic post-reactor is recommended, which is installed in the enlarged gas cooler casing (3). Reactor effluent gas is precooled in a gas cooler (3) before part of the PA vapor is condensed to a liquid in the precondensor (4) and is continuously discharged to the crude PA tank (5). The remainder of about 65 g PA/m^3 in the reaction gas is condensed as solid sublimate in switch condensors (6) on specially designed finned tubes. The switch condensors are periodically cooled and heated in a discontinuous operation of an automated switching cycle using heat transfer oil circuits. During the heating phase, solid PA is melted from the condensor tubes and discharged as a liquid to crude PA tank. Effluent gas is vented to the atmosphere after water scrubbing and/or incineration.

The crude PA is thermally pretreated (7) and then fed to the vacuum distillation system. Low boiling (LB) impurities are removed in the lights column (8) as LB residues. The high-boiling (HB) residue from the pure PA column (9) is sent to the residue boil-out vessel for PA recovery. Pure PA obtained as a distillate can be stored either in the molten state or flaked and bagged.

Catalyst: Special high-performance catalysts oxidize o-xylene as well as naphthalene and mixtures of both feedstocks in any proportions. All catalysts are ring-shaped.

Yield: 110–112 kg PA from 100 kg of pure o-xylene; 97–99 kg PA from 100 kg of pure naphthalene.

Economics: Excellent energy utilization and minimized offgas volume are due to high hydrocarbon/air ratio. Plants can be designed to operate independently of external power supply and export electric energy or HP steam.

Commercial plants: More than 110 plants with typical production capacities of 20,000–75,000 tpy, with a maximum capacity of 140,000 tpy, have been designed and built by Lurgi.

Licensor: BASF AG; **contractor:** Lurgi Oel Gas Chemie GmbH.

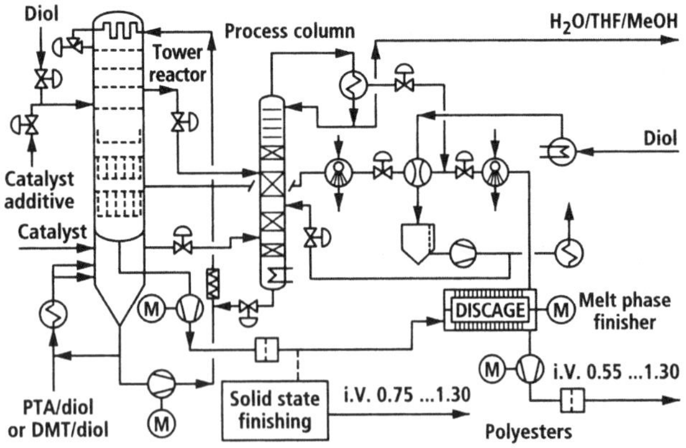

Polyalkylene terephthalates—PET, PBT, PTT, PEN

Application: New process to produce polyesters from the polyalkylene terephthalate family—from terephthalic acid (PTA) or dimethyl terephthalate (DMT) and diols using the IF proprietary two-reactor (2R)-process consisting of tower reactor ESPRE and DISCAGE finisher or alternatively a solid-state finishing.

Description: A slurry composed of a carboxylic acid or its methyl ester and a diol is prepared at a low mole ratio. The slurry is fed to the tower reactor's bottom where the main esterification or ester-interchange occurs under pressure or a vacuum at temperatures ranging between 170°C to 270°C. This reaction may be catalyzed or autocatalyzed.

Monomer is transferred via a pressurized pipe to the reactor top where reaction-side products are flashed out. Sudden postreaction occurs; thus raising conversion by 10%. Higher conversion rates (97–99%) are achieved by a sequential cascade of four to six reaction cups at decreasing pressures and increasing temperatures.

Stirring and intermixing are done by the reaction vapors while passing through the cups. A precondensate with i.V.s of 0.28 to 0.35 is obtained after surface-active-film evaporation—done as a twin assembly, under vacuum and higher temperature.

The prepolymer may be finished in the melt phase with IF's DISCAGE-reactor or in a solid-stating unit to obtain the required end-product features.

A process column separates side-reaction low boilers from the diol, which is then recycled back to reaction. Spray-condensers and vacuum units recover unreacted feedstock and recycle the diol; thus improving economics of this process.

Economics: This new process reduces processing costs by more than 25% as compared to conventional/historical processes due to its compact design, low-energy input, short-term reaction and agitator-less design. A product yield of more than 99.5% is attainable.

Energy costs can be reduced by more than 20%. Additionally, the end-product's quality is improved by eliminating intermediate product lines; it has a high uniformity of hold-up, as well as surface-intense, fast reaction.

Installations: Three commercial units with a total operating capacity of 600 mtpd and one pilot unit of 1 mtpd.

References: "Compact continuous process for high viscosity PBT," Polyester 2000, Fifth World Congress, Zürich.

Licensor: INVENTA-FISCHER.

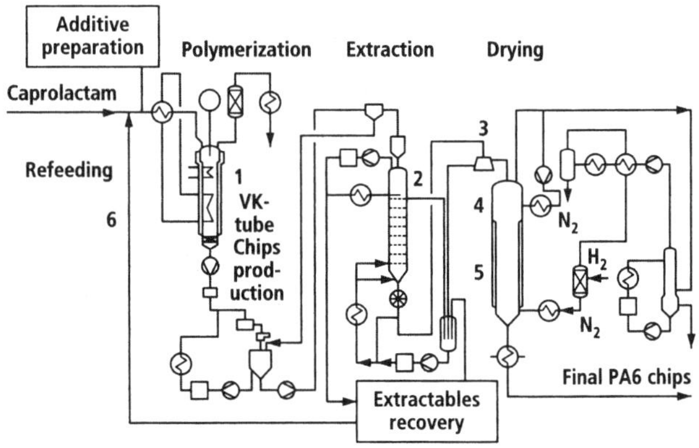

Polycaproamide

Application: INVENTA-FISCHER's VK-tube process polymerizes e-caprolactam (LC) monomer to produce polycaproamide (nylon-6) chips.

Description: Liquid LC is continuously polymerized in a VK-tube (1) in the presence of water, stabilizer and modifying additives at elevated temperatures. The polymerization process has proven to be very reliable, easy to operate and economical. Prepolymerization is available to reduce reactor volume for large capacity units. The polycaproamide chips are formed from the melt using strand cutters and are conveyed to the extraction column (2).

The chips—containing about 9% of monomer and cyclic oligomers—are treated with hot water in the extraction column. The extractables are, therefore, removed to a very large extent, to achieve a good polymer quality and high performance when processing further. Wet chips are sent to the centrifuge (3) and dried by hot, dry nitrogen in a two-zone dryer (4, 5). The nitrogen gas is regenerated in separate cycles. In the bottom zone of the dryer, the chips are cooled by means of a heat exchanger. The drying unit can be extended to a solid state postcondensation, i.e., drying and solid-state postcondensation take place in one process stage. Thus, high-viscosity chips for industrial yarns, films and foils can be produced.

Low utility and energy consumption are achieved by using closed circuits of water and nitrogen as well as by recovering heat. The recovery process for the recycling of the extractables reduces raw material cost. The extract water is concentrated and directly re-fed (6) to the polymerization. Alternatively, the concentrated extract is fed to a separate, specially designed, continuous re-polymerization unit.

Batch and continuous process units are available to meet all potential requirements regarding polymer grades as well as regarding flexibility in output rates and capacities. Special attention is devoted during plant design to attain minimal operating expenses for raw material, utilities and personnel.

Licensor: INVENTA-FISCHER.

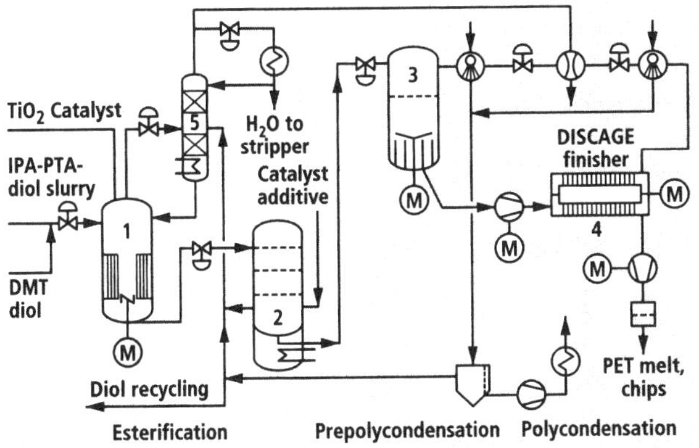

Polyesters (polyethylene terephthalate)

Application: To produce polyesters for resin and textile applications from terephthalic acid (PTA) or dimethyl terephthalate (DMT) and diols {ethylene glycol (EG) or others,} using the IF-proprietary four-reactor (4R)-process including DISCAGE-finisher.

Description: A slurry composed of PTA and EG, or molten DMT and EG is fed to a first esterification/ester-interchange reactor (1) in which main reaction occurs at elevated pressure and temperatures (200°C to 270°C). Reaction vapors—water or methanol—are sent to a low/high boiler separation column. High boilers are re-used as feedstock.

The oligomer is sent to a second cascaded, stirred reactor (2) with a lower pressure and higher temperature. The reaction conversion continues to more than 97%. Catalyst and additives may be added. Reaction gases are sent to the process column (5). The oligomer is then prepolymerized by a third cascaded reactor (3) under sub-atmospheric pressure and increased temperatures to obtain a degree of polycondensation >20. Final polycondensation to applicable intrinsic viscosities of up to i.V. = 0.9 is done in the DISCAGE-finisher (4). Pelletizing or direct-melt usage is optional.

EG is recovered by condensing process vapors at vacuum conditions. Vacuum generation may be either done by reaction gas as a motive stream or by the diol (EG). The average product yield may exceed 99%.

Economics: Typical utility requirements per mt of PET are:

Electricity, kWh	55.0
Fuel-oil, kg	61.0
Nitrogen, Nm³	0.8
Air, Nm³	9.0

Commercial plants: Thirteen lines with processing capacities ranging from 100 to 700 mtpd are operating; more than 50 Polyester-CP plants have been built worldwide. Presently, 700-mtpd lines are in operation as single-train lines, including a single finisher.

Licensor: INVENTA-FISCHER.

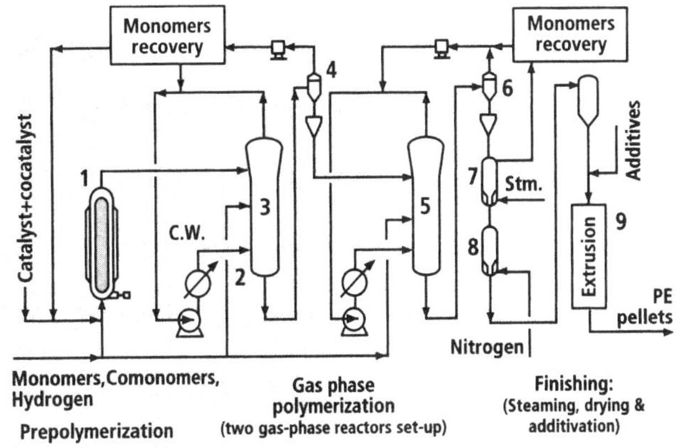

Polyethylene

Application: *Spherilene* gas-phase technology to produce very low-density polyethylene (VLDPE), linear-low-density polyethylene (LLDPE) as well as high-density polyethylene (HDPE) and other speciality types using a single Ziegler-Natta titanium-based catalysts family.

Description: Catalyst components are mixed and fed directly to prepolymerization (1) with a light inert hydrocarbon, where a first bulk polymerization occurs under mild controlled conditions. This step exploits the catalyst system potential in terms of morphology, mileage and complete reliability in the following gas-phase reaction sections. The slurry flows continuously into the first gas-phase reactor (GPR) (3). Reactor gas is circulated at high speed by a centrifugal compressor through a distribution grid. A cooler on the circulation gas loop (2) removes the reaction heat.

Polymer quality and reaction rate are controlled by gas composition, monomer feedrate and residence time. Product is continuously discharged from the first GPR via a proprietary device to a second GPR (5) with similar configuration. Resultant discharged gas is recovered, and no gas enters the second GPR, due to a proprietary "lock-hopper" system (4). Thus, an independent gas composition can be built up and maintained in each GPR, allowing growth to a different, if required, polymer within the polymeric matrix resulting from the first stage. Pressure and temperature in the GPRs are also independently controlled; no additional feed of catalytic components is required.

The polymer is then discharged in a receiver recovering the resultant gas (6) and to a proprietary unit for monomer stripping and catalyst deactivation in the polymer spheres (7). Residual hydrocarbons are stripped out and recycled to reaction, while the polymer is dried by a closed-loop nitrogen system (8) and, free from volatile substances, it is sent to liquid and/or solid additives incorporation step before extrusion (9).

Products: Product density is controlled from very-low-density polyethylene (ULDPE) < 900 g/ml to HDPE > 960 g/ml. Melt index (MI) capability ranges from 0.01 to > 100 g/10 min.

Because of the two GPR set-up, the *Spherilene* technology enables production of bimodal grades (MI, density) and speciality polymers based on new molecular models. For example, "high-performance" (HP) terpolymer grades guarantee superior properties to those of conventional butene LLDPE/VLDPE, and enhanced hexene HP "quattropolymer" grades are optimal economic choices to replace hexene/octene-based LLDPE. LDPE replacement grades for clarity and shrink film applications have also been developed.

Economics: Consumption, per metric ton of LLDPE:

Ethylene and comonomer, t	1.005
Electricity, kWh	410
Steam, kg	200
Cooling water, ΔT = 10°C, mt	150

Commercial plants: Licensed from 1992, eight plants using *Spherilene* process and technology have been licensed, with a total capacity of 1.8 million tpy. Single-line capacity ranges from 100 to 300 thousand tpy.

Licensor: Basell Polyolefins.

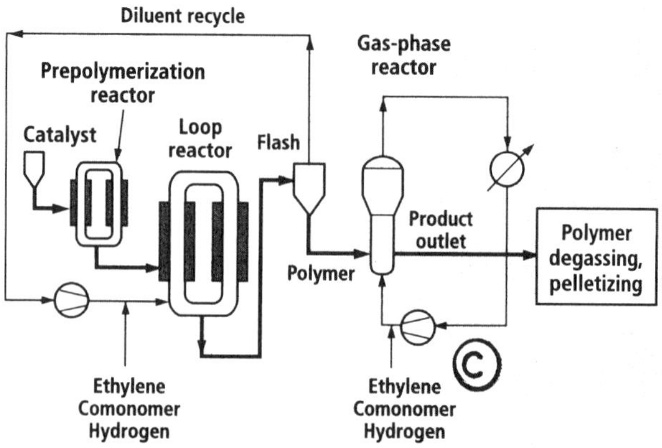

Polyethylene

Application: The Borstar polyethylene (PE) process can produce a broad range of bimodal and unimodal linear-low-density polyethylenes (LLDPE), medium-density polyethylenes (MDPE) and high-density polyethylenes (HDPE). The Borstar PE consists of a loop reactor and a gas-phase reactor in series.

Description: PEs with densities of 918–970 kg/m^3 and melt flowrate of 0.1–100 can be produced. Ziegler Natta catalysts are used, but single-site catalysts will be introduced for PE processing.

Catalyst is mixed with a propane diluent and fed to the compact pre-polymerization reactor. Cocatalyst, ethylene and hydrogen are also fed into this reactor. The pre-polymerized slurry is sent to the loop reactor, which is designed for supercritical conditions and operates at 75–100°C and 55–65 bar. Typically, lower molecular weight and higher density components of the bimodal polymers are produced in the loop reactor.

Diluent and unreacted components are separated from the polymer produced in the loop reactor in a flash tank. The loop reactor and gas-phase reactor are fully independent from each other, which enables easy control of reactor conditions and the flexibility to process a wide product range.

Polymer from the flash tank polymerizes further in a fluidized-bed gas-phase reactor. No fresh catalyst is needed. The polymer continues to grow on the same catalyst particles, resulting in a homogeneous polymer. The gas-phase reactor operates at 75–100°C and 20 bar. Fresh ethylene, comonomer and hydrogen are fed into the reactor. Typically, high molecular weight components are produced in this step, thus broadening the molecular weight distribution (MWD) and providing strength to the polymer. The production rate ratio between the reactors can be adjusted to meet the targeted product properties. After removing hydrocarbon residuals, the polymer powder is transferred to extrusion.

Products: A wide range of bimodal and unimodal products with a full control of comonomer distribution, can be produced, with densities ranging from 918 kg/m^3 to 970 kg/m^3 and melt flowrate from less than 0.1 to over 100. The MWD can be controlled from narrow to broad. Advanced properties are tailor-made for different applications such as pipe strength, film-bubble stability as well as high ESCR and stiffness in blow molding. Other special applications include extrusion coating and wire & cable.

Commercial plants: The first commercial scale Borstar PE plant is a 160,000-tpy facility operating at Borealis in Finland since 1995. A retrofitted HDPE Borstar plant at Stenungsund, Sweden, added 225,000-tpy capacity in 2000. In December 2001, Abu Dhabi Polymers Co. Ltd. (Borouge) brought onstream two lines producing 450,000 tpy in Abu Dhabi, the UAE. A fifth 250,000-tpy plant started up in 2002 in China for Shanghai Petrochemicals (SPC), thus increasing the total Borstar PE capacity to 1.085 million tpy. Maximum single-line design capacity is 350,000 tpy.

Licensor: Borealis A/S.

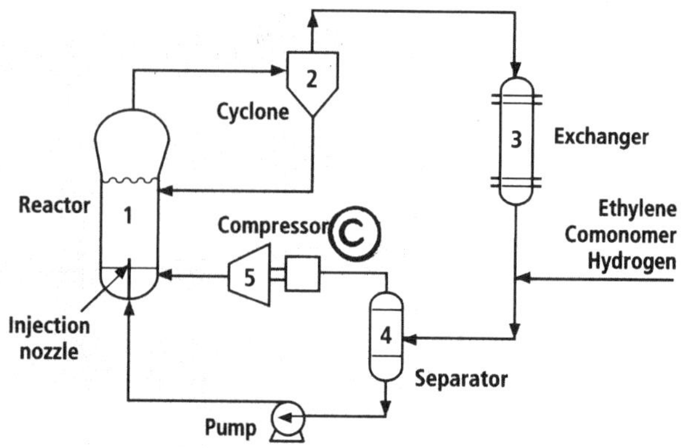

Polyethylene

Application: The Innovene gas phase process produces linear-low-density polyethylene (LLDPE) and high-density polyethylene (HDPE) using either Ziegler-Natta, chromium or metallocene catalysts.

Description: Ziegler-Natta and metallocene catalysts are directly injected into the reactor from storage whereas chromium catalysts are injected following activation of the catalyst via BP proprietary technology. The BP catalyst portfolio enables the production of a full-range of PE products with the same swing reactor using these three main catalyst families.

Accurate control of all the product properties, such as density and melt index, is achieved by continuous and automatic adjustment of the process gas composition and operating conditions. The reactor (1) is designed to ensure good mixing and a uniform temperature. Operating conditions within the bed are mild; the pressure is about 20 bar g, and the temperature between 75°C and 110°C. Polymer particles grow in the fluidized bed reactor where the fluidization gas is a mixture of ethylene, comonomer, hydrogen and nitrogen. Fine particles leaving the reactor with the exit gas are collected by cyclones, (2) which are unique to the Innovene gas-phase technology and recycled to the reactor. This feature ensures that fine particles do not circulate in the reaction loop, where they could foul the compressor, exchanger and reactor grid. The cyclones also prevent product contamination during transitions. Unreacted gas is cooled (3) and separated from any liquid (4), compressed (5) and returned to the reactor, maintaining the growing polymer particles at the desired temperature. Catalysts are incorporated into the final product without any catalyst removal step.

The reactor and almost all other equipment is made from carbon steel. Polymer powder is withdrawn from the reactor via a proprietary lateral discharge system and separated from associated process gas in a simple degassing stage using hot recirculating nitrogen. The powder is then pneumatically conveyed to the finishing section where additives are incorporated before pelletisation and storage.

Economics: The low-pressure technology and ease of operation ensures that the Innovene process is inherently safe, best-in-class environmentally and economically attractive with regard to both investment capex and opex.

Products: A wide range of LLDPE and HDPE products can be produced within the same reactor. LLDPE is used in film, injection molding and extrusion applications and can be made using either butene or hexene as the comonomer. Narrow molecular weight HDPE provides superior injection molding and rotational molding grades whereas broad molecular weight HDPE is used for blow molding, pipe, film and other extrusion applications.

Commercial plants: Thirty-five reactor lines are operating, in design or under construction worldwide representing around 6 MMtpy production with capacities ranging from 50,000 tpy to 350,000 tpy. Designs up to 450,000 tpy are also available.

Licensor: BP.

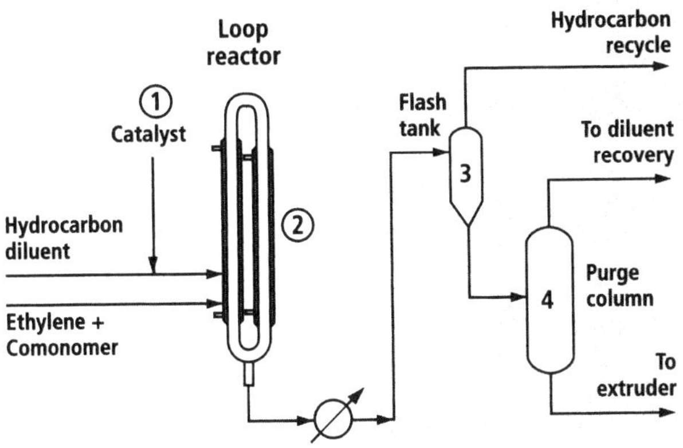

Polyethylene

Application: Produce linear polyethylene (LPE) using the Chevron Phillips Chemical Co., LP, LPE process.

Description: Polyethylene resins ranging in melt indices from less than 2 HLMI to greater than 50 MI, densities from slightly less than 0.920 to 0.970 gr/cc, and molecular weight distribution (MWD) from very narrow to very broad are produced by the Phillips LPE process. Polymerization occurs in an isobutane slurry using very high activity proprietary catalysts (1) in a loop reactor (2). Melt index and MWD are controlled by catalyst, operating conditions and hydrogen. Density is controlled by comonomer (hexene-1) incorporation.

High-activity catalysts eliminate the need for catalyst removal. No waxes or other byproducts are formed during polymerization, thereby minimizing environmental emissions.

Ethylene, isobutane, comonomer and catalyst are continuously fed to the loop reactor where polymerization occurs at temperatures lower than 100°C and pressures of approximately 40 kg/cm^2 and residence times of approximately one hour. Ethylene conversion exceeds 97% per pass. Reactor effluent is flashed to separate the solid resin from the gaseous stream (3). Polyethylene powder is purged (4) with nitrogen to remove traces of hydrocarbons and pneumatically conveyed to the extrusion area for stabilization and pelletizing. The gaseous stream is compressed, purified and recycled back to the reactor.

Products: Homopolymers and copolymers are produced for applications in film, blow molding, injection molding, roto molding, pipe, sheet and thermoforming, and wire and cable coating.

Economics: Representative of conditions for both homopolymer and copolymer production:

Typical raw material and utility requirements, per metric ton of pelleted resin:

Ethylene,* mt	0.99–1.005
Catalyst and chemicals, US$*	2.00–10.00
Steam, mt	0.25
Electricity, kWh	350
Water, cooling (circulating), mt	185
Nitrogen, Nm^3	30

*Depends on product slate

Commercial plants: Eighty-six reactor lines are either in operation or construction worldwide and account for 34% of worldwide capacity.

Licensor: Chevron Phillips Chemical Co., LP.

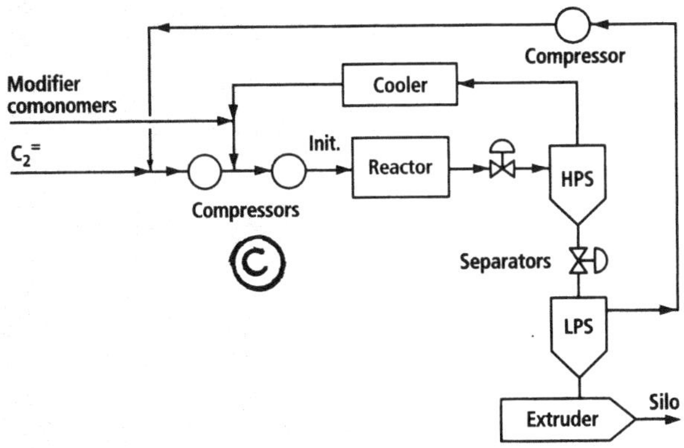

Polyethylene

Application: To produce low density polyethylene (LDPE) homopolymers and EVA copolymers using the high pressure free radical process. Large scale tubular reactors with a capacity in the range of 130 to 400 Mtpy, as well as stirred autoclave reactors with capacity around 100 Mtpy can be used.

Description: A variety of LDPE homopolymers and copolymers can be produced on these large reactors for various applications including films, molding and extrusion coating. The melt index, polymer density and molecular weight distribution are controlled with temperature profile, pressure, initiator and comonomer concentration. Autoclave reactors can give narrow or broad molecular weight distribution depending on the selected reactor conditions, whereas tubular reactors are typically used to produce narrow molecular weight distribution polymers.

Gaseous ethylene is supplied to the battery limits and boosted to 300 bar by the primary compressor. This makeup gas, together with the recycle gas stream, is compressed to reactor pressure in the secondary compressor. The tubular reactors operate at pressures up to 3,000 bar, whereas autoclaves normally operate below 2,000 bar. The polymer is separated in a high- and low-pressure separator; nonreacted gas is recycled from both separators. Molten polymer from the low-pressure separator is fed into the extruder; polymer pellets are then transferred to storage silos.

The main advantages for the high pressure process compared to other PE processes are short residence time and the ability to switch from homopolymers to copolymers incorporating polar comonomers in the same reactor. The high-pressure process produces long-chain, branched products from ethylene without expensive comonomers that are required by other processes to reduce product density. Also, the high pressure process allows fast and efficient transition for a broad range of polymers.

Products: Polymer density in the range 0.912 up to 0.935 for homopolymers; the melt index may be varied from 0.2 to greater than 150. Vinylacetate content up to 30 wt%.

Economics:

Raw materials and utilities, per metric ton of pelletized polymer:

Ethylene, ton/ton	1.008
Electricity, kWh	800
Steam, ton/ton	0.35
Nitrogen, Nm^3/t	5

Commercial plants: ExxonMobil Chemical Co. and its affiliates operate 23 high-pressure reactors on a worldwide basis with a capacity of approximately 1.7 MMtpy. Homopolymers and a variety of copolymers are produced. Since 1996, ExxonMobil Chemical Co. has sold licenses with a total installed capacity (either in operation or under construction) of approximately 1 million tpy.

Licensor: ExxonMobil Chemical Co.

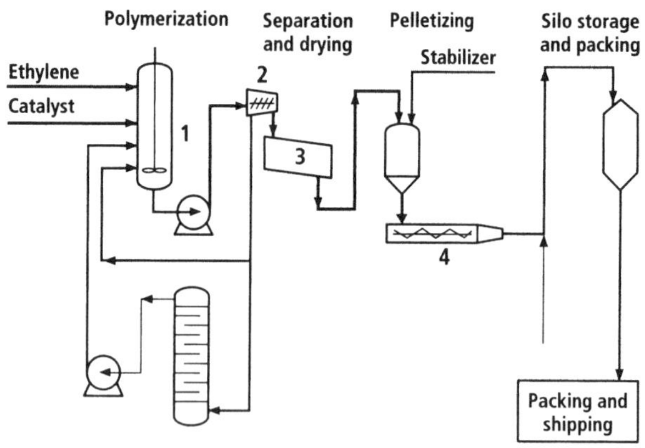

Polyethylene

Applications: To produce high-density polyethylene (HDPE) and medium density polyethylene (MDPE) under low-pressure slurry process—"CX process."

Description: The CX process uses two polymerization reactors in series and produces polymers having bimodal molecular-weight distribution (MWD). The MWD and composition distribution is easily controlled by adjusting the operating conditions of the two reactors without changing the catalyst types. This process produces a wide melt index range by applying innovative catalyst chemistry combined with a sophisticated polymerization process.

An all-round catalyst and simple polymerization operation provide easy product switch-over that results in short transition time and negligible amount of off-spec product. Mitsui has also developed new catalysts, which contribute better morphology of the polymer powder and ethylene consumption.

Ethylene, hydrogen, comonomer and super-high activity catalyst are fed into the reactors (1), and polymerization reaction occurs under slurry state. The automatic polymer property control system plays a very effective role in product-quality control. Slurry from the reactors is pumped to the separation system (2), and the wetcake is dried into powder in the dryer system (3). As much as 90% of the solvent separated from the slurry is directly recycled to the reactors without treatment. The dry powder is pelletized in the pelletizing system (4) along with required stabilizers.

Product: Broad range of homopolymer and copolymer can be produced.

Melt index	0.01 to > 50
Molecular-weight distribution Comonomer distribution	Freely controlled from narrow to very wide
Density	0.93 to 0.97

Economics: Typical consumption per metric ton of natural HDPE pellets:

Ethylene and co-monomer, kg	1,004
Electricity, kWh	345
Steam, kg	340
Water, cooling, t	190

Commercial plants: Thirty-eight reactor lines using the CX process are in operation or construction worldwide with a total production capacity of over 4 MMtpy.

Licensor: Mitsui Chemicals, Inc.

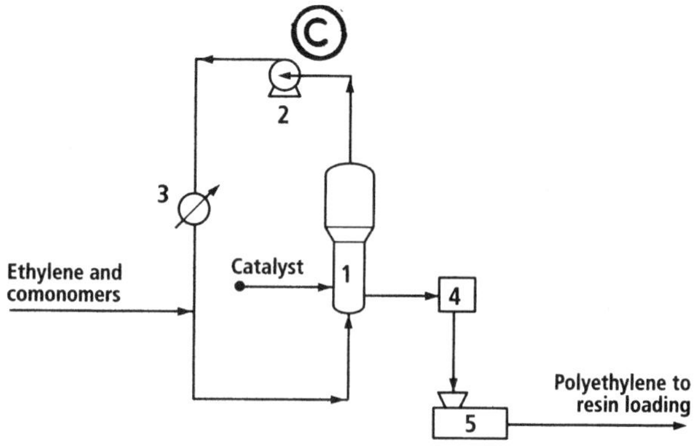

Polyethylene

Application: To produce linear low-density polyethylene (LLDPE) to high-density polyethylene (HDPE) using the low-pressure, gas phase UNIPOL PE process.

Description: A wide range of polyethylenes is made in a gas-phase, fluidized-bed reactor using proprietary solid and slurry catalysts. The product is in a dry, free-flowing granular form substantially free of fines as it leaves the reactor and is converted to pellet form for sale. Melt index and molecular weight distribution are controlled by selecting the proper catalyst type and adjusting operating conditions. Polymer density is controlled by adjusting comonomer content of the product. High productivity of conventional and metallocene catalysts eliminates the need for catalyst removal.

The simple and direct nature of this process results in low investment and operating costs, low levels of environmental pollution, minimal potential fire and explosion hazards, and easy operation and maintenance.

Gaseous ethylene, comonomer and catalyst are fed to a reactor (1) containing a fluidized bed of growing polymer particles and operating near 25 kg/cm^2 and approximately 100°C. A conventional, single-stage, centrifugal compressor (2) circulates reaction gas, which fluidizes the reaction bed, provides raw material for the polymerization reaction, and removes the heat of reaction from the bed. Circulating gas is cooled in a conventional heat exchanger (3).

The granular product flows intermittently into product discharge tanks (4) where unreacted gas is separated from the product and returned to the reactor. Hydrocarbons remaining with the product are removed by purging with nitrogen. The granular product is subsequently pelletized in a low-energy system (5) with the appropriate additives for each application.

Products: Polymer density is easily controlled from 0.915 to 0.970 g/cm. Depending on catalyst type, molecular weight distribution is either narrow or broad. Melt index may be varied from less than 0.1 to greater than 200. Grades suitable for film, blow-molding, pipe, roto-molding and extrusion applications are produced.

Commercial plants: Ninety-six reaction lines are in operation, under construction or in the design phase worldwide with single-line capacities ranging from 40,000 tpy to more than 450,000 tpy.

Licensor: Univation Technologies.

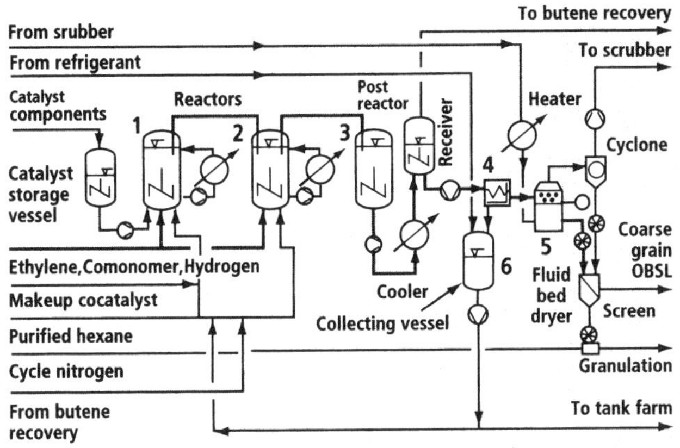

Polyethylene, HDPE

Application: To produce high-density polyethylene (HDPE) using the stirred-tank, heavy-diluent *Hostalen* process.

Description: The *Hostalen* process is a slurry polymerization method with two reactors parallel or in series. Switching from a single reaction to a reaction in cascade enables producing top quality unimodal and bimodal polyethylene (PE) from narrow to broad molecular weight distribution (MWD) with the same catalyst.

Polymerization occurs in a dispersing medium, such as like n-hexane, using a very high-activity Ziegler catalyst. No deactivation and catalyst removal is necessary because a very low level of catalyst residue remaining in the polymer. For unimodal-grade production the catalyst, the dispersing medium, monomer and hydrogen are fed to the reactor (1, 2) where polymerization occurs. In the case of bimodal grade production, the catalyst is only fed to the first reactor (1); the second step polymerization occurs under different reaction conditions with respect to the first reactor. Also ethylene, butene and further dispersing medium are fed to the second reactor (2). Reactor conditions are controlled continuously, thus a very high-quality PE is manufactured.

Finally, the HDPE slurry from the second reactor is sent to the post-reactor (3) to reduce dissolved monomer, and no monomer recycling is needed. In the decanter (4,) the polymer is separated from the dispersing medium. The polymer containing the remaining hexane is dried in a fluidized bed dryer (5) and then pelletized in the extrusion section. The separated and collected dispersing medium of the fluid separation step (6) with the dissolved co-catalyst and comonomer is recycled to the polymerization reactors. A small part of the dispersing medium is distilled to maintain the composition of the diluent.

Products: The cascade technology enables the manufacturing of tailor-made products with a definite MWD from narrow to broad MWD. The melt flow index may vary from 0.2 (bimodal product) to over 50 (unimodal product). Homopolymers and copolymers are used in various applications such as blow-molding (large containers, small bottles), extrusion molding (film, pipes, tapes and monofilaments, functional packaging) and injection molding (crates, waste bins, transport containers).

Economics: Consumption, per metric ton of PE (based on given product mix):

Ethylene and comonomer, t	1.015
Electricity, kWh	500
Steam, kg	450
Water, cooling water, $\Delta T = 10°C$, mt	175

Commercial plants: Twenty-six plants using the *Hostalen* process and technology are in operation or under construction, with a total licensed capacity of nearly 3.5 million tpy. Individual capacity can range up to 350,000 tpy for a single-line installation.

Licensor: Basell Polyolefins.

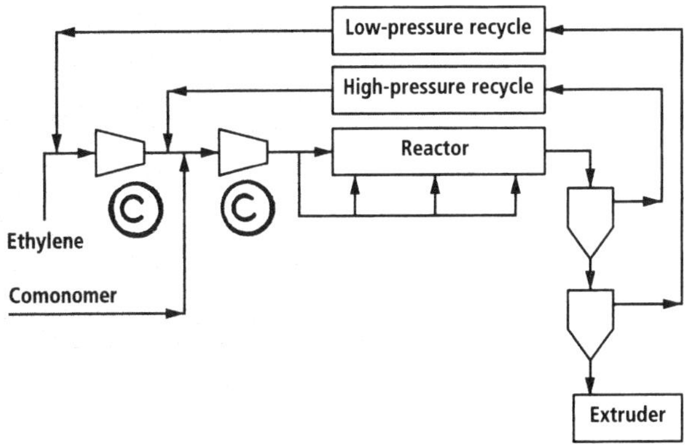

Polyethylene, LDPE

Application: The high-pressure *Lupotech* TS or TM tubular reactor process is used to produce low-density polyethylene (LDPE) homopolymers and EVA copolymers. Single-train capacity of up to 400,000 tpy can be provided.

Description: Ethylene, initiator and, if applicable, comonomers are fed to the process and compressed to pressures up to 3,100 bar before entering the tubular reactor. In the TS mode, the complete feed enters the reactor at the inlet after the preheater; in the TM mode, part of the gas is cooled and quenches the reactor contents at various points of injection.

The polymer properties (MI, ρ, MWD) are controlled by the initiator, pressure, temperature profile and comonomer content. After the reactor, excess ethylene is recovered and recycled to the reactor feed stream. The polymer melt is mixed with additives in an extruder to yield the final product.

A range of products can be obtained using the *Lupotech* T process, ranging from standard LDPE grades to EVA copolymers or N-butyl-acrylate modified copolymer. The products can be applied in (shrink) film extrusion, injection molding, extrusion blow molding, pipe extrusion, pipe coating, tapes and monofilaments.

There is no limit to the number of reactor grades that can be produced. The product mix can be adjusted to match market demand and economical product ranges. Advantages for the tubular reactor design with low residence time are easy and quick transitions, startup and shutdown.

Reactor grades from MI 0.15 to ≥50 and from density 0.917 to 0.934 g/cm^3, with comonomer content up to 30% can be prepared.

Economics: Consumption, per metric ton of PE:

Ethylene, t	1.010
Electricity, kWh	700–1,000
Steam, t	–1.2 (export credit)
Nitrogen, Nm^3	4

Commercial plants: Many *Lupotech* T plants have been installed after the first plant in 1955, with a total licensed capacity of 4.4 million tons. Basell operates LDPE plants in Europe with a total capacity of close to 1 million tpy. The newest state-of-the-art *Lupotech* TS unit at Basell's site in Aubette, France, was commissioned in 2000; with a capacity of 320 thousand ton, it is the largest single-line LDPE plant.

Licensor: Basell Polyolefins.

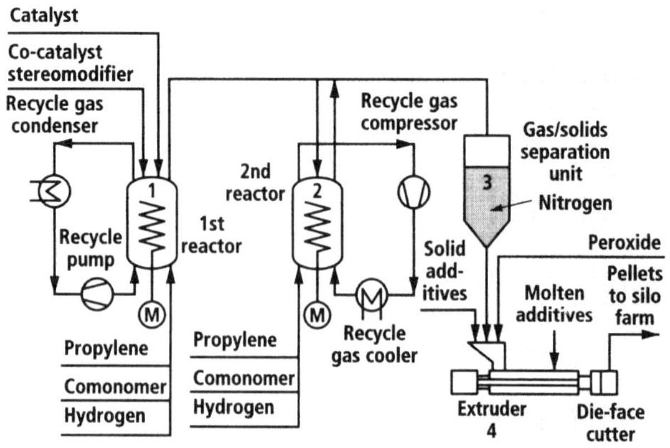

Polypropylene

Application: To produce polypropylene (PP) homopolymer, random copolymers and impact copolymers, including Metallocene PP in the Novolen process.

Description: In the Novolen process, polymerization is conducted in one or two gas-phase reactors, (1) and (2). The reactors contain a bed of polypropylene powder, which is agitated below the fluidization point by a helical agitator to keep the bed in motion and prevent powder agglomeration. A wide range of products can be produced with only two reactors connected in series, including super-high-impact copolymers. The second reactor is used either to incorporate rubber into the homopolymer matrix produced in the first reactor, or to increase the capacity while producing homopolymers or random copolymers.

Polymerization heat is removed from the reactors by external cooling circuits. Polymer powder is continually withdrawn from the reactors. The powder transfer from the first to the second reactor and from the second reactor to the gas/solids separation unit (3) is pressure driven. In this gas/solids separation unit polymer powder is separated from unreacted monomer and directly fed to the extruder (4) for pelletizing. The monomer is sent to the monomer recovery unit and recycled. Removal of catalyst residues or amorphous polymer is not required.

Products: The process can produce a wide variety of homopolymers, random copolymers including terpolymers and pentene copolymers, and impact copolymers with up to 50% rubber content. Product range also includes metallocene PP based on a simple drop-in technology.

Commercial plants: The capacities range from 60,000 to 400,000 mtpy for single lines. Forty-five production lines are in operation, engineering or under construction. The total licensed capacity worldwide for the Novolen process is in excess of 4.7 millions tons.

Licensor: Novolen Technology Holdings C.V.

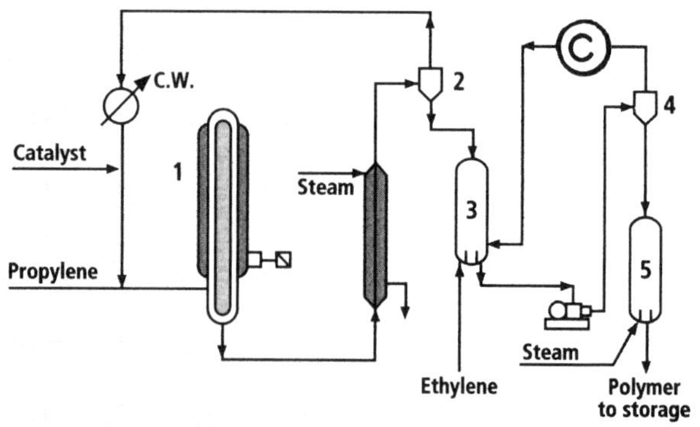

Polypropylene

Application: *Spheripol* process technology to produce propylene-based polymers including homopolymer PP and many families of random and heterophasic impact and specialty impact copolymers.

Description: In the *Spheripol* process, homopolymer and random copolymer polymerization takes place in liquid propylene within a tubular loop reactor (1). Heterophasic impact copolymerization can be achieved by adding a gas-phase reactor (3) in series.

Removal of catalyst residue and amorphous polymer is not required. Unreacted monomer is flashed in a two-stage pressure system (2, 4) and recycled back to the reactors. This improves yield and minimizes energy consumption. Dissolved monomer is removed from the polymer by a steam sparge (5). The process can use lower-assay chemical-grade propylene (94 %) or the typical polymerization-grade (99.5%).

Yields: Polymer yields of 40,000–60,000 kg/kg of supported catalyst are obtained. The polymer has a controlled particle size distribution and an isotactic index of 90–99%.

Economics: The *Spheripol* process offers a broad range of products with excellent quality and low-capital and operating costs.

Consumption, per metric ton of PP:

Propylene and comonomer, t	1.002–1.005
Catalyst, kg	0.016–0.025
Electricity, kWh	80*
Steam, kg	280
Water, cooling water, mt	90

* In case of copolymer production, an additional 20 kWh is required.

Products: The process can produce a broad range of propylene-based polymers, including homopolymer PP, various families of random copolymers and terpolymers, heterophasic impact and speciality impact copolymers (up to 25% bonded ethylene), as well as high-stiffness, high-clarity copolymers.

Commercial plants: *Spheripol* technology is used for about 50% of the total global PP capacity. There are 78 *Spheripol* process plants operating worldwide with installed capacity of about 12 million tpy. In addition, nine new *Spheripol* plants are under construction. Single-line design capacity is available in a range from 40,000 to 450,000 tpy.

Licensor: Basell Polyolefins.

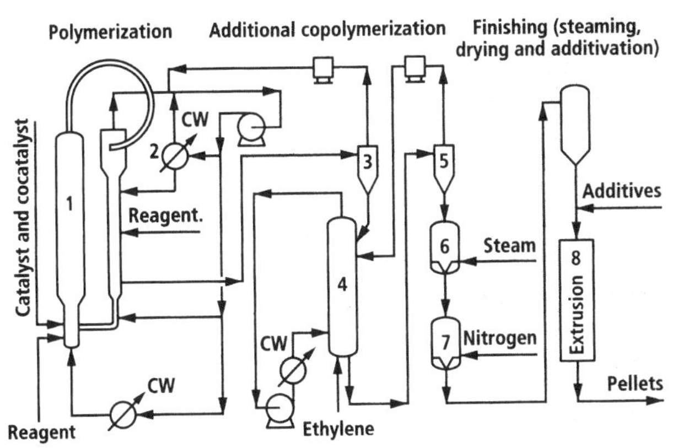

Polypropylene

Application: To produce polypropylene (PP)-based polymers using the *Spherizone* process technology.

Description: The *Spherizone* process is Basell's new proprietary gas-loop-reactor technology based on a multi-zone circulating reactor (MZCR) concept. Inside the reactor (1), the growing polymeric granules are continuously recirculating between two interrelated zones.

In the first zone (1a), the polymer is kept in a fast fluidization regime. When leaving this zone, the gas is separated, and the polymer crosses the second zone (1b) to a packed-bed mode and is then reintroduced in the first zone. A complete and massive solid re-circulation is obtained between the two zones.

The fluodynamic peculiar regime of the second zone, where the polymer enters as dense-phase in plug flow, allows altering the gas composition with respect to the chain terminator (hydrogen) and co-monomer. This is done by injecting monomers from the external system (2) in one or more points of the second zone (1b). Thus, two or more different polymers (MFR and/or comonomer type and content) can grow on the same granule.

While the granules recycle through the multiple zones, different polymers are generated in an alternate and cyclic way via continuous polymerization. Intimate mixing of different polymers is achieved; thus giving a substantial homogeneity of the final product.

Unreacted monomer is flashed at intermediate pressure (3) and recycled back to the loop reactor. Polymer can be fed to a fluidized-gas-phase reactor (4), operated in series (optional), where additional copolymer can be added.

From the intermediate separator/second reactor, the polymer is discharged to a receiver (5). The unreacted gas is recovered, while the polymer is sent to a proprietary unit for monomer steam stripping and catalyst deactivation (6). The removed residual hydrocarbons are recycled to the reaction, while the polymer is dried by a closed-loop-nitrogen system (7) and, free from volatile substances, sent to additives incorporation step (8).

Raw material and utility requirements per mton:

Propylene (plus comonomer for copolymers), kg	1,002–1,005
Catalyst, kg	0.025
Electricity, kWh	120*
Steam, kg	120
Water, cooling, m^3	85

* In case of high impact copolymer production, an additional 20 kWh is required.

Products: A broad range of propylene-based polymer can be produced including mono and bimodal (medium/wide/very wide MWD) homopolymer PP, high-stiffness homopolymers, random copolymers and terpolymers, high-clarity random copolymers as well as two compositions (homopolymer/random copolymer, twin-random copolymers or random/heterophasic copolymer). Conventional heterophasic impact copolymers can be produced with the second additional gas phase reactor, with ethylene/propylene rubber content up to 40%.

Commercial plant: A retrofitted 160,000-tpy plant is in operation at the Basell site in Brindisi since 2002.

Technology owner: Basell Polyolefins.

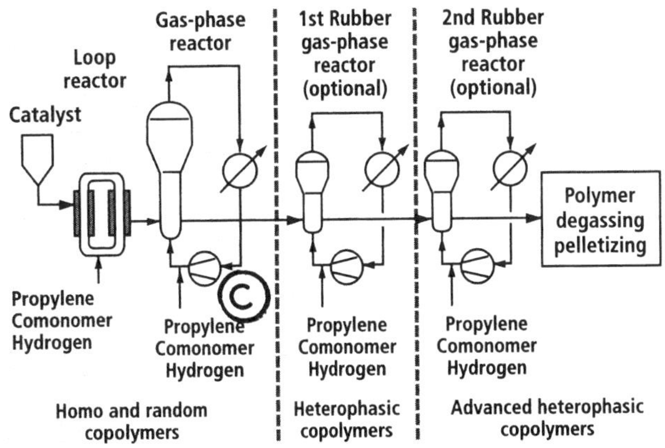

Polypropylene

Application: The Borstar polypropylene (PP) process can produce homopolymers, random copolymers, heterophasic copolymers, and very high-rubber-content heterophasic copolymers. It is a modular process; consisting of a loop reactor/gas-phase reactor combination.

Description: PP with a melt flowrate ranging from 0.1 to 1,200 can be produced. Borstar PP uses a Ziegler Natta catalyst, but single-site catalysts can be used in the future. When producing homopolymers and random copolymers, the process consists of a loop reactor and a gas-phase reactor in series. One or two gas-phase reactors are combined to manufacture heterophasic copolymers.

Propylene, catalyst, cocatalyst, donor, hydrogen, and comonomer (for random copolymers) are fed into the loop reactor; propylene is used as the polymerization medium (bulk polymerization). The loop reactor is designed for supercritical conditions and operates at 80–100°C and 50–60 bar. The propylene/polymer mixture exits the loop reactor and is sent to a fluidized-bed, gas-phase reactor, where propylene is consumed in polymerization. This reactor operates at 80–100°C and 25–35 bar. Fresh propylene, hydrogen and comonomer (in case of random copolymers) are fed into the reactor. After removing hydrocarbon residuals, the polymer powder is transferred to extrusion.

For heterophasic copolymers, polymer from the gas-phase reactor is transferred to another, smaller gas-phase reactor where the rubbery copolymer is made. After this processing step, hydrocarbon residuals are removed, and the powder is transferred for extrusion.

The basic module, loop/gas-phase reactor combination, enables high once-through conversion (minimized recycle), since unreacted monomer from loop reactor is consumed in the gas-phase reactor. Polymerization conditions in each reactor can be independently controlled, enabling production of both standard unimodal and broad-molecular-weight multimodal grades. The production rate ratio between the reactors can be adjusted to meet the targeted product properties.

Products: A wide range of polypropylenes with varying melt flowrates from 0.1 to 1,200, and from very stiff to very soft polymers are produced and can be tailored to customer needs. The products have reactor-made basic properties, thus minimizing additional compounding or other post-reactor treatment. Grades suitable for molding, film, fiber, thermoforming and pipe, as well as for engineering applications, are produced.

Commercial plants: The first Borstar PP plant has successfully operated since May 2000 in Schwechat, Austria, with 200,000-tpy capacity. Maximum single-line design capacity can achieve 400,000 tpy.

Licensor: Borealis A/S.

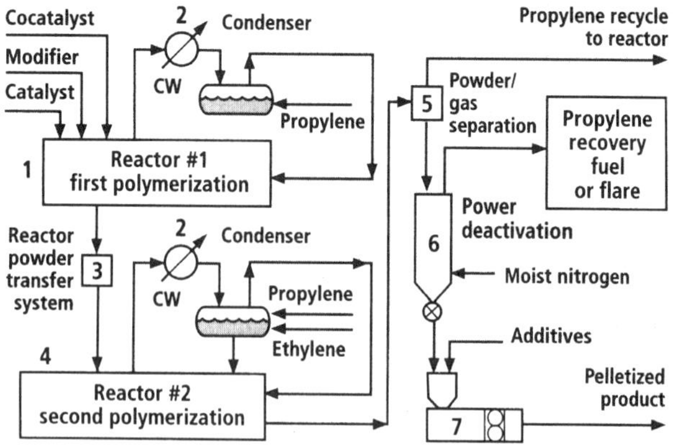

Polypropylene

Application: To produce polypropylene (PP) homopolymer, random copolymer and impact copolymer using the BP gas-phase process with proprietary 4th generation supported catalyst.

Description: Catalyst in mineral-oil-slurry is metered into the reactor together with co-catalyst and modifier. The proprietary supported catalyst developed by BP has control morphology, super-high activity and very high sterospecifity. The resulting PP product is characterized by narrow particle size distribution, good powder flowability, minimum catalyst residues, noncorrosiveness, excellent color and low odor.

The horizontal stirred-bed reactor (1) is unique in the industry in that it approaches plug-flow type of performance, which contributes to two major advantages. First, it minimizes catalyst bypassing, which enables the process to produce very high-performance impact copolymer. Second, it makes product transitions very quick and sharp, which minimizes off-spec transition materials. The reactor is not a fluidized bed, and powder mixing is accomplished by very mild agitation provided by a proprietary-designed horizontal agitator. Monomer leaving the reactor is partially condensed (2) and recycled. The condensed liquid together with fresh makeup monomer is sprayed onto the stirred reactor powder bed to provide evaporative cooling (remove the heat of polymerization) and control the bed temperature. Uncondensed gas is returned to the reactor.

For impact copolymer production, a second reactor (4) in series is required. A reliable and effective gas-lock system (3) transfers powder from the first (homopolymer) reactor to the second (copolymer) reactor, and prevents cross contamination of reactants between reactors. This is critically important when producing the highest quality impact copolymer. In most respects, the operation of the second reactor system is similar to that of the first, except that ethylene in addition to propylene is fed to the second reactor. Powder from the reactor is transferred and depressurized in a gas/powder separation system (5) and into a purge column (6) for catalyst deactivation. The deactivated powder is then pelletized (7) with additives into the final products.

Products: A wide range of polypropylene products (homopolymer, random copolymer and impact copolymer) can be produced to serve many applications, including injection molding, blow molding, thermoforming, film, extrusion, sheet and fiber. Impact copolymer produced using this process exhibit a superior balance of stiffness and impact resistance over a broad temperature range.

Commercial plants: Fourteen plants are either in operation or in design/construction worldwide with capacities ranging from 65,000 to 350,000 mtpy.

Licensor: BP.

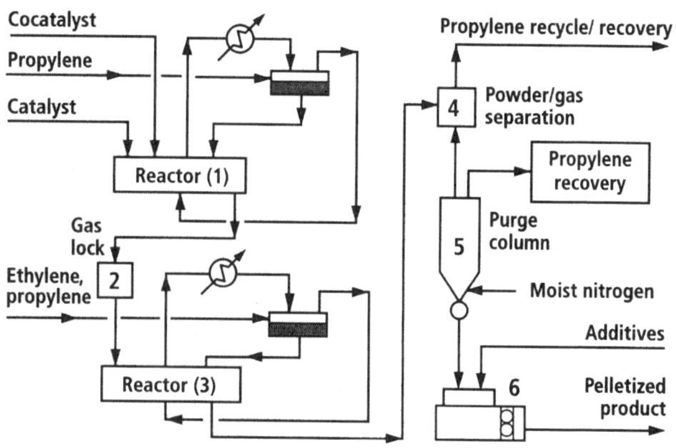

Polypropylene

Application: A process to produce homopolymer polypropylene and ethylene-propylene random and impact co-polymers using gas-phase polymerization in a special horizontal plug-flow reactor.

Description: The process features a horizontal agitated reactor and a high-performance catalyst specifically developed by the licensor. The catalyst has a controlled morphology, very high activity and very high selectivity. The process provides low energy consumption, superior ethylene-propylene impact co-polymer properties; minimum transition products, high polymer throughput and a high operating factor. Each process step has been simplified; consequently, the technology offers a low initial capital investment and reduced manufacturing costs while providing product uniformity, excellent quality control and versatile product capability.

Particles of polypropylene are continuously formed at low pressure in the reactor (1) in the presence of catalyst. Evaporated monomer is partially condensed and recycled. The liquid monomer with fresh propylene is sprayed onto the stirred powder bed to provide evaporative cooling. The powder is passed through a gas-lock system (2) to a second reactor (3). This acts in a similar manner to the first, except that ethylene as well as propylene is fed to the system for impact co-polymer production. The horizontal reactor makes the powder residence time distribution approach that of plug-flow. The stirred bed is well suited to handling some high ethylene co-polymers that may not flow or fluidize well.

The powder is released periodically to a gas-powder separation system (4). It is depressurized to a purge column (5) where moist nitrogen deactivates the catalyst and removes any remaining monomer. The monomer is concentrated and recovered. The powder is converted into a variety of pelletized resins (6) tailored for specific market applications.

Commercial plants: Eight polypropylene plants are in operation, with capacities ranging from 65,000 tpy to 360,000 tpy. Chisso offers processing designs for single-production with capacities reaching 400,000 tpy.

Licensor: Chisso Corp.

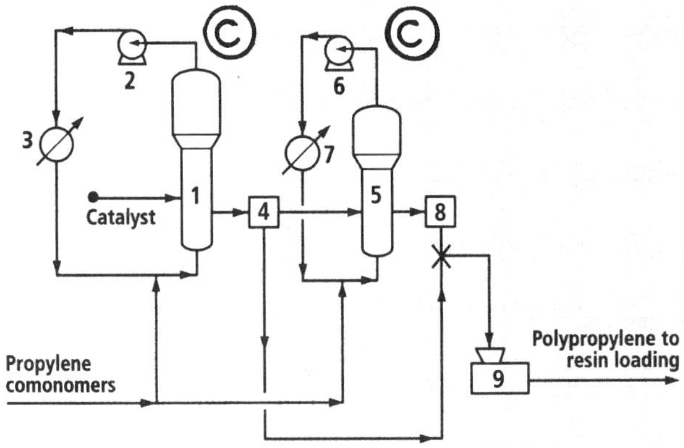

Polypropylene

Application: To produce homopolymer, random copolymer and impact copolymer polypropylene using the Dow gas-phase UNIPOL PP process.

Description: A wide range of polypropylene is made in a gas-phase, fluidized-bed reactor using proprietary catalysts. Melt index, isotactic level and molecular weight distribution are controlled by utilizing the proper catalyst, adjusting operating conditions and adding molecular-weight control agents. Random copolymers are produced by adding ethylene or butene to the reactor. Ethylene addition to a second reactor in series is used to produce the rubber phase of impact copolymers.

The UNIPOL PP process' simple, direct nature results in low investment and operating costs, low pollution levels, minimal potential fire and explosion hazards, and easy operation and maintenance. To produce homopolymers and random copolymers, gaseous propylene, comonomer and catalyst are fed to a reactor (1) containing a fluidized-bed of growing polymer particles and operating near 35 kg/cm^2 and approximately 70°C. A conventional, single-stage, centrifugal compressor (2) circulates the reaction gas, which fluidizes the reaction bed, provides raw materials for the polymerization reaction and removes the heat of the reaction from the bed. Circulating gas is cooled in a conventional heat exchanger (3). Granular product flows intermittently into product discharge tanks (4), unreacted gas is separated from the product and returned to the reactor.

To make impact copolymers, the polypropylene resin formed in the first reactor (1) is transferred into the second reactor (5). Gaseous propylene and ethylene, with no additional catalyst, are fed into the second reactor to produce the polymeric rubber phase within the existing polypropylene particles. The second reactor operates in the same manner as the initial reactor, but at approximately half the pressure, with a centrifugal compressor (6) circulating gas through a heat exchanger (7) and back to the fluid-bed reactor. Propylene product is removed by product discharge tanks (4, 8) and unreacted gas is returned to the reactor. Hydrocarbons remaining in the product are removed by purging with nitrogen. Granular products are pelletized in a proprietary low-energy system (9). Controlled rheology, high melt-flow grades are produced in the pelleting system through the addition of selected peroxides.

Products: Homopolymers can be produced with melt flows from less than 0.1 to 3,000 and isotactic content up to 99%. Random copolymers can be produced with up to 12 wt% ethylene or up to 21 wt% butene over a wide melt flow range (<0.1 to >100). A full range of impact copolymers can be polymerized with a good stiffness to impact balance. Products from narrow to broad molecular-weight distribution can be manufactured in grades suitable for film injection, molding, blow molding, extrusion and textile applications.

Commercial plants: More than 30 reaction lines are in operation, with capacities ranging from 80,000 tpy to 260,000 tpy. Total worldwide production with this technology is over five million tpy.

Licensor: The Dow Chemical Co. Univation Technologies is the licensor of the UNIPOL PE process.

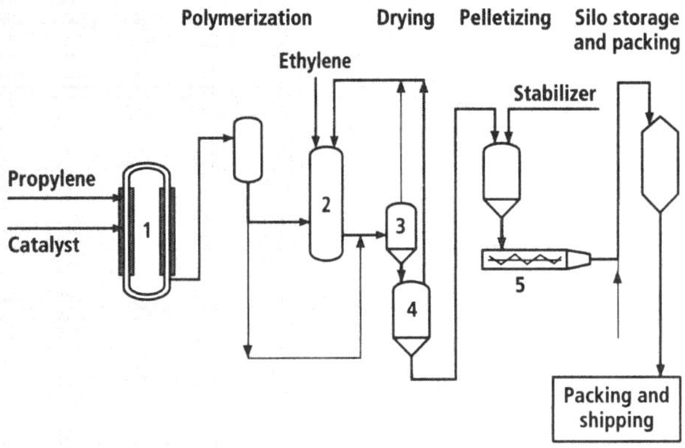

Polypropylene

Applications: To produce polypropylene (PP) including homopolymer, random copolymer and impact copolymer.

Description: The process, with a combination of the most advanced high-yield and high-stereospecificity catalyst, is a nonsolvent, nondeashing process. It eliminates atactic polymers and catalyst residue removal. The process can produce various grades of PP with outstanding product quality. Polymer yields of 20,000 to 100,000 kg/kg of supported catalyst are obtained, and the total isotactic index of polymer can reach 98% to 99%.

With new catalysts based on di-ether technology (5th generation catalyst, RK-Catalyst and RH-Catalyst), wider melt-index ranged polymers can be produced (compare with those produced with 4th generation catalyst) due to the high hydrogen response of RK/RH-Catalyst.

The reactor polymer has narrow and controlled particle size distribution that stabilizes plant operation and also permits easy shipment as powder. Due to the proprietary design of the gas-phase reactor, no fouling is observed during the operation, and, consequently, reactor cleaning after producing impact copolymer is not required. In addition, combination of the flexibility of the gas-phase reactor and high-performance catalysts allow processing impact copolymer with a high-ethylene content.

In the process, homopolymer and random copolymer polymerization occurs in the loop-type reactor(or vessel-type reactor) (1). For impact copolymer production, copolymerization is performed in a gas-phase reactor (2) after homopolymerization. The polymer is discharged from a gas-phase reactor and transferred to the separator (3). Unreacted gas accompanying the polymer is removed by the separator and recycled to the reactor system. The polymer powder is then transferred to the dryer system (4) where remaining propylene is removed and recovered. The dry powder is pelletized by the pelletizing system (5) along with required stabilizers.

Product: The process can produce a broad range of polypropylene polymers, including homo-polymer, random copolymer and impact copolymer, which become high-quality grades that can cover various applications.

Economics: Typical consumption per metric ton of natural propylene homopolymer pellets:

Propylene (and ethylene for copolymer),kg	1,005
Electricity, kWh	320
Steam, kg	310
Water, cooling, t	100

Commercial plants: Twenty-five reactor lines are in operation, engineering design or construction worldwide with a total production capacity of over 2.5 MMtpy.

Licensor: Mitsui Chemicals, Inc.

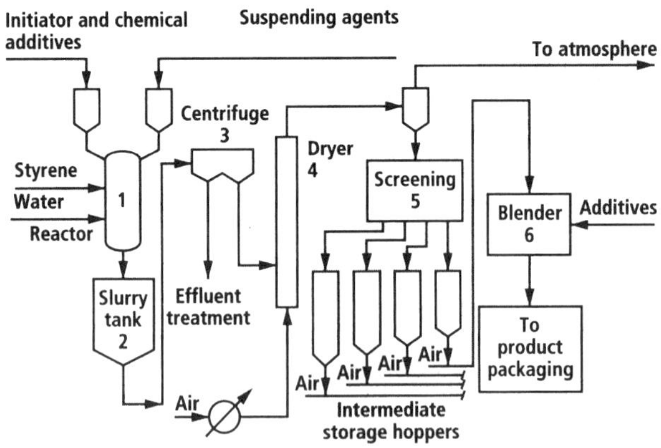

Polystyrene, expandable

Application: To produce expandable polystyrene (EPS) via the suspension process using BP Chemicals/ABB Lummus Global technology.

Description: The BP/Lummus styrene polymerization technology for the manufacture of regular and flame-retardant grades of EPS is a one-step batch suspension reaction followed by continuous dewatering, drying, and size classification.

Styrene monomer, water, initiators, suspending agents, nucleating agents, and other minor ingredients are added to the reactor (1). The contents are then subjected to a time-temperature profile under agitation. The suspending agent and agitation disperse the monomer to form beads. At the appropriate time, a premeasured quantity of pentane is introduced into the reactor. Polymerization is then continued to essentially 100% conversion. After cooling, the EPS beads and water are discharged to a holding tank (2).

From this point, the process becomes continuous. The bead/water slurry is centrifuged (3) where most of the "mother liquor" is removed. The beads are conveyed to a pneumatic dryer (4) where the remaining moisture is removed.

The dry beads are then screened (5) yielding as many as four product cuts. External lubricants are added in a proprietary blending operation (6) and the finished product is conveyed to shipping containers (7).

Economics: The BP/Lummus process is one of the most modern technologies for EPS production. Computer control is used to produce product uniformity while minimizing plant energy requirements. BP provides ongoing process research for product improvement and new product potential.

Raw materials and utilities, based on one metric ton of EPS:

Styrene and pentane, kg	1,000–1,015
Process chemicals, kg	25–49
Demineralized water, kg	1,000
Electricity, kWh	150
Steam, mt	0.42
Water, cooling, m^3	120

Commercial plants: Three commercial production units are in operation: one in France, one in Germany, and one in China for a total capacity of 200,000 metric tons.

Licensor: ABB Lummus Global/BP Chemicals.

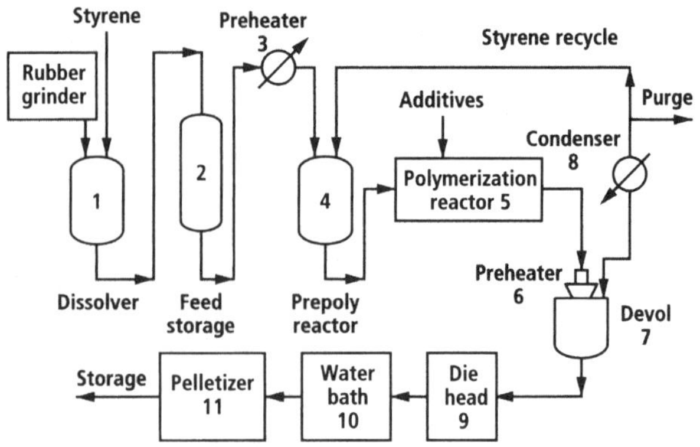

Polystyrene, high impact

Application: To produce a wide range of general purpose and high-impact polystyrenes (PS) via the bulk continuous process using the BP Chemicals/ABB Lummus Global technology.

Description: The production of general purpose PS (GPPS) and high-impact PS (HIPS) is essentially the same with the exception of the initial, rubber-dissolution step for HIPS.

The production of HIPS begins with the granulation and dissolving of rubber and other additives in styrene monomer (1) and then transferring the rubber solution to storage tank (2). For general-purpose product, controlled amounts of ingredients are fed directly to the feed preheater (3).

From this point on, the production steps for GPPS and HIPS are the same. The feed mixture is preheated (3) and continuously fed to the prepolymerizer (4) where the rubber morphology is established.

Following prepolymerization, the polymer mixture is pumped to the polymerization reactor (5) of proprietary design. At the exit of the reactor, the polymerization is essentially complete. The mixture is then preheated (6) in preparation for devolatization.

The devolatilizer (7) is held under a very high vacuum to remove unreacted monomer and solvent from the polymer melt. The residuals are condensed (8) and recycled back to the process. The polymer melt is then pumped through a die head (9) to form strands, a pelletization section (10 & 11) to form pellets and is screened to remove large pellets and fines. An external lubricant may be added at this time. The resultant product is air-conveyed to bulk storage and packaging facilities.

Economics: The BP/Lummus process offers one of the most modern technologies for GPPS and HIPS production. A broad product line is available with a consistently high product quality. BP provides ongoing process research for product improvement and new product potential.

Raw materials and utilities, based on one metric ton of polystyrene:

	GPPS	HIPS
Styrene and mineral oil, kg	1,011	937
Rubber, kg	–	73
Additives	1	2
Electricity, kWh	97	110
Fuel, 10^3 kcal	127	127
Water, cooling, m^3	46	26
Steam, LP, kg	6	6

Commercial plants: Plants in France, Germany, and Sweden are in operation with a total capacity of approximately 450,000 mtpy of GPPS and HIPS. Another 300,000 mtpy GPPS and HIPS unit will start up in China in 2005.

Licensor: ABB Lummus Global/BP Chemicals.

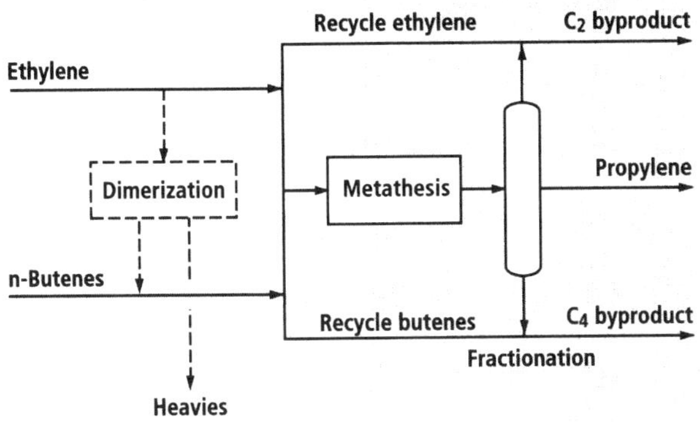

Propylene

Application: To produce propylene from ethylene and butenes using Lummus' olefin conversion technology (OCT). Other OCT process configurations involve interconversion of light olefins and production of C_2–C_5 mono-olefins.

Description: Ethylene feedstream (plus recycle ethylene) and butenes feedstream (plus recycle butenes) are introduced into the fixed-bed, metathesis reactor. The catalyst promotes reaction of ethylene and 2-butene to form propylene and simultaneously isomerizes 1-butene to 2-butene. Effluent from the metathesis reactor is fractionated to yield high-purity, polymerization-grade propylene, as well as ethylene and butenes for recycle and small byproduct streams. Due to the unique nature of the catalyst system, the mixed C_4 feed stream can contain a significant amount of isobutylene without impacting performance of the OCT process. A variation of OCT—Automet Technology—can be used to generate ethylene, propylene and the comonomer—hexene-1—by metathesis of n-butenes.

Yields: OCT process selectivity to propylene is typically greater than 98%. Overall conversion of n-butenes is 85–92%. Ethylene and butenes feedstreams can come from steam crackers or many refinery sources and in varying concentrations. Alternatively, butenes can come from ethylene dimerization, which is also licensed by Lummus.

In the Automet Technology, butenes yield about 10% ethylene, 38% propylene and 47% hexene-1. The balance is C_6 and heavier material.

Economics: Based on a 300,000-mtpy propylene plant, US Gulf Coast, mid-2000 (assuming 86% n-butenes in feedstream).

Investment, total direct field cost, US$	20.5 million
Utilities required per pound of product:	
Fuel gas (fired), Btu	340
Electricity, kWh	36
Steam, 50 psig saturated, Btu	704
Cooling duty, Btu	1,033
Nitrogen, scf	2.1
Catalyst, cost (est.) per yr, US$	325,000
Maintenance, per yr as % of investment	1.5

Commercial plants: Lyondell Petrochemical Co., Channelview, Texas, uses both the OCT technology and ethylene dimerization technology. Two other plants have used related technology, including a Phillips 66 Co. plant at Pasadena, Texas. Three other plants are under design or construction: a 690 MM lb/yr unit for BASF Fina Petrochemical in Port Arthur, Texas; a 350 MM lb/yr unit for the BP-Sinopec JV in Shanghai, China; and a 320 MM lb/yr unit for Mitsui Petrochemical in Osaka, Japan. The Automet Technology will go into operation on a semi-commercial scale at the Tianjin Petrochemical Co. in Tianjin, China.

Licensor: ABB Lummus Global.

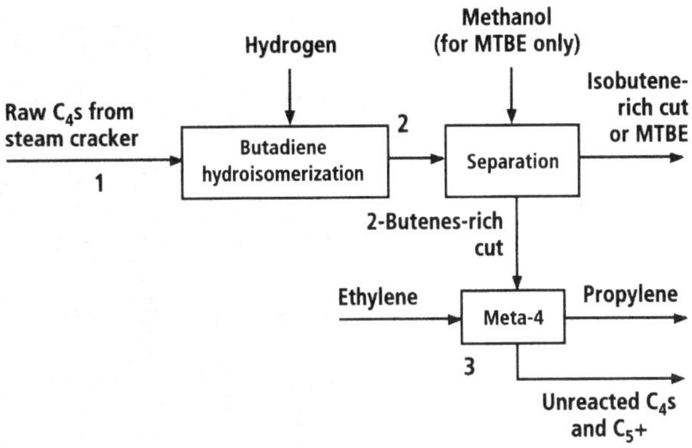

Propylene

Application: To produce polymer-grade propylene plus either an isobutylene-rich stream or MTBE by upgrading low-value pyrolysis C_4 cuts or butene-rich streams via selective hydrogen and Meta-4 processes. This process is particularly profitable when butadiene markets are weak and propylene demand is strong.

Description: Crude C_4 streams are converted into propylene and an isobutylene-rich stream in three IFP process steps: (1) butadiene and C_4 acetylenes selective hydrogenation and butenes hydroisomerization, (2) isobutylene removal via distillation or MTBE production and (3) metathesis (Meta-4).

The hydroisomerization step features: complete C_4 acetylenes and butadiene conversion to butenes, maximum 2-butenes production, flexibility to process different feeds, polymer-free product and no residual hydrogen. The second step separates isobutylene either by conventional distillation, or by reacting the isobutylene with methanol to produce MTBE.

The CCR Meta-4 process features are: a hard, highly active and robust catalyst, low catalyst inventory, low operating temperature and pressure, outstanding yields, liquid-phase operation, and continuous operation and catalyst regeneration.

Yields: Process selectivity to propylene is typically greater than 98%. Overall conversion of 2-butenes can reach 90%.

Economics: ISBL 2002 investment for a Gulf Coast location of a Meta-4 process producing 180,000 tpy propylene is US$ 19 million. Typical operating cost is $18 per metric ton of propylene.

Commercial plants: Over 100 C_4 hydrogenation units have been built using IFP technology. The CCR Meta-4 technology has been developed jointly with the Chinese Petroleum Corp., and demonstrated on real feedstock at Kaohsiung, Taiwan, industrial complex. The same type of moving-bed, continuous catalyst regeneration technology is industrially proven in Axens CCR Octanizing and Aromizing reformers.

Reference: Chodorge, J. A., J. Cosyns, D. Commereuc, Q. Debuisschert, and P. Travers, "Maximizing propylene and the Meta-4 process," Oil Gas 2000.

Licensor: Axens, Axens NA.

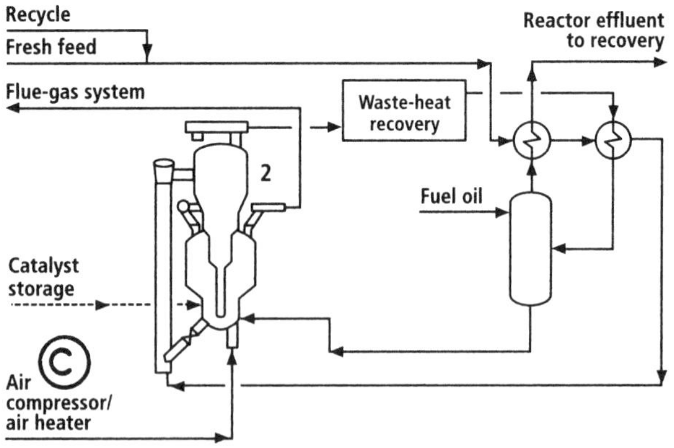

Propylene

Application: To produce propylene and ethylene from low-value, light hydrocarbon streams from ethylene plants and refineries in the carbon number range of C_4 to C_8, such as steam cracker C_4/C_5 olefins, cat-cracker naphthas, or coker gasolines.

Description: The SUPERFLEX process is a proprietary technology patented by ARCO Chemical Technology, Inc. (now Lyondell Chemical Co.), and is exclusively offered worldwide for license by Kellogg Brown & Root. It uses a fluidized catalytic reactor system with a proprietary catalyst to convert low-value feedstocks to predominately propylene and ethylene products. The catalyst is very robust; thus, no feed pretreatment is required for typical contaminants such as sulfur, water, oxygenates or nitrogen. Attractive feedstocks include C_4 and C_5 olefin-rich streams from ethylene plants, FCC naphthas or C_4s, thermally cracked naphthas from visbreakers or cokers, BTX or MTBE raffinates,C_5 olefin-rich streams removed from motor gasolines, and Fischer-Tropsch light liquids.

The fluidized reactor system is similar to a refinery FCC unit and consists of a fluidized reactor/regenerator vessel, air compression, catalyst handling, flue-gas handling and feed and effluent heat recovery. Using this reactor system with continuous catalyst regeneration allows higher operating temperatures than with competing fixed-bed reactors so that a substantial portion of the paraffins, as well as olefins, are converted. This allows for flexibility in the amounts of paraffins in the feeds to SUPERFLEX and the ability to recycle unconverted feed to extinction.

The cooled reactor effluent can be processed for the ultimate production of polymer-grade olefins. Several design options are available, including fully dedicated recovery facilities; recovery in a nearby, existing ethylene plant recovery section to minimize capital investment; or processing in a partial recovery unit to recover recycle streams and concentrate olefin-rich streams for further processing in a nearby plants.

Yields: The technology produces up to 70 wt% propylene plus ethylene, with a propylene yield about twice that of ethylene, from typical C_4 and C_5 raffinate streams. Some typical yields are:

Feedstock	FCC LCN	Coker LN	Pyrolysis C_4s	Pyrolysis C_5s
Ultimate yield, wt%*				
Fuel gas	13.6	11.6	7.2	12.0
Ethylene	20.0	19.8	22.5	22.1
Propylene	40.1	38.7	48.2	43.8
Propane	6.6	7.0	5.3	6.5
C_6^+ gasoline	19.7	22.9	16.8	15.6

* Ultimate yield with C_4s and C_5s recycled.

Commercial plants: The first SUPERFLEX licensee with a propylene production of 250,000 mtpy is Sasol Technology. Engineering is underway and completion of the unit in South Africa is scheduled for 2005.

Licensor: Kellogg Brown & Root, Inc.

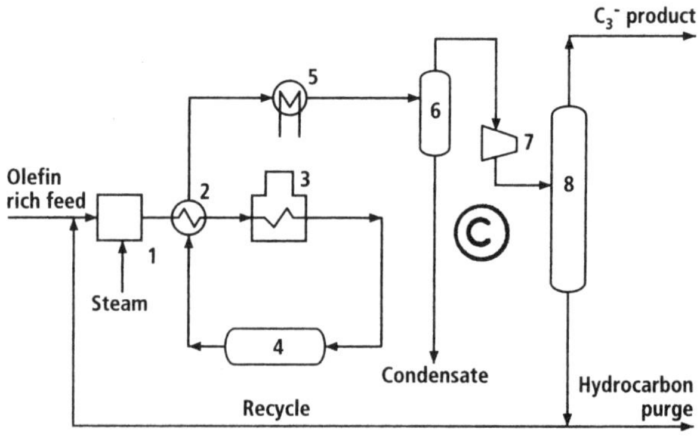

Propylene

Application: To produce predominantly propylene beside ethylene from low-value, olefin-rich feeds ranging from C_4 to C_8. The technology is suitable for conversion of C_4 and C_5 olefins from ethylene plants as well as light hydrocarbons from refineries independent from the olefin composition and fits into the environment for both.

Description: The Propylur process is a proprietary technology patented by Lurgi Oel Gas Chemie GmbH. The reaction system is a fixed-bed reactor with a proprietary catalyst, which is very robust and not affected by typical catalyst poisons. The process offers high selectivity toward propylene and excels by a simple process control.

The feed stream is mixed with a recycle from the downstream fractionation. Process steam is added in a saturator (1). The resulting steam/hydrocarbon mixture is superheated by the hot reactor effluent in a heat exchanger (2). Reaction temperature is adjusted in a fired heater (3) before entering the reactor (4). Passing through the catalyst layer, the conversion occurs at a slight temperature decrease.

Heat exchange with the feed and further cooling in a heat exchanger (5) results in a partial condensation. The condensate—mainly water—is separated from the organic vapors in a condensate separator (6). The water is evaporated and fed to the saturator (1) and reused as process steam. By recycling the process condensate, the water balance is closed.

Hydrocarbon vapors are compressed (7) and sent to a product separation column (8). The organic vapors are split into a crude propylene fraction in the top and the heavier hydrocarbons in the bottom of the column. The bottom fraction is split into a purge stream and a recycle stream. The recycle is sent back and mixed with the fresh feed to raise the overall conversion of higher olefins into propylene.

Yields: The technology produces 75 wt% propylene plus ethylene based on the olefin content in the feed, whereas the propylene yield is 60 wt%. The achieved yield and product spectrum is independent from the olefin composition in the feed.

Typical product slate from a Propylur unit (related to the olefins in the feed):

Propylene, wt%	60
Ethylene, wt%	15
Butenes, wt%	10
Gasoline, wt%	15

Economics: Using economic models for a Propylur unit to convert low-value olefin streams into propylene, the pay back time is between 2 and 3 years depending on the assumed value of the feed.

Commercial plant: A demonstration plant is operated at BP Köln in Worringen; it converts an olefin-rich C_5 cut from BP's steam cracker.

Licensor: Lurgi Oel Gas Chemie GmbH.

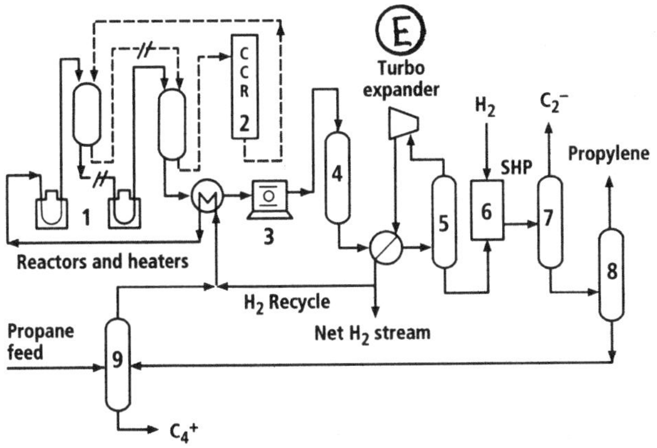

Propylene

Application: To produce polymer-grade propylene from propane using the UOP Oleflex process in a propylene production complex.

Description: The complex consists of a reactor section, continuous catalyst regeneration (CCR) section, product separation section and fractionation section. Four radial-flow reactors (1) are used to achieve optimum conversion and selectivity for the endothermic reaction. Catalyst activity is maintained by continuously regenerating catalyst (2). Reactor effluent is compressed (3), dried (4) and sent to a cryogenic separation system (5). A net hydrogen stream is recovered at approximately 90 mol% hydrogen purity. The olefin product is sent to a selective hydrogenation process (6) where dienes and acetylenes are removed. The propylene stream goes to a deethanizer (7) where light-ends are removed prior to the propane-propylene splitter (8). Unconverted feedstock is recycled back to the depropanizer (9) where it combines with fresh feed before being sent back to the reactor section.

Yields: Propylene yield from propane is approximately 85 wt% of fresh feed. Hydrogen yield is about 3.6 wt% of fresh feed.

Economics: US Gulf Coast inside battery limits are based on an Oleflex complex unit for production of 350,000 mtpy of polymer-grade propylene. The utility summary is net utilities assuming all light ends are used as fuel.

Inside battery limits investment, \$ million	145
Total project investment, \$ million	210
Typical net utility requirements, per ton of propylene product	
Electricity, kWh	200
Water, cooling, m^3	50
Net fuel gas, MMkcal (export credit)	1.2
Catalyst and chemical cost, \$/metric ton product	14

Commercial plants: Nine Oleflex units are in operation to produce propylene and isobutylene. Four of these units produce propylene. Two additional Oleflex units are under construction or in detailed design. These units will produce propylene. Over 1.3 million mty of propylene will be produced using Oleflex technology by 2001.

Licensor: UOP LLC.

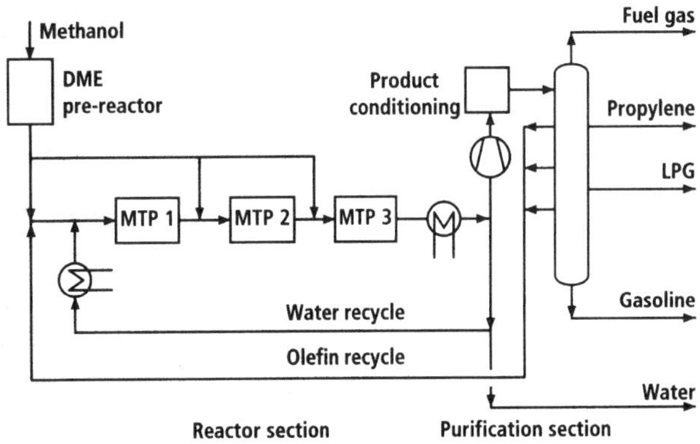

Propylene—methanol to propylene (MTP)

Application: To produce propylene from natural gas via methanol. This route delivers dedicated propylene from nonpetroleum sources, i.e., independently from steam crackers and FCCs.

Description: Methanol feed from a MegaMethanol plant is sent to an adiabatic DME pre-reactor where methanol is converted to DME and water. The high-activity, high-selectivity catalyst nearly achieves thermodynamic equilibrium. The methanol/water/DME stream is routed to the first MTP reactor stage where the steam is added. Methanol/DME are converted by more than 99%, with propylene as the predominant hydrocarbon product. Additional reaction proceeds in the second and third MTP reactor stages. Process conditions in the three MTP reactor stages are chosen to guarantee similar reaction conditions and maximum total propylene yield. The product mixture is then cooled, and product gas, organic liquid and water are separated.

The product gas is compressed, and traces of water, CO_2 and DME are removed by standard techniques. The cleaned gas is then further processed yielding chemical- or polymer-grade propylene as specified. Several olefin-containing streams are recycled to the MTP reactor as additional propylene sources. To avoid accumulation of inert materials within the loop, a small purge removes light-ends and the C_4/C_5 cut. High-grade gasoline is produced as a byproduct.

Water is recycled to steam generation; excess water from the methanol conversion is purged. This process water can be used for irrigation after appropriate and inexpensive treatment.

Economics: Current studies and projects are based on a combined MegaMethanol/ MTP plant with a capacity of 5,000 mtpd of methanol (1.667 mtpy), yielding approx. 519,000 mtpy of propylene and 143,000 mtpy of gasoline.

Based on a natural gas cost of 0.5 \$/MMBtu, net production cost for propylene will be 166 \$/mt. (Including owner's cost, capitalized interest and depreciation, assuming a moderate credit of 160 \$/mt for the byproduct gasoline.)

Technology status: Since January 2002, a demonstration unit has been operating at the Statoil methanol plant at Tjeldbergodden, Norway. This unit has confirmed the lab results. The catalyst is commercially available. Lurgi offers the process on commercial terms.

References: Rothaemel M., and H-D. Holtmann, "MTP, Methanol To Propylene—Lurgi's Way," DGMK-Conference "Creating Value from Light Olefins—Production and Conversion," Hamburg, Oct. 10–12, 2001.

Koempel, H., W. Liebner and M. Wagner, "MTP—An economical route to dedicated propylene," Second ICIS-LOR World Olefin Conference, Amsterdam, Feb. 11–12, 2003.

Licensor: Lurgi Oel Gas Chemie GmbH.

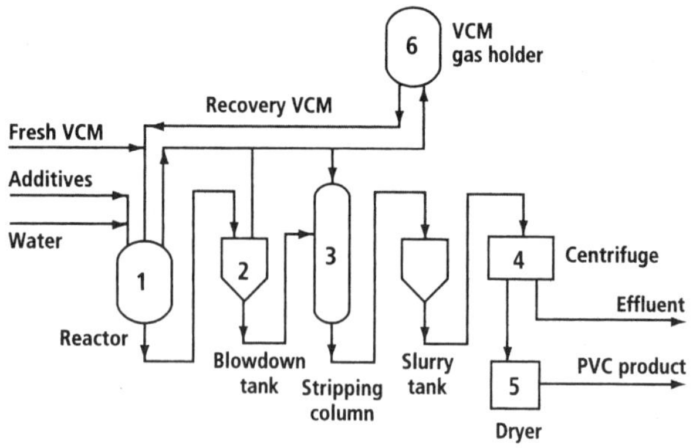

PVC (suspension)

Application: A process to produce polyvinyl chloride (PVC) from vinyl chloride monomer (VCM) using suspension polymerization. Many types of PVC grades are produced including: commodity, high K-value, low K-value, matted type and co-polymer PVC. The PVC possesses excellent product qualities such as easy processability and good heat stability.

Description: PVC is produced by batch polymerization of VCM dispersed in water. Standard reactor sizes are 60, 80, 100 or 130 m^3.

The stirred reactor (1) is charged with water, additives and VCM. During polymerization reaction, the temperature is controlled at a defined temperature depending on the grade by cooling water or chilled water. At the end of the reaction, the contents are discharged into a blowdown tank (2) where most of the unreacted VCM is flashed off. The reactor is rinsed and sprayed with an anti-fouling agent, and is ready for the following batch.

The PVC slurry containing VCM is continuously fed to the stripping column (3). The column has a proprietary design and effectively recovers VCM from the PVC slurry without any deterioration of PVC quality. After stripping, the slurry is de-watered (4), and dried effectively by the proprietary dryer (5). It is then passed to storage silos for tanker loading or bagging.

Recovered VCM is held in a gas holder (6), then compressed, cooled and condensed to be reused for the following polymerization batch.

Economics:

Raw materials and utilities, per ton of PVC:

VCM, t	1.003
Electricity, kWh	160
Steam, t	0.7
Additives, for pipe grade, $U.S.	12

Commercial plants: The process has been successfully licensed 12 times worldwide. Total capacity of the Chisso process in the world is more than 1 million tpy. In addition, Chisso VCM removal technology has been licensed to many PVC producers worldwide.

Licensor: Chisso Corp.

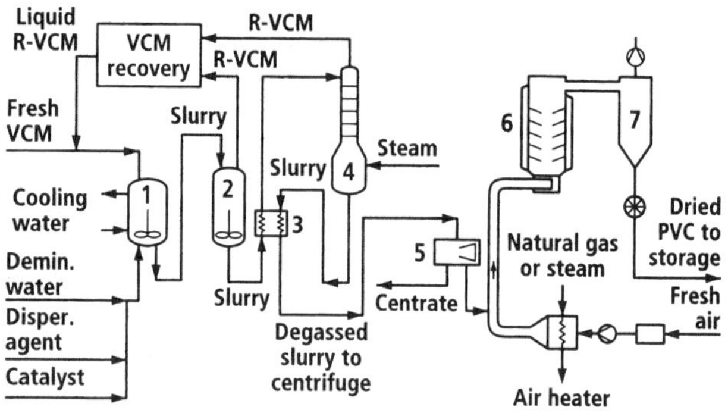

S-PVC process polymerization and degassing

S-PVC process drying

PVC (suspension)

Application: Production of suspension polyvinyl chloride (PVC) resins from vinyl chloride monomer (VCM) using the Vinnolit process.

Description: The Vinnolit PVC process uses a new high-performance reactor (1), which is available in sizes up to 150 m^3. A closed and clean reactor technology is applied; thus, opening of the reactors is not necessary, except for occasional inspections. Equally important, high-pressure water cleaning is not necessary. All process operations of this unit are controlled by a distributed process control system (DCS).

The batch-wise polymerization occurs in the following operation sequence:

- Prepare the reactor, which includes applying a highly effective antifouling agent
- Charge reaction solutions including dispersing agents, additives, chemicals, VCM and water into the reactor
- Exothermic conversion from VCM to PVC
- Discharge of the PVC slurry into the blowdown tank
- Flush the reactor internals.

The PVC slurry and unreacted VCM from the polymerization reactors are fed to the blowdown tank—the intermediate buffer between the discontinuous polymerization and the continuous degassing and drying unit.

In the blowdown tank (2), unreacted VCM is flashed out of the PVC slurry. From the blowdown tank, the slurry is fed through heat recuperator (3) to the sieve-tray type Vinnolit degassing column (4). VCM is stripped out with steam. The VCM concentration of the slurry leaving the degassing column is less than 1 ppm. The unreacted VCM is liquefied in the VCM recovery unit and charged back to polymerization. After dewatering the suspension in the centrifuge (5), the wet PVC cake is fed in the Vinnolit cyclone drying system (6). The solid particles and air are separated in the cyclone separator (7).

Economics: Chilled water for polymerization is not required. High productivity is achieved by using an inner-cooler reactor.

Raw materials and utilities, per ton of PVC:

VCM, t	1.001
Steam, t	0.8
Electricity, kWh	170
Additive costs, for pipe grade US$	14
Productivity, t/m^3/y	up to 600

Commercial plants: Vinnolit is producing up to 650,000 PVC tpy. Total capacity of the Vinnolit process in the world is about one million tpy. Vinnolit cyclone dryer has been licensed to many PVC producers worldwide.

Licensor: VinTec GmbH; **Contractor:** Uhde GmbH.

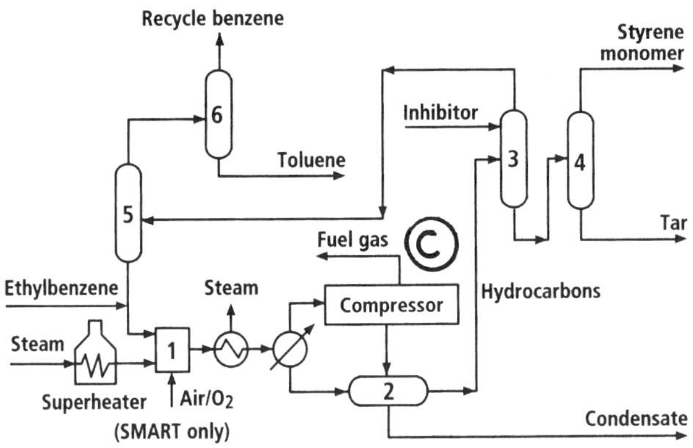

Styrene

Application: To produce polymer-grade styrene monomer (SM) by dehydrogenating ethylbenzene (EB) to form styrene using the Lummus/UOP "Classic" styrene process for new plants and the Lummus/UOP SMART process for revamps involving plant capacity expansion.

Description: In the Classic SM process, EB is catalytically dehydrogenated to styrene in the presence of steam. The vapor phase reaction is carried out at high temperature under vacuum. The EB (fresh and recycle) is combined with superheated steam, and the mixture is dehydrogenated in a multistage reactor system (1). A heater reheats the process gas between stages. Reactor effluents are cooled to recover waste heat and condense the hydrocarbons and steam. Uncondensed offgas—containing mostly hydrogen—is compressed and is used as fuel. Condensed hydrocarbons from an oil/water separator (2) are sent to the distillation section. Process condensate is stripped to remove dissolved aromatics.

A fractionation train (3,4) separates high-purity styrene product, unconverted EB, which is recycled, and the relatively minor byproduct tar, which is used as fuel. Toluene is produced (5,6) as a minor byproduct and benzene (6) is normally recycled to the upstream EB process.

Typical SM product purity ranges from 99.85% to 99.95%. The process provides high-product yield due to a unique combination of catalyst and operating conditions used in the reactors and the use of a highly effective polymerization inhibitor in the fractionation columns.

The SMART SM process is the same as Classic SM except that oxidative reheat technology is used between the dehydrogenation stages of the multistage reactor system (1). Specially designed reactors are used to achieve the oxidation and dehydrogenation reactions. In oxidative reheat, oxygen is introduced to partially oxidize the hydrogen produced over a proprietary catalyst to reheat the process gas and to remove the equilibrium constraint for the dehydrogenation reaction. The process achieves up to about 80% EB conversion per pass, eliminates the costly interstage reheater, and reduces superheated steam requirements. For existing SM producers, revamping to SMART SM may be the most cost-effective route to increased capacity.

Economics: (Classic) (500,000 mtpy, ISBL, US Gulf Coast)

Investment, US$ million	78
Ethylbenzene, ton/ton SM	1.055
Utilities, US$/mton SM	29

Commercial plants: Presently, 36 operating plants incorporate the Lummus/UOP Classic Styrene technology. Five operating facilities are using the SMART process technology. Many future units using the SMART process are expected to be retrofits of conventional units, since the technology is ideally suited for revamps.

Licensor: ABB Lummus Global and UOP LLC.

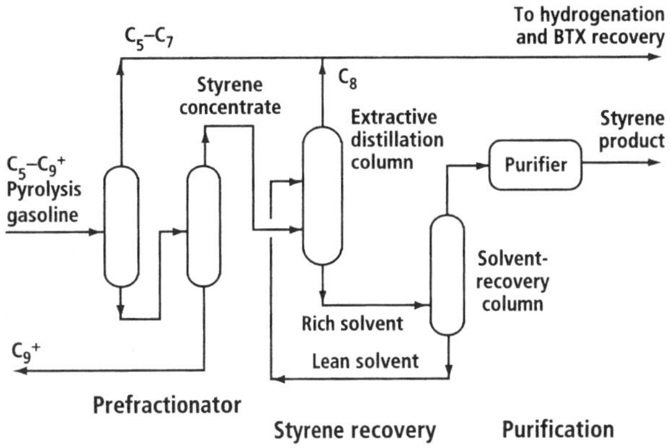

Styrene

Application: To directly recover styrene from raw pyrolysis gasoline derived from steam cracking of naphtha, gas oils and NGLs using the GT-Styrene process.

Description: Raw pyrolysis gasoline is prefractionated into a heart-cut C_8 stream. The resulting styrene concentrate is fed to an extractive-distillation column and mixed with a selective solvent, which extracts styrene to the tower bottoms. The rich solvent mixture is routed to a solvent-recovery column, which recycles lean solvent to the extractive-distillation column and recovers the styrene overhead. A final purification step produces a 99.9% styrene product containing less than 50 ppm phenyl acetylene.

The extractive-distillation column overhead can be further processed to recover a high-quality mixed xylene stream. A typical world-scale cracker could produce approximately 25,000 tpy styrene and 75,000 tpy mixed xylenes from pyrolysis gasoline.

The styrene is a high-purity product, suitable for polymerization, at a very attractive cost compared with conventional styrene production routes. If desired, the mixed xylenes can also be extracted from the pygas, upgrading their value as chemical feedstock. The process is economically attractive for typical pygas and supplemental feeds, which contain 15,000 tpy or more styrene.

Traditional pygas processing schemes destroy styrene in the first-stage hydrogenation unit. Hydrotreated pygas is then fractionated to extract benzene and toluene. With the GT-Styrene process, this fractionation is made upstream of the hydrotreaters, which avoids some hydrogen consumption and catalyst fouling by styrene polymers. In many cases, most of the existing fractionation equipment can be re-used in the styrene-recovery mode of operation.

Economics: Styrene recovery (considering styrene upgrade only) Basis: 25,000-tpy styrene capacity

Typical US GC capital cost, $MM:	20
Styrene value in pygas, $/t	180
Styrene product sales value, $/t	525
Processing cost, $/t	50
Gross margin, $MM/yr	7.37
Pretax ROI, %	37

Commercial plants: One license has been placed.

Reference: "Generate more revenues from pygas processing," *Hydrocarbon Processing,* June 1997.

Licensor: GTC Technology Inc.

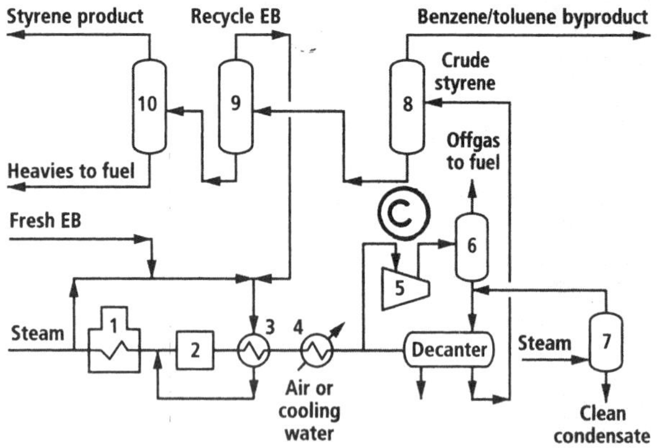

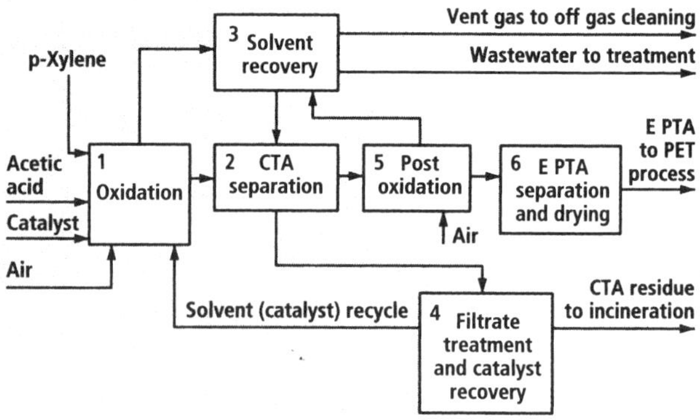

Styrene

Application: Process to manufacture styrene monomer (SM) by dehydrogenating ethylbenzene (EB) to styrene. Feedstock EB is produced by alkylating benzene with ethylene using the Mobil/Badger EBMax process.

Description: EB is dehydrogenated to styrene over potassium promoted iron-oxide catalyst in the presence of steam. The endothermic reaction is done under vacuum conditions and high temperature. At 1.0 weight ratio of steam to EB feed and a moderate EB conversion, reaction selectivity to styrene is over 97%. Byproducts, benzene and toluene, are recovered via distillation with the benzene fraction being recycled to the EB unit.

Vaporized fresh and recycle EB are mixed with superheated steam (1) and fed to a multi-stage adiabatic reactor system (2). Between dehydrogenation stages, heat is added to drive the EB conversion to economic levels, typically between 60%–75%. Heat can be added either indirectly using conventional means such as a steam heat exchanger or directly using a proprietary Direct Heating Technology developed by Shell Oil.

Reactor effluent is cooled in a series of exchangers (3) to recover waste heat and to condense (4) the hydrocarbons and steam. Uncondensed offgas—primarily hydrogen—is compressed (5) and then directed to an absorber system (6) for recovery of trace aromatics. Following aromatics recovery, the hydrogen-rich offgas is consumed as fuel by process heaters. Condensed hydrocarbons and crude styrene are sent to the distillation section, while process condensate is stripped (7) to remove dissolved aromatics and gases. The clean process condensate is returned as boiler feedwater to offsite steam boilers.

The distillation train first separates the benzene/toluene byproduct from main crude styrene stream (8). Unconverted EB is separated from styrene (9) and recycled to the reaction section. Various heat recovery schemes are used to conserve energy from the EB/SM column system. In the final purification step (10), trace C_9 components and heavies are separated from the finished SM. To minimize polymerization in distillation equipment, a dinitrophenolic type inhibitor is co-fed with the crude feed from the reaction section. Typical SM purity ranges between 99.90%–99.95%

Economics:

Ethylbenzene consumption, per ton of SM	1.052
Net energy Input, kcal per ton of SM	1.25
Water, cooling, m^3 per ton of SM	150

Note: Raw material and utility requirements presented are representative; each plant is optimized based on specific raw material and utility costs.

Commercial plants: The technology has been selected for use in over 40 units having design capacities (single train) ranging from 32–78 Mmtpy. The aggregate capacity of these units exceeds 8 MM mtpy.

Licensor: The Badger Technology Center of Washington Group International, and Fina Technology Co.

Terephthalic acid (E PTA)

Application: Production of polymer-grade terephthalic acid (E PTA) is an excellent raw material to produce polyethylene terephthalate resin (PET), which is used for engineering plastics, packaging materials—bottles and other food containers—as well as films. E PTA is also used as raw material to produce various polyester fibers used in the textile industry.

Description: The general flow diagram to produce E PTA using Eastman Chemical's proprietary process, comprises three different main sections—crude terephthalic acid (CTA), polymer-grade terephthalic acid (E PTA) and catalyst recovery.

Crude terephthalic acid (1,2,3): CTA is produced by the catalytic oxidation of p-xylene with air in the liquid phase using acetic acid as a solvent (1). The feed mix—p-xylene, solvent and catalyst—together with compressed air is continuously fed to the reactor, which is a bubble-column oxidizer. It operates at moderate temperature and offers an extremely high yield. The oxidizer product is known as crude terephthalic acid (CTA) due to the high level of impurities contained. Many impurities are fairly soluble in the solvent. In the CTA separation step (2), impurities can be effectively removed from the product by exchanging the reaction liquor with lean solvent from the solvent-recovery system. The reactor overhead vapor, mainly reaction water, acetic acid and nitrogen is sent to the solvent-recovery system (3), where water is separated from the solvent by distillation. After recovering its energy, the offgas is sent to a regenerative thermal oxidation unit for further cleaning.

Polymer-grade terephthalic acid (5,6): The crude acid is purified to obtain E PTA in a post-oxidation step, at elevated temperature conditions. The post oxidizers serve as reactors to increase conversion of the partially oxidized compounds to terephthalic acid. The level of 4-carboxy benzaldehyde (4-CBA) p-toluic acid (p-TA)—the main impurities in terephthalic acid—is significantly lowered. In a final step (6), E PTA is separated from the solvent and dried for further processing in the polyester-production facilities.

Catalyst recovery (4): After exchanging the liquor in the CTA separation, the suspended solids are separated and removed as CTA residue, which can be burned in a fluidized-bed incinerator or, if desirable, used as land fill. The soluble impurities are removed from the filtrate within the filtrate treatment section, and the dissolved catalyst is recycled.

Economics: The advanced Eastman E PTA technology uses fewer processing steps. In combination with the outstanding mild-oxidation technology, this technology leads to considerable capital cost savings and lower production cost than in other technologies.

Commercial plants: Commercial plants are operating in the US, Europe and Asia Pacific. The latest plant was constructed in Europe during 1998. Total worldwide installed capacity is 1.5 million mtpy.

Licensor: Lurgi Oel Gas Chemie GmbH.

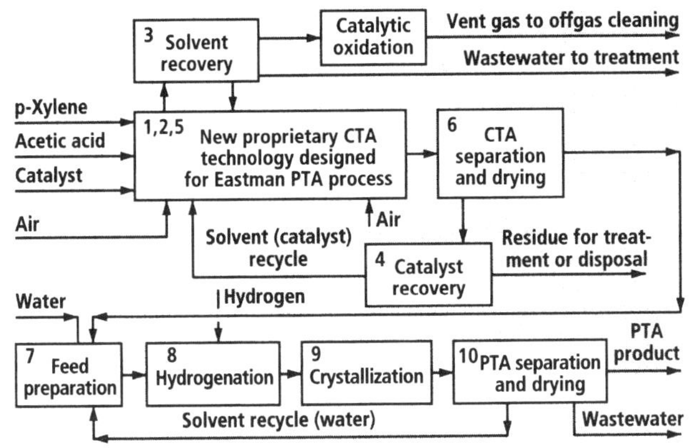

Terephthalic acid (PTA)

Application: Produce fiber-grade terephthalic acid (PTA). PTA is the market standard raw material to produce all types of polyester fibers used in the textile industry.

Description: The general flow diagram to produce PTA using Eastman Chemical's new proprietary process comprises two main production lines—production of crude terephthalic acid (CTA), followed by hydrogenation to produce fiber-grade terephthalic acid (PTA).

Polymer-grade terephthalic acid (1,2,3,4,5,6): To produce PTA, the same process steps are applied to process polymer-grade terephthalic acid. However, some modifications have been made, which significantly reduce capital cost, as well as, optimize the properties of E PTA when it is used as feed material for the hydrogenation process. This modified E PTA quality is known as CTA.

Fiber-grade terephthalic acid (7,8,9,10): To produce a market-typical fiber-grade quality, the CTA undergoes a further purification step. In the feed preparation system (7), the CTA is dissolved in water at elevated temperatures and sent to the hydrogenation reactor (8) in which it is catalytically treated to remove impurities. The hydrogenation catalyst is palladium based. The purified terephthalic acid is crystallized by stepwise depressurizing in a unique crystallizer train (9).

A heat-integation system is attached. The combination of the CTA quality and unique crystallization system generates less waste. In a following step, the PTA crystals are separated from the mother liquor and finally dried for further processing in the polyester production system.

Economics: Although the PTA purification of terephthalic acid by hydrogenation is well known from other technologies, the integrated advanced Eastman E PTA technology requires fewer process steps—including the purification. Such design offers considerable capital-cost savings and lowers production expenses.

Commercial plants: Commercial E PTA plants are operated in the US, Europe and Asia Pacific, with a worldwide installed capacity of 1.5 million mtpy. The hydrogenation section at the new PTA process has been developed from Eastman's operating isophthalic acid process, which is regarded as one of the best worldwide.

Licensor: Lurgi Oel Gas Chemie GmbH.

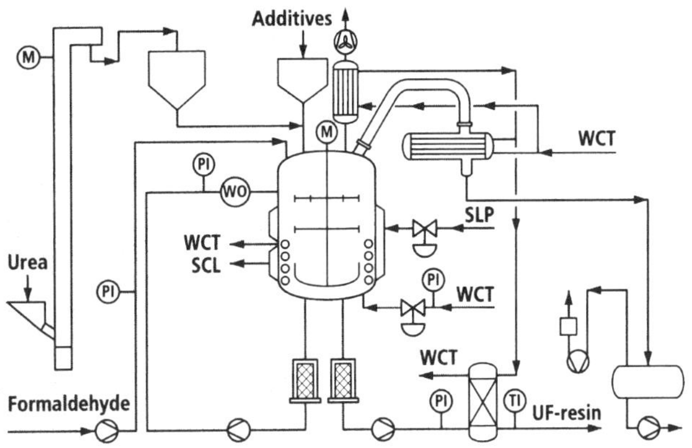

Urea-formaldehyde

Application: Urea-formaldehyde resins are used as adhesives in the wood working industry and are typically used in the production of plywood and particle board. They are available as concentrated solutions or in powder form as a result of the spray-drying process.

Description: The reaction mechanisms of the major components formaldehyde and urea are by polyaddition:

$$H_2N{-}CO{-}NH_2 + CH_2O \rightarrow H_2N{-}CO{-}NH{-}CH_2OH$$
$$\Delta h = -24 \text{ kJ/mol}$$

whereby the hydroxymethyl compounds undergoes further slow reaction by polycondensation

$$H_2N{-}CO{-}NH_2 + H_2N{-}CO{-}NH{-}CH_2OH \rightarrow$$
$$H_2N{-}CO{-}NH{-}CH_2{-}NH{-}CO{-}NH_2 + H_20$$

which is also responsible for the viscosity increase during the storage. The formation of methylene bridges will be accelerated by raising storage temperatures. The technology is based on batchwise production of the aqueous solution, short intermediate storage and continuously operating spray drying in a connected stage.

The formaldehyde is charged into the reactor and the pH value has to be adjusted. Urea is then added and the mixture will be heated to the reaction temperature. The reactor consists of a stirred kettle based on load cells and is equipped with heating jacket and internal cooling coils. Heating is carried out with low-pressure steam. An efficient anchor agitator ensures good heat transmission and uniform reaction conditions. The reaction proceeds in several steps at 70°–100°C.

To control the reaction rate, the pH value is measured continuously, which is readjusted by adding formic acid. When the end-point of the reaction is reached, determined by a solubility test with water, the reaction mixture is brought to pH 7 with caustic soda lye. The condensation is complete and the resin is evaporated under reduced pressure. The vapor is condensed in a tubular condenser and collected in a receiver.

After cooling the resin in the reactor, the resin is pumped to the buffer tank of the connected spray dryer plant. Usually, the complete batch processing takes 4–5 h. The urea-formaldehyde resin solution can be dried in a spray dryer based on the co-current flow principle.

This process cost-effectively produces high-quality glues at large quantities. The product is a low formaldehyde resin adhesive, suitable for veneering, plywood and particle board production by hot pressing process. The quality of the bonding complies with the requirements of *DIN 68705, Part 2* respectively to *DIN 68763-V20.* For particle board, a perforate value according to *DIN 120* of less 10 mg HCHO/100 g dry board will be maintained.

Licensor: INVENTA-FISCHER.

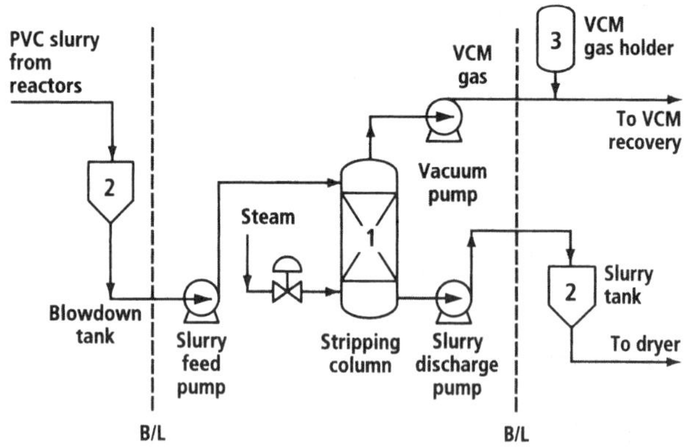

VCM removal

Application: Adding a stripping column to existing polyvinyl chloride (PVC) plants to remove vinyl chloride monomer (VCM) from PVC slurry. The recovered VCM can be reused in the PVC process, without any deterioration of PVC polymer quality.

Description: PVC slurry discharged from reactors contains significant amounts of VCM (> 30,000 ppm) even after initial flashing. This process effectively removes the remaining VCM so that the monomer is recovered and reused. Recycling of raw materials drastically reduces VCM emissions from the following dryer. There is no significant change in PVC quality after stripping. Residual VCM level in the PVC product can be lowered below 1 ppm, and in some cases below 0.1 ppm.

The PVC slurry, containing VCM, is continuously fed to the stripping column (1). The slurry passes counter-currently to steam, which is fed into the base of the column. The proprietary internals of the column are specially designed to ensure intimate contact between the steam and the PVC slurry and to ensure that no PVC particles remain inside the column. All process operations, including grade change, are automatically done in a completely closed system.

While steam stripping is widely used, this proprietary technology, which involves sophisticated design and know-how of the column, offers attractive benefits to existing PVC plant sites.

The process design is compact with a small area requirement and low investment cost. The size of the column is 2.5 to 30 t/h.

Economics:

Steam	130 kg/t of PVC

Commercial plants: Chisso has licensed the technology to many PVC producers worldwide. More than 70 columns of the Chisso process are under operation or construction, and total capacity exceeds 4 million tpy of PVC.

Licensor: Chisso Corp.

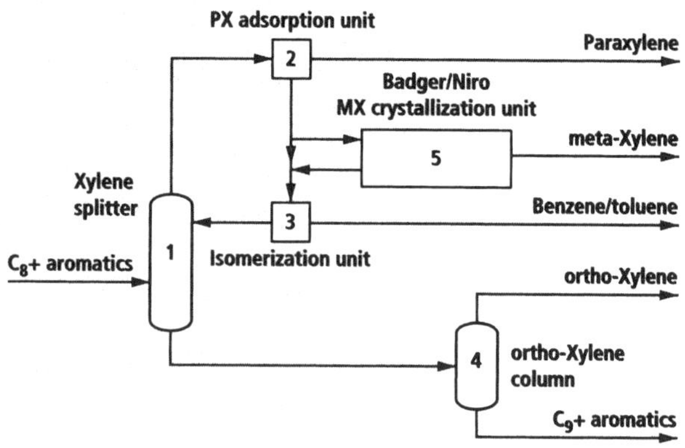

m-Xylene

Application: To recover and purify meta-xylene from C_8^+ aromatics streams using the Badger/Niro meta-Xylene crystallization process.

Description: As shown in the above figure, the integration of a Badger/Niro meta-Xylene crystallization unit into an existing aromatics plant's xylene section. It consists of a xylene splitter (1), paraxylene adsorption unit (2), isomerization unit (3) and optionally an o-xylene column (4). For an existing aromatics plant producing 300,000-mtpy paraxylene, over 100,000 mtpy of high-purity m-xylene production can be added, with no reductions in the paraxylene and o-xylene productions, by installing a Badger/Niro Unit (5) along with other relatively minor changes to the xylene section.

The raffinate from the adsorption unit (2) is fed to the crystallization section of the Badger/Niro Unit, which uses proprietary suspension crystallizers to create a slurry with high-purity m-xylene crystals. This slurry is fed to Niro wash column system where the crystals are separated from the mother liquor and melted for the final product. The wash column is a near-perfect separator—nearly all of the crystals fed to it become product in one pass. The crystals are washed countercurrently to remove liquid impurities; the used wash is recovered internally to the column with none of the capital or operating costs associated with the wash recovery when centrifuges or filters are used. Because of the high-purity of the crystals and the wash column separation efficiency, product purity is greater than 99.7%.

With this technology, the utilities consumption is minimal because the m-xylene is crystallized only once, the crystallization unit feed needs little or no pretreatment. Crystal wash recovery is done without additional refrigeration or other utilities; consequently, the cost for utilities is the lower than for any other licensed m-xylene technology. Furthermore, due to the efficiency of the process, capital cost should be substantially lower than for competing technologies.

Economics: Utility consumptions for the crystallization unit including refrigeration, per mt of product:

Electrical power, kWh	213
Water, cooling, m^3	37

Commercial plants: This technology has been used successfully for paraxylene purification. The first two paraxylene commercial units started up in 1999 and 2000 and met design performance requirements.

Licensor: The Badger Technology Center of Washington Group International, Inc., and Niro Process Technology B.V.

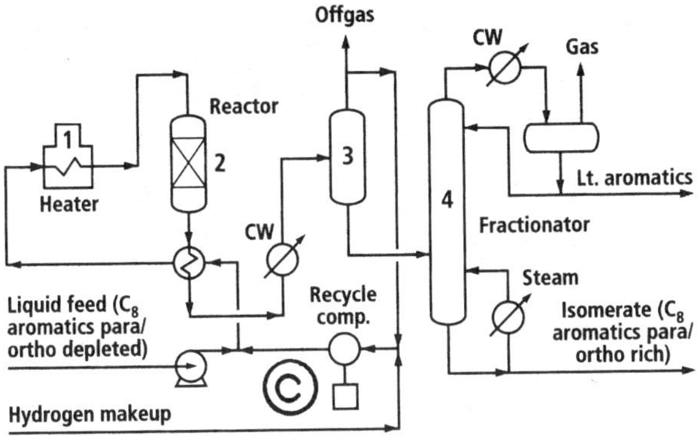

Xylene isomerization

Application: To selectively isomerize a paraxylene depleted-C_8 aromatics mixture to greater than equilibrium paraxylene concentration using ExxonMobil Chemical's XyMax and Advanced MHAI processes. Simultaneously, ethylbenzene (EB) and nonaromatics in the feed are converted to benzene and light paraffins, respectively. Conversion of EB is typically 60–80%.

Description: The para-depleted liquid C_8 aromatics raffinate stream from the paraxylene separation unit, along with hydrogen-rich recycle gas are pumped through feed/effluent exchangers and the charge heater (1) and into the reactor (2). Vapor then flows down through the fixed, dual-bed catalyst system. Dealkylation of EB and cracking of non-aromatics preferentially occurs in the top bed. The bottom bed promotes isomerization of xylenes, while minimizing loss of xylenes from side reactions. The reactor effluent is cooled by heat exchange and the resulting liquid and vapor phases are separated in the product separator (3). The liquid is then sent to a fractionator (4) for recovery of benzene and toluene from the isomerate.

Two enhanced isomerization catalyst technologies have been developed by ExxonMobil Chemical. The first technology, referred to as Advanced Mobil High Activity Isomerization (AMHAI), provides higher selectivity and lower operating costs compared to isomerization processes used in the past. The AMHAI technology offers increased operating flexibility in terms of a greater range of EB conversion and a lower temperature requirement. The second technology, referred to as XyMax, further increases yield performance and debottleneck potential. This technology can operate at even higher EB conversion, with higher selectivity and significantly lower xylene loss.

Operating conditions: XyMax and AMHAI units operate with a high-space velocity and a low hydrogen-to-hydrocarbon ratio, which results in increased debottleneck potential and decreased utilities costs. By converting a high portion of EB in the feed, these technologies can provide significant savings in associated paraxylene recovery facilities. Both technologies offer long operating cycles.

Commercial plants: The AMHAI process was first commercialized in 1999. Five AMHAI units are currently in operation. The first commercial unit using XyMax technology was brought onstream in 2000. Since then, two additional applications of the XyMax technology have been licensed. Including other ExxonMobil xylene isomerization technologies, there are a total of 22 units in operation.

Licensor: ExxonMobil Chemical (retrofit applications); Axens, Axens NA (grassroots applications).

References and Additional Reading

NUMBERED REFERENCES

1. Bloch, H. P. and J. J. Hoefner, *Reciprocating Compressors: Operation and Maintenance,* 1996, Gulf Publishing Company, Houston, TX, ISBN 0-88415-525-0.
2. Bloch, H. P., *Improving Machinery Reliability,* 3rd Edition, 1998, Gulf Publishing Company, Houston, TX, ISBN 0-88415-661-3.
3. Bloch, H. P. and F. Geitner, *Machinery Failure Analysis and Troubleshooting,* 3rd Edition, 1997, Gulf Publishing Company, Houston, TX; ISBN 0-88415-662-1.
4. Bloch, H. P. and C. Soares, *Process Plant Machinery,* 2nd Edition, 1998, Elsevier Publishing, Stoneham, MA, ISBN 0-7506-7081-9.
5. Bloch, H. P., *Practical Lubrication for Industrial Facilities,* 2000, Fairmont Press, Lilburn, GA, ISBN 0-88173-296-6.
6. Bloch, H. P. and A. Budris, *Pump User's Handbook: Life Extension,* 2nd Edition, 2005, Fairmont Press, Lilburn, GA, ISBN 0-88173-452-7.
7. Bloch, H. P.; "Solve Your Hydro-Cracker Compressor Problems," *Hydrocarbon Processing,* August 2000.
8. Bloch, H. P. and F. Geitner, *Major Process Equipment Maintenance and Repair,* 2nd Edition, 1997, Gulf Publishing Company, Houston, TX, ISBN 0-88415-663-X.
9. Brown, R. N., *Compressors, Selection and Sizing,* 2nd Edition, 1997, Gulf Publishing Company, Houston, TX, ISBN 0-88415-164-6.
10. Eisenmann, R. C. Sr. and R. Eisenmann, Jr., *Machinery Malfunction Diagnosis and Correction,* 1997, Prentice-Hall, Englewood Cliffs, NJ, ISBN 0-13-240946-1.
11. Price, S. M., "Development and Use of a Real-time Acquisition, Monitoring and Analysis System for a Process Control Environment," in *Proceedings of National Instruments User Symposium,* Austin, TX, March 28–30, 1993.
12. Continuous Monitoring of High Speed Data Improves Machinery Analysis, Diagnosis. Staff Report, *Gas Machinery Journal,* Second Quarter, 2004, pp. 13–14.
13. Haight, Brent, "Compressor Monitoring and Control," in *Proceedings of Compressor Tech Two,* September, 2004, pp. 24–26.
14. Compressor Controls Corporation, Des Moines, Iowa, Marketing Literature, 2005.
15. Elliott Company, Jeannette, Pennsylvania, "Compressor Refresher" Training Course, 1996.
16. Bloch, H. P., Continuous Torque Measurement Boosts Machinery Efficiency, *Hydrocarbon Processing,* September 1977.
17. Boyce, M. P., *Gas Turbine Engineering Handbook,* 1982, Gulf Publishing Company, Houston, TX, ISBN 0-87201-878-4.
18. Mitchell, J. S., *Physical Asset Management Handbook,* 3rd Edition, 2002, Clarion Technical Publishers, Houston, Texas.
19. Gresh, T., *Compressor Performance: Aerodynamics for the User,* 2nd Edition, 2004, Elsevier Publishing, Stoneham, Massachusetts, ISBN 0-7506-7342-7.
20. American Society of Mechanical Engineers, Power Test Code Department, New York, NY, 1997.
21. Bloch, Heinz P., *Machinery Reliability Improvement,* Gulf Publishing Company, Houston, TX, 1982, also 1988 and 1998, ISBN 0-88415-661-3.
22. Bloch, et al., *Compressors and Expanders,* Marcel Dekker, New York, 1982, ISBN 0-8247-1854-2
23. Bloch, H. P., and F. Geitner, Machinery Component Maintenance and Repair, 3rd Edition, 2004, Gulf Publishing Company, Houston, TX, ISBN 0-87201-781-8, also 0-7506-7726-0.
24. Godse, A. G., "Predict Compressor Performance at New Conditions," *Hydrocarbon Processing,* June 1989.
25. Godse, A. G., "Rotating Machinery Shop Testing," *Hydrocarbon Processing,* January, 1990.
26. Godse, A. G., "Improve Reliability of Liquid Film Shaft Seals," *Hydrocarbon Processing,* Auust 1995.
27. Godse, A. G., "Understand Dry Gas Seals," *Hydrocarbon Processing,* February 2000.
28. Bloch, H. P., *Practical Guide to Steam Turbine Technology,* 1996, McGraw-Hill, New York, ISBN 0-07-005924-1.
29. Bloch, H. P., and C. Soares, *Turboexpanders and Process Appli-*

Compressors and Modern Process Applications. By Heinz P. Bloch.

cations, 2001, Gulf Professional Publishing, Houston, TX, ISBN 0-88415-509-9.

ADDITIONAL READING

Harrison, W. J., "The Hydrodynamical Theory of Lubrication of a Cylindrical Bearing."

Anonymous, Under Variable Load and of a Pivot Bearing, *Transactions of the Cambridge Philosophical Society,* Vol. 22, 24 April 1919, pp. 373–388.

Evans, W. R., *Control-System Dynamics,* McGraw-Hill, New York, 1954.

Cannon, R. H., *Dynamics of Physical Systems,* McGraw-Hill, New York, 1967.

Externally Pressurized Bearings, *Proceedings of The Institution of Mechanical Engineers,* a Joint Conference arranged by the Tribology Group of the Institution of Mechanical Engineers and the Institution of Production Engineers, 17–18 November 1971, The Institution of Mechanical Engineers, London, England, 1972.

Cameron, A., *Basic Lubrication Theory,* 3rd Edition, Wiley, New York, 1981, p. 177.

Rowe, W. B., *Hydrostatic and Hybrid Bearing Design,* Butterworth & Co. Ltd., 1983.

Fuller, D. D., *Theory and Practice of Lubrication for Engineers,* 2nd Edition, Wiley, New York, 1984, p. 73.

Bently, D. E., and A. Muszynska, Why Have Hydrostatic Bearings Been Avoided as a Stabilizing Element for Rotating Machines?, in *Proceedings of the Symposium on Instability in Rotating Machinery,* Carson City, Nevada, June 1985, NASA Conference Publication, No. 2409, 1985.

Bently, D. E., and A. Muszynska, Role of Circumferential Flow in the Stability of Fluid-Handling Machines, Rotors, in *Texas A&M Fifth Workshop on Rotordynamics Instability Problems in High Performance Turbomachinery,* 16–18 May 1988, College Station, Texas, pp. 415–430.

Bassani, R., and B. Piccigallo, *Hydrostatic Lubrication,* Elsevier Science Publishers, 1992.

Bently, D. E., and C. T. Hatch, Root Locus and the Analysis of Rotor Stability Problems, *Orbit,* Vol. 14, No. 4, Minden, NV, Bently Nevada Corporation, December 1993.

Bently, D. E., and C. T. Hatch, *Fundamentals of Rotating Machinery Diagnostics,* Bently Pressurized Bearing Press, 2002.

Thomas, G. R., A Brief Review of Fluid Film Bearing Lubrication Principles, in *Vibration Institute Seminar,* February 2004.

Bently, D. E., D. W. Mathis and R. G. Thomas, "Externally Pressurized Bearing—A Tool for Rotordynamic Machinery Management," Submitted for publication at the ISROMAC10 Conference, March 2004.

Bloch, H. P. and A. Shamim, *Oil Mist Handbook: Practical Applications,* 1998, Fairmont Press, Lilburn, GA, ISBN 0-88173-256-7.

Bloch, H. P. and F. Geitner, *Introduction to Machinery Reliability Assessment,* 2nd Edition, Elsevier Publishiing Company, Stoneham, Massachusetts, 2005, ISBN 0-88415-172-7.

Lobanoff, V. S. and R. R. Ross, *Centrifugal Pumps: Design and Application,* 2nd Edition; 1992, Gulf Publishing Company, Houston, TX, ISBN 0-87201-200-X.

Karassik, I. J. et al, *Pump Handbook,* 2nd Edition, 1986, McGraw-Hill, New York, NY, ISBN 0-07-033302-5.

Index